Sustaining Our Social and Natural Capital

Proceedings of the 12th ANZSYS Conference

ISBN: 978-0-9791688-8-8

Edited by

Roger Attwater & John Merson

Published and distributed by:

ISCE Publishing

17947 W Porter Ln
Goodyear, AZ 85338

Sustaining Our Social and Natural Capital:
Proceedings of the 12th ANZSYS Conference
Library of Congress Control Number: 2007936998
ISBN: 978-0-9791688-8-8

Printed in the United States of America

Contents

THEME FOUR
Systems Thinking and Dynamics
John Barton, Barton Consulting
Bob Cavana, Victoria University of Wellington

THEME FIVE
Social Sustainability and Community

Wendy Gregory, Environmental Science & Research
Jack W. Meek, University of La Verne

THEME SIX
Livelihood, Innovation and Opportunity

Ian Knowd, University of Western Sydney, Australia

THEME SEVEN
Language, Learning and Transdisciplinary Practice

Susan Goff, University of Western Sydney, Australia

EDITORIAL
Sustaining Our Social and Natural Capital

Roger Attwater & John Merson

Our efforts to sustain our communities, and the natural environments that support them, are challenged by our ability to communicate effectively between our different forms of knowledge. Respect for diversity and difference, drawing upon all our methods of inquiry, advocacy, and learning to find common ground, are all part of the integrative approach needed to address the complexity of the challenges we face. This conference was an opportunity for practitioners from broad ranging traditions to share their experiences regarding integrative and innovative approaches that can make a difference.

Systems thinkers from across the globe were invited to participate in the conference through submitting a paper, contributing to workshops, and involvement in the range of planned and emergent activities. Papers were invited, on any aspects of relevance to the conference themes, and were peer reviewed prior to the conference by contributors within each thematic panel. The intent of this was to assist a process of dialogue between contributing authors. The conference themes included:

- Resilience, ecology and governance;
- Cultural knowledge and heritage;
- Creativity, organization and complexity;
- Systems thinking and dynamics;
- Social sustainability and community;
- Livelihood, innovation and opportunity;
- Language, learning and transdisciplinary practice;
- Engineering systems and infrastructure design.

The following chapters reflect these thematic plenary panels in which papers were presented and discussed. Each chapter begins with a brief overview from the chairperson(s) of each theme. The structure of these themes was also followed through into a conversation mapping exercise which followed the plenary panels of papers. A brief outline of the conversation mapping exercise is provided in Appendix 1, along with a summary of some of the key issues and discussion points which emerged. As well as this dialogue around thematic issues, planned activities included a 'circle of elders' discussion, workshops, and field trips.

The 'circle of elders' term was an after dinner round table discussion including a number of respected systems practitioners. The conversation was moderated by Stuart Hill, and included Richard Bawden, Bob Dick, Kay Fielden, Gerald Midgley, and Ray Ison. Workshops were contributed over the last two days of the conference by Bob Dick, Andrew Gaines, Susan Goff Janet McIntyre, and Jim Sheffield. A local field trip on the last morning of the conference enabled delegates to visit the local landscape, drawing upon local government initiatives with Geoffrey Smith and Jasmine Payget, and indigenous cultural perspectives with Chris Tobin. Within the conference venue, an exhibition of artwork reflecting cultural and indigenous perspectives was organized by Cheryle Yin-Lo, Tracey Allen and Rosie Forsyth-Jackson. Other exhibits included photographs of Californian coastal issues by Kathy Pompe, local conservation initiatives by the Blue Mountains City Council, and the East Timor Sisters, all of which contributed to a vibrant venue. We would also like to acknowledge the support of the Blue Mountains City Council, along with funding from Sydney Catchment Authority. Finally, we'd like to thank our colleagues at the Blue Mountains World Heritage Institute for their involvement.

THEME ONE
Resilience, Ecology and Governance

Rosalie Chapple, Blue Mountains World Heritage Institute
Peter Ampt, Future of Australia's Threatened Ecosystems (FATE) program, University of New South Wales.

The six papers in this theme dealt with topics ranging from environmental degradation of the River Ganges and of the north Katoomba catchment, to sustainability accounting, local govt sustainability, coastal planning, and the project of Gross National Happiness in Bhutan. Three of these papers dealt directly with ecological management issues, and the other three dealt more generally with governance for sustainability. All of the papers pointed toward failure when solutions don't take a whole-of-system approach. All papers also shared in common a fundamental concern with the values held by people, and the question of how these values can be aligned with sustainability. The papers demonstrated that introducing an external driver of change mostly fails, until personal values are used as the internal driver of change and action.

Values are inextricably bound up in culture, which is where values are formed and expressed. The fact that 2 of the 6 papers were focused on issues in less developed countries, namely India and Bhutan, helped in the comparison of the key role played by values in determining whether a society can move toward more sustainable practices. The paper about Bhutan raised the question of whether sustainability can be possible in the west without a foundation based more explicitly on morals/ethics/spirituality. The coastal planning paper demonstrated the case of a society failing to take individual responsibility for an environmental problem – and the difficulty inherent in bringing about change when externally applying a new model. The case studies in India demonstrated how those values embedded in Indian culture and religion need to be explicitly addressed in order to bring about the changes sought – in this case specifically in relation to 'cleaning and waste discharge' – in a culture where 'cleaning' was not an activity that one was proud to be seen doing - and to water conservation – in the example presented in Rajasthan, water conservation was achieved by directly appealing to religious values. The paper demonstrated that when those values that govern a certain behavior are explicitly addressed, then it is possible to bring about a desired change which previously had failed.

In summary, these papers pointed to the fact that we can't just apply a new model, or a blanket of sustainability, over an old system. People need to internalize the need for change, from their own personal values and culture. The papers also point to the need for the dawn of a new era of governance in which there is movement from traditional scientific management in tune with a centralised hierarchical bureaucracy, to systems of natural resource management which are societal-driven, from the ground up - a more inclusive and integrated approach in which the local community is fundamentally engaged rather than marginalized from a central bureaucracy.

Personal values are the drivers of change for sustainability – for the change to be owned by an individual or by an organization, these values may need to be made explicit – what processes are effective for values to be expressed and brought in alignment with the change sought? To what extent do we need a religion to facilitate this? Can't "culture" achieve what we want without appealing to religious values?

Papers from this theme (and others) illustrate a range of ways in which systems practitioners engage with 'projects' or attempt to improve situations. What can we say about the role of the systems practitioner as an agent for change?

In particular, to what extent should we:

- Be transparent about our own agendas/values and how are we best to work with the agendas/values of participants?
- Engage with and participate in identifying and framing problems?
- Provide expertise/ offer an agenda/ suggest a vision?
- Become engaged in the process and influence the direction or just observe and report?
- Challenge assumptions around linear progression and assist to navigate through the maze?
- Monitor the process/illustrate the journey/ document the process and outcomes/ be accountable?
- Maintain contact after the project has ended?

The papers in this theme implied that the authors have different positions on these questions.

For example, in the beach erosion paper (Love & Cooper) the change agent challenges the science underlying orthodox interpretations of beach erosion events. On the basis of their hypothesis they reframe the problem and advocate a radically different approach to coastal development planning. They recognise the political nature of this conclusion but as systems practitioners have not as yet attempted to go further and examine the culture and values of the system in which they are advocating change. If their hypothesis is championed it may generate sufficient momentum to impact on planning policy and practice but this is unlikely if it remains in the scientific realm.

In a similar way, the sustainability accounting paper (Osborne) analyses the rhetoric and practice around the subject and draws a conclusion that the field is in a mess, with no clearly dominant and thoroughly elucidated approach or methodology gaining widespread acceptance. The paper works towards developing a philosophical and theoretical framework to give shape and structure to the mess. The author remains in an academic position. For the field to become coherent and to generate positive activity will require robust intellectual dialogue that is grounded in reality from a critical mass of participants, and for that endeavour to link with policy and practice.

Both of these papers essentially criticise present practice and advocate new directions from an academic viewpoint. As such they may be part of the stimulus that nourishes an epistemic community that ultimately influences the situation. By contrast, the other 4 papers describe past or current actions to improve problem situations. These actions achieved contrasting degrees of success and the authors have played varying roles. As leader of the North Katoomba Catchment project, the author (Smith) has evidently played an active role in generating local participation with very positive consequences and strong prospect for this to be a sustainable effort. By contrast, the author of the Liverpool Council paper (Parissi) was a participant/observer in a broader Council initiative essentially as an outsider. From that perspective he was able to observe the culture of the council as being at best unhelpful or at worst incompatible with the Partnership Project.

The final two papers embrace non-linear systems perspectives more apparent in India and Bhutan and contrast them with progressivist perspectives that in one paper (Kumar and Sankaran) have stimulated actions that are incompatible with local cultural practices and values. This paper suggests the existence of an epistemic community that is building on theory and practice to generate culturally appropriate actions with positive environmental consequences, with systems practitioners in both industry and in academia being active participants in and informed commentators on the process. In the paper on Bhutan's Gross National Happiness Project, the author (Lamberton) is an outside observer who is scrutinising western sustainability rhetoric through the lens of a country that remained apart from western development pressure and is making a genuine attempt to retain key Buddhist philosophies and apply them to a small nation steering an independent course in an era of globalisation. The prospect of impacting on the currently dominant neo-classical paradigm is daunting, but bringing alternative views into public dialogue is clearly critical.

The lesson I draw from these papers is the critical importance of having a strong epistemic community generating systems perspectives to influence culture and values actively within key organizations.

Giving Shape and Structure to the Mess of Sustainability Accounting

Dick Osborn
Centre for the Public Awareness of Science, The Australian National University, AUS

Pidd (1996) defines an unstructured situation where disagreement exists as to what needs to be done and why as a mess, and where it is impossible therefore to say how it should be done. Finding shape and structure are the means towards resolving a mess. This work-in-progress establishes sustainability accounting as a mess. New ideas transmitted from a mess are likely to inhibit behavioral change. The Innovation Adoption Curve is used to test this proposition. Adoption rates of ISO 14001 and the Global Reporting Initiative Guidelines cannot reach the critical mass interval on the Adoption Curve within any reasonable planning horizon. Adoption rates for an accounting tool for communities of place, such as Local Agenda 21, are significantly higher. Three hierarchical models are then aligned to offer a structure for going beyond the messy consequences from past initiatives in sustainability accounting. Daly's capitals hierarchy is aligned with an Australian hierarchy of communities-of-place, and with the UN's hierarchical classifications of human, natural, and built capital stocks.

1. Introduction

This work-in-progress argues for simplifying the communication and organizational learning tasks of managing for sustainability transition. It advocates decision-makers monitor progress within their community-of-place by assessing change in its capital stocks, doing so across multiple levels with existing metrics and institutional arrangements. Argument is presented in the following three sections.

Section 2 uses a definition from management science to establish the collective efforts in sustainability accounting over past decades as a mess. Widespread behavioral change towards sustainability seems unlikely when messages promoting its accounting practices are transmitted from a mess. Section 3 supports this view by providing examples of progress along the shape of the Adoption Curve for popular innovations in sustainability accounting. Resolving the mess and its consequences is essential, since managing for sustainability transition depends on engaging most, if not all, decision-makers (e.g., Carver, 2001; Walker, *et al.*, 2002; Backstrand, 2003; Siebenhuner, 2004). Section 4 identifies integrated assessment of change in a place's capital stocks as an effective means to engage many decision-makers, and to transmit information across the multiple levels of sustainability governance. A framework for structuring the process and content of an integrated assessment on community capital are presented.

2. Establishing sustainability accounting as a mess

A mess is an unstructured situation where disagreements exist on what needs to be done and why, and therefore where it is impossible to say how it should be done (Pidd, 1996). Giving shape and structure precedes resolving a messy situation (Richey, 2002; Mackenzie, *et al.*, 2006). Two entry points seem useful for the mess considered here: finding a definition of accounting to accommodate the interests of accountants and non-accountants, and establishing when the communication of new ideas on sustainability accounting began. Two channels are used to connect transmitter and receiver in establishing a mutual understanding on the worth of new ideas: interpersonal channels and mass media channels (Rogers,1995: 17-18). Considering when sustainability accounting began must rely on evidence from transmitters using one-to-many channels.

Peskin (1998) defines accounting as providing decision-makers at household, business, and government levels with a structured body of information, where movements in a system's inputs and outputs during an accounting period are described relative to its state of balance. Peskin applies this definition to financial accounting, and to accounting for environmental sustainability.

The US National Erosion Reconnaissance Survey of 1934 meets Peskin's accounting definition as a structured body of information describing the extent and nature of imbalances in a land use system (Natural Resources Conservation Service, 2001). Its pioneering methods were repeated in Australia a decade later (Rural Reconstruction Commission, 1944), and have been adopted and adapted since by many agencies. A research literature on social and environmental accounting practice in the corporate sector began in the early 1970s (Mathews, 1997). In 1973, the US Water Resources Council (WRC) mandated selected federal agencies apply common principles and accounting standards when evaluating and reporting proposals for water and land-related project. The US WRC regulations require the construction of four accounts: national economic development, environmental quality, regional economic development, and social well-being (Water Science and Technology Board, 2004). Some member countries in the OECD (including Australia) compiled and published State-of-Environment-Reports in the 1970s (OECD, 1979).

The evidence shows mass media channels were being used between 40-70 years ago to communicate innovations in sustainability accounting to practitioners at multiple scales of decision-making. Even so, significant disagreements as to the why and what of sustainability accounting practice remain. Examples include:

- *It [the sustainability debate] is prone to inclusive political correctness and the accounting profession should not attempt to respond to all the different agendas and expectations* (Institute of Chartered Accountants in England and Wales, 2002) v. *There are no other binding commitments remotely achievable at the World Summit on Sustainable Development that could be more valuable than a commitment to creating the means to authoritatively assess progress to sustainable development* (Hales & Prescott-Allen, 2002).

- *At the heart of accounting is the measurement of financial transactions which are transfers of legal property rights made under contractual relationships. Non-financial transactions are specifically excluded due to conservatism and materiality principles* (Wikipedia 2006) v. *The task of the accounting profession in relation to intangible assets and knowledge-based enterprise is less about counting than it is about giving an account – telling the story of both tangible and intangible assets in meaningful ways, for both managers and markets* (Lambe, 2002).
- *Correct accounting is good economics and good economics accounts correctly... Green accounting does not provide a method for accounting for sustainability and cannot be massaged, manipulated or extended to do so* (Cairns, 2006) v. *Sustainability accounting desegregates the internal accounts to show costs and benefits relating to economic, social, and environmental performance. It also extends the accounting boundary to consider the monetary value of external impacts* (Forum for the Future, 2003).
- *Developing and using information on environmental performance and conditions is critical to any environmental management framework and must be incorporated at all levels of decision-making. There is a need to develop, agree upon, and apply common metrics for measuring and reporting environmental performance of products, households, services, firms, facilities and the economy* (President's Council for Sustainable Development, 1999). v. *675 tools applicable to the assessment of sustainability in urban development were identified, with 165 of these undergoing evaluations against a set of criteria identified as important for the integrated assessment of urban sustainability. Additionally, stakeholders, including urban decision-makers and tool developers were canvassed on the strengths and weaknesses of current assessment tools and future user requirements. The results of this work confirmed that there is no tool currently capable of simultaneously covering all assessment criteria* (Walton, *et al.*, 2005).

3. Using the Innovation Adoption Curve to illustrate the mess's consequences.

The communication research field of innovation diffusion began in 1903 (Rogers, 1995: 39-40). It now contributes, for example, to evidence-based policy and practice (Nutley, *et al.*, 2002); and to policy design for dealing with complex adaptive systems (Rogers, *et al.*, 2005). Progress in adopting accounting innovations can be estimated by combining a generic Innovation Adoption Curve with secondary data on the number of adopters, and on the number of decision-making units at saturation point (Osborn, *et al.*, 2002).

3.1 The generic Innovation Adoption Curve

The cumulative rate at which decision-makers adopt innovations within a given social system generally follows an S-curve over time (Rogers, 1995: 11). The take-off interval on the generic curve (Figure 1) is predicted to occur when some 10-20% of all decision-makers within the social system under consideration have adopted the innovation. Original work on the strength of weak ties by Granovetter (1973) provides the basis for this prediction, and for much of present understanding on how social networks operate. For example, Granovetter's work has popular acceptance through Gladwell's (2000) description of epidemic diffusion thresholds.

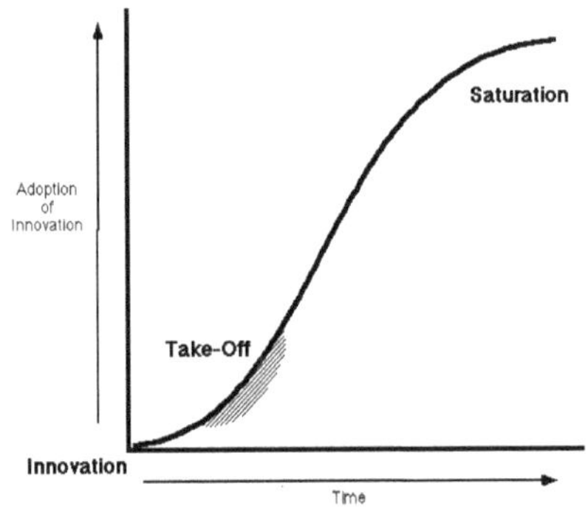

Figure 1 *The innovation adoption curve (adapted from Rogers, 1995)*

3.2 Mandatory versus voluntary adoption in the US

The generic Innovation Adoption Curve of Figure 1 illustrates the cumulative rate of adoption over time where the decision made to adopt or reject an innovation is voluntary. Adoption can also occur through an authority-innovation decision (Rogers, 1995: 28-30). Political judgements determine the extent and nature of adopter engagement within a social system as part of an authority innovation decision.

Differences between mandatory and voluntary innovation adoption provide one illustration of consequences from the mess of sustainability accounting. Figure 2 shows adoption by US establishments of two sustainability accounting practices. Trends in adoption rates are expressed relative to the known number of establishments at the saturation point of the Innovation Curve.

Some differences between the two practices are as follows:

1. The Toxic Release Inventory (TRI) Program began in 1987 as a response to the Bhopal disaster (US EPA, 2006). Establishments operating above specified thresholds within the manufacturing, oil refinery, and public utilities industries are required to disclose their location coordinates, plus provide data on their release of specified toxic chemicals and other wastes. Establishment numbers within targeted industries declined slightly from 410, 000 to 400,000 during 1990-2002 (OECD, 2006). Of these, close to 6% qualified each year in that interval as TRI Reporters (US EPA, 2006).

2. Certification of performance under ISO's Standard 14001: Environmental Management Systems. In 1996, the International Standards Organization (ISO) released a standard set of procedures for any organization in any in-

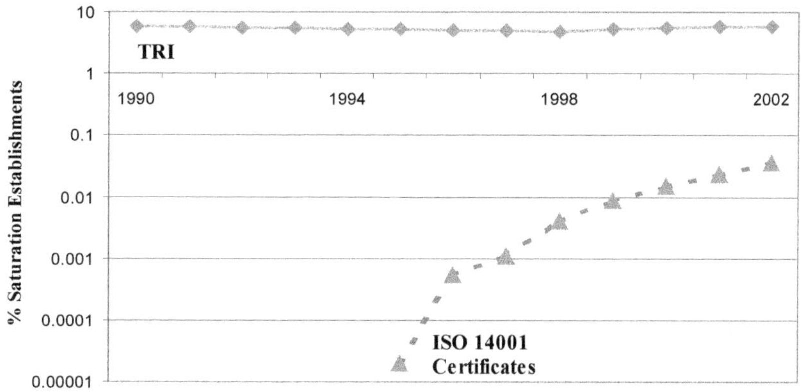

Figure 2 *Mandatory v. voluntary adoption of sustainability accounting in the US (sources: ISO 2000, ISO 2004, OECD 2006, US EPA 2006)*

dustry to follow in forming, implementing, evaluating, and communicating its environmental management system (EMS). OECD data indicates the size of the US social system meeting the 'any organization, any industry' eligibility criteria went from some 6.1M private establishments in 1995 to 7.1M in 2002. ISO 14001 procedures follow the Plan-Do-Check-Act steps in the management cycle of continuous improvement (ISO 2006). Should an organization decide to communicate its performance to external stakeholders, it seeks a certificate on its EMS from an accredited auditor. Global surveys on the number of ISO 14001 Certificates issued are conducted annually. Results illustrated in Figure 2 reflect growth in ISO 14001certifications went from 0.00003% to 0.01015% of saturation between 1995 and 2002.

3.3 Comparing ISO 14001 Certifications among some OECD members

The current inventory of the ISO contains nearly 15, 000 standards, with its Standard 9000: Total Quality Management and Standard 14001: Environmental Management Systems being by far the most popular (ISO 2004). Figure 2 data sources are used again to provide another example of a sustainability accounting tool's progression along the Innovation Adoption Curve. Good sustainability policy built on an understanding of the Curve will use the instruments available to get adopters within a targeted social system to the critical mass interval or tipping point within the shortest possible time, and with the least economic and political costs. Table 1 is built on a no-policy change scenario. It assumes average annual growth in ISO 14001 adopter numbers during 1998-2002 will continue into the future. The assumption is then used to estimate the time required to reach an ISO 14001 tipping point within each of the twenty-four countries in the OECD sample. Estimates shown in Table 1 for reaching critical mass range between 480 to 5, 500 years from present.

3.4 Comparing adoption in Australasia between community-of-practice and community-of-place

As with the US TRI Program and Bhopal, the Global Reporting Initiative (GRI) originated as a response to an environmental disaster: in this case the Exxon Val-

Global Region	Countries in sample (n)	Establishments in Critical Mass Target (n)	Years to Critical Mass
Asia	2	471, 000	5, 500
Oceania	2	363, 000	3, 400
Americas	2	1, 200, 000	1, 800
Europe	18	2, 500, 000	480

Table 1 *Years to critical mass for ISO 14001 adoption - estimates for 24 OECD countries (sources: ISO 2000, ISO 2004, OECD 2006)*

	GRI	LA21
Organizations at Saturation Level	2.4M	920
Organizations @ Critical Mass (15% Saturation)	360, 000	138
Adopters as % Saturation in 'most recent' year (1)	0.0028%	23%
Adopters as % Saturation in ('most recent' - 5)	0.0010%	5%
Years from 'most recent' to Critical Mass	43, 000	-2
(1) 'most recent' for GRI = 2006; for LA21 = 2001		

Table 2 *Progress in GRI and LA21 adoption - Oceania circa 1996-2006*

dez oil spill of 1989 (Consumer Protection Working Group, 2002). The GRI is identified as the best seed on which to grow a uniform global framework for reporting any organization's performance in relation to sustainability (e.g., Ranganathan, 1999; UN Division for Sustainable Development, 2002). Adoption and adaptation of GRI is driven by seeking consensus among stakeholders, and involves organizations located in sixty countries (GRI, 2006). The GRI is probably the most well known among many hundreds of sustainability accounting tools, where significant efforts are made to create a community-of-practice.

Local Agenda 21 Planning (LA21), is probably the most well known among many hundreds of sustainability accounting tools, where significant efforts are made to sustain a community-of-place. Its origins can be traced back to preparing for the 1992 Earth Summit, since Section 28.28 of Agenda 21 requires local authorities *undertake a consultative process with their populations and achieve a consensus on a 'local Agenda 21' for the community* (UN Division for Sustainable Development, 1993). LA21 therefore shares with GRI the process of consensus seeking among stakeholders. LA 21 differs from GRI by using accounting to assess environmental, economic, and social conditions within a community-of-place built from many organizations, rather than one. LA21 thereafter uses performance indicators for tracking progress towards needs in sustainability conditions for a community-of-place.

Supplemented with data on adopter numbers and on the number of local authorities, OECD statistics on private establishments across all industry sectors again provides a platform for comparing the adoption of two sustainability accounting tools. Table 2 shows results from this comparison. Adoption of GRI and LA21 increased some three or four fold over the five years of most recent data available. Differences between the two practices on time remaining before reaching critical mass on their respective Innovation Adoption Curves remain substantial. If growth in LA21 adoption rates between 1996-2001 continued, then around 2004 all local authorities in Australia and New Zealand could be engaged in this form of sustainability accounting. In Oceania, reaching critical mass with GRI does not seem possible within any reasonable planning horizon, should growth achieved under its arrangements and institutional settings of 2002-2006 continue.

Differences between GRI and LA21 with respect to interpersonal communication channels could be one reason for differences in their progress along the Innovation Adoption Curve. Rogers (1995: 82) discusses the significant differences between interpersonal and mass-media channels, with many small-scale empirical studies showing the former to be more effective in achieving a positive adoption decision. Community Innovation Surveys conducted in Europe, however, provide the most robust evidence. For example, a significant majority (>70%) of some 201, 000 enterprises adopting innovations during 1998-2001 did so through interpersonal communication channels within their own industry and market (Eurostat, 2004). A minority (<10%) identified governments, universities and research institutions as important sources of information in innovation adoption. The ratio between non-adopter and adopter organizations also provides some understanding of the interpersonal communication task facing promoters of sustainability accounting. In Oceania, for GRI this ratio is around 36, 000:1, compared with a ratio of 4:1 for LA21.

3.5 Summarizing results

Some 40-70 years of using mass-media channels have failed to resolve disagreements between scientists and practitioners as to the why and what of sustainability accounting practice. The resulting mess transmits conflicting information, increasing the uncertainty that always accompanies the decision to adopt or reject an innovation (Rogers, 1995: 6). Section 3 reflects this uncertainty by combining a generic Innovation Adoption Curve with secondary data on relatively well-known examples of sustainability accounting practices. LA21 in Oceania is the only case among examples considered where the critical mass benchmark is achieved. The activity thresholds of the US TRI Program limit disclosure requirements to one third of the critical mass benchmark within the social system it targets. The large social systems they target means voluntary adoption of ISO 14001 and GRI is unlikely to reach critical mass within any reasonable planning horizon. Results support Peskin's (1998) assertion that communication failure contributes to poor adoption of accounting for sustainability.

4. Resolving the mess's consequences

Giving shape and structure are necessary steps in dealing with a mess. Peskin's 1998 definition sees any form of accounting as a structured body of information describing change in a system's inputs and outputs relative to its state of balance. Anielski's definition of sustainability accounting specifies the state of balance sought: *In the Genuine Wealth model, sustainability is being achieved when the overall integrity of the five core capital assets of a com-*

munity or organization (human, social, natural, built and financial capital) are non-declining in their overall physical and qualitative 'condition'. In this model, the conditions of the capital stocks and flows can be reported both in physical/ qualitative terms (composite indices) and monetary (full cost) accounting terms (Anielski 2005).

Concepts and case studies on a capitals approach for tracking progress toward sustainable development are common at many levels of decision-making. Examples for businesses include Dyllick and Hockerts (2002), Sigma Project (2003); for local communities include Kretzmann and McKnight (2005); Roseland (2005); for engaging community stakeholders in qualitative assessments include Grosskurth and Rotmans (2005), Osborn and Macfarlane (2006); and for national governments include UN Statistics Division (2005), World Bank (2006). Anielski's Genuine Wealth model provides therefore an overarching framework for accommodating many initiatives. What could a structured body of information look like when combining the Genuine Wealth model with multiple levels of sustainability governance? Accepting the evidence on adoption rates presented elsewhere pushes policy design toward promoting existing ideas and common practices.

Hierarchical structures are used frequently as mental models in learning for sustainability transition (Meadows 1998). Figure 3 aligns three existing hierarchies to create a structured body of information consistent with Anielski's Genuine Wealth model, and to realize the opportunities it provides for convergence between many initiatives. The hierarchy of capital stocks, where nature provides the means for achieving human well-being was proposed by Daly in 1973 (Meadows, 1998:41), and accepted by many as a guiding principle for structuring sustainability information. Each country has its own statistical geography where communities-of-place exist at different levels, and with their own sizes in terms of decision-making units. An Australian example is provided. The classification systems established as standards and guidelines by the UN's Statistical Commission provide hierarchical structures for codifying natural, built and human capital stocks with consistent, credible and comparable metrics. Data collection and dissemination will vary over time and space, but regularly engage many decision-makers through census and sample collections.

The shape of the Innovation Adoption Curve, and the body of information possible from aligning three hierarchical structures, provide insights into the why, what, and how of sustainability accounting.

References

Anielski M (2005), Measurement, in Office of Research and Development, US EPA *A Forum on Sustainability, Well-Being, and Environmental Protection: What's an Agency to Do?* www.epa.gov/sustainability/econforum/pdfs/2005_Econ_Forum_Panelist_Precis_Papers.pdf.

Australian Bureau of Statistics, Catalogue # 1209.0.55.001. *Draft Mesh Blocks Australia.* www.abs.gov.au.

Australian Bureau of Statistics, Catalogue # 3236.0. *Household and Family Projections Australia 2001 to 2026.* www.abs.gov.au.

Australian Bureau of Statistics, Catalogue # 8161.0.55.001. *Business Register Counts of*

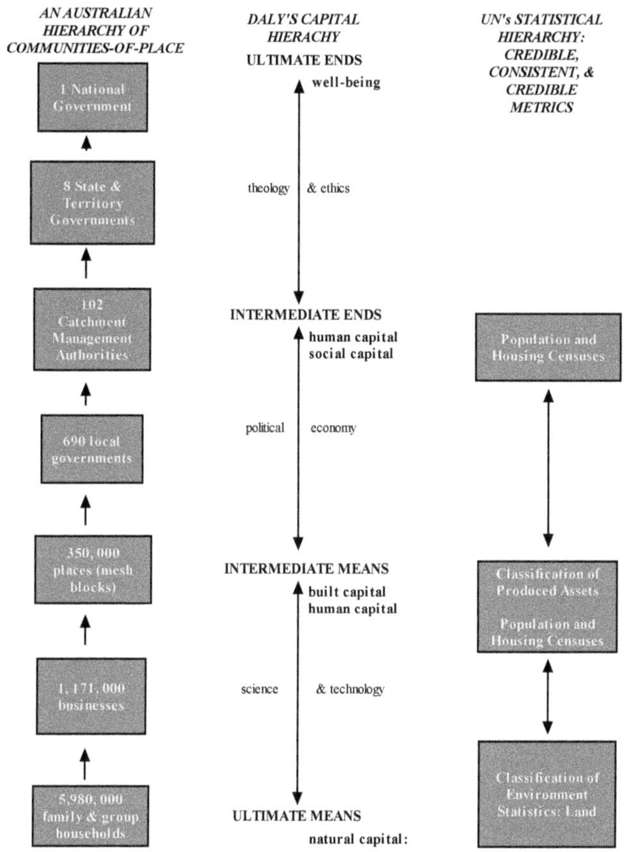

Sources: ABS 1209.0.55.001; ABS 8161.0.55.001; ABS 3236.0; Meadows 1998; National Office of Local Government 2006; National Land & Water Resources Audit 2002; UN Statistics Division 2004; UN Statistics Division 2006a; UN Statistics Division 2006b.

Figure 3 *Aligning Three Hierarchies to Structure Sustainability Accounting*

Businesses, Summary Table 2004. www.abs.gov.au.

Backstrand K (2003), Civic Science for Sustainability: Reframing the role of experts, policy-makers and citizens in environmental governance. *Global Environmental Politics* 3 (4): 24-41.

Cairns RD (2006), *Accounting for the Environment or for Sustainable Development?* Minerals Council of Australia. www.minerals.org.au/_data/assets.pdf_file/5984/24-2CairnsRobert.pdf.

Carver S (2001), *Participation and Geographic Information: a position paper*. ESF-NSF Workshop on Geographic Information and Participatory Approaches. www.shef.ac.uk/~scgisa/spoleto/carver.pdf.

Consumer Protection Working Group (2002), *The Desirability and Feasibility of ISO Corporate Social Responsibility Standards: Report from the Consumer Protection Working Group to the ISO Consumer Policy Committee (COPOLCO)*. www.jisc.go.jp/policy/pdf/Final_report_on_the_feasibility_and_desirability_of_ISO_CSR_standards.pdf.

Dyllick T and Hockerts K (2002), Beyond the business case for corporate sustainability. *Business Strategy and the Environment*, 11: 130-141.

Eurostat (2004), *Innovation in Europe: Results for the EU, Iceland and Norway*. 2004 Edition. ISBN 92-894-7262-6.

Forum for the Future (2003), *Accounting for Sustainability: Guidance for Higher Education Institutions*. www.forumforthefuture.org.uk/docs/publications/185/accounting%20for%20sustainability.pdf.
Gilbert s (2003), *The GRI – An Institution and Emerging Standard*. www.bcsd.org.tw/images/doc/701/20030818/01.PPT.
Gladwell M (2000), *The tipping point how little things can make a big difference*. Boston: Little Brown. ISBN 0-316-31696-2.
Global Reporting Initiative (2006), About GRI. www.globalreporting.org/about/.
Granovetter MS (1973), The Strength of Weak Ties, *American Journal of Sociology*, 78: 1360-1380.
Grosskurth J and Rotmans J (2005), Getting a grip on sustainable development in policy making, *Environment, Development and Sustainability*, 7 (1): 135-151.
Hales D and Prescott-Allen R (2002), Flying Blind: Assessing Progress Toward Sustainability, in Esty D and Ivanova MH (eds), *Global Environmental Governance: Options and Opportunities*.
International Council for Local Environmental Initiatives (2002), *Second Local Agenda 21 Survey: Background Paper 15*. www.iclei.org/documents/Global/final_document.pdf.
International Standards Organization (2001), *The ISO Survey of ISO 9000 and ISO 14000 Certificates: Tenth Cycle*, Geneva, ISO.
International Standards Organization (2004), *The ISO Survey of ISO 9001:2000 and ISO 14001 Certificates – 2003*. Geneva, ISO.
International Standards Organization (2006), *ISO 9000/ISO 14000*. www.iso.org/iso/en/iso9000-14000/ims/ims.html.
Institute of Chartered Accountants in England and Wales (2002), Quotation on p8, in Adams R, *Accounting: Industry as a partner for sustainable development*. London. Association of Certified Chartered Accountants and UN Environment Programme. ISBN: 92-807-21747-7.
Kretzmann JP and McKnight JL (2005), *Discovering Community Power: A guide to mobilizing assets and your organization's capacity*. The ABCD Institute.
Lambe P (2002) Accounting for Knowledge Management, Presentation to 16th World Congress of Accountants. www.greenchameleon.com/thoughtpieces/account.pdf.
Mackenzie A, Pidd M, Rooksby J, Sommerville I, Warren I, and Westcombe M (2006), *Wisdom*, decision support and paradigms of decision making, *European Journal of Operational Research* 170: 156-171.
Mathews MR (1997), Twenty-five years of social and environmental accounting research: Is there a silver jubilee to celebrate? *Accounting, Auditing and Accountability Journal*, 10, 4: 481-531.
Meadows D (1998), *Indicators and Information Systems for Sustainable Development: A Report to the Balaton Group*. Hartland Vermont. The Sustainability Institute.
National Office of Local Government (2006), *Local Government National Report: 2004-05 Report on the operation of the Local Government (Financial Assistance) Act 1995*. www.dotars.gov.au/local/publications/reports/2004-2005/pdf/Local_report.pdf.
National Land and Water Resource Audit (2002), *Australian Catchment, River and Estuary Assessment 2002*. http://audit.ea.gov.au/ANRA/coasts/docs/estuary_assessment_Est_Ass_Catchment.cfm.
Natural Resources Conservation Services (2001), *Natural Resources Inventory: Background*. www.nrcs.usda.gov/TECHNICAL/NRI/1997/docs.background.pdf.
Nutley S, Davies H, and Walter I (2002), Conceptual Synthesis 1: Learning from the Diffusion of Innovations. *Working Paper Series: 10*, ESRC UK Centre for Evidence-Based Policy and Practice.
www.st-andrews.ac.uk/~ruru/Learning%20from%20the%20Diffusion%of%20Innovations.pdf.

OECD (1979), *The State of the Environment in OECD Member Countries*, Paris.OECD.

OECD (2006), Number of establishments and enterprises in OECD member countries. www.oecd.org/statsportal/.

Osborn D, Savage D, Reyes MF and Muradyan T (2002) *Images of effectiveness, equity and efficiency in the diffusion of environmental management accounting*. www.emawebsite.org/documents/emaric_316.pdf.

Osborn D and Macfarlane M (2006), Sustaining communities by learning from integrated assessments of place, in Petheram RJ. and Johnson RC (eds.) *Practice change for sustainable communities: Exploring footprints, pathways and possibilities*: APEN 2006 International Conference, La Trobe University, Beechworth, Victoria, Australia, 6 – 8 March 2006. Published online at www.regional.org.au/au/apen/2006. The Regional Institute. ISBN 1 920842 31 4.

Pidd M (1996), *Tools for Thinking: Modelling in Management Science*, John Wiley and Sons, Chichester.

Peskin H (1998), Alternative resource and environmental accounting approaches and their contribution to policy, in Uno K and Bartelmus P (Eds) *Environmental Accounting in Theory and Practice*. Kluwer Academic Publishers Dordrecht/Boston/London ISBN 0-7923-4559-2.

President's Council for Sustainable Development (1999), *Towards a Sustainable America: Advancing Prosperity, Opportunity, and a Healthy Environment for the 21st Century*. http://clinton2.nara.gov/PCSD/Publications/tsa.pdf.

Ranganathan J (1999), Signs of Sustainability: Measuring Corporate Environmental and Social Performance, in Bennett M and James P (eds), *Sustainable Measures: Evaluation and Reporting of Environmental and Social Performance*. Greenleaf Publishing, Sheffield. ISBN 1874719160.

Richey T (2002), *Modelling complex socio-technical systems using morphological analysis*. www.swemorph.com/.

Rogers EM (1995), *Diffusion of Innovations*, Fourth Edition. New York. Free Press. ISBN 0-020926671-8.

Rogers EM, Medina UE, Rivera MA, and Wiley CJ (2005), Complex Adaptive Systems and the Diffusion of Innovations, *The Innovation Journal: The Public Sector Innovation Journal*, 10(3), article 30. www.innovation.cc/volumes-issues/rogers-adaptivesystem7final.pdf.

Roseland M (2005), *Toward Sustainable Communities*. Third Edition. Vancouver, Canada. New Society Publishers. ISBN: 0-86571-535-1.

Rural Reconstruction Commission (1944), *Land utilization and farm settlement*, The Commission's Third Report. Canberra. Australian Government Printer.

Siebenhuner B (2004), Social learning and sustainability science: which role can stakeholder participation play? *International Journal for Sustainable Development* 7(2): 146-163.

Sigma Project (2003), *The Sigma Guidelines: Putting sustainable development into practice – a guide for organisations*. London, British Standards Association. ISBN: 0-580-42448-0.

UN Commission for Sustainable Development (1997), *A study of responses by local authorities and their national and international associations to Agenda 21: prepared by ICLEI and UNDPCSD*. www.un.org/documents/ecosoc/cn17/1997/background/ecn171997-1.rept1.htm.

UN Division for Sustainable Development (1993), *Agenda 21: Earth Summit – The United Nations Programme of Action from Rio*. ISBN 9211005094.

UN Division for Sustainable Development (2002), *Johannesburg:World Summit on Sustainable Development: Plan of Implementation*. www.un.org/esa/sustdev/documents/WSSD_POI_PD/English/POIToc.htm.

UN Statistics Division (2003), Integrated Environmental and Economic Accounting 2003

(SEEA 2003). http://unstats.un.org/unsd/envAccounting.seea.htm.
UN Statistics Division (2004), *UN Statistics Division/UN Environment Programme Questionnaire 2004 on Environment Statistics: Land.* http://unstats.org/unsd/environment/q2004land.pdf.
UN Statistics Division (2006a), *Population and Housing Censuses.* http://unstats.un.org/unsd/demographic/sources/census/census/censusmethods.htm.
UN Statistics Division (2006b), *Classification of Produced Assets.* http://unstats.un.org/unsd/Sna1993/tables/table_Annex_13.pdf.
US Environmental Protection Agency (2006), *Toxic Release Inventory (TRI) Program.* www.epa.gov/tri/.
Walker, B., S. Carpenter, J. Anderies, N. Abel, G. S. Cumming, M. Janssen, L. Lebel, J. Norberg, G. D. Peterson, and R. Pritchard. (2002). Resilience management in social-ecological systems: a working hypothesis for a participatory approach. *Conservation Ecology* **6**(1): 14. www.consecol.org/vol6/iss1/art14/.
Walton JS, El-Haram M, Castillo NH, Horner RMW, Price ADF, and Hardcastle C (2005), Integrated assessment of urban sustainability, *Engineering sustainability* 158 (2): 57-65.
Water Science and Technology Board, Oceans Studies Board (2004), *Analytical Methods and Approaches for Water Resources Project Planning*, Washington DC. National Academies Press.
Wikipedia (2006), *Accountancy.* http://en.wikipedia.org/wiki/Accountancy.
World Bank (2005), *Where is the wealth of nations? Measuring capital in the 21st century*, Washington DC. siteresources.worldbank.org/INTEEI/214578-1110886258964/20748034/All.pdf.. Accessed 1st February 2006.

Acknowledgements

Suggestions from an anonymous reviewer on an earlier draft are gratefully acknowledged, as are those from Sue Stocklmayer and Leonie Pearson. The paper remains a work-in-progress. Comments on further improvement are most welcome.

Richard (Dick) Osborn's research addresses the communication challenges of sustainability science. It is a place-based, integrative, and civic science able to engage stakeholders from multiple decision-making scales in designing local futures. His work builds on some four decades as an information broker. He began as an agricultural extension officer, providing a communication channel between cattle husbandry researchers and producers. Then as broker between scientists from many disciplines and research institutions, and with local stakeholders, working on integrated natural resource management. That work extended over time from the Darling Downs, through the Upper Darling Basin, to the Murray-Darling Basin. As Senior Policy Analyst within the Australian Local Government Association, he then worked in all areas of public policy, at national scale, and with actors in each of the three government spheres in Australia's federal system. His responsibilities in the early 1990s led to concentrating on sustainability issues. They included contributions to drafting and implementing Australia's National Strategy for Ecologically Sustainable Development, its Inter-Government Agreement on the Environment, and, at global scale, on the role of local authorities in Agenda 21. He has been working on accounting design to sustain communities since the mid-1990s. Much of this work is within Australia. It also includes networking internationally, particularly by invitation in an Expert Working Group on Environmental Management Accounting established by the UN's Division for Sustainable Development.

A Systemic Understanding of Environmental Degradation of the River Ganga

Madhu Ranjan Kumar & Shankar Sankaran
Indian Railways, IND
Faculty of Design, Architecture and Building, University of Technology, Sydney, AUS

This paper explains the pollution of the River Ganga as arising out of three reasons:

1. **'Energy' as the source of improvement of the Ganga basin;**
2. **Lack of communication among different worldviews about the Ganga;**
3. **Lack of regulation to the fast pace of 'improvement' brought about by 'energy'.**

Then this paper indicates one instance of improved waste water management, one instance of improved fresh water management and a case study of improved cleanliness in a factory which was earlier a polluter of the Ganga.

1. Introduction

"The Ganga is the most important river system in India with almost one-tenth of the world's population living within its basin. It is also one of the most polluted rivers in the country" (Markendya & Murty, 2000). The Ganga Action Plan (GAP) started in 1985, aimed to clean the river. It involved among others, cleaning the sewage discharge and treatment of the industrial affluent before these are released into the river and construction of electric crematorium for burning of dead bodies. The result of 13 years of pollution control is shown in Table 1.

This shows that the handling of the affairs of pollution control of the Ganga has produced mixed results. This brings us to first develop a generic understanding of how we have handled our affairs of environmental management.

2. Two approaches to environmental management

It has been argued that human affairs display the interaction of two processes – the goal seeking process and the process of generation of multiple and often conflicting standards for appreciating and regulating ongoing relationships (Vickers, 1970: 202). On these lines, this paper argues that the approach to environmental management can be looked in two ways. The dominant one is based on the Western rational-scientific ethos. The other approach can be called a social approach which aims at appreciating different and constantly changing worldviews. Here actions emanate from an understanding of the different worldviews. In trying to understand the present threat to the River Ganga, let us begin with the goal seeking process of the world.

S.No	Location	1986 (summer)		1999 (summer)	
		DO(mg/lt)	BOD(mg/lt)	DO(mg/lt)	BOD(mg/lt)
1.	Haridwar	8.1	1.8	8.6	1.2
2.	Allahabad	6.6	15.5	7.9	3.2
3.	Patna	8.1	2.2	7.8	2.4
4.	Calcutta at Uluberia	6.9	2.1	6.6	2.2

Table 1: *The level of pollutants in Ganga with respect to pollution parameters (Source: http://haridwar.nic.in/gap.htm).*

Note:
DO: Dissolved Oxygen. The desired standards fixed by the Central Pollution Control Board of Ganga is 5 mg/lt (minimum).
BOD: Bio-chemical Oxygen Demand. The desired standards fixed by the Central Pollution Control Board of Ganga is 3 mg/lt (maximum).

2.1 Social systems as goal seeking

The understanding of human behaviour as goal seeking gave rise to the problem solving approach to the issues of the mankind. This affected our response to many issues facing the world including that of environmental management. Before the industrial age, 'protestant ethics' – the foundation of Western civilisation believed in expending energy (by being industrious) and then relaxing. In contrast, the Indians are said to believe in *aram* culture. '*Aram* means rest and relaxation without being preceded by hard work' (Sinha, 1995: 101). It is not surprising that since the advent of industrial age, technology – which again is energy driven - has been the most accepted means of bringing about improvement much in line with the problem solving approach. However there is an inherent flaw in accepting improvement in life as energy driven: The second law of thermodynamics says that entropy – a measure of wasteful heat in a system - keeps on increasing if we try to change the system. This means the more we change a system the more its entropy increases. This results in an increase in the percentage of unusable heat in the system. The present problem of global warming can be understood from this basic argument. Further, one can argue based on the principles of thermodynamics that because of the innate irreversibility in any change process, the more we try to improve a system the worse it could become. This is congruent with the traditional Indian philosophy which is systemic in nature (Kumar & Sankaran, 2006) and the ideas of the systems school by system dynamics. From this point of view, our eco-system has to perforce move away from where it was say a hundred years ago to a more disordered eco-system thanks to the 'progress' we have made in this period. Thus what was once a natural system called the 'Ganga basin' became a manmade system of 'water utilization in the Ganga basin'.

Returning back to energy as the prime mover of improvement, this mindset affected the way of tackling the environmental issues. Therefore, sewage treatment was conceived in a manner which needed electricity which however is not available in the entire Ganga basin for almost half a day throughout the year. It needs proper maintenance of the sewage treatment plant for which the cash strapped state government had made no provisions. Thus sewage treatment became a costly affair – it costs about Rs 5 million to treat a million litres per day of sewage before it is discharged into the Ganga (Dasgupta, 2005). Do we see the second law of thermodynamics asserting itself in these downstream 'disorders'?

2.2 Social systems as appreciating systems

If the Ganga is polluted today, besides the goal seeking, energy driven approach to improvement, another reason that could be attributed is our inability to appreciate and reconcile different worldviews about the Ganga and a lack of communication among these different 'worldviews'. What are the different worldviews about the Ganga? – Is it a large open drain in which to discharge the pollutants of a factory or a flow of sacred water brought from Heaven in which one must bathe so as to cleanse one of all his/her sins or a major river in an irrigation system or a open pipe line to transport sweet dinking water from the Himalayas to the plains of India, or a river whose annual flooding bring rich alluvial soil or is it a river whose beautiful *ghats* (banks) can be used to attract tourists?

These worldviews show the problem of pollution of the Ganga through the second process of generation of multiple and often conflicting standards for appreciating the world. To paraphrase Vickers (1970: 203) our fears and aspirations about the Ganga have generated standards so exacting and so conflicting as to pose insoluble problems of multi-valued choice.

3. The problem of rate of change

However, a larger problem that continues (Vickers, 1970: 203) is the rate of change both in the course of events and in the categories and standards of our appreciation. In the process we have created a system which is neither stable nor regulable (Vickers, 1970: 19). Today, we are facing the problem of lack of ecological, political and economic regulation partly because we are not able to set widely acceptable and internally self-consistent standards with which we can interpret our experiences. This has resulted in three threats: the environmental threat – the water, the air – the threat to the Ganga falls in this category; the institutional threat – political, economic and social institution do not know how to adapt themselves to the energy driven 'improvements', and the appreciative threat – the need to revise and revalue, more rapidly than time permits the way we understand the events of the world which is the most serious threat (Vickers, 1970: 20).

This understanding tells us to shift our attention from the more physical and therefore the more compelling 'failure to keep the Ganga clean' to the failure to develop institutions and the failure to modify appreciation which could keep pace with the 'improvements'.

So there are three systemic issues with respect to the pollution of the Ganga:

1. 'Energy' as the source of improvement of Ganga basin;
2. Lack of communication among different worldviews of the Ganga;
3. Lack of regulation to the fast pace of 'improvement' brought about by 'energy'.

4. Alternate approach to environment management

However, there are examples of environmental rejuvenation in pockets of India which are not based on energy and in which there are ample communication between different worldviews about pollution. This paper quotes two of them and then discusses a case study which aimed at in-depth penetration of each others mind-set and communication among participant's worldviews through action science.

(i) Spread over an area of 3500 hectares, the wetland on the eastern fringe of Calcutta has been using the wastewater of Calcutta not as a pollutant but as a resource for breeding of fish (Ghosh, 1998). Using the natural eastward slope for the flow of wastewater from Calcutta, the fishermen have used their traditional knowledge of the ecosystem called '*bheries*' (wetland) to clean the wastewater. Ghosh (1998) adds:

"*Retention of wastewater in the ponds, before the initial stocking of fish, allowed bacteria to act upon the organic matter in the sewage and decompose the organic waste. The growth of these beneficial bacteria was supported by the algae that thrived in these shallow ponds under the ample sunshine. The algae also provided food for the fish. This ecosystem provides a natural kidney for the city's organic wastewater. So a double boon was taking place: The organic sewage was being treated by the natural ecosystem, and was producing rich quantities of fish food on which the local varieties thrive.*"

This is an example of approaching water-treatment in terms of resource recovery (in contrast to energy expending), 'regulation' of water quality in place of improving water quality and community participation instead of lack of communication.

(ii) In the semiarid state of Rajasthan, Laporia village was described as barren with highly saline landscape where 40% of population would migrate to adjoining districts in summer. Further, the government policies removed the community ownership of forests and water resources and therefore their responsibilities for these resources. The environment deteriorated, and water became scarce. (Sen, 2002). After a local initiative, Laporia turned into an oasis and its water table rose from 60 feet in 1991 to 15 feet in 2002. This was achieved by integrating the natural system of water cycle into the human social system. Religion was used to 'regulate' the use of water and conserve it. Earlier, holding values such as respect for natural resources was a way of life. But such basic respect for environment along with the need to share and conserve it took a back seat as the population exploded. Laxman who spearheaded the effort to conserve water combined the forgotten customs and rituals of the region with traditional methods of water conservation said, "I therefore placed all natural resources on a pedestal in the form of deities to preserve the sanctity of the environment" (Sen, 2002).

Both examples described above show instances of communication between multiple worldviews. However, it does not show them in detail. The case study narrated next shows how difficult it is to establish communication between different worldviews and then arrive at a consensual action. The case study is available in detail in Kumar and Sankaran (2005).

4.1 Environment management through and action science

The organisation where the study was done is a large manufacturing entity in the Indian public sector engaged in the manufacturing of internal combustion engines. It is located in Banaras city where the Ganga Action Plan was launched in 1986.

The manifest problem in the organisation was that there was no organized method of collection or disposal of the packing material which were received with goods that arrived at the warehouse. They were treated as waste and thrown here and there around the work place.

The initial problem for the improvement in work practice was thus defined as: *How to have an organized system for disposal of waste and cleanliness in the factory?*

The first initiative looked at the issue from mode 1 of soft systems methodology which is pictorially shown in Figure 1. A participatory approach where everyone was encouraged to give his/her point of view was adopted in the soft system methodology so that different worldviews can be understood.

However, after some months there were reports of packing waste lying

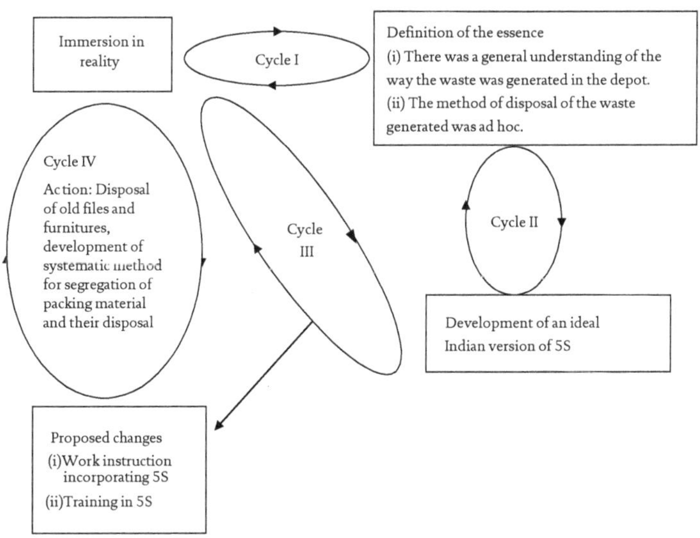

Figure 1 *Summary of different cycles in terms of SSM (Source: Based on Tay, 2003)*

in the warehouse for days. They were not being dumped in the bins earmarked for them. Sweeping of the warehouse was reported to have become irregular. If not checked, the system ran the risk of getting reverted to the old state.

Reassessing the same situation from the point of view of action science showed that there were differences between the 'espoused theory' and 'theory-in-use' of the employees because of the culture of India which considers cleaning a menial activity.

It was clear to the action research team that any effort for effective implementation of cleanliness was required to address and tackle this mind-set against cleaning. The strategies till now contributed only to the existing norms of interaction – it could be called a rational approach to housekeeping. It made no attempt to discover, much less change, the underlying values which governed the behaviour of the workshop managers and staff. It was akin to what Argyris, *et al.* (1985: 281) have called a 'culture of protectionism'- strategies that contribute to norms of interaction that inhibit learning.

Reflecting further on the situation, it gradually became clear that the intervention has to go beyond the conventional methodologies of SSM or action science. Edmondson says that Schein offers an approach for assessing organisation specific cultural patterns (Edmondson, 1996: 591). The insight provided by action science about employees' defensive behaviour inhibiting change was combined with Schein's idea of culture being sum of shared tacit assumptions and with Indian's propensity of being guided by social pressure (Derne, 1992). The revised action strategy was based on these three theoretical moorings. The revised action is shown in Figure 2.

Figure 2 shows that the earlier governing values-in-use of the group become nested in the more fundamental governing values of dignity of labour, hierarchy and power distance. When the dormant value of dignity of labour was invoked there was an imbalance in the relative magnitude of different governing values nested in each other. This disturbed the existing applecart of hierarchy which went against the power distance value of the group as a whole. Thus to bring back the values of these three governing values to the earlier level (which was the normal level), the revised behaviour of following the dictums of cleaning became necessary for everyone. This therefore triggered the revised action strategy shown in Figure 2. It is to be noted here that the feedback loop was not used to change the action strategy. It was instead used to change the governing value.

5. Discussion

The human predicament is produced by the interaction of two streams of processes (Vickers, 1970: 202) discussed in this paper. The case studies show how enduring improvement in environment is possible which need not be energy driven and where developing a common apparatus for interpreting the incoming communication of the world around and also the world within was a crucial enabler. Therefore, if our earlier ecosystem was a better ecosystem, one way to induce minimal change in it is to think beyond considering energy as the prime mover of improvement. This calls for a radically different approach

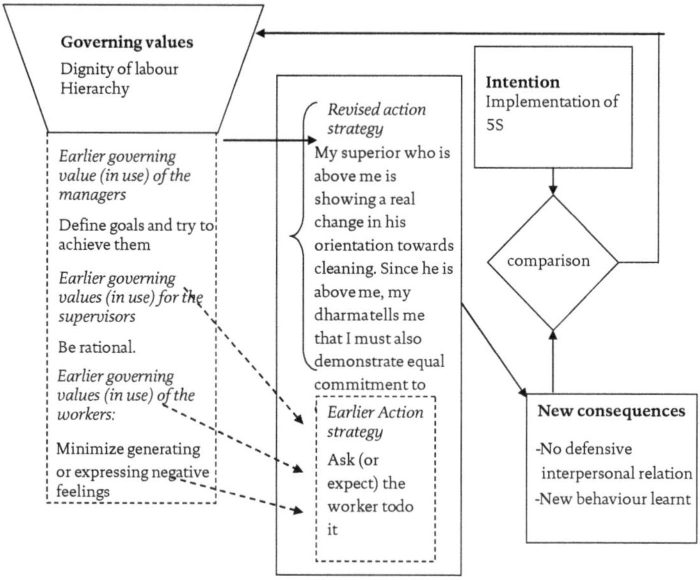

Figure 2 *Nesting of earlier governing values in more fundamental governing values and resulting action strategy and consequence of the action strategy (Source: developed for this work. Note: dashed boxes show earlier governing values and action strategies.)*

to environmental management than what has hitherto been attempted in the Ganga Action Plan. We modified our earlier ecosystem because - thanks to the domination of protestant thinking – we felt that we could control the environment to our 'preferred' standard. 'The great emancipation which the power of technology brought soon resulted in wastage of resources, pollution of the environment and multiplication of populations so as to defeat even their own productivity' (Vickers, 1970). 'As God disappeared from the Western world, there was nothing to stop us from befouling our own nest... Our belief in progress or ... in general our future orientation made us challenge the present situation, the social environment as much as the physical one... As they say, the rest is history' (Loy, 1993). This is the trap we are in (Vickers, 1970). The source of this trap is the energy driven model of improvement. Market economy – as the economic institution, parliamentary democracy – as the political institution and scientific temper or rationalism –as the social institution were supposed to provide stability to the output of the energy driven world. But what about mechanisms to regulate this energy driven world? None. The aspect of regulation cannot come when you are constantly trying to 'improve'. Homeostasis will then be dismissed as a throwback to an earlier state.

References

Argyris, C., Putman, R. and Smith, D.M. (1985). *Action science, concepts, methods and skills for research and intervention*, San Francisco: Jossey-Boss, ISBN 0875896650.

Dasgupta, S. (2005). "The desecration of the holy Rivers", available http://www.

dayafterindia.com/june205/pollution.html as retrieved on 27 May 2006.
Derne, S. (1992). "Beyond institutional and impulsive conceptions of self: Family structure and the socially anchored real self", *Ethos*, ISSN 0091-2131, 20, 259-288.
Edmondson, A.C. (1996). "Three faces of Eden: The persistence of competing theories and multiple diagnoses in organisational intervention research", *Human Relations*, ISSN 0018-7267, 49(5): 571-595.
Ghosh, D. (1998). "The Calcutta wetlands: Turning bad water into good", *Changemakers*, available: http://www.changemakers.net/journal/98october/ghosh.cfm as retrieved on 29 May 2006.
Kumar, M.R. (2006). "Total quality management as the basis for organisational transformation of Indian railways – A study in action research", DBA thesis, Southern Cross University, Australia.
Kumar, M.R. and Sankaran, S. (2005). "Implementation of 5S through action research: An Indian context", presented at the 11th International Conference on Quality and Productivity, I.I.T. Delhi, India.
Kumar, M.R. and Sankaran, S. (2006). "Theories of action - A comparison between action science and *Gita*", presented at the World Congress on Action Research, Action Learning and Process Management (ALARPM), The Netherlands.
Loy, R. (1993). "Transcendence East and West", *Man and World*, ISSN 0025-1534, 26(4): 403-427.
Markandya, A. & Murty, M.N. (2000). *Cleaning up the Ganges: A cost benefit analysis of the Ganga Action Plan*, Delhi, Oxford University Press, ISBN 0195649451.
Sen, P. (2002). "Reviving customs and rituals to deliver a people from drought", *Changemakers*, February 2002.
Sinha J.B.P. (1995) *The cultural context of leadership and power*, Delhi, Sage Publications, ISBN 0803992319.
Tay, B.H. (2003), "Using action research to develop a social technical expert system for an industrial environment", PhD Dissertation, Graduate College of Management, Southern Cross University, Lismore: Australia.
Vickers, G. (1970). *Freedom in a Rocking Boat, Changing Values in a Unstable Society*, London, The Penguin Press, ISBN 0140212051.

Mahdu Ranjan Kumar is a mechanical engineer and works in Indian Railways in a senior management capacity. His areas of interest include organisational learning, organisational change and knowledge management. He has completed a Doctor of Business Administration from Southern Cross University implementing Total Quality Management in the Indian Railways using action research to help implement change in work practices. Madhu has published refereed papers in international journals such as Systemic Practice and Action Research and the TQM Journal.

Shankar Sankaran teaches Project Management at University of Technology Sydney and supervises doctoral students researching in project management, knowledge management applying systems thinking and action research approaches. Shankar has published/presented refereed papers in conferences and journals. He is a member of the editorial advisory board of Action Research and the editorial boards of Action Research International, Australasian Journal of Business and Social Inquiry and the founding editor of the International Journal of Health and Ageing Management. Shankar and his doctoral students have presented papers covering knowledge management, systems thinking and action research in several international conferences. Shankar is the Vice President (International) of the Action Learning, Action Research and Process Management Association of Australia.

North Katoomba Catchment Restoration and Action Project

Geoffrey Smith
Blue Mountains City Council, AUS

The North Katoomba Catchment Restoration and Action Project is an innovative holistic approach to catchment management that is being adopted as a model for how Blue Mountains City Council sustainably manages its natural resources. The North Katoomba area was selected as a priority sub-catchment under the NSW State Government's Urban Runoff Control Program (URCP) and Councils Catchment Assessment and Prioritization Program (CAPP) due to a number of factors including riparian degradation, severe in-stream erosion, channel instability, sedimentation, habitat loss, weed encroachment, presence of threatened species, an active community of volunteers engaged in long running environmental activities, sites of Aboriginal significance, urban and industrial activity and the surrounding Blue Mountains World Heritage Area and Grose Wilderness Area downstream. Through a multifaceted approach, the project has successfully engaged the local community and using their vision for the area was able to pull together state government authorities, local businesses and other agencies to work collaboratively to create a range of sustainable on-ground outcomes and promote collaboration between all stakeholders. Through its whole of catchment and multi-disciplinary methodology this project demonstrates how to deliver significant and long term outcomes in natural resource management, which can be transferable to not only other priority catchments within the Blue Mountains local government area, but with other sub-catchments within the Hawkesbury-Nepean System and throughout NSW. The key lesson from NKCRAP was that achieving sustainable and lasting on-ground environmental outcomes and improved catchment health (sustaining the natural capital) was only possible through actively supporting and empowering the local community (sustaining the social capital).

Introduction

The North Katoomba Catchment Restoration and Action Project (NKCRAP) started as a standard catchment improvement scheme but evolved into an innovative holistic approach to natural systems management that is being adopted as a model for how Blue Mountains City Council sustainably manages its catchments.

North Katoomba was selected as a priority sub-catchment under the State Government's *Urban Runoff Control Program* (URCP) and Councils *Catchment Assessment and Prioritization Program* (CAPP). The CAPP was designed as a GIS tool for Council to better allocate and strategically target the limited resources available for natural resource management. The CAPP identifies the key natural resource issues that need to be addressed to achieve improved catchment health. Some of the key issues identified in the North Katoomba catchment included extensive riparian degradation, severe in-stream erosion, channel instability, sedimentation that was infilling pools and affecting stream hydrology, habitat loss, weed encroachment, presence of threatened species, sites of Aboriginal significance, recent bushfires throughout the catchment, residential development and industrial activity including a Waste Management Facility, a decommissioned STP and the SCA Greaves Dam system that provides drinking water to the Upper Blue Mountains. All these key issues in the catchment have the potential to impact the surrounding Blue Mountains World Heritage Area and Grose Wilderness Area downstream.

The NKRAP was designed to follow on from the on-ground work in the North Katoomba catchment carried out under the URCP which was run between 1998 and 2003. The aim of the URCP was to implement a range of on-ground catchment improvement actions to address some of the key environmental issues related to stormwater and urban runoff. Unfortunately many of the proposed solutions actually created new or bigger problems that would have gone undetected if not for the local community that have been actively involved in volunteer activities and had intimate knowledge of the changes in their catchment. It soon became apparent that without working collaboratively with the local community the project would be far less effective and much less sustainable.

The ultimate success of the North Katoomba Catchment Restoration and Action Project was only achieved through the engagement of an active community of volunteers in the catchment including 3 Bushcare groups, 2 Living Streams Streamwatch groups, the Community Gardens, the Aboriginal Cultural Resource Centre, and the K@toomba Men's Shed. The key lesson from NKCRAP was that achieving sustainable and lasting on-ground environmental outcomes and improved catchment health (sustaining the natural capital) was only possible through actively supporting and empowering the local community (sustaining the social capital).

Background

Several of the longest running Bushcare groups in the Blue Mountains exist in the North Katoomba catchment (over 15 years) and a number of their members have lived in the catchment for many more years and have observed marked and often rapid changes in the local environment.

The community in North Katoomba became mobilized when the 'Blue Pool' at the bottom of Minihaha Falls once known as the bottomless pool, became totally infilled with sand and sediment from the catchment. As recently as 1975 a Council officer lowered a brick tied to a rope and reached over 30 metres without touching the bottom. It had been used as a popular recreation area by the community and many locals reported jumping off the waterfall without any risk of touching the bottom because it was so deep. This pool has traditionally been a special area for the Dharug people with particular significance to the women, and has been visited by Aboriginal elders from around Australia.

In the last decade Council and a number of government agencies started undertaking a variety of programs and on-ground works designed to improve the North Katoomba catchment and most undertook some form of community consultation prior to commencing works. Unfortunately this level of community consultation proved not to be sufficient to ensure long term sustainable outcomes and in some cases actually resulted in a scenario where one problem was 'fixed' but another and potentially bigger problem was created. These 'bigger' problems that arose were often only brought to Council's attention by the local community, particularly those involved in long term environmental volunteer work such as Bushcare, who have an intimate knowledge of their catchment and were able to observe the environmental changes that were taking place.

Despite all of the on-ground 'improvement' works that were being carried out in the catchment, it was only when the local community became involved, engaged and empowered did the program start achieving sustainable results. Through a multifaceted approach, the NKRAP started engaging the local community through a number of community catchment activities. A focus group was formed and their vision was to restore the natural pools and improve the overall health of the North Katoomba catchment. This focus group was essential in pulling together Council, state government authorities, local businesses and other agencies to work collaboratively to create a range of sustainable on-ground outcomes and promote collaboration between all stakeholders.

The following are some examples of well intentioned catchment improvement programs for North Katoomba which the local community was able to point out subsequent consequences and identified environmental impacts. Once the community was brought on board as collaborators as part of the KKRAP many of these problems were rectified and preventative measures put in place.

One objective of the URCP North Katoomba catchment improvement program, was to reduce the amount of sediment entering the local streams. Through the use of a rainfall simulator the amount of sediments coming off unformed roads was calculated and this information was used as justification for an extensive road sealing program with curb and guttering. Unfortunately by creating impervious surfaces and concentrating the stormwater by piping it directly into the local streams, the increased flows and velocities of the stormwater caused massive in-stream erosion. This resulted in many more tonnes of sediment being eroded and exported down into the surrounding World Heritage National Park than would have come of the unformed roads in the first place. The local volunteers working along the Glen Road tributary of Yosemite Creek first identified a massive instream headcut that resulted in over 15 eucalypts falling into the creekline. As a result, funding was sought from the Local Government Advisory Group (LGAG) of the Hawkesbury Nepean Catchment Management Authority (HNCMA) and used to start the NKRAP. Some of this funding was used to rock line the most active instream erosion sites identified by the local community, and the creekbanks were stabilized and the riffle zones restored improving the biodiversity.

Sydney Water through its Upper Blue Mountains Sewerage Improvement Program was upgrading the sewerage system through North Katoomba using directional tunnelling technology to minimise the environmental damage to the local environment. Unfortunately several directional bores in North Katoomba penetrated through an aquifer resulting in the groundwater levels dropping several metres. This affected the nearby Marmion Hanging swamp which the local Marmion Rd Bushcare group reported to be dying out in areas as well as the loss of spring and stream flow to the swamp. As a result of direct action from this group and the Blue Mountains Conservation Society (ConSoc), Sydney Water agreed to undertake a hydrological study of the swamp and have planned to re-grout the bores to restore the water tables and rehabilitate the adjacent hanging swamp.

The RTA has been upgrading the Great Western Highway through the Blue Mountains to improve traffic flows to the west. Unfortunately in the North Leura/Katoomba section, storm events resulted in tonnes of sediment being exported into local streams which was brought to the attention of authorities by the local community which traced it downstream into the surrounding WHNP. Through their lobbying, the RTA agreed to install an instream sediment basin to capture sediments making their way down Govett's Creek and to fund a multi-year riparian weed removal program.

The RFS and NPWS through its fire mitigation program designed to improve the safety of the North Katoomba residents, back-burned extensive areas of the catchment including some hanging swamps in North Katoomba, resulting in a much higher erosion risk and die-back of the hanging swamp vegetation, home to rare and endangered species including the Blue Mountains water skink. Local conservation groups were successful in getting the upper Blue Mountain hanging swamp communities listed as Federally listed endangered ecological communities thus changing the back burn practice so that they now are protected.

BMCC though the operation of its Waste Management Facility (WMF) and South Street Depot was also responsible for the tonnes of sediment making its way into Yosemite Creek and an adjacent hanging swamp. Once again by listening to the concerns of the local community Council improved its WMF and lined the new cells preventing leachate and sediment from entering the creek system and introduced ISO 14001 protocols for staff and built sedimentation basins and set up erosion control works at its South Street Depot. Soft engineering solutions were trialled to restore the hanging swamp which was used as a model to obtain an additional $150k to run similar hanging swamp restoration programs throughout the Mountains including the hanging swamps in North Katoomba affected by bushfires.

It was only by tapping into local knowledge and the groundswell of community action, that Council found ways to add value to the works that they were doing and tackle some of the larger challenges.

For example along sections of the creeks where the volunteers were working, Council sought funding to remediate sections with severe instream erosion and active headcuts that were beyond the capacity of the community volunteers to restore. External grant funding was sought to implement these larger works. One of the key factors in the success in obtaining State and Federal grant funding was being able to demonstrate the large amount of in-kind volunteer time and effort. Having active community groups to carry out some of the ongoing maintenance of the sites once the contract work had finished also proved to be very influential.

The community groups were also successful in attracting funding through programs such as the Environmental Trust Grants. Local volunteers in North Katoomba put together a comprehensive weed removal program for the catchment specifically targeting scotch broom whose seeds can remain dormant for up to 80 years but were snapped into germination following the 2002 bushfires. The community was also successful in attracting grant funding for projects in the North Katoomba Community Gardens and the K@toomba Men's Shed. The K@toomba Men's Shed works with men and youth in the area in a range of hand-on mentoring programs including environmental restoration projects. This is particularly important as North Katoomba has some of the lowest socio-economic indicators in the Blue Mountains and Western Sydney.

The following aims, objectives, methodology were developed for the NKCRAP with input from the focus group. The large number of outcomes were only made possible through the ongoing involvement of the local community.

Aim

The overall aim of the NKCRAP is to involve the North Katoomba community and other key stakeholders in a holistic catchment restoration project that delivers long term environmental outcomes (sustaining the natural capital) while exploring how to continually build community and live sustainably within the North Katoomba catchment (sustaining the social capital).

Objectives

- To engage the community in all aspects of the project through the formation of a representative focus group;
- Creation of new and support of existing volunteer groups including Bushcare, Living Streams Streamwatch, the North Katoomba Community Gardens, the Aboriginal Cultural Resource Centre and the K@toomba Men's Shed;
- To help build a sense of place in the North Katoomba catchment by conducting a range of community catchment events and hands-on workshops;
- Engage the Aboriginal community in the long-term restoration of key sites including the pools within the catchment that have been identified as significant sites to the Dharug people, and the planting of a 'Bush Foods' garden within the existing community garden that caters for the indigenous community living in Katoomba;
- Conduct a living sustainably within a catchment campaign to promote the adoption of sustainable practices for household and businesses within the

catchment including the sealing or stabilization of driveways and other bare areas, the installation of rainwater tanks and preventing the spread of weeds from private properties into the surrounding bushland;

- Run a sediment source control campaign with residents, builders and businesses to prevent sediment export at or close to the source;
- To identify priority issues in the catchment and implement sustainable remediation works designed to conserve local scheduled and vulnerable, threatened or endangered ecological communities and species (Blue Mountains Hanging Swamp complex; Blue Mountains Riparian Complex; Blue Mountains Water Skink *Eulamprus leuraensis;* Giant Burrowing Frog *Helioporus australicus*) and increase biodiversity by improving habitat connectivity and wildlife corridors;
- Create and implement a North Katoomba Weed Management Plan and hold targeted Scotch Broom (Cytissus scoparius) weeding days in the catchment following the germination of thousands of plants after the bush fires of 2002;
- To promote a sense of community within the catchment by building on the connectivity established during the bushfires of 2002.

Methods used to implement the project included

- The formation of a focus group to create a vision for the catchment and to identify and recommend on-ground works;
- The establishment of community groups to monitor and work on rehabilitating the catchment as well as increasing the level of community involvement;
- Inter-agency collaboration to jointly deliver programs and on ground works to achieve greater value for money;
- Community catchment days to educate and involve residents in on ground activities and actions;
- Catchment pledges engaging the community in making changes on their own property and adopting sustainable practices that will benefit the catchment;
- 'Broom Blitz' events engaging the community in targeted weeding days, which fosters a feeling of achievement by the residents as they work toward an achievable goal;
- Additional support for existing community projects and programs including Bushcare, the Community Gardens and the K@toomba Men's Shed.

The NKCPAP has been running for four years and has demonstrated sustainability by delivering the following triple bottom line outcomes through improved catchment health, an enhanced local economy and an active and engaged local community.

Economic outcomes

- State agencies including the RTA, Sydney Water and the URCP have invested over $5m in the past decade in the North Katoomba catchment;
- Council has been successful in obtaining over $500k in state and federal grant funding for the NKCRAP;
- Council Green Team staff (many who live in the Katoomba catchment) and local bush regenerator contractors (based in North Katoomba employing locals from the North Katoomba community) used 'best practice' techniques when removing weeds and sourced local progeny plants from Katoomba based community nursery 'Wild Plant Rescue' to repair riparian zones and improve habitat connectivity;
- Contractors/Council used NKCRAP grants totalling over $175k for weeding and revegetating riparian zones throughout the catchment;
- Contractors/Council completed over 350m of instream channel stabilisation works;
- Over 30 driveways were sealed or stabilized by local residents to stop erosion;
- 35 household in the catchment installed rainwater tanks as well as 3 rainwater tanks at the Community Gardens and 2 tanks at the K@toomba Men's Shed;
- Lithgow Green Corps spent a day calculating sediment loads in the pools along Yosemite Creek and under Minnehaha Falls, a pool once over 30m deep now filled with sediment. They calculated that over 54,000 tonnes of sediment had been deposited in the pool;
- Council's work crews reviewed work practices with the challenge of improving on "best practice" and attaining ISO 14001 accreditation;
- Council's depot was upgraded to include stormwater retention basins, improved sediment controls and specialised storage for unconsolidated materials;

Environmental outcomes

- Installation of six new stormwater quality improvement devices (SQIDs) and improved maintenance schedules for existing devices;
- Volunteers/Council removed target weed species and rehabilitated over 15km of riparian corridor;
- 350m of instream stabilisation work using rock lining which restored riffles improving biodiversity;
- Over 5km of eroding dirt road edge was sealed;
- 15km of unused trails closed and rehabilitated in Mike Eades Reserve, North Katoomba;
- 3km of unvegetated road verges mulched/turfed;
- Illegal vehicular access to reserves blocked at 15 points to prevent damage from 4WDs;
- Joint Council/Sydney Water(SW) initiative results in 35 houses installing rainwater tanks reducing the impacts of stormwater on the local creek system;

- Several severely degraded walking tracks were restored in Mike Eades Reserve reducing erosion and sedimentation;
- Australian Trust for Conservation Volunteers closed off and brush matted the length of an eroded unformed road leading into a hanging swamp in North Katoomba;
- Over 3,000 native species were planted in riparian zones by contractors and Bushcare groups;
- The North Katoomba Weed Management Plan was devised and implemented. Over 6210 hours of work has been completed by volunteers and contractors;
- Two disused Council quarries were rehabilitated in North Katoomba and North Leura;

Social outcomes

- Formation of a North Katoomba focus group to give direction for ongoing catchment improvement works;
- Three Blue Mountains Sustainable Living courses were conducted with over 75 residents in North Katoomba which provided hands on training in how to live more sustainably and reduce their ecological footprint. Participants were given training in actions they could undertake within their own homes to minimise waste, recycle, build earthworm farms and compost heaps, reduce water usage through no dig gardens and mulching and replacing weed species with natives. Participants were offered follow-up visits by Council's Community Weeds Officer to develop a property plan and receive one on one training in how to remove unwanted weed species and replace with natives. Vouchers for local providence native plants from the local community nursery Wild Plant Rescue were provided as well.
- Two new Streamwatch groups were established for Yosemite and Katoomba Creeks;
- 25 households have joined the Bush Backyards program and prepared property plans;
- Conducted three catchment crawls with local school students and residents identifying catchment issues and engaging them in monitoring activities;
- Delivered Catchment Newsletters to all residents in the catchment, which contained maps, identified source points of pollutants and detailed the remediation efforts and the outcomes;
- Conducted three community catchment days to promote sustainable living in the North Katoomba catchment;
- Held two special Broom Blitz events with over 90 Bushcare volunteers removing broom seedlings along creeklines 5km down into the Grose Wilderness;
- Volunteers spent 2324hrs weeding and revegetating riparian zones in 2004-05

Figure 1 *The 'bottomless' pool at Mininhaha Falls in the 1960s and in 2006*

- Over 200 catchment pledges were signed by residents outlining a range of sustainable actions they were willing to implement;
- A waterwise gardening demonstration sites is being established at the Community Gardens with rainwater tanks, drip irrigation, no dig gardening and mulching techniques, permaculture and organic gardening;
- The Aboriginal community is establishing a Bush Foods Botanic Garden with support from the World Heritage Institute and the Aboriginal Cultural Resource Centre;
- Three Festivals of Joy at the Community Gardens were held with over 1500 participants with a range of workshops and entertainment with an environmental and water theme;
- The Katoomba Men's Shed was successful in obtaining a community water grant to install rainwater tanks at the Shed and Community Gardens;
- The heritage walking track to Minihaha Falls was restored:
- Increased and improved partnerships with all stakeholders including local residents, community groups, State government authorities (including Sydney Water, NPWS, & DIPNR), other groups (URCP, LGAG HNCMA), Aboriginal community, and local business.
- Possibly the best long-term outcome for this project, is the high level of involvement of the North Katoomba community and the ongoing community engagement in natural resource management.

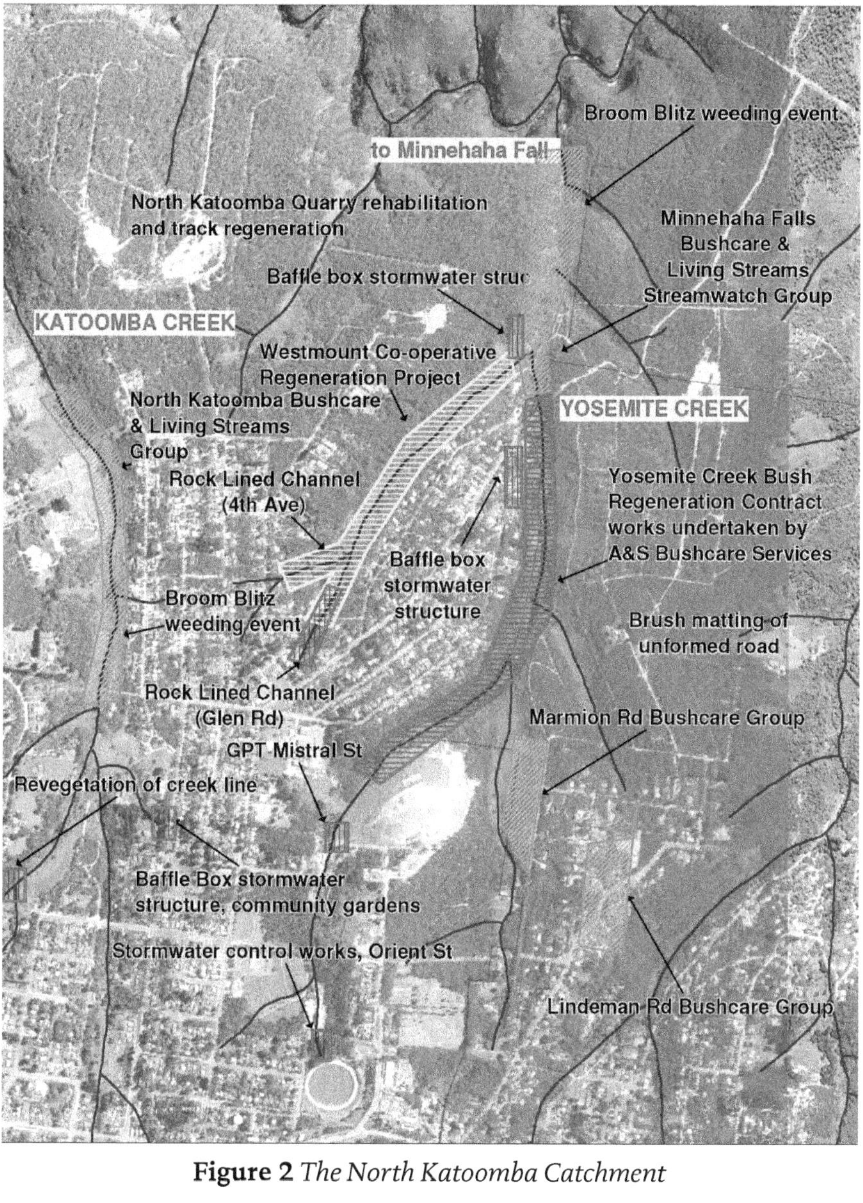

Figure 2 *The North Katoomba Catchment*

Conclusion

Through its whole of catchment approach and multi-disciplinary methodology this project is being adopted as a model for how Blue Mountains City Council sustainably manages its natural resources. The high level of involvement by catchment residents and stakeholders (businesses and industries, indigenous groups, Council staff, State government authorities) in planning and implementing the project has turned what was often confrontational into a collaborative approach to natural resource management. This collabora-

Figure 3 *Examples of instream and riparian works including new Stormwater Quality Improvement Devices (SQIDs)*

Figure 4 *Community volunteers and community catchment activities in North Katoomba*

tive approach has promoted a sense of place and community by actively engaging the community and key stakeholders in adopting sustainable behaviours and practices that will benefit the catchment into the future. On ground works undertaken as part of this project has also provided Council's work crews with new skills and innovative ways to perform work, and as a result have changed some of the old practices.

The North Katoomba Catchment Restoration and Action Project has been deemed a huge success for many reasons including the: number and scope of on-ground works that were undertaken; partnerships and associations that were created; collaboration between all stakeholders; level of community engagement and the subsequent education and empowerment of the community. Official recognition of the KKCRAP was achieved by winning the NSW Local Government Award 2004 for Excellence in the Environment, winning the National Award for Local Government for Integrating Biodiversity Conservation into Planning and Management in 2005, and being selected for the Department of Environment and Conservation's (DEC) 'Leading Practice Showcase' in 2006.

Through its whole of catchment and multi-disciplinary methodology this project demonstrates how to deliver significant and long term outcomes in natural resource management, which can be transferable to not only other priority catchments within the Blue Mountains LGA, but with other Councils within the Hawkesbury-Nepean Catchment and throughout NSW. The North Katoomba Catchment Restoration and Action Project focussed on grass roots reporting of biodiversity and sustainability outcomes, which was developed for specific target audiences (eg. school groups, general public, interest groups, government agencies). By informing the local community of the continual on-ground improvements being done by Council, they were more motivated and inspired to continue their own onground volunteer efforts.

The NKCRAP has benefits on a number of scales. Globally, as it deals with catchment restoration in the only city in the southern hemisphere totally surrounded by a World Heritage Area. All improvement works done in this catchment will have beneficial flow on effects in the Grose Valley Wilderness Area and the Greater Blue Mountains National Park and will assist Australia in meeting its international obligations through UNESCO. On a national scale, the project is beneficial due to its work restoring threatened species habitats, sensitive vegetation communities and endangered ecological communities including the recently federally listed upland hanging swamps, and improving biodiversity and habitat connectivity. Regionally the project is beneficial, as it will provide a model for sustainable catchment management. All on-ground works undertaken within the catchment will have a beneficial effect on the Hawkesbury Nepean (HN) river system and address key targets in the HN Catchment Blueprint and Catchment Action Plan. Locally, the project improved the water quality in Yosemite and Katoomba Creeks, restored the riparian habitat and increased the biodiversity, enhanced the sense of community and place, and fostered partnerships between the local community, Council, State government authorities and other key stakeholders including local businesses.

The success of the NKCRAP demonstrates that one of the best ways to obtain long term environmental outcomes (sustaining the natural capital) is to engage and empower the local community (sustaining the social capital), which ultimately reduces costs of future environmental restoration through increased care and reduced impacts.

Geoffrey Smith holds a BSc with Honours from the University of Florida, USA and a Masters of Environmental Science from Monash University, Victoria, Australia. Geoffrey is an environmental scientist and educator with a passion for looking after the world's waterways. He has been actively involved in establishing community water quality monitoring programs and groups throughout Sydney, NSW, Australia and many countries throughout the world through his work with Streamwatch, NSW Waterwatch, Waterwatch Australia and OzGREEN (Global Rivers Environmental Education Network, Australia). He is a founding member of OzGREEN and Friends of the Ganges, Australia and had been working with the Swatcha Ganga (Clean Ganges) campaign since 1995. Geoffrey has been particularly interested in establishing the natural link between art, spirituality and the environment. He feels that by providing people with an opportunity to become actively involved in looking after their natural environment, or creating artwork associated with the environment, it helps reconnect them with their spirituality as they rediscover the sacredness of nature. Geoffrey Smith is currently working for Blue Mountains City Council as the Natural Systems Program Leader. Following the declaration of the Blue Mountains World Heritage listing, Geoffrey saw a unique opportunity for the community in the Mountains to become more actively involved in a stewardship role in looking after the waterways and created the Living Streams Streamwatch water quality monitoring and catchment health assessment program. There are currently over 20 active Living Streams Streamwatch groups in the Blue Mountains. Geoffrey attended the World Summit on Sustainable Development in Johannesburg in 2002 where he became the Global Facilitator for the International Rainwater Harvesting Alliance, a network of over 60 countries committed to using alternative sustainable technologies for providing fresh water to the 1.2 billion people on the planet without access to safe drinking water and the 2.5 billion people without access to adequate sanitation. In addition, he attended the Global Mountain Summit in Bhiskek Kyrzgyzstan, the World Water Forum in Kyoto, Japan and the Eco-Sanitation Symposium in Lubek Germany, to promote rainwater harvesting and ecological sanitation as a sustainable alternatives to addressing the Millennium Development Goals for freshwater and sanitation which aim to reduce by 50% the proportion of people without access to fresh drinking water and sanitation.

Bhutan's Gross National Happiness Project: Lessons from the East

Geoffrey Lamberton
School of Commerce & Management, Southern Cross University, AUS

The small kingdom of Bhutan nestled in the eastern Himalayas has declared Gross National Happiness (GNH) as its national goal, clearly distinguishing its developmental path from nations fervently pursuing economic growth measured by increases in Gross National Product. Bhutan has a long history of preserving its natural capital, resisting the temptation to exploit rich mineral and forest resources, investing in the preservation of cultural traditions, social welfare and economic independence. Underpinning the Bhutanese objective of GNH is a strong tradition of Buddhist philosophy and practice leading to an alternative economic path which has much to offer the rest of the world in the global search for sustainability. This paper presents a proposal to examine the Buddhist and ethical underpinnings of GNH together with the accounting process of measuring progress towards this objective, leading to a discussion of whether the Western version of sustainability is inherently flawed due to the lack of a virtuous moral basis.

1.0 Gross national happiness in Bhutan

Bhutan, a small Himalayan Kingdom of less than one million people and blessed with abundant natural resources is treading an economic development path which maintains its separateness and independence from the global economy. The Fourth King of Bhutan, His Majesty King Jigme Singye Wangchuck, in the late 1980s declared *Gross National Happiness* (GNH) the national goal of Bhutan, reflecting the prioritization of social welfare ahead of the accumulation of economic wealth.

Recognizing the shortcomings of gross national product as a measure of welfare, GNH proposes a form of economic development which is consistent with values prescribed by the unique Bhutanese culture which is profoundly influenced by Buddhist teachings. The Bhutanese view of *happiness* is that it is much more than material wealth.

"Happiness is not simply sensory pleasure, derived from physical comfort. Rather, happiness is an innate state of mind, which can be cultivated through spiritual practice, overcoming mental and emotional states, which induce suffering" (Tideman, 2004: 222).

This concept of happiness reflects the essence of Buddhist teachings which connects human suffering with attachment to and grasping for sensory desire caused by greed and selfishness. GNH also displays an equivalent foundation to the multi-dimensional objective of sustainability. The conceptual foundations of GNH are preservation of culture; environmental conservation; good governance; and socio-economic development (Hershock, 2004: 71), equivalent to the social, environmental and economic dimensions of sustainability.

The similarity between GNH and sustainability forms a central theme to this research, providing an opportunity to compare progress toward GNH within Bhutan, with progress toward sustainability within an economically developed country such as Australia. In light of continuing failure to make genuine progress toward the global objective of sustainability, the relevance of the Buddhist ethical framework which underpins national progress toward GNH is examined and compared to the ethics which underpin market economic systems.

1.1 Buddhist influences over GNH

Bhutanese culture is profoundly influenced and informed by Buddhist tradition, specifically the Tibetan school of Mahayana Buddhism. Buddhist tradition is founded upon the teaching of *Siddhartha Gautama* who, according to earliest Buddhist scripture, lived the affluent life of a young prince, until his observations of human suffering led to disenchantment with his noble life, and he embarked on a spiritual journey ending in a profound experience that led to his awakening as a *Buddha* (Gethin, 1998). Buddism is often referred to as the *Middle Way*, as Gautama rejected both his affluent life and the extreme fasting and ascetic life that he experienced on the path to enlightenment.

According to Buddhist teachings attachment to sensory indulgence (including material wealth) is the root cause of suffering and is removed by adherence to moral precepts in daily life, as well as meditation and mindfulness practice. Buddhist teachings clearly deny a link between the acquisition of material goods and lasting happiness, which can only be achieved by *the cultivation of values within one's mind, such as insight, compassion, tolerance* (Tideman, 2004: 224) and equanimity.

The ethical basis underpinning Buddhist teachings is expressed in the doctrine of *causing no harm to other beings*. This principle is applied in various ways across the many different cultural and regional interpretations of Buddhism. Examples of the *no harm doctrine* include adhering strictly to a vegetarian diet, adoption of low environmental impact lifestyles avoiding violence toward people, animals and the natural environment.

Within Christianity,

"The individual is considered to be a totally unique being, possessed of a unique soul, and to be the primary actor....in the drama of his or her own salvation" (Mancall, 2004: 26).

Reification of each individual influences western views of economic development which based on the principle of self interest, encourage individualistic activities such as wealth creation and consumerism in the false belief these behaviors can be optimal for society as a whole.

Market economics is underpinned by the logic of individual self interest and is inherently unsustainable as it fails to represent its true nature as a mere partial (economic) representation of a complex living reality (Saravanamuthu, 2005). The neoclassical economic model conflicts inherently with the need to be in equilibrium within *oneself, with other people, with society and with nature* (Tideman, 2004: 234).

In contrast to Christianity, Buddhism denies the existence of a separate self and enduring soul. This distinction has critical implications for both human relationships and humankind's relationship with the natural environment. Buddhist teachings emphasise the link between internal happiness and contributing directly to the well being of the whole community in recognition that the perceived separateness between humans and other humans (as well as humans and the natural environment including all cohabiting species) is an illusion. Humans only exist in the context of their relationships with the *other*, and it's this *other* that requires nurturing to enable a state of happiness.

2.0 GNH and the path to sustainability

Bhutan's GNH project is underpinned by their strong Buddhist ethic and driven by their observation of environmental and social degradation outside Bhutan. It is also a recognition that gross national product is a crude measure of welfare counting social negatives such as expenditure on environmental disasters, crime, imprisonment to name a few examples as increments to growth (Dixon, 2004: 105), whilst ignoring many indicators of strength within a society (volunteer work, home duties, leisure time, quality of education).

Bhutan's vision is to achieve a high level of welfare for all of its citizens providing inspiration to the rest of the world to find solutions to problems caused by unsustainable lifestyles. GNH includes the elimination of income disparity (Mancall, 2004), as well as the socially debilitating force of unemployment as the State acts as employer of last resort. GNH also requires the control of urbanization by promoting rural village lifestyles and providing incentives for people to remain in rural villages.

Furthermore a (Buddhist) economic core value recognizing global interdependence is built on the concept of mutual help (Van Willensward, 2004), that is the need to contribute to the welfare of the global community through commercial trading activities which benefit all stakeholders.

Recognizing and fixing negative environmental and social impacts of economic transactions requires limits to economic growth. One source of unsustainable growth is the creation of excess patterns of demand using creative marketing techniques. Advertising often creates unnecessary demand fuelling unsustainable consumption patterns and attachment to excessive consumption of material goods, leading to restrictions on advertising within Bhutan (McDonald, 2006).

3.0 Measuring progress toward GNH & sustainability

An important aspect of Bhutan's GNH project is measurement of progress toward this objective. The measurement methodology adopted involves a performance indicator based approach. Indicators are being developed which measure progress toward all round wellbeing throughout the Bhutanese society which include measures of the maintenance and enhancement of natural capital, measures of generosity, for example monitoring the strength of the volunteer sector, as well as measures of security and:

"Education, health care, housing, clothing, food and nutrition, shelter, environmental and habitat protection, parents spending time with and raising children, arts, business practices, infrastructure, legal and regulatory issues, reported levels of happiness..." (Dixon, 2004: 113).

GNH is not the aggregate of individual happiness (Mancall, 2004), and within the GNH measurement framework there is no attempt to measure individual happiness. Integral to Buddhist philosophy is the interdependence between each subunit and the whole. It is also recognised that accountability requires society's values to be made explicit.

"Every measure of progress...is based on values...What we...measure...reflects our deepest social values" (Coleman & Sagebien, 2004: 253)

The measurement of GNH using a broad set of performance indicators has much in common with efforts to measure corporate sustainability using the sustainability accounting guidelines provided by the United Nation's Global Reporting Initiative (GRI). As with measurement of GNH, the GRI provides an evolutionary step in the identification of indicators relevant to measuring sustainability (Tideman, 2004). The *GRI Sustainability Accounting Guidelines* (GRI, 2002) identify a wide array of indicators relevant to measuring performance toward the goal of sustainability. The Guidelines draw on the three dimensional definition of sustainability using a series of performance indicators to measure each of the economic, environmental and social dimensions.

A common concern amongst critical accounting theorists is a perception that accounting even in its environmental or sustainability forms is incapable of contributing positively to the restoration of the natural environment. Accounting has developed to meet the objectives of capitalism and its obsession with wealth creation and continuing economic growth. Critical accounting theory warns against the potential for accounting processes to subvert attention from the actual goal toward manipulation of the metrics believed to measure progress toward the goal (Cooper, 1992; Lehman, 1995).

The next section explores a relatively obscure section of the accounting literature which suggests that accounting can be transformed to contribute to spiritual objectives.

3.1 Accounting and spirituality

A question critical to the pursuit of GNH is whether the accounting measurement process can contribute to what is inherently a spiritual path?

Ezzamel (2003) considered the symbolic role of accounting which can emerge out of functional practices such as inventory and expenditure accounts. Ezzamel examined inscriptions made on the walls of Egyptian temples which together with linguistic texts, pictorial scenes and religious monuments produced a powerful discourse connecting the physical and spiritual realms (Ezzamel, 2003: 6) through the orderly and coherent ritualistic accounting process, including lists of offerings made to the gods.

Gallhofer and Haslam (2004) identify the contribution of accounting to the relentless pursuit of profit and wealth and the consequential loss of spiritual focus and understanding. They envisage accounting in an emancipating role reflecting social community values and environmental stewardship. Exploring different narrative forms (*interviews, oral histories, self-reflection and conversations,* Gallhofer and Haslam, 2004: 399) combined with images and pictorial forms has potential to reflect a holistic vision which includes the spiritual.

Sustainability accounting attempts a more holistic picture by the separate measurement of specific environmental, social and economic performance indicators. Saravanamuthu (2005) believes that to be in harmony with a holistic vision such as sustainability requires the ability to transcend its perceived separate dimensions, identifying the Vedic approach for individuals *to become sensitive to the needs of the other.*

Linking accounting to spirituality uncovers the potential for accounting to develop as a tool which lays transparent the broader impacts of the global economy's relentless pursuit of material wealth. A more spiritually and less materially focused accounting could be based on values determined by society-at-large, which must necessarily include a cohesive set of social, cultural and environmental values, where economic goals are subsumed within overriding social imperatives. Ultimately the relevance of accounting will depend on whether it contributes to society's broader goals (Funnell, 2004).

To make a positive contribution to a spiritual goal such as GNH the accounting process will need to be underpinned by a broad set of societal values woven into a holistic vision such as sustainability, as well as being equipped with a diverse array of qualitative and quantitative measurement and reporting tools (Hargens, 2002) including narratives in various forms, images and pictorial forms, and sophisticated analysis such as life cycle analysis, together with performance indicator scorecards suggested by GRI and GNH measurement systems.

4.0 Proposed direction of this project

Fundamental to this research is the contrast between the influence of Buddhist values along the path to GNH and the influence of the dominant market economic view that is driving unsustainability.

The primary research objective is to explore the different moral basis underpinning the Bhutanese goal of GNH compared to the Western goal of sustainability, and to determine whether the path to sustainability is blocked by the absence of an appropriate ethical foundation.

This research will continue with a comparison of the Bhutanese Government's energy policy, identifying the role of Buddhist ethics in the formation of policy, with the energy policies of the Australian government. It is anticipated that energy policy will provides a compelling contrast given both countries have extensive mineral reserves, but different sources of generation and patterns of consumption.

The impact of energy policies within Bhutan and Australia on consumption, welfare and the environment will be compared using social, environmental and economic performance indicators relevant to energy management at the national level.

References

Coleman, R. & Sagebien, J. (2004). "Towards Gross National Happiness Measuring Genuine Progress - Indicators for Enlightened Society", in Karma Ura & Karma Galay (Eds), *Gross National Happiness and Development, Proceedings of the First International Seminar on Operationalization of Gross National Happiness,* The Centre for Bhutan Studies, Thimphu, Bhutan, pp 252-259.

Cooper, C. (1992). "The Non and Nom of Accounting for (M)other Nature", *Accounting, Auditing and Accountability,* ISSN 0951-3574, vol 5 no 3, pp 16–39.

Dixon, F. (2004). "Improving Unsustainable Western Economc Systems", in Karma Ura & Karma Galay (Eds), *Gross National Happiness and Development, Proceedings of the First International Seminar on Operationalization of Gross National Happiness,* The Centre for Bhutan Studies, Thimphu, Bhutan, pp 105-120.

Ezzamel, M. (2003). "Accounting, the Divine and the Symbolic: the Memorial Temples of Ancient Egyypt", *Cardiff Business School Working Paper,* Cardiff, March, pp 1-59.

Funnell, W. (2004). "Accounting and the Pursuit of Utopia: the Possibility of Perfection in Paraguay", *Accounting Historians Journal,* ISSN 0148-4184, vol 31 no 1, pp 57-91.

Gallhofer, S. and Haslam, J. (2004). "Accounting and Liberation Theology, Some Insights for the Project of Emancipatory Accounting", *Accounting, Auditing & Accountability Journal,* ISSN 0951-3574, vol 17 no 3, pp 382-407.

Gethin R. 1998. *The Foundations of Buddhism*; Oxford University Press: Oxford, ISBN 0-19-289223-1.

Global Reporting Initiative (GRI). (2002). *Sustainability Reporting Guidelines.* Boston.

Hargens, S. (2002). "Integral Development: Taking 'The Middle Path Towards Gross National Happiness", *Journal of Bhutan Studies,* ISSN 1608-4128, vol 6, pp 24-87.

Hershock, P. (2004). "Trade, Development, and the Broken Promise of Interdependence: A Buddhist Reflection on the Possibility of Post-Market Economics" in Karma Ura & Karma Galay (Eds), *Gross National Happiness and Development, Proceedings of the First International Seminar on Operationalization of Gross National Happiness,* The Centre for Bhutan Studies, Thimphu, Bhutan, pp 51-76.

Lehman, G. (1995). "A Legitimate Concern for Environmental Accounting", *Critical Perspectives on Accounting,* ISSN 1045-2354, vol 6, pp 393–412.

Mancall, M. (2004). "Gross National Happiness and Development: an Essay", in Karma Ura & Karma Galay (Eds), *Gross National Happiness and Development, Proceedings of the First International Seminar on Operationalization of Gross National Happiness,* The Centre for Bhutan Studies, Thimphu, Bhutan, pp 252-259.

McDonald, R. (2006). "Selling Desire and Dissatisfaction: Why Advertising should be Banned from Bhutanese Television", *Journal of Bhutan Studies,* ISSN 1608-4128, vol

14, pp 1-22.
Saravanamuthu K. (2005). "Emancipatory Accounting and Sustainable Development: a Ghandian-Vedic Theorization of Experimenting with Truth", *Sustainable Development*, ISSN 1447-3599, in press.
Tideman, S. (2004). "Gross National Happiness: Towards a New Paradigm in Economics", in Karma Ura & Karma Galay (Eds), *Gross National Happiness and Development, Proceedings of the First International Seminar on Operationalization of Gross National Happiness,* The Centre for Bhutan Studies, Thimphu, Bhutan, pp 222-246.
Van Willensward, H. (2004). Will 'Middle Way Economics' Emerge from the Gross National Happiness Approach of Bhutan? in Karma Ura & Karma Galay (Eds), *Gross National Happiness and Development, Proceedings of the First International Seminar on Operationalization of Gross National Happiness,* The Centre for Bhutan Studies, Thimphu, Bhutan, pp 214-221.

Geoff Lamberton was a corporate and chartered accountant before entering academia. He teaches ethics and sustainability, business finance and sustainable business management. His research interests include sustainability accounting and the ethical change required for the transition to a sustainable society. His major publications are in the areas of sustainability accounting and the link between sustainability and ethics.

Between Council and Community: Public Participation and Sustainability in Local Government

Cesidio Parissi
School of Natural Sciences, University of Western Sydney, AUS

Sustainability is an important process and objective of society, although not supported by all. In any case, legislation and practice increasingly emphasises public participation in the context of environmental and sustainability issues. Legislation in NSW links the progress of environmental sustainability, partly, to the actions of local governments through their budgeting plans. Whatever its importance and legislative imperative, the sustainability project has been less than successful in two crucial ways. Firstly, the effects of sustainability practice in local government have not been completely successful and, secondly, while practitioners have refined the techniques of public participation, public participation has had little impact on decision making. The gap between theory and practice in both cases seems wide and is explored here. The study broadly fits within an ethnographic approach and primarily used participant observation as the method of data gathering. This paper presents some preliminary findings of a PhD-in-progress that examines the case of Liverpool City Council and its attempt to institute a new model of public participation to directly link the Council's overall strategic budgeting document, the Corporate Plan, to renewed efforts at direct public participation. The new model is presented and some of the problems of directly linking sustainability to the Corporate Plan are discussed.

Introduction

This paper is a brief account of *Creating Our Future Together Partnership Project* initiated by Liverpool City Council, a Western Sydney local government authority, in June 2004. And of my participation, as a PhD research student, in the Partnership Project (the Project) throughout its first 19 months, to January 2006. My involvement encompassed 'Stage 1' of an iterative process. This paper comes from that research: the preliminary findings of a work in progress. This project was initiated in the wake of the dismissal of all elected Councillors by the State Government due to the Oasis scandal and the subsequent appointment of an Administrator in April 2004 (Daly, 2004). The Project set out to initiate a profound change program within the community and Council of Liverpool. The Project was originally located within a change-framework of facilitating Liverpool into becoming a community with a sustainable future.

My study is grounded in several underlying concepts. One of these concepts describes the integrity of life on this planet as not only being in a poor condition, but in decline, and that changing this situation requires action (NRC,

1999; WRI, 2003; Brown, *et al.*, 2005; Preston and Jones, 2006). Furthermore it is argued that this parlous situation affects both humanity and nature with triple-bottom-line (TBL) social/environmental/economic effects (McMichael, 2001; Diamond, 2005). TBL consequences include: the dissolution of community (Roseland, 1998) and its disempowerment (Ristock & Pennell, 1996; Bookchin, 2005); degradation of aboriginal societies (Sinatra & Murphy, 1999); effects on the natural environment, such as loss of biodiversity, pollution and loss of ecosystems (Shiva, 1993; WRI, 2003); the skewing of human ethics (Singer, 2002); impacts on women (Sen, 1994); weakening our economic structure (Hawken, *et al.*, 2000; Albert, 2003); adverse bearings on human health (McMichael, 2001; Brown, *et al.*, 2005); and a worsening 'North-South' dichotomy (Anderson, 2000; WRI, 2003).

Sustainability discourse moves beyond an environmental approach that simply aims to describe and correct humanity's past errors, to a position of developing future-oriented theories and processes. A critical aspect of the sustainability discourse is that it places an emphasis on action at the local level, in these social formations: government authorities, corporations, Non Government Organisations (NGOs) and citizens (Brown, 1997; Harris & Chu, 2001; Andrew & Robottom, 2005). Notions of civil and civic governance come into play as the above social formations interact, with the term 'governance' meaning all forms of decision-making, not just those of governments or courts. Andrew and Robottom (2005) emphasise that how governance is viewed and how it is expressed depends largely on how the more powerful social formations (governments, corporations) include lesser-powered actors (NGOs, citizens) in decision-making processes. Brown and Pitcher (2005) stress that, however well intentioned, engaging all 'interested' parties in the decision making process should be seen as a matter for negotiation.

Public participation is a critical part of the sustainability project, as advocated by the United Nation's two *Earth Summits* at Rio in 1992 and Johannesburg in 2002 (Buckingham, 2003; WRI, 2003). This theme was explored at the first United Nations conference that focused on public participation, the *International Conference on Engaging Communities* at Brisbane in 2005 (UNO, 2005). This emphasis is supported by NSW State government legislation through the Local Government Act 1993 (Farrier and Stein, 2006). Furthermore according to Edwards (2005) increased public participation in sustainability issues is the focus of many of the citizens of the world. For Western Sydney, the literature has described only one model that has been developed specifically for the region, although several councils have incorporated elements of other models (Parissi, 2006). The model identified is Brown's (2001) Pressure-State-Response-Potential Model of sustainability reporting for the *Western Sydney Regional State of the Environment Report 2000* (WSROC, 2000; Parissi, 2003, 2005).

The focus of this paper is to explore the imperative of sustainability in its location at the local level. One of the questions that my research explores and is addressed here is, 'how does the practice of governance respond to sustainability in relation to public participation?'. Or, as I have put it elsewhere, *"How does this impetus for positioning sustainability at the local level manifest itself in terms of praxis, that is, in the terrain where theory, practice and effect interact?"* (Parissi, 2005). This is the terrain where our journey with Liverpool's Partnership Project is undertaken.

Packing for the journey: My approach

My research is broadly framed within a qualitative research tradition (Creswell, 1994). This study is ethnographic in approach as it examines real people within their natural settings. Subsequently my research looks upon the community of Liverpool in relation to the Partnership Project as a window *"...to gain a comprehensive and complete picture of a social group"* (Fetterman, 1998: 19). This approach was adopted for, as Robson (1993: 149) states, an ethnographic case study allows emergent qualities to unfold, where *"...the main purpose is exploratory"*. This is important as there was no effective model for Western Sydney, as the other model dealt with environmental reporting. A new model came about during the research. Secondly, as the core of the Project was to foment change, then aspects of Critical Ethnography came into play. Furthermore, due to the Project's iterative nature, the approach of Action Research also plays a part in the story (Creswell, 1994).

Although not an employee of Council, I was not an impartial observer, but participated in planning processes of engagement within and between community and council, in developing concrete outputs, and in aspects of evaluation. Besides conducting numerous interviews, the principal methods for data collection were participant observation, interview, and document analysis. In all, my participation included about 140 meetings (planning, data processing, and reflective-evaluation), 16 community visioning workshops, 10 corporate planning workshops and 6 public meetings. Document analysis consisted of examining the output of the steering committee, newspaper articles, council papers and State government records.

The journey's setting

Figure 1 shows the Liverpool local government area (LGA) extending for about 30 kilometres, along the east-west axis of the Sydney Basin, covering about 305km^2 (Keating, 1996). 'Blaxland Ridge' divides the area in two, with rainfall heavier in the east, as is population density. About 25% of the region is wilderness (mostly the south-eastern Holsworthy Military area), 40% is rural (west) and 35% is urban (east) (WSROC, 2005a).

In 2005 the LGA had an estimated 170,192 residents, this was the second youngest population of any LGA in NSW (ABS, 2006). Over 40% were born in a non-English speaking country, conversing in one or more of 125 different languages (LCC, 2004). Already the fastest growing council in NSW (LCC, 2004), the State government has designated Liverpool to be one of five growth centres, one which will support 115,000 new dwellings to 2031; over that time Liverpool will absorb about 20% of Sydney's population growth (GCC, 2006).

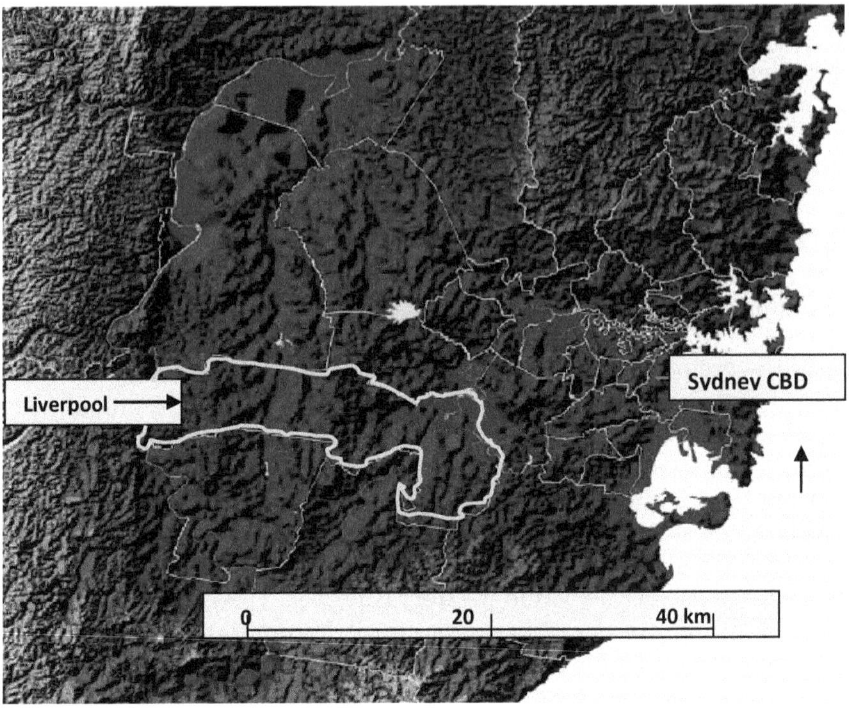

Figure 1 *Sydney Basin and its Surrounds, Showing Topographic Relief and Local Government Area Boundaries (Source: Western Sydney Regional Organisation of Councils, modified by author, WSROC 2005b)*

A journey undertaken: The partnership project

Following the Public Inquiry into Liverpool City Council, the Partnership Project was an important component of the Administrator's plan to improve the governance of Liverpool after the Inquiry identified this as being deficient (Kibble, 2004). A small Strategic Planning group within Council was delegated to organise what was to become the Project, and volunteers were called for from within the organization. The steering committee 'the Team' first met on the 15th June 2004 and I joined it on the 29th.

These initial meetings set the foundation for the Project, with Team members stating that they wanted something different, ethically grounded, with meaningful public participation. They wanted to have a significant affect on how 'Liverpool works', and have the community produce an 'effective plan'. At one of the first meetings, some of the comments were *"It's our strategic plan ... it will mean a lot and not just be nice ...* [but] ... *something to have faith in..."*, *"Something to make decisions by..."*, *"...the journey doesn't stop, there will be future renewal, keep the people informed even after the plan is completed and put out..."*, so the Team included an ongoing iterative/reflective element to the plan. It was to have an inclusive ethos, with contributors *"...from the cross-section of our population, involve the Administrator, residents and others..."*. A consistent aim of the Team was to produce a plan that was "Community owned, and Council led".

In addition to this 'idealistic' enthusiasm, there was also a deep realism, with constraining factors noted, such as *"...budgets, State government decisions...inertia, status quo..."* that would necessitate the Council officers needing to *"...take the vision for Liverpool and build solutions mediated by all sorts of practical, institutional, political, etc constraints..."*. There was also trepidation, for nothing like this had been tried at Liverpool. It was commented that, considering the nature of the community and of the recent Oasis scandal, *"...people are cynical and apathetic towards Council"*. One of the tasks was to regain the trust of the community in Council processes which meant that work had to be done to change both that culture of 'apathy' and the organizational culture within Council. These objectives were to be achieved through a framework that included an ongoing community engagement structure linked to the outcomes of the Project (as distinct from existing Neighbourhood Forums).

The Team reported to the Executive Board, the General Manager and the Administrator/Council and organized a series of sub-groups for publicity, public-workshop planning, corporate-change planning, and data analysis. The Project's initial timetable was for six months, but due to the small budget, the part-time nature of the work, the need to negotiate with higher levels of Council, and many ad hoc organizational problems, delays were experienced. Two main streams of activity were initiated: the development of a Community Vision for Liverpool and an internal Council Culture Change program.

Community engagement begins

Following eight months of planning, in February of 2005 Stakeholder Briefing Sessions took place. Another two months of effort produced the main facilitated Visioning workshops for all who *'lived, learned, worked or played in Liverpool'*, between May and July, heralded by a flurry of promotional activity. A pilot workshop was held on 27th May for 21 Council workers, most of who lived in the area. The month of June was taken up by nine publicly-open events and two for special-interest groups ('youth' and 'newly arrived migrants'), followed by a disabled-persons workshop on 4th July. In all, 13 sessions of three and a half hours were held at different times of the day, on week-days and week-ends, across the length and breadth of the LGA, for a total of 79 participants. In addition there were 110 interviews, 80 submissions from a Schools Program, and 30 involved in the Stakeholders Briefing. Including the Pilot workshop, 320 participants provided qualitative data. This data was supported by information derived from externally-conducted quantitative surveys in 2004 and 2005 that together elicited 1,170 responses from 5,000 randomly selected residents.

Several drafts of the document, called "Liverpool Directions 2006-2016", were constructed and approval of the document was negotiated with Senior Council. *Liverpool Directions 2006-2016* was endorsed by Council on 10th April, 2006. This represented the conclusion of Stage 1 of the *Creating Our Future Together Partnership Project,* a process planned for repetition in subsequent years.

The second stream of activity was the Council Cultural Change Program. This began with an all-staff launch in October 2004 conducted by the Administrator, and was soon followed by 'Corporate Planning Workshops'. These workshops introduced department managers and other staff to the Project, and began a process of *'Integrated Strategic Planning'*. One council officer stated that this process was moving from a position *"...of having managers sitting at their separate desks formulating their section's budget and then Council stapling all these together as the Corporate Plan. Each department silo is talking to itself, one section can't understand the jargon of the other; what hope is there for the average resident to understand?´* The aim was to move to where, *"We have all the departments sit at a round table and be informed by the community's priorities"*. The first Corporate Planning Workshops cleared the path by getting managers together, rationalizing the Corporate Plan document, simplifying language and adopting a unified structure for the document. Although much work was done in the meanwhile, the next workshops were held a year later, in 2005 and incorporated the preliminary outcomes of the community Visioning Workshops into a first attempt at an integrated Strategic Plan.

The Liverpool model

From the research I have identified seven major elements of the Liverpool Model:

- A Community Visioning process and document;
- A Corporate Plan/Council Cultural Change program (Integrated Strategic Planning);
- A Community Cultural Change process, with both the council and the community forming an evolving team where both are learner/teacher;
- A framework of the Triple Bottom Line;
- Linking of the Community Vision to the Strategic Plan;
- An on-going community engagement structure, and
- A staged iterative process of review and development.

Unpacking the journey

Most features of the Liverpool Model have been put into effect, although not all, and not all optimally. The Vision (Directions) document has been created and validated by the community; it has been published and is available on the Council website (LCC, 2006). The Corporate Planning/Cultural Change process has several successes and is progressing well towards its next incarnation for the 2006/07 budget. An extensive internal evaluation has been undertaken, including a celebratory lunch where we exchanged initial impressions of the Project. Finally, considerations for the next stage of the iterative process has begun as a part of the next Corporate Plan and as part of the reflective evaluation process.

Several limitations of the Project have surfaced. These include the lack of an ongoing engagement structure that was distinct from the current Neighbourhood Forums. The new structure was to include a capacity to monitor the uptake of the Directions document into the workings of Council, and be a vehicle for the community to progress the outcomes of the Project. A Consultation Toolkit was developed, exhibited and adopted by Council. Although a positive and necessary outcome in itself, this approach limits the engagement process to a toolkit that is the province of Council officers and not the community.

Another issue is that a TBL/Sustainability framework was not adopted for the Directions document. TBL issues were acknowledged as points within the document only. Sustainability was sent back to its council silo of the Sustainable Environment and Health department when it was decided that this department's development of a Sustainability Policy for Council would replace the previous approach. This limited the capacity of the Project to foster sustainability as part of its cultural change program. Also, an external evaluation was not undertaken, rather, an internal review process was adopted; this, however valuable, was not the same thing. The internal process did not allow for the establishment of a benchmark analysis of where Liverpool was with regard to the aims of the Project, limiting its ability to assess progress flowing from the Project.

Tentative conclusions

The development of the Liverpool Model shows a possible way forward for local governments that seek to encourage an active and engaged community. An improvement in efficiency was achieved, but what of attempting fundamental change in the governance procedures of Liverpool? Had a TBL/ Sustainability framework been adopted for the Directions document and then for the future incarnations of the Project, and had a specific community engagement structure emerged, it would have given greater potency to the process. Both Council and community would have been in a better position to challenge each other in a collaborative effort to further the aims of achieving a more sustainable Liverpool based on improvements in its TBL. These outcomes are yet possible as the process of continuous review and its ongoing iterative nature, promote new opportunities and open paths to innovative solutions.

References

ABS (2006). *Regional population growth, Australia, 2004-05*, Canberra: Australian Bureau of Statistics. Available at http://www.abs.gov.au/ausstats/abs@.nsf/Latestproducts/3218.0Main%20Features22004-05?opendocument&tabname=Summary&prodno=3218.0&issue=2004-05&num=&view=, accessed on 14/08/2006.

Albert, M. (2003). *Parecon: life after capitalism*, London: Verso, ISBN 185984698X.

Anderson, S. (2000). *Views from the south: the effects of globalization and the WTO on third world countries*, Chicago: Food First Books and The International Forum on Globalization, ISBN 093502882X.

Andrew, J. and Robottom, I. (2005). "Communites' self-determination: whose interests count?", in M. Keen, V.A. Brown and R. Dyball (eds.), *Social learning in environmental*

management: towards a sustainable future, London: Earthscan, ISBN 1844071839.
AtKisson, A. (1999), *Believing Cassandra: an optimist looks at a pessimist's world,* Melbourne: Scribe, ISBN 0908011458.
Bailey, B. (ed) (1995). *The true state of the planet,* New York: The Free Press, ISBN 0028740106.
Brown, V.A. (1997). "Managing change: the new manager," in V.A. Brown (ed.), *Managing for local sustainability: policy, problem-solving, practice and place,* Canberra: National Office of Local Government, ISBN 0642264643.
Brown, V.A. (2001). *Producing a Regional State of the Environment Report, Companion Report to the Western Sydney Regional State of the Environment Report 2000,* Richmond, NSW: University of Western Sydney Regional Integrated Monitoring Centre. The author has a copy.
Brown, V.A., Grootjans, J., Richie, J., Townsend, M. and Verrinder, G. (eds.) (2005). *Sustainability and health: supporting global ecological integrity in public health,* Sydney: Allen & Unwin, ISBN 1741144426.
Brown, V.A. and Pitcher, J. (2005). "Linking community and government: islands and beaches," in M. Keen, V.A. Brown and R. Dyball (eds.), *Social learning in environmental management: towards a sustainable future,* London: Earthscan, ISBN 1844071839.
Buckingham, S. (2003). "Building alliances for local environmental sustainability," in S. Buckingham and K. Theobald (eds.), *Local environmental sustainability,* Cambridge: Woodhead Publishing Limited, ISBN 1855736853, pp. 1-16.
Creswell, J.W. (1994). *Research design: qualitative and quantitative approaches,* Thousand Oaks: Sage Publications, ISBN 0803952554.
Daly, M. (2004). *Liverpool City Council Public Inquiry: Primary Findings and Initial Report,* Sydney: Department of Local Government, ISBN 1920766081. Available at: http://www.dlg.nsw.gov.au/liverpool/documents/Information/Liverpool_Inquiry_Report_final.pdf accessed on 03/08/2004.
Diamond, J. (2005). *Collapse: how societies choose to fail or survive,* Camberwell, Victoria: Allen Lane, ISBN 0713998628.
Edwards, A.R. (2005). *The sustainability revolution: portrait of a paradigm shift,* Gabriola Island: New Society Publishers, ISBN 0865715319.
Farrier, D. and Stein, P. (2006). *Environmental law handbook: planning and land use in New South Wales,* 4th edn., Sydney: UNSW Press, ISBN 0947205853.
Flannery, T. (1994). *The future eaters: an ecological history of the Australasian lands and people,* Sydney: Reed New Holland, ISBN 1876334215.
Fetterman, D.M. (1998). *Ethnography,* 2nd edn., Thousand Oaks: Sage Publications, ISBN 076191384X.
GCC (2006).*South-west growth centre,* Sydney: Growth Centres Commission. Available at: http://www.gcc.nsw.gov.au/the-growth-centres/south-west-growth-centre.aspx, accessed on 10/08/2006.
Hamilton, C. (2003). *Growth fetish,* Crows Nest, NSW: Allen & Unwin, ISBN 1741140781.
Harris, N. and Chu, C. (2001). "The development of indicators of sustainability at the local level: methodology and key issues," *Environmental Health,* ISSN 14445212, 1(2): 39-52.
Hawken, P., Lovins, A.B. and Lovins, L.H. (2000). *Natural capitalism: the next industrial revolution,* London: Earthscan, ISBN 1853837636.
Keating, C. (1996). *On the Frontier: a Social History of Liverpool,* Sydney: Hale & Ironmonger, ISBN 0868065862.
Kibble, G. (2004). "Survey to put city on right path," *Liverpool City Champion,* Wednesday, 20 Oct., p.9.
LCC (2004). *Social Plan 2004-2005,* Liverpool, NSW: Liverpool City Council. Available

at: http://www.liverpool.nsw.gov.au/scripts/viewoverview_contact.asp?NID=3182, accessed on 10/08/2006.

LCC (2006). *Liverpool City Council Website,* Liverpool, NSW: Liverpool City Council. Available at: http://www.liverpool.nsw.gov.au/default-liverpool.asp, accessed on 28/08/2006.

Lomborg, B. (2001). *The sceptical environmentalist: measuring the real state of the world,* Cambridge: Cambridge University Press, ISBN 0521010683.

McMichael, T. (2001). *Human frontiers, environments and disease: past patterns, uncertain futures,* Cambridge: Cambridge University Press, ISBN 0521004942.

Metro Strategy (2005). *City of cities, a plan for Sydney's future: metropolitan strategy supporting information,* Sydney: Department of Planning, ISBN 0734756550. Available at: http://www.metrostrategy.nsw.gov.au/dev/uploads/paper/introduction/index.html, accessed on 10/08/2006.

NRC (National Research Council) (1999). *Our common journey: a transition towards sustainability,* Washington: National Academy Press, ISBN 0309067839.

Parissi, C. (2003). "Lessons from local sustainability monitoring: outcomes and aftermath of the Western Sydney Regional State of the Environment Report 2001", *Environmental Health,* ISSN 14445212, 3(3): 54-66.

Parissi, C. (2005). "Public participation in local government: a case study of regional sustainability monitoring in Western Sydney", *Environmental Research Event 2005 Conference Proceedings, Environmental Change: Making it Happen,* Hobart, 29th November – 2nd December, ISBN 0864593864.

Parissi, C. (2006). *Between council and community: public participation and sustainability in local government,* work in progress, unpublished PhD Thesis, University of Western Sydney, School of Natural Sciences, Hawkesbury Campus, Richmond, Australia.

Pezzoli, K. (1997). 'Sustainable development: transdisciplinary overview of the literature', *Journal of Environmental Planning and Management,* ISSN (online) 13600559, 40(Sept): 549-574.

Preston, B.L. and Jones, R.N. (2006). *Climate change impacts on Australia and the benefits of early action to reduce global greenhouse gas emissions,* CSIRO: A Consultancy Report for the Australian Business Roundtable on Climate Change. Available at: http://www.csiro.au/csiro/content/file/pfbg,,.html, accessed on 04/08/2006.

Ristock, J.L. and Pennell, J. (1996). *Community research as empowerment: feminist links, postmodern interruptions,* Toronto: Oxford University Press, ISBN 0195410807.

Robson, C. (1993). *Real world research: a resource for social scientists and practitioner-researchers,* Oxford: Blackwell Publishers, ISBN 0631176896.

Roseland, M. (1998). *Toward sustainable communities: resources for citizens and their governments,* Gabriola Island: New Society Publishers, ISBN 086571374X.

Sen, G. (1994). "Women, poverty and population: issues for the concerned environmentalist", in W. Harcourt (ed.), *Feminist perspectives on sustainable development,* London: Zed Books and Society for International Development, ISBN 1856492443.

Shiva, V. (1993). *Monocultures of the mind: perspectives on biodiversity and biotechnology,* London: Zed Books and Third World Network, ISBN 1856492184.

Sinatra, J. and Murphy, P. (1999). *Listen to the people, listen to the land,* Melbourne: Melbourne University Press, ISBN 0522848613.

Singer, P. (2002). *One world: the ethics of globalization,* Melbourne: Text Publishing, ISBN 1877008451.

UNO (2005). *International conference on engaging communities,* Brisbane, 14-17 August. Available at http://www.engagingcommunities2005.org/, accessed on 10/10/2006.

WSROC (2000). *Western Sydney Regional State of the Environment Report*

2000, Blacktown: Western Sydney Regional Organisation of Councils, ISBN 1862712654.
WSROC (2005a). *Regional Maps: Rural Urban Fringe Map*, Blacktown, Australia: Western Sydney Regional Organisation Of Councils. Available at: http://www.wsroc.com.au/regionalfutures/pdfs/14_H4_Rural_Urban_Fringe_A4.pdf, accessed on 01/08/2005.
WSROC (2005b). *Regional Maps: Topography of Western Sydney*, Blacktown, Australia: Western Sydney Regional Organisation of Councils. Available at: http://www.wsroc.com.au/regionalfutures/pdfs/897_H3_Topography_Map_A4.pdf, accessed on 01/08/2005.
WRI (World Resources Institute) (2003). *World resources 2002-2004*, Washington: World Resources Institute, United Nations Development Program, United Nations Environment Program, World Bank, ISBN 1569735328.

Acknowledgements

Firstly, I wish to acknowledge the original inhabitants of the Liverpool region, the peoples of three aboriginal tribes, the Darug, Tharawal and the Gandangara. In addition, my thanks go to the Liverpool City Council officers who so warmly accepted me into the Partnership Project, and the community of Liverpool who so generously gave of their time to the project and accepted me as a participant.

Cesidio Parissi attained a Bachelor of Applied Science in Environmental Health, Honours Class 1, with University Medal. He is currently a doctoral candidate at the University of Western Sydney, in the School of Natural Sciences and is particularly interested in trans-disciplinary and inter-disciplinary approaches to research with a local area focus. Specifically, areas of active research have been in examining regional State of the Environment Reporting with local governments of Western Sydney and sustainability engagement processes of a particular Western Sydney council, namely, Liverpool City Council.

Causal Model of Beach Erosion with Inland Sand Transport: Political and Urban Planning Implications

Terence Love[1] & Trudi Cooper[2]
[1]Curtin University, AUS
[2]Edith Cowan University, AUS

This paper describes first stage findings from systems-based research into beach erosion and accretion of the coast near Perth in Western Australia. It presents an alternative explanation that views the coastal hinterland as an ocean-ward travelling dune system with windborne sand transported from inland. The role of built environment in blocking sand transport is included. Issues of scale, politics and urban planning are significant. The majority of Western Australia's populace live in the coast near Perth, and over half this area of coast has similar conditions to those described. Politically and legally, the model redirects responsibility and liability for the costs of coastal repair away from natural events. It suggests that in part, the beach erosion, and its subsequent costs and liabilities, result from local development, particularly of large houses in the close to beach coastal strip. This has liability implications for developers, building owners and planners. In urban planning terms, this implies a different urban planning topology if natural windborne sand transport from inland across the coastal dune system is to be maintained. One possibility is to arrange urban development in east-west strips alternating with significant bands of dune system.

Introduction

This paper presents a systems-based, alternative explanation of beach erosion and accretion of the beaches north of the city of Perth in Western Australia. Currently, serious beach erosion problems threaten coastal housing in the City of Wanneroo with similar geological conditions found in over half of the developed coastline of the City of Perth (Brachmanis, 2002). The problem is significant in population terms because the home for most of Western Australia's population with this north of city coastline one of the fastest growing regions in Australia.

Traditional explanations of beach erosion and accretion phenomena in this coastal region centre on the wave-beach interface and assume erosion and accretion primarily is explained by classic longshore drift and storm event models (see, for example, FirePlan WA & M. P. Rogers & Associates, 2001; M P Rogers and Associates, 2002). This paper presents an alternative explanation via a system causal loop model that includes wind-driven sand transport mechanisms feeding the coastal dune system from inland. This approach regards the extensive inland dunes as a westward travelling dune system that feeds the beach edge dune, the beach, and the beach/surf interface.

A significant aspect of the proposed system dynamic model is the inclusion of the effect of housing development on the coastal dunes. Housing, fences and other built environment infrastructure are likely to act in a similar manner to the fence technologies developed for stopping the movement of travelling dunes (Zandonella, 2003). Anthoni (2000) for example, noted similar adverse effect on beach health from buildings that create effects on sand transport to distance of around 30 x their height above local ground level. The proposed model therefore includes the built environment in this role.

This project is divided into three phases. The first phase has established causal systems models based on observation and literature review. The second phase identifies numeric values to enable system dynamic modelling of the sand transport mechanisms. The third phase involves using the system models to alternative interventions. This paper outlines preliminary findings from the first and part of the second phases.

Background

The coastal area that is the subject of this paper comprises an extensive complex of sand dunes along the predominately North-South coast and extending several kilometres inland. The underlying geology is karstic overlaid with sand dunes of various heights. Offshore for several nautical miles the seabed is primarily the same sand with limestone reefs.

To date, the main strategies for managing coastal erosion have been first to allow erosion to threaten houses and destroy the dune system. Then local government has used the ensuing media storm, community pressure and marine engineering consultants recommendations to draw on funding from local, State and Federal government to pay for protection and remediation. Typically, protection comprises stone groynes perpendicular to the beach and a program of sand replenishment.

To date, marine engineering consultants have assumed beach erosion phenomena are fully explained by longshore, littoral, drift processes and storm erosion (see, for example, FirePlan WA & M. P. Rogers & Associates, 2001; M P Rogers and Associates, 2002; Tremarfon Pty Ltd, 1997). The line of coast above and below Perth is North-South with onshore winds during the day in summer approach from the South and West and in winter, wild storms with heavy winds mainly from the North West (EPA, 2005). The days of winter storms are very clearly days of major beach erosion. Both wind conditions are oblique to the coast, typical conditions for longshore drift.

Systems causal models

There are several factors, however, that suggest the situation is not so simple. The front 'row' of dunes is typically between 5 and 50 meters high and covered with light coastal vegetation. After a major storm and erosion event, a new sand cliff is exposed. Like a cake, clearly visible on this exposed face are thin and widely separated 'layers' marking the soil and vegetation surface at previous times. Many of these layers are high above the waterline and meters below the current top surface of the dune. They indicate significant

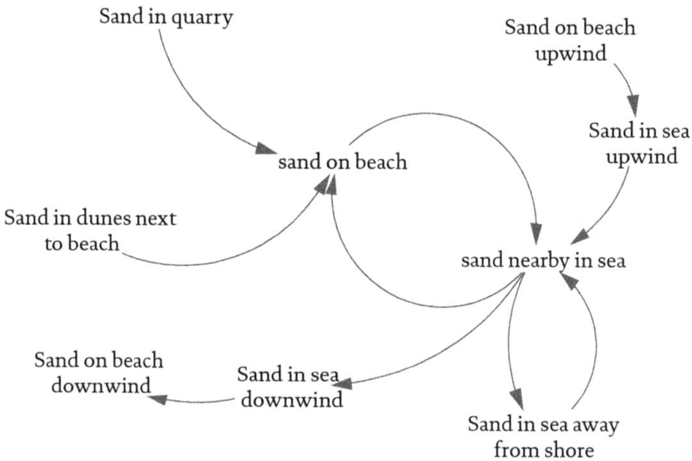

Figure 1 *Causal model of longshore drift system of erosion*

amounts of dune building at heights that are not easily reachable by saltation in the beach zone which maximum height of sand movement, even in storm conditions rarely reaches 2m. Omitted from the commonly held picture of wind behavior are the overnight strong easterly wind conditions found most of the year. In addition, year on year rates of coastal beach erosion appear relate directly to the amount of built environment located directly inland. The beach facing edge dune appears to be a dynamic system fed by sand from both inland and from the beach. It cascades sand from its foot onto the beach, which is eroded at times by wave action and at other times built up by deposition of sand from beaches elsewhere. The dynamic state of the beach-edge dune system is reflected in dune shape. Moving dunes have a shallow slope on the rear and are steeper on the front face. In this region, dunes with steep beachside faces, such as those near Tamala Park are associated with low levels of development whereas dunes with shallow beachside faces are associated with inland built development.

The marine consultant models takes weather as a variable, mainly surrogated by wind, and matches it to wave-driven sand transport at the surf line via longshore drift along the littoral region (see, for example, Cooper & Pilkey, 2004; M P Rogers and Associates, 2002). Intermittent sand replenishment occurs because of re-deposition of sand in the waves from sand eroded further down along the coast and blown up beach when it has dried out via saltation processes. Additional sand replenishment is via sand trucked onto the beach. In systems terms, this is represented in Fig 1.

Longshore drift modelling is not exact. In the case of Quinns Rocks, there are substantial differences between survey estimates and models predictions of sand movement in different reports (FirePlan WA & M. P. Rogers & Associates, 2001; M P Rogers and Associates, 2002; Tremarfon Pty Ltd, 1997). This may be due to stochastic variation of weather. Of greater concern, however, is the high sensitivity of outcomes of these models to 'fudge factors' decided by the consultants (Cooper & Pilkey, 2004).

From observation, wind is a significant transport mechanism of sand. Three examples that indicate this are:

- Beach movement occurs under action of afternoon onshore breeze. It can be painful to walk against the high levels of windborne sand, which is clearly visible in the air as a form of mist. A hand held up to face the wind accumulates sand particles. Cars parked close to the beach have their paintwork abraded, and in some cases removed, by the sand.
- Strong easterly winds from inland during the night and early morning smell and taste of sand. Rainfall in these conditions results in sand staining on vehicles, windows, pathways and clothing hung to dry.
- New building development in which the light saltbush typical of the area is removed, immediately results in very high levels of windborne dust deposited over gardens of nearby houses and high levels of sand dust inside houses at a distance form the development. This occurs to the extent that developers are required to surround new excavations with cloth fences to catch windblown sand.

The wind borne transport of sand is clearly high. Potentially it is of a similar order to that needed to supply sand from inland at a level that can fully compensate for longshore drift erosion of the beach. If the dune systems acts as a 'travelling dune', then the amount of windborne sand per m of coast width from inland required to feed beach fore dune is distributed over 20 km depth of the dune system from the coast to inland. The amount of sand required to be transported by inland wind to compensate for beach erosion is many orders lower than that observed in beachside windborne sand transport and windborne sand drift from building developments. This offers support for the hypothesis that the erosion and dune feeding system was in dynamic homeostasis until land development. It does this by providing an additional mechanism of supply of sand to the dune system, face dune and beach independent of the longshore drift source of sand deposition to beach in the surf zone.

The authors have developed an alternative causal systems model that hypothesises that dune replenishment, and thus beach replenishment, also occurs from inland, and that coastal housing, large trees and fencing block dune movement reduce sand transport in a similar manner to the technologies developed for controlling sand in coastal and central Africa. The inclusion of this additional sand transport mechanism is based on direct observations of the dune and beach system. It is also supported by wind analysis and sand transport theory (AUSGEO, 2003; EPA, 2005; Liu, Skidmore, Wagner, & Tatarko, 2002; Mangimeli, n.d.). In sand transport terms, the easterly wind blows strong and long (EPA, 2005) at a rate and duration of an order that would be expected to result in strong sand flow westward. In fact, although not readily observed during the working day, the dominant wind is easterly in contrast with the more easily observable peak storm winds that occur from the west (AUSGEO, 2003).

A causal system representation that includes both coastal and inland wind and wave driven processes as described above is shown in Fig 2. below.

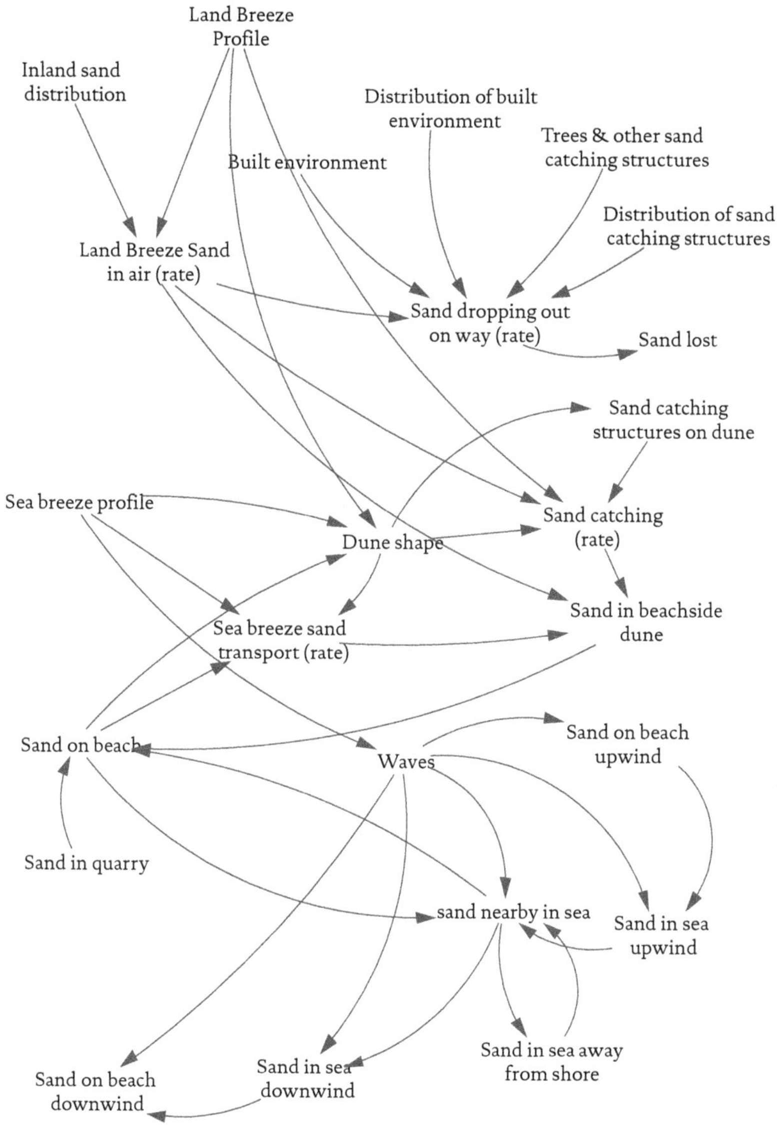

Figure 2 *A model of sand transport for beach erosion and accretion that includes sand transport from inland and the effects of built environment on reducing dune movement*

Political considerations

The marine engineering modelling of longshore drift and erosion dynamics, although widely used, is open to criticism because of the lack of a tight orthogonal relationship between all model variables and their direct connection with the real physical processes. Many variables cannot easily be accurately quantified because of the high dynamic complexity. Perhaps most important, , predicted outcomes are highly sensitive to small changes in equation parameters chosen on the basis of consultants judgements (Cooper & Pilkey, 2004). This provides a means by which political considerations may and can strongly influence modelling outcomes, predictions and implementation decisions in this situation.

The differences between the traditional and proposed explanations for the erosion dynamics of the coastal region north of Perth have potentially strong political and economic consequences. The traditional explanation of beach erosion, based solely on longshore drift and storm events, serves liability on God, chance and weather. By default, it leaves *all* citizens responsible for loss, and establishing and maintaining coastal protection through rates and taxes. The approach justifies a position that beach erosion is relatively independent of urban planning, or rather that urban planners must allow for the possibility that beaches may erode and plan to accommodate that possibility usually by buffer zones to the sea front.

In contrast, the model outlined in this paper includes the effect of houses and other built environment elements on the delicate balance between erosion and renewal of the beach line. It, therefore, locates liability and potential cost for beach remedial action with those responsible for building the structures or providing planning permission for them. It suggests that, in part, the beach erosion, and its subsequent costs and liabilities, result from local development, particularly of large houses in the close to beach coastal strip. This results in liability implications for developers, building owners and planners.

If the above inland sand replenishment model is correct, then it has implications for urban planning. It suggests that urban form must be arranged in order to minimise the flow of sand westward towards the coast. It also suggests a moratorium on extensive building within the dune system close to the coast. If natural windborne sand transport from inland across the coastal dune system is to be maintained, it implies the need for a different urban planning topology. One possible approach might be to arrange urban development in east-west strips alternating with significant bands of dune system.

Conclusion

This paper has presented a model of sand erosion and accretion mechanisms that includes additional factors. The model has political and economic implications in terms of relocation of liability for beach erosion and remediation

References

Anthoni, J. F. (2000). Disappearing beaches: engineering solutions, from http://www.seafriends.org.nz/oceano/beachsav.htm

AUSGEO. (2003). A Blow to Severe Storm Costs. Ausgeo News(70), 13-14.

Brachmanis, J. (2002). A Regional Review of Coastal Hazards along the South West Coast of Western Australia, from https://www.ga.gov.au/image_cache/GA2059.pdf

Cooper, J. A. G., & Pilkey, O. H. (2004). Longshore drift: Trapped in an expected universe. Journal of Sedimentary Research, 74(5), 599-606.

EPA. (2005). Wind Roses using data between May 1944 and Dec 2005 for Perth Airport, from http://www.epa.wa.gov.au/docs/keysbrook/volume1/PER_keysbrook_App1.pdf

FirePlan WA, & M. P. Rogers & Associates. (2001). Quinns Beach - Erosion Risk Managent Study Report 2001. Wanneroo: City of Wanneroo.
Liu, L. Y., Skidmore, E., Wagner, L., & Tatarko, J. (2002). Dune Sand Transport as Influenced by Direction, Magnitude and Frequency of the Erosive Winds, Ordos Plateau, China. In J. A. Lee & T. M. Zobeck (Eds.), Proceedings of ICAR5/GCTE-SEN Joint Conference, International Center for Arid and Semiarid Lands Studies: Texas Tech University, Lubbock, Texas.
M. P. Rogers and Associates. (2002). August 2002 City of Wanneroo Quinns Beach Erosion and COastal Protection. Osborne Park, Western Australia: M P Rogers and Associates.
Mangimeli, J. (n.d.). Geology of Sand Dunes. Retrieved July 2006, 2006, from http://www.nps.gov/whsa/Sand%20Dune%20Geology.htm
Tremarfon Pty Ltd. (1997). Quinns Coastal Processes Study. Wanneroo, Western Australia: City of Wanneroo.
Zandonella, C. (2003). Shifting Sands. New Scientist(28 June), 40.

Acknowledgments

The City of Wanneroo supported this project with copies of consultant reports that underpinned decisions for coastal erosion interventions.

THEME TWO

Cultural Knowledge and Heritage

Shaun Hooper, Blue Mountains World Heritage Institute

How we connect to and respect our past is very important to self and community identity. Ways of living today are firmly rooted in the past and are impacted by many change agents. This theme provides a basis for some of the issues that challenge conservation of our connection to and understandings of the past. These challenges provide focus for research and development of culturally appropriate tools and concepts in the protection of what makes us who we are and what we value from our past and present.

Sustainability of our cultures knowledge systems and its heritage is also an important challenge we face today. In the world of the global village, cultural groups need to balance the demands of everyday life and economics with the challenge of inter-generational equity. Providing a strong cultural identity with places and stories that our children can connect to is an essential element in sustainable cultural communities.

Indigenous communities in the current time are especially susceptible to loss of cultural identity through government policy and research epistemologies that do not allow for the expression of their unique cultural knowledge and heritage. Action at all levels needs to be guided by Indigenous people themselves to allow for culturally appropriate knowledge systems to be developed that support their unique perspectives of our world.

New Zealand Communities Forever: Integrating Sustainability, Education and the Treaty of Waitangi

Sylila Monteiro & Rashika Sharma
Applied Technology Institute, UNITEC New Zealand, NZ

To sustain the natural and social capital in New Zealand it is essential to integrate sustainability, education and the Treaty of Waitangi. The Treaty, the core of New Zealand governance was signed in 1840 between Māori and the Crown. The Treaty recognizes the unique relationship of Māori with their environment and promises Māori full rangatiratanga (chieftainship) of their lands and taonga (treasures). The concept of managing New Zealand's natural and social capital is underlined by the Treaty which plays a central role in sustainability education. The Māori concepts of hauora (total well-being and balance with nature) and rāhui tapu (conservation) are significant reflections of sustainability. The Treaty itself is a fundamental theme in New Zealand's resource management legislation - the Resource Management Act 1991, whose main objective is to support and maintain sustainable management of natural and physical resources. Unitec, New Zealand is a tertiary institute that incorporates its commitment to the Treaty of Waitangi and its partnership with Māori through an agreement: "Te Noho Kotahitanga". The Unitec Applied Technology Institute (UATI) offers a Sustainable Technologies course where students are required to design sustainable technological solutions in authentic settings. The compulsory requirement to integrate the Treaty into sustainable technology issues ensures that students learn to appreciate the cultural and social importance of the environment to Māori. This paper through a case study attempts to illustrate how these elements have been incorporated effectively into the learning and teaching context at Unitec Applied Technology Institute (UATI) both within this course and into the wider degree programme.

Introduction

Sustainability is a means of meeting the needs of the present without compromising the ability of future generations to meet their own needs. Education and awareness is essential in order for communities to plan for a sustainable future. In New Zealand, to achieve a sustainable future the integration of technology and traditional Māori values into environmental legislation and education is crucial for the development and management of natural resources. This paper attempts to demonstrate how this integration is possible and can be replicated by other global communities. It must be emphasized from the start that the Māori terminology used in this paper has several connotations with regional variations. To clarify the meaning of Māori terminology used by the authors a glossary is included.

Why education for sustainability?

Education is a sociological phenomenon focused on what educators do to facilitate learning in others. It is a very powerful medium through which information can be transferred and improved practice can be effectuated. Education can initiate a change in mind-set, attitudes and habits. Learning generally takes place within social contexts though it can be argued that individuals can learn independently. Education for sustainability directs people in society to live in sustainable ways. Environmental education, the precursor to education for sustainability, concentrated on conservation and protection rather than effective ways to reduce environmental degradation. Education for sustainability takes a more proactive approach. Instead of merely cleaning up the symptoms of environmental problems it aims to do things differently in the same place through social indoctrination.

The Earth Summit in 1992 highlighted the importance of education for sustainability. The United Nations called on all countries to develop strategies for its implementation. The Earth Summit defined education for sustainability as:

"...critical for achieving environmental and ethical awareness, values and attitudes, skills and behavior consistent with sustainable development and for effective public participation in decision making (UNCED, 1992)."

Education for sustainability was first prioritized by the New Zealand Parliamentary Commissioner for the Environment in 1997. The awareness of education for sustainability has been slow therefore New Zealand has reaffirmed its commitment by declaring the Decade of Education for Sustainable Development from 2005 to 2014 (PCE, 2004: 37). Changes had to be made so that people could live in sustainable ways.

Education for sustainability necessitates people to critically reflect on their own values and surroundings for community well-being. It aims to create an integrated approach whereby people and institutions can share their ideas and use their expertise to make sustainability a reality. More than a mere immediate social awareness, the ultimate objective is changing social beliefs that inculcate in an individual and in the community the practice of sustainability.

Tertiary education is vital in influencing New Zealand society and community. Conventionally the acquisition of knowledge at tertiary level was restricted to the specializations undertaken. Specialist students have difficulty developing knowledge of issues from different perspectives including sustainability. Many students graduate without the knowledge of sustainability and are unable to apply the concept in practice. Education today needs to focus on integrating principles of sustainability within all courses so that not only environmental graduates but graduates from all disciplines become sustainable practitioners.

The New Zealand education environment is unique as compared to those of other countries. To encourage sustainability in New Zealand special attention needs to be paid to the political and social aspects that directly and indirectly influence education. In bi-cultural New Zealand society, the Treaty of Waitangi and the Resource Management Act 1991, encourages dialogue between Māori (*tangata whenua*) and Non-Māori. This dialogue enhances the implementation of education for sustainability in locally relevant and culturally appropriate ways.

The Treaty of Waitangi

The Treaty of Waitangi signed on 6 February 1840, is the founding document of present day New Zealand and the closest thing this country has to a written constitution. The Treaty recognizes the unique relationship of Māori with their environment and promises Māori full *rangatiratanga* (chieftainship) of their lands and *taonga* (treasures).The Treaty ensures that Māori were able to preserve and uphold their way of life yet were able to draw on the positive aspects of European culture, technology, skills and material wealth.

Rangatiratanga is the key term in the Treaty. In essence it is the working out of a moral contract between a leader, his people and his god. According to the Ngawha Geothermal Resource Report it is a dynamic not static concept, emphasising the reciprocity between the human, material and non material worlds. Realistically, it means the wise administration of all the assets possessed by a group for that group's benefit: in a word, trusteeship (Waitangi Tribunal, 1993: 5).

Rangatiratanga is a birthright, the exclusive control of *taonga* for the benefit of the tribe: those living and those yet to be born. Without this crucial authority the tribal base is threatened socially, culturally, economically and spiritually. This entails stewardship to maintain the tribal base for succeeding generations. The Treaty ensures that Māori retain the right to protect and guard the natural environment. They have a strong tie with the land and have *mana whenua* (power and authority) to produce a livelihood from the land and its natural resources. Māori protect the land from despoliation by careless exploitation and ensure that sustainability is always maintained (Mutu, 2004: 166)

NZ Resource Management Act 1991

The Treaty is the central theme of the Resource Management Act (RMA) 1991 which was designed to promote sustainable management of New Zealand's natural and physical resources. This Act provides a legal framework for making decisions about the use, development and protection of renewable and non-renewable resources. Section 5(2) of the act defines this as:

"'managing the use, development, and protection of natural and physical resources in a way, or at a rate, which enables people and communities to provide for their social, economic, and cultural wellbeing and for their health and safety; while sustaining the potential of natural and physical resources (excluding minerals) to meet the reasonably foreseeable needs of future generations; and safeguarding the life-supporting capacity of air, water, soil and ecosystems; and avoiding, remedying or mitigating any adverse effects on the environment" (Statistics NZ, 2004: 288).

To achieve sustainability, the Act obliges persons with authority and power to consider *kaitiakitanga* (the ethic of stewardship), the appropriate development of natural and physical resources, the protection of heritage, the maintenance and the enhancement of environmental quality by proper management of finite resources (BIZ, 2004).

Consultation with the *iwi* (tribe) is the core of the Treaty signed in 1840 and the Resource Management Act 1991 legalizes this consultation making it relevant for the present. Therefore liable authorities such as the local councils have to consult *tangata whenua* (people of the land) wherever Māori are directly affected by a proposal concerning the natural resources and the environment. The Resource Management Act 1991 requires resource consents to be granted for any development and these take the form of specific approvals through submissions made by affected parties to the local councils.

Traditional environmental knowledge is important in environmental management hence the incorporation of the traditional values of the Treaty finds current application through the Resource Management Act. Immaculate understanding of their environment has supported Māori efforts to maintain and sustain their families and communities for many centuries. The Resource Management Act recognizes Māori as being *kaitiaki* (caretakers or guardians) and responsible for maintaining and preserving the life sustaining environment. The Māori concepts of *hauora* (total well-being and balance with nature) and *rāhui tapu* (conservation) are significant reflections of sustainability. Māori have been great believers in a lifestyle that preserves their natural resources so that the sea and earth would not be harmed. They have *mana whenua* (authority) to produce a livelihood for family and tribe from the land and its natural resources (Mutu, 2003: 167).

In Māoridom, all the elements are the world's natural resources- the *taonga*- that are greatly treasured and respected. If *taonga* is violated *mana* (power) is removed and power to sustain life is lost (Mutu, 2003: 167). Preserving the life force of *taonga* ensures that the life essence is healthy and strong. The judicious use of *taonga* guarantees healthier communities and a sustainable lifestyle. Māori have always practised sustainable environmental management and this traditional knowledge is vital in future resource management in New Zealand.

UNITEC New Zealand and *Te Noho Kotahitanga*: The Partnership

UNITEC's mission is to inspire students to discover and apply their intellectual and creative potential and contribute responsibly to their societies and cultures (UNITEC, 1994: 11). UNITEC as a tertiary institute accepts the responsibility to educate New Zealanders and acknowledges the fact that all New Zealanders need to be informed about the Treaty of Waitangi. UNITEC believes that only from an informed and educated understanding of each party's history, culture, values and beliefs that everyone can begin to fully appreciate and value the cultural richness of New Zealand and each other's uniqueness and sameness (UNITEC, 1994: 17). "*Te Noho Kotahitanga*", a partnership document of 2001, expresses UNITEC's commitment to this uniqueness and to the Treaty of Waitangi. This commitment is implemented through the integration of the Treaty in most courses, to empower the students with an appreciation of the bi-cultural richness of Aotearoa.

UNITEC's Applied Technology Institute (UATI) offers wide range of industry-related education and training programmes designed for tradesman and technologists who help shape the future. In all aspects of society the training and education of tradesman and technologists is important so they can apply knowledge and skills to solve practical problems and help make the world a better place to live.

UATI's Bachelor of Applied Technology (BAT) programme is a generic degree for the technologists of tomorrow. UATI recognizes the important role technologists play in developing society. With proper training students not only specialize in their disciplines but also on wider issues such as sustainability. The BAT programme incorporates the principle of sustainability within its applied technology courses for students to comprehend the principle and apply it within their own trades and make a difference tomorrow.

UATI's Sustainable Technologies course enables students to investigate global and local aspects of sustainable technologies and to integrate this knowledge within their trade. Students articulate the rationale of sustainable technologies. They design creative solutions for sustainable practices in individual, community and workplace settings. This course aims at teaching the principle of sustainability to students so that they can make a difference to the environment by taking a step closer to making their trade more sustainable. In designing sustainable solutions, students have to integrate the implications of the Treaty of Waitangi into their trade areas. The real world scenario given below provides a typical classroom example of the application of the Treaty as incorporated within the delivery of the Bachelor of Electrotechnology programme.

"Your project team is working on the feasibility of a geothermal power station in central North Island. The design has been developed and submitted to the local council for resource consent. You have now been contacted by the council officials who require evidence of consultation with the local iwi (tribe) to ensure that the principles of the Treaty of Waitangi have been addressed".

Students are given a time limit to research and come up with relevant solutions to the above problem. During this process students will deem it necessary to study the implications of the Treaty of Waitangi and the effect it will have on the local *iwi* and the environment. Each group then presents their findings at the appointed time and a discussion of the various issues create a forum of learning about sustainable practice in New Zealand.

Educating 'tomorrow's sustainable technologists' today!

Technologists play a significant decisive role in advancing and transforming our society. Tomorrow's developments are in the hands of today's technologists therefore it is imperative that today's technologists are educated about sustainability. This ensures that every New Zealander can enjoy the same resources in the future as we have today. Technologies act as a double-edged sword and regularly contribute to further pressures on sustainability that may have been unforeseen (PCE, 2004: 42). These unforeseen pressures can be minimized by educating technologists about sustainability. The Sustainable Technologies course at UATI seeks to achieve this goal by educating the technologists of tomorrow today so that they have a better understanding of sustainability and can help make the future more liveable.

Integration of sustainability, education and the treaty

The integration of the Treaty of Waitangi in the Sustainable Technologies course ensures that students comprehend the critical link between the technical environment and the natural environment. All technologies impact the environment in one form or another which in turn impacts on society. Therefore students are encouraged to appreciate the link between technology and the Treaty to ensure that the correlation is maintained to achieve sustainability. This appreciation will inevitably inculcate sustainable practice in the individual and result in educating the communities and the wider society.

Once again it is vital to emphasize that education for sustainability is critical for realizing environmental and ethical awareness. Values and attitudes, skills and behavior consistent with sustainable development are essential for effective public participation in decision making to foster social change. Learning is multi-layered and inter-generational. By incorporating the Treaty of Waitangi, UATI's Sustainable Technologies course aims at successfully educating the whole person rather than specifically equipping students for a career. New Zealand is unique in that traditional Māori values are intrinsic with current principles of sustainable practice. Boyer's (1990) *scholarship of integration* advocates making connections across disciplines, and placing the specialities in larger contexts. Sustainability fits into all learning areas as a cross-curricula theme. Education that endorses this integration, consciously and unconsciously creates in the individual the awareness and sensitivity to the environment and related issues. Knowledge and understanding of the environment and its impact on people, takes on practical stances that reflect in the attitudes, values and feelings in individuals that in turn permeate into the community.

A sense of community responsibility develops through participation and action as individuals or members of groups (*whanau* or *iwi*) address environmental issues. With knowledge of their heritage, their language and their culture together with tertiary education, New Zealand communities can handle the world at large with confidence and self- determination and be assured of a sustainable future for succeeding generations establishing New Zealand communities forever.

As in the case of New Zealand further investigation into the integration history, culture, society and education in other countries could lead to further insightful research. Such heuristic research can indoctrinate various communities creating sustainable global communities making the world a better place in the future. The fusion of new technologies with traditional practices can build a sustainable future with global tangible long-term benefits. Specific behaviors that support a healthy, humane, and environmentally friendly world would become globally coherent and compelling. By living in ways that are deeply fulfilling and at the same time socially and environmentally responsible, global sustainability can become a changing social reality.

References

Boyer, E. L. (1990). *Scholarship Reconsidered: Priorities of the Professoriate*. Princeton, NJ Carnegie Foundation for the Advancement of Teaching.

Boyer, E. L. (1990). The Scholarship of Engagement. *Journal of Public Service and Outreach*, 1(1)11-20. ISSN 1086-2447

Business Information Zone. (2004). *Resource Management Act (RMA)*. Retrieved April 19 2006 from http://www.biz.org.nz/public/content.aspx?SectionID=36&ContentID=550#

Cavanagh, P. (Ed.). (2004). *New Zealand Official Year Book 2004* (104th ed.). Auckland: David Bateman Ltd, ISBN 0-478-26925-0, pp. 288. http://www.un.org/esa/sustdev/documents/agenda21/english/agenda21chapter36.htm

Mutu, M.(2003). *Te Whanau Moana*. Auckland: Reed Publishing NZ Ltd, ISBN 0-7900-0839-4, pp. 166-168.

Parliamentary Commissioner for the Environment (PCE). 2004. *See Change: Learning and Education for Sustainability*. Wellington: PCE, ISBN 1-877274-12-7, pp 37-67

Ministry of Environment (2006). *The Resource Management Act 1991*. Retrieved 26 September 2006 from http://www.mfe.govt.nz/rma/public/rma-terms.html

UNCED. 1992. *Agenda 21: Promoting Education, Public Awareness and Training*. Retrieved May 14 2006 from http://www.un.org/esa/sustdev/documents/agenda21/english/agenda21chapter36.htm

UNITEC. 2004. The Bachelor of Applied Technology. Auckland: Author, pp. 11.

Waitangi Tribunal. 1993. *Ngawaha Geothermal Resource Report* WAI 304. Wellington, Brooker and Friend, pp. 5.

Glossary

Hauora- total well-being and balance with nature

Iwi- group of hapu tribal groupings

Kaitiaki- spiritual minders of the elements of the natural worlds; group with responsibilities of *kaitiakitanga*

Kaitiakitanga- the responsibilities and *kaupapa* passes down from the ancestors for *tangata whenua* to take care of the places, natural resources and other *taonga* in their *rohe* and the *Mauri* of those places, resources and *taonga*.

Kaupapa- fundamental principles, plan, tactics, strategies and methods.

Mana- power, authority, ownership, status, influence, dignity and respect derived from the gods.

Mana whenua- power of the people

Mauri- essential life force, spiritual power and distinctiveness that enables each thing to exist as itself.

Rāhui tapu- conservation

Rangatiratanga- chieftainship including sovereignty, rights of self-determination, self government, the authority and power of *iwi* or *hapu*, to make decisions and to own and control resources.

Rohe- geographical territories of a tribe
Tangata whenua- people of the land
Taonga- valued resources, assets, prized possessions both material and non-material

Sylila teaches a wide range of communication and integrated practice papers on Certificate, Diploma and Degree programmes offered across UNITEC. She previously taught communication in Zambia Institute of Technology in Kitwe, Zambia. She specializes in business document translation services for French and Portuguese organizations engaged in international communication. She also provides training for New Zealand Army personnel in Portuguese and French as preparation for overseas assignments. Her consultancy work involves Industry Training on Communication components of Management Development Programmes for Unitec Enterprise Unit.

Rashika teaches a wide range of integrated practice courses for the School of Applied Technology at Unitec New Zealand. She is actively involved in delivery of communication, health and safety, sustainable technologies, and societal context papers for the institute. She previously taught environmental science at the Fiji Institute of Technology in Suva, Fiji. Rashika is specifically interested in education for sustainability research.

Recording My Mum's Songs

Fiona Brady & Tina Asela

This paper draws on a current project, recording hymns and choruses in the local language, to explore some of the issues with language maintenance and revitalization programs in a remote Indigenous community. I use the work of Walter Ong and David Crystal to underpin the discussion about nature of oral cultures, technology and their implications for language loss and revival. I identify a mosaic of existing initiatives and positive factors for language survival and I outline the new and difficult legal considerations the community must work through as they create language resources with new media.

Introduction

The wealth of a country may be indicated in the richness and diversity of its natural and social environments. Australia has a wealth of Indigenous languages, over 250 with 600 dialects, however, all these languages are under threat (Nathan, 1999).

This paper describes a small language support project in a remote Indigenous community in Far North Queensland, Australia. Using this project as a lens we discuss some of the broader issues of language maintenance, and how technology can support a minority language. We are able to identify both the mosaic of people, skills and initiatives that support language use in this community; and the gaps in our knowledge and procedures, particularly in regard to legal issues.

Project description

My name is Tina Asela. I applied for, and received, a small arts grant to record the hymns and choruses sung in the local language, Kalaw Kawaw Ya (KKY), in the Church of the Holy Cross on my home island, Dauan. Dauan Island is one of the most remote and isolated islands in the Torres Strait, approximately 150 people live here and we are 180km from the regional centre.

I am recording hymns sung on the four special church days on Dauan Island: Easter; Coming of the Light day in July; Church day in September, this celebrates the anniversary of the consecration of our church; and Christmas. My mother, Naina Asela, is an elder of Dauan Island; she leads the singing of the hymns at community events and teaches them to the community. She first learned these songs as a girl of 16; her auntie encouraged her to join the church choir because there were few other activities then. She is a very good singer, she used to work on cruise ships with my father when they were young, she would sing while my father performed traditional dances. This is not the first time she has been recorded, about 20 years ago a visitor to the island took a video of her singing.

My mother has an old handwritten copy of the hymns that was given to her by Mrs Sarbie Mooka. I refer to this copy when I am printing up hymn sheets for church days. In past times the church service was conducted in English and the hymns were in KKY. Priests on Dauan continue to compose new songs in KKY for the church and recently they have started to translate some of the services into KKY as well.

The burubur [drum] is traditionally used to accompany the hymns, however, sometimes other instruments such as guitar, bass and keyboard are used by the younger men in church services. I am making these recordings so Dauan people can listen to them and become more familiar with the hymns. I am also making hymn books in KKY for the church so we do not have to depend on my mother's very old copy. The hymn recordings, the accompanying interviews, which were conducted in English, Creole and KKY, together with a hymn book with an English translation, will be deposited with the State Library of Queensland so this knowledge will not be lost.

Project need

One of the objectives of this project is to provide a resource that is linguistically accurate. There is a need for this sort of resource because, as language use has declined and people are dispersed, errors are creeping into the language. At the recent Coming of the Light celebration in the regional centre of Cairns, a diverse group of Islanders performed a composite re-enactment that included songs from a number of islands. (Each island has its own particular celebration and hymns). A Dauan Island song was included but it was not sung correctly and this caused pain to the Dauan Island visitors to the event.

Description of the language

These songs are in Kalaw Kawaw Ya (KKY) a dialect of Kalaw Lagaw Ya, (KLY) the language of the Central and Western Islands of the Torres Strait. KKY is spoken on Dauan, Saibai and Boigu Islands; it may also be used to talk on the telephone to family members who have moved away to the mainland or to other islands. Nowadays it is mostly people over 35 years of age who speak KKY. The majority of Islanders are bi or tri-lingual, Torres Strait Creole and English are the other main languages.

Modern Dauan Island culture is predominantly oral. The major source of texts in KKY is the Summer Institute of Linguistics (SIL). Wikipedia describes the Institute as a "Christian linguistic service organization which studies lesser-known languages primarily to provide the speakers with Bibles in their native language". The other published resources in or about KLY/KKY are either academic or religious, although, a small number of story and song books have been produced by the local primary school.

Song is a major reservoir of KKY language. Song is highly regarded: Torres Strait Islanders have a strong musical tradition, from the song sagas of traditional times to hymns and popular songs. Songs are disseminated orally rather than written down. For example, Father Aragu may compose a new hymn and sing it over the telephone to teach someone on another island, or he may record his service on cassette and mail it to distant singers so they can prepare for an event. There is an active modern music scene and musicians on Dauan, Saibai and Boigu Islands compose songs in KKY and in Creole and produce their own CDs. Younger people may sing in KKY although they do not regularly speak the language. Elsewhere in the region CDs of secular and religious songs are produced in a combination of languages. Such CDs are widely distributed and sold through the networks of family and friends.

We should not be surprised that religion and song are strong carriers of the language. Walter Ong (1982), one of the foremost scholars in the field of orality and literacy, describes how music may help to "fix" narrative, so it can be recalled verbatim. Religion also has many aspects which can support local language: religion is held in high regard throughout the region; each Island has its own church or churches so there is local ownership; the repetition of hymns and the gravity of ritual help people memorise the words.

Christianity has a strong oral tradition, the bible itself is a written record of oral stories and religious organizations have recognized the value of using the local language and oral learning techniques to introduce Christianity to people in developing areas.

Theoretical framework

The loss of language is a world wide issue. Languages are dynamic and just as species have become extinct in the natural world so languages have died, however, we are now in a situation where the majority of the world's languages are under threat (Crystal, 2000).

David Crystal (2000) identifies two main causes of language loss: physical danger to the speakers through natural disaster, disease, or destruction of habitat, these lead to the death or displacement of people to other language communities; and assimilation into a dominant culture through invasion or colonization. He describes the sequence of assimilation as firstly; pressure to speak the dominant language, followed by a phase of emerging bilingualism that rapidly declines into the final phase where the younger generation becomes more proficient in the dominant language (Crystal, 2000). In this phase language may be lost very quickly. A recent example of this would be where one island in the region lost its three remaining fluent KLY speakers in one year (Edwards, 2006).

Language is intricately connected with culture, it is a means of group identification, and it contains the social values and world view of the speakers (A matter of survival, 1992). The loss of language will affect the culture and it has been argued that the health of a culture is directly linked to a community's economic and social wellbeing (The Crossing Boundaries National Council, 2005). Nickerson (2005) states that if there is a starting point to cultural continuity and community renewal it lies in the preservation and learning of Aboriginal languages.

Crystal (2000) recommends a "healthy bilingualism", he argues that the dominant language is outward looking and necessary as a bridge between the two worlds, whereas, the local language is inward looking and used to express identity, family ties, social relationships and history.

Language revitalization

Languages have been successfully revived, Maori and Welsh are recent examples. A language is living when it is used. Government recognition and support are part of the revitalization process but there are too few linguists, and funds, available to do all the language work required to develop language programs for all the endangered languages. However, language groups themselves can take action to facilitate language use. Successful strategies include individuals using the language at home and especially teaching the children; the community can choose to use language in religious and cultural practice, in schools, and in the local media; resources in and about the language can be made, these could include dictionaries, religious materials, academic writings, teaching and entertainment CDs and videos.

Community control and community involvement in language work underpin the long term success of language revival exercises (Crystal, 2000). Through their participation in language projects, local people have an opportunity to develop knowledge and skills, which could enable them to manage their own language programs. Participation also reinforces local ownership of the language and the relationship between language and land (Nathan, 1999).

Technology is identified as an ideal tool for language resource projects (Nathan, 1999) although there is some concern that the very ease of technology can lead to a less than appropriate implementation (Villa, 2002).

Language and technology

Technology has been identified as a factor in the demise of languages: modern transport and communications technology have reduced the isolation of many people exposing them more and more to the dominant language and culture. (Crystal 2000; A matter of survival, 1992). However, technology may also be important in their revival. (Crystal, 2000; Ong, 1982; McHenry, 2002). Modern technological methods of oral and visual communication will transmit any language. In the past, writing was the main means of communicating over a distance, thus profoundly disadvantaging oral communicators and those whose language had not yet been written. Communications technologies are working increasingly like natural language; this makes it a "good fit" for oral language and culture (Ong, 1982). The earliest examples are the telephone and radio, now there is widespread use of mobile phones, TV, video, multimedia computers with internet, audiovisual recorders and players.

People who are predominantly oral can acquire information technology skills using a range of strategies when they have a specific purpose for so doing (Brady, 2005), and it has been suggested that there is a congruence between IT and Indigenous learning styles.

Furthermore, technology offers the opportunity to create and control representations of self and culture. Self publishing could enable Indigenous people to define themselves and in some way balance historical, written, anthropological definitions of their culture. Marcia Langton states that Indigenous self publishing demonstrates a "cultural identity rooted in tradition yet open to modernity" (UNESCO, 2001). It may start to redress situations such as the under-representation of Indigenous Australians in conventional media. (Nathan, 1997). This lack of representation has hampered the legitimation of cultural knowledge and recognition of Indigenous people as experts on their own environment (UNESCO, 2001).

IT is a modern, relevant and highly valued area (Nathan, 1999) so it can enhance the status and motivation of language work. The very existence of materials sends a message that the language is important (Nathan, 2005).

Audio visual resources can be very rich; video may record the surroundings, the events, the facial expressions and gestures of participants as well as the song or story. This makes them important learning tools particularly for dispersed communities. Working with IT can also generate cross generational, cross cultural and collaborative links amongst peers. (Auld, 2002; UNESCO, 2001). These activities themselves strengthen networks and communities. We should not restrict our interest to traditional culture materials, multimedia is being used for creative works that have strong economic potential and may engage more of the community, and enhance their self image.

ICT offers opportunity for cultural renewal but also threat of cultural hybridization or homogenization (Nickerson, 2005) and loss of linguistic diversity (Auld, 2002). Dauan, in part because of its remoteness, has its own distinct lifestyle, a visitor from neighboring Badu Island referred to Dauan as being "more cultural" than her Island. It will be a challenge to retain this given the ease of production and the much greater market for resources in Creole.

Legal issues

One of the difficulties with any recording or oral culture/history is that the act of recording fixes something that is "very changeable" (Ong 1982). In lived culture the event is: singing at the church. It is conducted in accordance with the local protocols and adapted to the audience. These protocols are dynamic reflecting the people and circumstances of this particular event. A recording idealizes this version, separating the singing from the context which governs it, so new types of protocols are needed. Rules for appropriate access, attribution and use of recorded material are needed urgently. Technology has de-professionalized media production and materials can be distributed virtually without cost across the globe via the Internet, this is increasing the potential for abuse by others and for the distribution of inaccurate or sensitive material (Australia Council, 2002).

There is currently no legislation in Australia that specifically protects Indigenous cultural heritage. The Federation for Aboriginal and Torres Strait Islander Languages (FATSIL) has produced a set of protocols for Indigenous communities and their consultants who are jointly producing language materials at local level. These protocols are designed to give a general guide to the types of issues involved in publishing language materials and are aligned with the *Our Culture: Our future* report by Terri Janke (1998) and the AIATSIS

(2000) *Guidelines for Ethical Research.* It includes a model agreement, as they state formal written agreements are important for all parties. The agreement defines the rights and roles of communities as owners and custodians of their languages and cultures based on the linguistic rights of ATSI described by Australian Linguistic Society.

In this project verbal agreements and permissions have been sought from performers and community. The performers would like to be acknowledged as performers with the right to sing the hymns. Royalties on any sales will be paid to performers. Tina would like to make money from the sale of the CD. The IP and Copyright has been given to the Holy Cross Church of Dauan. This was the first time any of the participants had had to think about these concepts and try to synthesize with their own cultural practice.

Conclusion

This project shows part of a large network of individuals and organizations who have contributed to KKY language over a long period of time in a variety of ways. This study indicates the range of positive factors already in place but not necessarily directed at language revitalization such as: local initiation of projects; participation in the project by elders; the existence of a small pool of local people who have undertaken linguistic studies; the local music production; the increasing support of the church for language use; the young people's involvement in the church; that there is existing interest in and support for products in KKY language; the collaborative arrangements to structure and acquit projects; existing local technology skills; and availability of technology privately and in the council and school.

We have used this study to reflect on our current position, to recognize and value work that is already being done, to fit our project into an overall picture of language revitalization and to identify gaps to be addressed. Language is an important aspect of the unique culture of Dauan Island and as individuals and community we have the responsibility to keep it strong.

References

"A matter of survival". (1992). Report of the Inquiry into Aboriginal and Torres Strait Islander Language Maintenance. Retrieved June 5 2006 from http://www.austlii.edu.au/au/special/reproject/rsjlibrary/

AIATSIS. (2000). Guidelines for ethical research in Indigenous studies. Retrieved August 16, 2004 from http://www.aiatsis.gov.au/corp/docs/ethicsguidea4.pdf

Auld, Glenn. (2002). "The role of the computer in learning Ndjebbana" *Language Learning & Technology*. ISSN 1094-3501. Vol 6 No. 2 May 2002, pp41-58.

Brady, F. (2005). "Learning to Internet Bank". Chapter in Dyson, E., Hendriks, M., and Grant, S. (Eds.), *Information Technology and Indigenous people.* Idea Group Reference. ISBN 1-59904-298-3.

Crystal, D. (2000). *Language Death.* Cambridge: Cambridge UP. ISBN -13; 9780521653213.

Edwards, Ron. (2006). Personal comment. Cairns QLD July 2006.

FATSIL. (2004). *Guide to community protocols for Indigenous language projects.* Retrieved December 10 2005 from http://www.fatsil.org.au

Janke, T. (2002). *New media culture: protocols for producing Indigenous Australian new*

media. Aboriginal &Torres Strait Islander Arts Board, Australia Council. ISBN 0 642 47242 4. Retrieved December 2 2005 from http://www.ozco.gov.au
Mc Convell, P. & Thieberger, N. (2001). *State of Indigenous languages in Australia – 2001*. Australia State of the Environment Technical Paper Series (Natural and Cultural Heritage), Series 2. Canberra: *Department of the Environment and Heritage*. ISBN 0 642 54871 4.
McHenry, T. (2002). "Words as big as the screen: native American languages and the internet." *Language Learning & Technology*. ISSN 1094-3501. Vol 6 No. 2 May 2002. pp102-115.
Nathan, D. (1997). "Plugging in Indigenous knowledge: connections and innovations." Paper read at the 1997 Fullbright symposium: Indigenous people in an interconnected world. Retrieved December 10 2005 from http://www.uiowa.edu/-anthro/fulbright/abstracts.nathanful.htm
Nathan, D. (2005). "The spoken Karaim CD: Sound, text, lexicon and "Active morphology" for language learning multimedia." Paper. Retrieved December 10 2005 from http://www.dnathan.com/papers/karaimcd2.pdf
Nathan, D. (1999). "Language support with IT: not a high wire act". Paper presented at Learning IT Together, Brisbane April 1999. Retrieved December 10 2005 from http://www.it.usyd.edu.au/~djn/papers/NotHighWire.htm
Nickerson, M and Kaufman, J. (2005). "Aboriginal culture in the digital age". *Policy, Politics and Governance*. Vol 10 July 2005. Retrieved December 10 2005 from http://www.crossingboundaries.ca/aboriginalvoice
Ong, W.J. (1982). *Orality and literacy: The technologizing of the word*. Methuen: London. ISBN 0415281294 (2002).
The crossing boundaries national council. (2005). *Volume 5 Aboriginal voice national recommendations: from digital divide to digital opportunity*. ISBN 0 9738193-4-0. Retrieved December 10 2005 from http://www.crossingboundaries.ca
UNESCO. (2001). "Indigenous identities: Oral, written expressions and new technologies". Symposium at UNESCO (Paris):15 to 18 May 2001.
Villa, D. "Integrating technology into minority language preservation and teaching efforts: an inside job." *Language Learning & Technology*, ISSN 1094-3501. Vol 6 No. 2 May 2002, pp92-101.

Knowledge and Wiradjuri: Who is Telling the Truth? A Wiradjuri Perspective

Yalmambirra
School of Environmental & Information Sciences, Charles Sturt University, NZ

One must question if it is possible to portray Indigenous peoples in a correct light if one should follow the writings of some sociologists, anthropologists, archaeologists, and historians. Much of their respective findings have provided false depictions of peoples, places, languages, cultures, origins and meanings. Wiradjuri peoples for instance have been informed by some historians and others of dubious intent that they never existed as Wiradjuri until someone came along and named them that! The danger here lies in how many contemporary researchers take the writings of those gone before as gospel, and in doing so contribute to the false depiction of Indigenous peoples. This false depiction adds to the knowledge base inherent in studies and research on Indigenous peoples. An unethical and inappropriate knowledge base is a major concern for my peoples, Wiradjuri peoples. There is much confusion in the broader community in regards to whose knowledge is correct and who should be believed. Readers of this paper will have the opportunity in many instances, to decide for themselves what to make of it all. This has been done in a deliberate way. Firstly it provides an opportunity to get an idea of what has been written about Wiradjuri and by whom, albeit briefly. Secondly, readers are given every opportunity to glimpse other perspectives of events and to also understand the thoughts, ideas, and concerns of a Wiradjuri man and whilst I am not a 'Black Dictionary', surely this is an entirely ethical and appropriate approach to take.

Introduction

"Knowledge is everything. It tells us where we came from, who we are, why we are here, what to eat, what to use for medicines, what is woman's business and what is men's business, where we can go and where we are not allowed to venture. Knowledge is about land, law and ceremony, and was given to us by our creator" (Wiradjuri Elder, 2004).

The knowledge that the above Elder speaks of relates to all things. This knowledge was passed down from Elder to young, through countless generations (Walsh & Mitchell, 2002), knowledge that came with creation and the creator himself; Baiame (Grant & Rudder, 2005). This was our education, the environments our classrooms (Walsh & Mitchell, 2002). According to knowledge passed down, Wiradjuri peoples have always been, always existed, and our country and our languages have always belonged to us (Clarke, 2003). We never questioned this knowledge (Broome, 2001) and respected those who gave it to us (Tonkinson, 1991). Our knowledge was passed down ethically and appropriately where ethically means 'honestly / truthfully' and appropriately meaning 'the right way'.

As a current PhD student, I am undertaking a number of face-to-face interviews with Wiradjuri peoples in order to ascertain their perspectives on their peoples and cultures. During the interview process, the participants stress time and again that Wiradjuri knowledge has been misinterpreted in so many ways that they themselves live in 'confusion country'. The participants have also stated that in contemporary times one of the major concerns is that non-Wiradjuri people depict Wiradjuri.

It has been my experience as a Wiradjuri Elder, that I am expected by most non-Indigenous people to have an enormous amount of knowledge pertaining to Wiradjuri peoples and or issues, as well as having knowledge about other Indigenous peoples and cultures; I am not a 'black dictionary'; referring to a 'knowledgeable one' according to Tonkinson (1991). I have never been initiated and given secret and sacred knowledge by Elders as defined by Myers (1991) and so in reality I do not have all the knowledge that I should have had by rights of birth; I am culturally poor (Keen, 1994). But knowledge grows when partnerships are formed, and this collective knowledge can be and often is, a very powerful tool (Walsh & Mitchell, 2002).

This paper also highlights how non-Indigenous people have depicted Wiradjuri; how our cultures and ownership of our own history, has been 'taken away' from us through writings that could be considered ambiguous and contradictory. This line of thought includes in part, an examination of what has been written about Wiradjuri peoples and how those writings have contributed to knowledge that has led, and still does, to the misconceptions and stereotyping that has become a part of our (Wiradjuri) daily lives.

It is only in the last few decades that the true pictures of Wiradjuri have come to light and in many instances only because we have put pen to paper (Macdonald, 1994; Keen, 1994). Perhaps it is time for others to let Wiradjuri speak for Wiradjuri. Surely our version of history should be accepted, or at the very least considered..."We want to represent ourselves, and we want to be heard" (Kinoshi, 1998: 5).

Discussion

It is a given that major research has been undertaken on Wiradjuri; the literature testifies to this. Some of this research has looked at Wiradjuri from every aspect of their lives (see Read, 1983; White, 1986) others have looked at specific customs or languages only (see Hale, 1846: Mathews, 1897: 1900; Holmer, n.d.) whilst others still, have only paid Wiradjuri a cursory glance (see Gribble, 1884; Heaton, 1879). But there are anomalies in the combined works of people such as these.

On the issue of origin for example, our knowledge is in direct contrast to many non-Indigenous people who saw us as immigrants as suggested by Laidlaw (1990); Rickard (1992); Berndt & Berndt (1999) and Clarke (2003). In David Loewenthal's 'The past is a foreign country', is written that "every feature of the landscape is filled with creatures of their own imagination...The present-day natives are on the whole uninspired preservers of a great and interesting tradition" (Strehlow cited in Lowenthal, 1985: 379). One wonders if Strehlow has ever seen a kangaroo, emu or goanna; they are not from our imaginations! One must also ask whether Strehlow has seen our sites, our art, our fish traps and been witness to ceremonies; we were always 'inspired' to preserve our cultures and spiritual beliefs.

Bates (cited in McGregor, 1997:128) suggested that we were all cannibals... [That] "human meat was our favorite food" and in so saying was commended for her examination of 'life in the primitive state'.

How does one go about discerning what is right or wrong in relation to writings of this sort? What constitutes knowledge in this area? Who is telling the truth? According to Wikipedia (2005), a rather dubious and dynamic website, Indigenous peoples had no name for themselves pre-invasion. This is a fairly ambiguous suggestion as it can be taken from two different perspectives; that no single clan group had a name for themselves; or that Indigenous peoples had no collective name for themselves. Then there is the statement by Windschuttle (2003) that Wiradjuri did not exist in the 1820's as the term "Wiradjuri" was invented. Windschuttle quotes Tindale (1974: 156) in stating that Wiradjuri were one of those tribes whose name was given by white people in the 1890's. If Wiradjuri were not Wiradjuri until the 1890's then who were we?

Certainly Wiradjuri do not follow the same line of thinking as those of Wikipedia, Windschuttle and Tindale. Wesson (2000) in discussing traditional boundaries cites numerous sources in her historical atlas of Indigenous peoples and names (Robinson (1840); Barber (1841); Lane (1859); and Smyth (1878) as examples of authors who have mentioned 'Wiradjuri' prior to the 1890s.

Knowledge that is 'tainted' has impacted upon Wiradjuri in many ways. Firstly it takes away our identity. Secondly, knowledge is directly attributed to 'wealth' (Myers, 1991); we were rich in culture because of that knowledge, but we are now the poorer, and thirdly, 'ownership' as defined by Keen (1994) has been taken from us to such an extent that we are no longer owners of our own histories.

The problem however is not how, but what non-Wiradjuri write about us. An ethnocentric approach such as this according to Ferraro *et al.,* (1994) emanated from the innate belief that their (European) cultures were far better, far superior, and were cultures that were most wanted above those of Indigenous peoples; they were the experts. Ethnocentrism meant that non-Indigenous people looked at Wiradjuri through non-Indigenous eyes, from their own cultural perspectives and did so "with a false sense of superiority" stated Broome (2001:8). However Lampert (2005) suggests that whilst non-Indigenous Australians could be classed as experts in relation to a number of things, they cannot be so when it comes to Indigenous peoples.

Learning white language has provided us with a tool in order to deliver our own histories but if Wiradjuri history was an oral one, and it was and still is to a certain extent, then our history has been written by mostly anthropologists; a "false history" (Walley, 1990: 69).

Therein is the heart of the problem. The majority of research undertaken by non-Indigenous people on Indigenous peoples has been tainted simply because it has not been sanctioned by the very peoples themselves. Wiradjuri and by association, other Indigenous peoples, have been presented from the invaders perspectives; "who we were, who we are, and who we could be" (Phillips, 2005:25). They have not consulted Indigenous peoples before respective findings are presented to the world, with the end result being that Wiradjuri have become stereotyped; that we are a homogenous peoples; all do the same thing, carry out ceremony the same way, eat the same foods, have the same spiritual beliefs as other Indigenous peoples as suggested by the Australian Information Service (1979).

The thoughts of an anthropologist add to the confusion that exists in relation to the issue of 'knowledge'. Maddock (1983: ix) inform readers that "Anthropology is important because it is the *main* (emphasis added) source of knowledge about traditional Aboriginal society, *including* (emphasis added) the nature of relations to land". There are a number of things wrong with this statement. Maddock does not give credence to Indigenous peoples; their respective levels of knowledge are not considered. In using the words "traditional Aboriginal society", Maddock does not take into account the diversity of Indigenous peoples and infers that we are homogenous, but then states that there are "cultures" (:3).

Maddock goes further and confuses even more. He states that he began research in the Northern Territory in 1964 among "traditionally oriented Aborigines"... [That] "they knew a great deal about their ancestral culture". Readers could take this statement as suggesting that Indigenous peoples that lived elsewhere were not "traditionally oriented" or that they had no knowledge of culture.

'Many past writers or observers of Wiradjuri history fail to advise that knowledge is not distributed equally among clans or communities, that appropriate ceremony had to be undertaken in order to acquire knowledge, as highlighted earlier (ATSIC, 1999); and that, if an informant has not undertaken such, then information passed on may well be corrupted'. Talking to *any* Indigenous person about Indigenous issues may well provide an answer to a question, but the information may be incorrect and if used, can be the catalyst for additional stereotyping or misconceptions.

Interviews conducted during my PhD research highlight the need for Elders to ensure that knowledge is passed on honestly; ethically (in the right manner); appropriately (with the right peoples); and to ensure that the knowledge is correct. The participants all agree with Chalmers (2005:163) that "only Indigenous peoples know who they are". Chalmers has suggested that Indigenous peoples must be allowed to "represent themselves in such a way that does not restrict or limit how they wish to represent themselves". Wherever and whenever possible the teachers of history must be those who have lived it, or at the very least be "written by and be under the control of Aboriginal people" (Keen, 1994:21). In order to reduce misconceptions and stereotyping, this I believe, must emanate from perspectives that are inclusive of country; Wiradjuri, Walpiri, Gamilaroi, Yorta Yorta etc.

One only has to look at the issue of 'culture' to see how confusing things can be. Kroeber and Kluckhohn (cited in Voget, 1975:383) report that "Between 1903 and 1916, six definitions appeared – two by sociologists, two by a chemist, and two by an anthropologist". Otterbein (1972: 2) suggested that "the use of the term culture to refer to both a group of people and to their way of life entered the vocabulary of anthropologists after 1900". Voget (1975) had written that approximately 157 definitions of culture existed between the years 1940 and 1950. Regardless of who first initiated the term, White (1968) puts it into perspective when stating that there aren't too many people that would need approximately three hundred definitions of the meaning of culture. Three hundred!!! Readers can be left to wonder at the brilliance of the scientific mind!

Whilst this paper has highlighted, albeit briefly, some areas of research that have impacted upon Wiradjuri, and here I write of the literature, it is important however that we look at what has been written and classed as 'knowledge', as resources that should not be thrown out with the garbage. Each of these 'tells' a different story and whilst each may add to 'confusion country' they can be a very powerful educational tool, now and in the future. The writings of Maddock (1983) for example can be utilised to highlight how 'tainted' thoughts have permeated Indigenous history.

This must be done however with the thoughts of a Wiradjuri Elder in mind:

"*We must not burn the books...* [We] *need to utilise these resources, these 'offerings of knowledge'...* [We] *need to utilise them in an appropriate manner and in direct consultation with those who may have the true knowledge; Indigenous peoples"* (Wiradjuri Elder, 2004).

The formation of partnerships, as alluded to earlier in this paper, infers partnerships between Wiradjuri and non-Indigenous people. Whilst this may be a giant step in the right direction as suggested by Davies (cited in Baker, *et al.*, 2001; Liddle (2001) other partnerships must also be forthcoming. Wiradjuri peoples, especially the Elders, need to work in partnership with each other. In this way their collective knowledge can become a 'pool' whereby those who would research and study Wiradjuri, and by association other Indigenous peoples, could cross-reference the written words of the past with the oral history of those living in the present and provide awareness among the broader community (Lands, 1990).

It is not hard to ascertain from the literature that in the majority of times, the opinions of the sociologist, anthropologist, archaeologist and other historians differed; and differed greatly and often. The problem with the historical record is that in the context of this paper, it is corrupted. There are only a few researchers that identify as their primary resource; the peoples themselves, others rely on the written word of their peers and other writers for information as suggested by Shaw (1992), and this is also cause for concern. The writings of others may well be as corrupted as their own!

The impacts to Wiradjuri via corrupted writings are many. There is the obvious of course. Those who are not Wiradjuri take the words of these people as gospel; that their words are correct in every instance, therefore that is what has happened; that is what Wiradjuri peoples are and do. The danger in this can be found in those who utilise the resources in an unethical (in the wrong way) and inappropriate (with the wrong peoples) manner in the education of others. Those who do this are teaching about Wiradjuri or any other Indigenous peoples, in such a way as to often aid and abet the misconceptions and the stereotyping that seems to go hand-in-hand when any issue Indigenous is taught, debated and/or argued.

The way in which history has been written has taken ownership or custodianship away from Wiradjuri, and by association, other Indigenous peoples. Changes are now happening however. Attwood's (1996) comments that we are 'reclaiming our right to provide our own perspectives of history by claiming the right of ownership and custodianship of those histories' are a timely reminder that our histories belong to no other than us...we are willing to share our knowledge, but we are it's keepers.

Conclusion

It is my belief that the history that belongs to Wiradjuri be returned to Wiradjuri; we are the rightful custodians of knowledge that belongs to us and is ours by right of birth. Those who have written about us have not spoken with us; they have no real understanding of how we feel, what our thoughts, ideas and concerns are. It is my belief that it does not take courage to write, but it does take courage to write the truth. You see, the truth my friends, is where our knowledge and history lies.

References

ATSIC, (1999). *As a matter of fact: Answering the myths and misconceptions about Indigenous Australians.* Second Edition. Canberra, Australia. Aboriginal and Torres Strait Islander Commission, ISBN 0-642-32022-5.

Attwood, B. (ed) (1996). *In the age of Mabo: History, Aborigines and Australia*, St Leonards, NSW. Australia: Allen & Unwin, ISBN 1 86373 841 X.

Australian Information Service, (1979). *The Australian Aboriginals.* Reference Paper. Canberra, Australia: The Australian Government Publishing Service.

Berndt, R. M. and Berndt, C. H. (1999). *The world of the first Australians: Aboriginal traditional life: past and present.* Canberra. Australia: Aboriginal Studies Press, ISBN 0 85575 184 3.

Broome, R. (2001). *Aboriginal Australians: Black responses to white dominance. 1788-2001.* Third Edition. Crows Nest, NSW, Australia: Allen & Unwin, ISBN 1 86508 755 6, p. 8.

Clarke, P. (2003). *Where the ancestors walked: Australia as an Aboriginal landscape.* Crows Nest, NSW. Australia: Allen & Unwin, ISBN 1 74114 070 6.

Chalmers, G. (2005). "The repercussions of representation," in Phillips and Lampert, (eds) (2005). *Introductory Indigenous studies in education: The importance of knowing.* Frenchs Forest. NSW, Australia: Pearson Education Australia, ISBN 0 7339 7266 7.

Baker, E., Davies, J. and Young, E. (eds) (2001). *Working on country: Contemporary Indigenous management of Australia's lands and coastal regions.* South Melbourne, Victoria, Australia: Oxford University Press, ISBN 0 19 551217 0.

Ferraro, G., Trevathan, W. and Levy, J. (1994). *Anthropolgy: an applied perspective.* New

York. USA: West Publishing Company, ISBN 0-314-02879-x (alk. paper). File://P:\Data\PHD\Sydney%20Line.htm [Accessed 13/07/2004].
Grant, S. and Rudder, J. (2005). *A first Wiradjuri dictionary: English to Wiradjuri, Wiradjuri to English*. ACT, Australia: Restoration House, ISBN 0 86942 131 X.
Gribble, J. B. (1884). *'Black but comely:' or, Glimpses of Aboriginal life in Australia*. London: Morgan & Scott.
Hale, H. (1846). *United States exploring expedition. During the years 1838, 1839, 1840, 1841, 1842. Ethnography and Philology*. USA: Lea & Blanchard, Library of Congress Catalog Card Number 68-56969
Heaton, J. H. (1879). *Australian dictionary of dates and men of the time: containing the history of Australia from 1542 to May, 1879*. Sydney. Australia: George Robertson.
Holmer, N. M. (n.d.). in Liljegren, S. B. (ed) (n.d.). *Australian essays and studies on the history and structure of the Australian languages*. Copenhagen: Upsala. http://www.fact-index.com/a/au/australian-aborigine.html
Ingold, T., Riches, D. and Woodburn, J. (eds) (1991). *Hunters and gatherers: Property, power and ideology*. Volume 2. Oxford, UK: Berg, ISBN 0-85496-155-0, ISBN 0-85496-735-4 (pbk).
Keen, I. (1994). *Being black: Aboriginal cultures in 'settled' Australia*. Canberra, Australia: Aboriginal Studies Press, ISBN 0 85575 185 1, p. 21.
Kinoshi, S. (ed) (1998). "Introduction: The University of Sydney Aboriginal students' black on black," An anthology of Aboriginal writings on Aboriginal issues. *Aboriginal Studies Journal*, Volume 1. ISSN 1440-2955, p. 5.
Laidlaw, R. (1990) *Discovering Australian history to 1900: an evidence based approach*. Melbourne, Victoria. Australia: Edward Arnold, ISBN 0 7131 8340 3.
Lampert, J. (2005). "The teaching that matters: merging our personal beliefs with our professional practice in the classroom," in Phillips and Lampert, (eds) (2005). *Introductory Indigenous studies in education: The importance of knowing*. Frenchs Forest. NSW, Australia: Pearson Education Australia, ISBN 0 7339 7266 7, p. 163.
Lands, M. (1990). "Author and publicist," in Thompson, L. (1990). *Aboriginal voices: Contemporary Aboriginal artists, writers and performers*. Brookvale. NSW, Australia: Simon & Schuster Australia, ISBN 0 7318 0165 2.
Liddle, L. (2001). "Bridging the communication gap: transferring information between scientists and Aboriginal land managers," in Baker, E., Davies, J. and Young, E. (eds) (2001). *Working on country: Contemporary Indigenous management of Australia's lands and coastal regions*. South Melbourne, Victoria, Australia: Oxford University Press, ISBN 0 19 551217 0.
Macdonald, G. (1994). "A Wiradjuri fight story," in Keen, I. (1994). *Being black: Aboriginal cultures in 'settled' Australia*. Canberra, Australia: Aboriginal Studies Press, ISBN 0 85575 185 1, p. 21.
Maddock, K. (1983). *Your land is our land: Aboriginal land rights*. Ringwood, Victoria. Australia: Penguin Books, ISBN 0 14 022505, pp. ix-3.
Mathews, R. H. (1897). The burbung, or intitation ceremonies of the Murrumbidgee tribes. Read before the Royal Society of New South Wales, July 7. 1897. *Journal of the Royal Society of NSW*.
Mathews, R. H. (1900). "The burbung of the Wiradthuri tribes," Read before the Royal Society of Queensland, April 7, 1900. *Social Anthropology* – Papers. Volume 1. American Anthropology.
McGregor, R. (1997). *Imagined destinies: Aboriginal Australians and the doomed race theory, 1880-1939,* Carlton South, Victoria, Australia: Melbourne University Press, ISBN 0 522 84762 5, p. 128.
Myers, F. (1991). "Burning the truck and holding the country: property, time and the negotiation of identity among Pintupi Aborigines," in Ingold, T., Riches, D. and

Woodburn, J. (eds) (1991). *Hunters and gatherers: Property, power and ideology.* Volume 2. Oxford, UK: Berg, ISBN 0-85496-155-0, ISBN 0-85496-735-4 (pbk).
Otterbein, K. F. (1972). *Comparative cultural analysis: an introduction to anthropology.* Sydney. Australia: Holt, Rinehart & Winston, ISBN 0-03-084417-7, p. 2.
Phillips, J. (2005). "Indigenous knowledge: making space in the Australian centre," in Phillips and Lampert, (eds) (2005). *Introductory Indigenous studies in education: The importance of knowing.* Frenchs Forest. NSW, Australia: Pearson Education Australia, ISBN 0 7339 7266 7, p. 163.
Read, P. J. (1983). *A history of the Wiradjuri people of New South Wales 1883 – 1969.* A thesis submitted for the degree of Doctor of Philosophy of the Australian National University. Canberra, Australia.
Rickard, J. (1992). "Aborigines," in Whitlock, G. and Carter, D. (eds) (1992). *Images of Australia: An introductory reader in Australian studies.* Queensland. Australia: University of Queensland Press, ISBN 0 7022 2447 2.
Shaw, B. (1992). *When the dust come in between: Aboriginal viewpoints in the east Kimberley prior to 1982.* Canberra. Australia: Aboriginal Studies Press, ISBN 0 85575 232 7.
Lowenthal, D. (1985). *The past is a foreign country.* Oakleigh. Melbourne, Australia: Cambridge University Press, ISBN 0 521 22415 2 hardback / ISBN 0 512 29480 0 paperback, p. 379.
Tindale, N.B. (1974). "Aboriginal tribes of Australia: their terrain, environmental controls, distribution, limits, and proper names," in Windschuttle, K. (2003). *The Sydney Line.* Submission to the review of the National Museum of Australia. (Online) [Accessed 13/07/2004].
Tonkinson, R. (1991). "Ideology and domination in Aboriginal Australia: a Western desert test case," in Ingold, T., Riches, D. and Woodburn, J. (eds) (1991). *Hunters and gatherers: Property, power and ideology.* Volume 2. Oxford, UK: Berg, ISBN 0-85496-155-0, ISBN 0-85496-735-4 (pbk).
Voget, F. W. (1975). *A history of ethnology.* Sydney, Australia: Holt, Rinehart & Winston, ISBN 0-03-079665-2.
Walsh, F. and Mitchell, P. (2002). *Planning for country: Cross-cultural approaches to decision-making on Aboriginal lands.* jukurrpa books. Alice Springs. NT, Australia: ISBN 1 86465 037 0.
Walley, R. (1990). "Director and Playwright," in Thompson, L. (1990). *Aboriginal voices: Contemporary Aboriginal artists, writers and performers.* Brookvale. NSW, Australia: Simon & Schuster Australia, ISBN 0 7318 0165 2.
Wesson, S. (2000). *An historical atlas of the Aborigines of eastern Victoria and far south-eastern New South Wales.* Monash Publications in Geography and Environmental Science. Number 53. School of Geography and Environmental Science, Monash University: Melbourne. Australia, ISSN 0313 8410. ISBN 0 909685 68 1.
White, I. (1986). *Dimensions of Wiradjuri.* Thesis. Canberra, Australia: Australian National University.
White, L A. (1968). "On the concept of culture," in Manners, R. A. and Kaplan, D. (eds) (1968). *Theory in anthropology.* Illinois. USA. Aldine Publishing Company: Library of Congress Catalog Card Number 67 – 17606.
Whitlock, G. and Carter, D. (eds) (1992). *Images of Australia: An introductory reader in Australian studies.* Queensland. Australia: University of Queensland Press, ISBN 0 7022 2447 2.
Wikipedia, (2005). *Australian Aborigine.* (Online) [Accessed March 2005].
Windschuttle, K. (2003). *The Sydney Line.* Submission to the review of the National Museum of Australia. (Online) File://P:\Data\PHD\Sydney%20Line.htm

Acknowledgements

Thanks go to Rik Thwaites, Jim Birkhead and Don Wyatt for their continued encouragement and support.

Yalmambirra left school at age eleven and entered University at age 45. He is now the Koorie Academic (Wiradjuri) at Charles Sturt University, Albury and Thurgoona Campuses. Yalmambirra has a degree in Bachelor of Applied Science (Parks, Recreation & Heritage), an Honours Degree (Class 1), and is currently undertaking research for his PhD. Although relatively new to writing papers, Yalmambirra has had published, a number of papers on many issues that are inherent in research on and with Indigenous (Wiradjuri) peoples. Yalmambirra is currently undertaking a PhD. His research is oriented toward cultural change and adaptation in relation to Wiradjuri peoples.

Towards a Metadesign Approach for Building Indigenous Multimedia Cultural Archives

Laurel Evelyn Dyson & Mike Leggett
Faculty of Information Technology, University of Technology, Sydney, AUS

Indigenous Australian cultures have experienced a revival in recent years as communities realize the value of recording their cultural heritage. Computer systems designed for Indigenous cultural archives, can play a role in storing Indigenous culture in multimedia formats such as video, sound recordings, photographs and text, and then making them available back to the community. This paper proposes a new approach to the design of Indigenous multimedia archives. Practice-based approaches to metadesign can be combined with Indigenous protocols to develop better systems for Indigenous communities. Using this approach it is possible to avoid "top-down" design methods and, instead, partner with Indigenous people in building authoring software with which they can construct their own cultural systems. By this means Indigenous Australians can gain control of a system which is suited to their needs, cultural traditions and protocols.

Introduction

The preservation and revitalization of Indigenous cultures is of major interest in Australia today. The process of colonization and assimilation prevented the passing on of traditional knowledge from generation to generation in many Indigenous communities and resulted in the extinction of many cultural practices and languages. Today there are concerns that globalization may contribute to this process still further through the colonizing influences of dominant cultures on the Web (McConaghy, 2000).

However, technological advances provide opportunities to redress this issue. In particular, multimedia technologies offer a way of storing and displaying video, photographs and sound recordings and so fit with many characteristics of traditional Indigenous culture (Dyson, 2003). The latter is not a written culture, but founded in oral and audio practices (story-telling, ceremony, song and music) and also in pictorial cultural forms (paintings, sculptures and carvings). In recent years multimedia has extended to the development of Indigenous living cultural archives (also referred to as Indigenous multimedia knowledge systems). These allow culture and knowledge to be stored and then made available back to the community (Dyson & Underwood, 2006). Furthermore they provide a method of repatriation of digital copies of items held in museums and other collections back to communities (Hunter, *et al.*, 2003).

A number of Indigenous multimedia archives have been implemented in Australia as well as overseas (e.g., Hunter, 2002; Injie & Haintz, 2004; Hughes & Dallwitz, 2007). There are also several systems currently under development, still in the prototype phase (e.g., Leggett, 2005; Leavy, 2007; Chesselet, n.d.). Some of these systems have been more successful than others in addressing the various requirements of Indigenous communities. A major barrier is the complexity of designing new systems which fit well with Indigenous knowledge concepts and protocols. Secondly, there is the cost of bringing in outside development teams to implement systems in remote communities, with further costs for ongoing maintenance and upgrades. Thirdly, some systems are hard-coded and so not easily adaptable from one community to another.

In this paper we propose a new approach to designing Indigenous multimedia cultural archives which, we believe, will address these issues. Metadesign is at the cutting edge of current design methodologies: it seeks to aid communities to solve their own design problems by providing them with the support to do this (Fischer, 1999). A metadesign approach will involve partnering with Indigenous communities in the design of authoring software which will allow Indigenous people themselves to create their own systems for storing and viewing their culture and knowledge. This will give Indigenous people control over their multimedia archives and produce systems which more closely reflect their cultures and needs.

In this paper a brief overview of existing systems is given in order to highlight the complexities that have to be addressed in the design of Indigenous multimedia archives. Then we explain how practice-based approaches to metadesign can be combined with Indigenous protocols to develop better archiving systems for Indigenous communities.

Indigenous cultural archives in Australia

The main requirements of Indigenous cultural archive systems can be summarized as:

- Appropriate to Indigenous culture, particularly its oral and graphical strengths;
- Robust enough to withstand the harsh environments where many remote communities live;
- Acknowledging Indigenous knowledge protocols, security concerns over who has access to secret or sacred knowledge, and intellectual property issues;
- Easy to use and navigate (given low computer literacy levels in many communities);
- Cost-effective (given the poverty of many communities);
- Allowing for diversity of communities and cultural evolution over time;
- Able to be placed outdoors at the locus of creative practice;
- Providing community control over contents and over design, development, implementation and maintenance.

None of the existing Indigenous cultural archives has managed to address all of these needs although some have succeeded very well with certain issues. For example, Hunter's Indigenous Rights Management System has implemented Indigenous knowledge concepts of access and intellectual property into its database through XrML (XML-based rights markup language) (Hunter, 2002). Another system, Ara Irititja has screens in colors appropriate to the central Australian deserts of the Anangu people who commissioned it, and is displayed on a mobile workstation which is heat-proof, dust-proof, mice-proof, has an uninterruptible power source, and can be placed on the back of a truck and taken to wherever the community wish to use it (Hughes & Dallwitz, 2007). The Digital Songlines prototype has screens which are almost totally graphical and has moved away from the widespread "desktop" design metaphor to one which acknowledges the strong link between Indigenous people and the land (Leavy, 2007). Likewise the PathScape prototype uses video files of the land and has the advantage of simple navigation by gesture control of the screen cursor (Leggett, 2005).

One major challenge is placing control of the system in the hands of the Indigenous community. To date the most successful at achieving this are simple, cheap systems based on FileMaker Pro, where text-based lists of available multimedia items are presented to the user. Nyirti, used at the Wangka Maya Centre in the Pilbara, is an example of this (Injie & Haintz, 2004). However, these raise obvious problems when the population who will be using them come from a non-text-based culture.

A new approach to creating an indigenous multimedia cultural archive

The overall approach which we propose for the design of Indigenous cultural archives is practice-based. Practice-based methodology has resulted in highly successful outcomes and is well described in the literature. It consists of "...acquiring knowledge using sensitive methods for gathering and analysing necessary data" and, most importantly, must be deeply rooted in the actual context and experiences of the participants (Candy & Edmonds, 2002: 39-43). It is a "...creative production process [that] is self-conscious, rational and reflective" (Scrivener & Chapman, 2004). It is an approach employed frequently in the end user development of software tools and multimedia.

We propose focusing upon the early design stages of practice-based methodology, using metadesign as a conceptual framework. Metadesign is "... another species of design, where the artefacts being designed are themselves interfaces for designing – hence meta-design' (Lieberman, 2005). Metadesign takes up many known principles of good design within human-computer interaction, providing options for guiding the users' design process.

The methodology of metadesign has been developed from theoretical and practical investigations in several international centres by a range of researchers (Lazarev, 1994; Maturana, 1997; Fischer, 2003; Fischer & Giaccardi, 2004). Metadesign is characterized by objectives, techniques and processes for creating interactive media environments. This has been achieved by researchers working closely and sympathetically with community groups that encourage participants to engage in the collaborative construction of artefacts and activities meaningful to the context of their production and usefulness. The relationship between the designers and the community group is symbiotic – researchers, skilled in electronic tools and their use, learn with community members how to organise cultural knowledge in the form of artefacts, oral narrative, performance, etc. The collaborative project becomes a "... creative process defin[ing] a 'seed' able to generate endless variations recognizable as belonging to the same idea but open to change" (Giaccardi, 2005: 345).

In Australia, Indigenous communities across the continent, whilst sharing protocols of behavior, exhibit wide variations in cultural forms and knowledge. Responding to this diversity, the design approach is a seeding process rather than planning exercise. Indigenous communities need a system open to adaptation and capable of responding to developments in technology and evolution in their own cultural practices. A new "praxis of design" will enable communities to develop their own computer-based systems as an aid to sustaining, growing, preserving and transmitting their culture to successive generations.

Using this approach "top-down" design methods are avoided and, instead, Indigenous people can become partners in building authoring software with which to construct their own cultural systems. By this means Indigenous Australians can gain control of the design process and achieve a system which is suited to their needs, cultural traditions and protocols. Providing the authoring tool is sufficiently flexible and able to be used by any community, it will have the power to build systems which truly reflect the varying cultures and knowledges across the country.

The metadesign approach will incorporate Indigenous protocols within the authoring system. These protocols are well documented and cover both traditional cultural protocols, and more recent media and archiving protocols (e.g., Janke, 2002; Museums Australia, 1998; Byrne, *et al.*, 1995). The incorporation of protocols into Indigenous multimedia archives is absolutely essential, given cultural sensitivities regarding the display of traditional knowledge and artefacts. Indigenous people have concerns over who has the right to knowledge and do not wish unauthorized members of even their own community, let alone outsiders, gaining access to material that is seen as sacred or secret, viewable only by the initiated or by people of a certain gender (Radoll, 2004).

The proposed metadesign operational methodology is anticipated to be characterized specifically by:

- User-centred and participatory design techniques;
- Rapid prototyping of coded modules representing the core structure and processes;
- Iterative development;
- Fluid techniques like seeding, building and "tearing down of strawmen";
- Use of APIs for end users to customise and capture their participatory decisions;
- Free tagging processes rather than predetermined metatags.

The authoring system will remain open to participation, evolution and emergence, recognizing that Indigenous cultures and needs are not immutable, and that Indigenous expectations of what the system can do may well change over time.

Conclusion

The significance of the approach suggested by this paper is that we are concerned with the design *process*: how, by pursuing a metadesign approach integrated with Indigenous knowledge protocols, we can arrive at a system which fits better with Indigenous culture. It is innovative in that we wish to move a step back from the design of the Indigenous multimedia knowledge system itself, and instead consider the design of authoring software which will allow Indigenous people themselves to create their own systems for storing and viewing their culture and knowledge. This is radically different from the way that existing Indigenous multimedia knowledge systems have been implemented, where the development has adopted a more traditional systems design methodology and centred on the end product: the multimedia knowledge system itself. By placing the *authoring tool* into the hands of Indigenous people, each community will be able to create an Indigenous multimedia knowledge system which truly reflects their particular culture and their community's specific needs. Moreover, this approach will permit flexibility in recording and managing their culture so that the cultural archive continues to reflect the community out of which it grew.

References

Byrne, A., Garwood, A., Moorcroft, H. and Barnes, A. (1995). *Aboriginal and Torres Strait Islander Protocols for Libraries, Archives and Information Services: The Protocols.* Available at: www.CDU.edu.au/library/protocol.html (accessed 6 Feb 2006).

Candy, L. and Edmonds, E. A. (2002). *Explorations in art and technology*, London: Springer, ISBN 1852335459.

Chesselet, J. (n.d.). *The Living Archive Project: Mapping time and space*, Cape Town, SA: Doxa Productions.

Dyson, L. E. (2003). "Indigenous Australians in the information age: Exploring issues of neutrality in information technology," in C. Ciborra, R. Mercurio, M. De Marco, M. Martinez, and A. Carignani, (eds.), *New Paradigms in Organizations, Markets and Society: Proceedings of the 11th European Conference on Information Systems (ECIS), Naples, Italy, 19 – 21 June 2003*, pp. 1-12.

Dyson, L. E. and Underwood, J. (2006). "Indigenous people on the Web", *Journal of Theoretical and Applied Electronic Commerce Research*, ISSN 0718-1876, 1(April): 65-76.

Fischer, G. (2003). "Meta-Design: Beyond user-centered and participatory design," in J. J. a.C. Stephanidis (ed.), *Proceedings of HCI International 2003*, Crete, Greece, pp. 88-92.

Fischer, G. (1999). "Symmetry of ignorance, social creativity, and metadesign," in E. Edmonds and L. Candy (eds.), *Third Conference on Creativity & Cognition*, ISBN 1581130783, New York: ACM Press, pp. 116-123.

Fischer, G. and Giaccardi, E. (2004). "Meta-Design: a framework for the future of end-user development", in H. Lieberman, F. Paterno and V. Wulf (eds.), *End user development,*

Dortrecht: Kluwer Academic Publishers.
Giaccardi, E. (2005). "Metadesign as an emergent design culture", *Leonardo*, ISBN 8809212185, 38 (4): 343-349.
Hughes, M. and Dallwitz, J. (2007). "Ara Irititja: Towards culturally appropriate IT best practice in remote Indigenous Australia," in L. E. Dyson, M. Hendriks and S. Grant, (eds.), *Information technology and Indigenous people*, Hershey, PA: Information Science Publishing, ISBN 1-59904-299-1, pp. 146-158.
Hunter, J., (2002). "Rights Markup Extensions for the Protection of Indigenous Knowledge," *Global Communities Track, WWW2002, Honolulu, May* 2002. Available at: www2002.org/CDROM/alternate/748/ (accessed 29 Jan. 2004).
Hunter, J., Koopman, B. and Sledge, J. (2003). "Software tools for Indigenous knowledge management," *Museums and the Web 2003 Conference*, Charlotte, March 2003, ISBN 188625274.
Injie, L. and Haintz, F. (2004). "The natural development of Wangka Maya into the direction of a knowledge centre." *Indigenous Studies – Sharing the cultural and theoretical space: AIATSIS Conference 2004, Canberra, Australia.*
Janke, T. (2002). *New media cultures: Protocols for producing Indigenous Australian new media*, Sydney: Australia Council, ISBN 0642472424.
Lazarev, Y. N. (1994). "The art of metadesign", *Leonardo: Prometheus: Art, science and technology in the former Soviet Union* (Special Issue), ISBN 8809212185, 27(5): 423-425.
Leavy, B. (2007). "Digital songlines – Digitising the arts, culture and heritage landscape of Aboriginal Australia," in L. E. Dyson, M. Hendriks and Grant (eds.), *Information technology and Indigenous people*, Hershey, PA: Information Science Publishing, ISBN 1-59904-299-1, pp. 159-169.
Leggett, M. (2005). "Indexing audio-visual digital media: the PathScape prototype," *Scan*, 2(2). Available at: http://scan.net.au/scan/journal/index.php (accessed 20 July 2006).
Lieberman, H. (2005). "Preface", in H. Lieberman, F. Paterno and V. Wulf (eds.), *End-user development*, Kluwer/Springer.
McConaghy, C. (2000). "The Web and today's colonialism," *Australian Aboriginal Studies*, 1 & 2: 48-55.
Maturana, H. (1997). *Metadesign*. Available at: http://www.hum.aau.dk/~rasand/Artikler/metadesign.htm (accessed 20 July 2006).
Museums Australia (1998). *Taking the time – Museums and galleries, cultural protocols and communities: A resource guide*, Brisbane.
Radoll, P. J. (2004). "Protecting Copyrights on the Internet: A cultural perspective from Indigenous Australia," in F. Sudweeks and C. Ess (eds.), *Proceedings of the Fourth International Conference on Cultural Attitudes towards Technology and Communication (CATaC)*, Karlstad, Sweden, Murdoch University, Murdoch, Australia, ISBN 0869058622, pp. 339-348.
Scrivener, S. and Chapman, P. (2004). "The practical implications of applying a theory of practice based research: A case study", *Working Papers in Art & Design*, ISSN 14564917, 3.

Laurel Evelyn Dyson, BSc (Hons), BA (Hons), PhD, CELTA, GradDipABE, GradDipInfTech (Distinction), MInfTech, is a lecturer in Information Technology at the University of Technology, Sydney, Australia, where she teaches computer ethics and information systems courses. Her many years of teaching experience have included computer education programs for Indigenous Australians, senior citizens, adult literacy students and prisoners Her main research interests are in the fields of Australian cultural studies, and the cultural, ethical and design issues which impact on Indigenous people's adoption and use of Information Technology. She has recently co-edited the book Information Technology and Indigenous People.

Mike Leggett, MFA, PolyDipPhot, has been working with media across the institutions of art, education, IT, cinema and television since the early-70s. Currently he is the recipient of an Australian Postgraduate Award and works within the PhD program of the Creativity &Cognition Studio in the Faculty of Information Technology, University of Technology Sydney. He has film and video work in archives and collections in Europe, Australia, North and South America. He has curated exhibitions of interactive multimedia including the Museum of Contemporary Art in Sydney. He writes and lectures about media art, contributing to journals (Leonardo; Continuum), magazines (World Art), online 'zines (FineArt Forum).

Integrating Indigenous Cultural Traditions in the Management of Protected Marine Resources: The Fiordland example

James M. Mize
Environment and Earth Sciences, Victoria University of Wellington, NZ

Indigenous communities that rely upon natural resources for their cultural practices depend on the ecosystem health for such traditions to continue. Yet efforts to protect ecosystem integrity by "locking up" resources in reserves risk the disruption of traditions compatible with sustainability. This tension can be seen in the Fiordland area of New Zealand. Statutes governing the management of adjacent marine resources incorporate Māori values of kaitiakitanga (guardianship). This paper traces the evolution of recognition of Māori guardianship practices in the management of Fiordland marine resources, culminating in the inclusion of representation of the local iwi (tribe), Te Runanga o Ngāi Tahu, on the Fiordland Marine Guardians management authority established in the Fiordland (Te Moana O Atawhenua) Marine Management Act of 2005. In conclusion, this paper makes recommendations for application of similar statutory tools in situations where protective management measures may threaten traditional cultural use.

Increasingly, natural resource managers recognise limits of marine ecosystems under their jurisdiction, and move to adopt or promote conservation measures to protect these resources. One tool favoured to protect biodiversity is the designation of marine reserves, spatially explicit marine management areas where no take of life is allowed. Marine reserves have been shown to be effective at rebuilding depleted fish stocks and restoring ecosystem health, a broad scientific consensus supports the use of marine reserves for improved ecosystem integrity outcomes (Roberts & Hawkins, 2000).

New Zealand is a leader in adoption of marine reserves as a protective measure, with its early adoption of the Marine Reserves Act of 1971 and its consistent record of accomplishment of marine reserve designations under this act (Sobel & Dahlgren, 2004). New Zealand's signing of the Convention on Biological Diversity (CBD) in 1992 reaffirms this commitment, as the government reviews the Act to "better provide for the protection of marine biodiversity" (Department of Conservation and Ministry for the Environment, 2000). However, the CBD requires not only that the contracting nation establish networks of protected areas, but also that they "respect, preserve, and maintain knowledge, innovations, and practices of indigenous and local communities embodying traditional lifestyles relevant for the conservation and sustainable use of biological diversity" [1].

While indigenous communities that rely upon natural resources for their cultural practices and livelihoods depend on the health of the ecosystem for their traditional practices to continue, government efforts to protect ecosystem integrity by "locking up" resources in "no-take" reserves risk disrupting such practices. Marine reserves address environmental conservation goals, yet incautiously applied they defeat parallel goals of sustaining indigenous traditions and knowledge. The Te Wāhipounamou / South West New Zealand (Fiordland) area illustrates this tension (see Figure 1), where frictions between increased conservation focus and traditional customary uses of marine resources have resulted in new solutions to resolving the debate.

Fiordland, its indigenous peoples and marine resources

The Fiordland marine area is unique. Heavy rainfall washes tannins down from the vegetation on the land to the deep and confined fiords, which forms a layer of dark water that filters sunlight, allowing deepwater corals and fishes to thrive within 40 metres of the surface (Department of Conservation, 2005). The area is known both for its abundance and for diversity of marine species and is a tourist destination popular with divers and boaters as well as a fishing hotspot for both recreational and commercial fishers (Warne, 2000).

From before the arrival of Europeans, the South Island Māori *iwi* (tribe) [2] Ngāi Tahu have relied upon the natural resources within this unique area, from the native *pounamou* (greenstone) not found elsewhere to the various and plentiful *kai moana* (seafood), the plentiful fish and shellfish found in its waters (Waitangi Tribunal, 1992). After the arrival of Europeans, this reliance continued but became subject to relations between Ngāi Tahu and the Crown of England set forth in the Treaty of Waitangi, signed by *Rangatira* (chiefs) of Ngāi Tahu, which described the nature of affairs between the Crown of England and Māori (Boast, 1989). While the treaty was signed in both English and Māori language versions, generating much scholarship over proper interpretation and harmonization between the two, this paper only considers the understanding expressed by Ngāi Tahu in the Māori version. In Article I of the Treaty, Māori agreed to allow the crown *kawanatanga* (governance) over New Zealand in exchange for assurances in Article II that the Crown confirm and guarantee Ngāi Tahu the *tino rangatiratanga* (full tribal chieftainship and authority) of their lands, estates, forests, fisheries and other *taonga* (treasured possessions) (Boast, 1989). Taonga is an encompassing term, and includes both tangible possessions such as the actual resource itself and intangible qualities of the resource, such as sustainability achieved through maintenance of its natural environment (Goodall, 1997).

For well over a century, the Crown failed to honour guarantees expressed in Article II of the Treaty; however, in the 1970s the government began to acknowledge its treaty obligations (Robinson, 2002, Boast, 1989). The passage of the Treaty of Waitangi Act in 1975 aimed to "provide for the observance, and confirmation, of the principles of the Treaty of Waitangi" and set up the Waitangi Tribunal to hear claims over breaches of the treaty's provisions (Robinson, 2002).

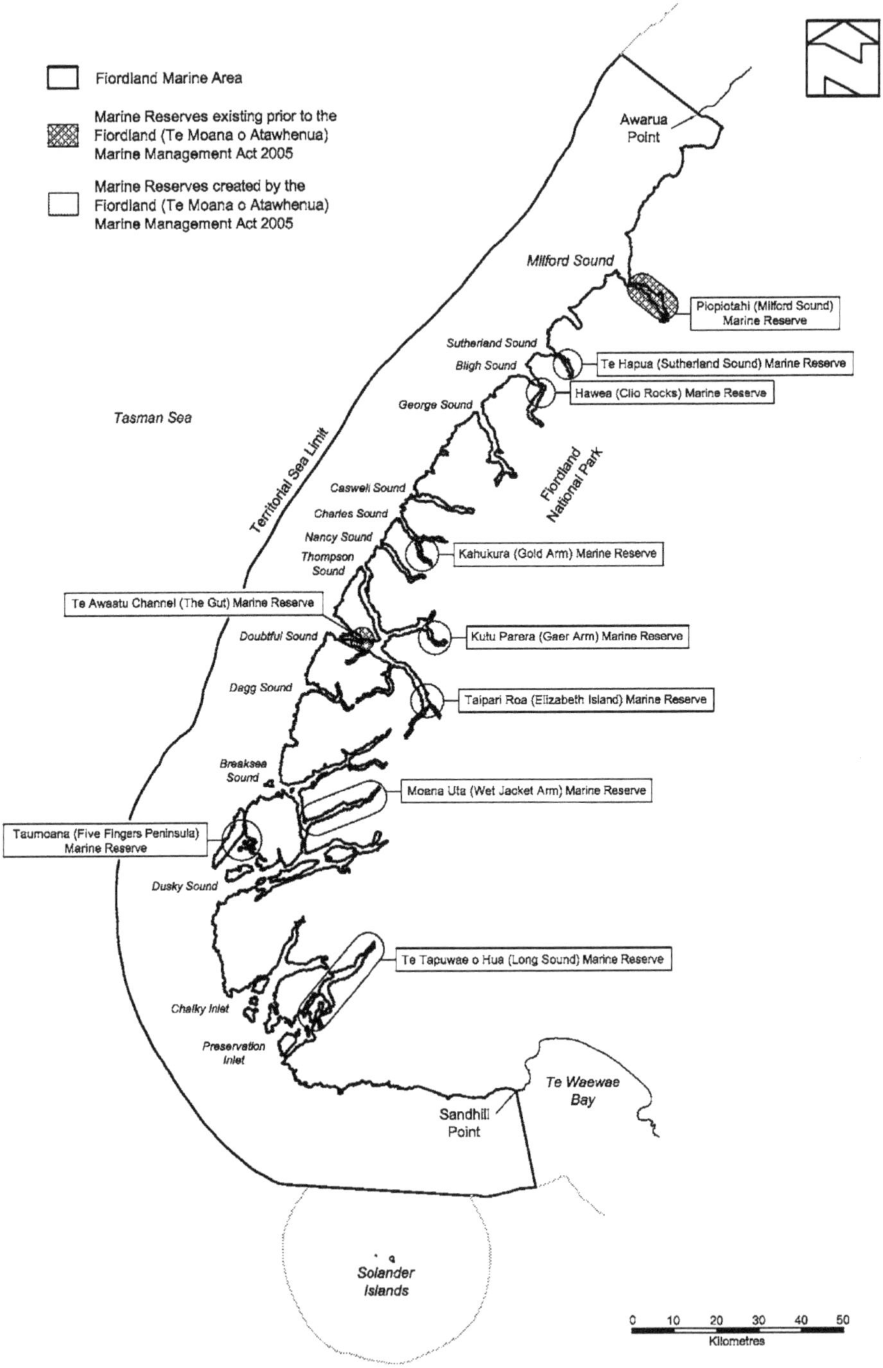

Figure 1 *Indicative map of Fiordland (Te Moana o Atawhenua) marine area and marine services. (Source: Fiordland - Te Moana o Atawhenua - Marine Management Act 2005, Schedule 1)*

The importance of recognition of treaty principles was brought into the common law in the case of *New Zealand Māori Council v. Attorney General* in 1987 (Goodall, 1997). However, customary use of marine resources was not explicitly recognized until 1989, when in the case of *Te Weehi v. Regional Fisheries Officer* the court acknowledged the existence of, and the Crown's requirement to protect, Māori customary fishing practices (Wickliffe, 1995; Munro, 1994; Dawson, 1992). Subsequent legislation in the Treaty of Waitangi (Fisheries Claims) Settlement Act of 1992 separated traditional Māori customary fishing practices between commercial and non-commercial, settled the commercial claims and clarified and made explicit the nature and extent of non-commercial fishing rights (Wickliffe, 1995; Dawson, 1992). Further legislation in the Fisheries Act of 1996 set forth requirements for the Crown to consult with representatives of Māori and to provide for input of *tangata whenua* (people of the land) with affected customary fishery interests when enacting fishery sustainability measures [3], and clarified the nature and extent of customary (non-commercial) fishing practices [4]. Critics note that this separation itself does not respect traditional Māori practices and question whether the non-commercial rights are accurately portrayed (Wickliffe, 1995). Others suggest that regulatory embodiment of traditional non-commercial practices offers an opportunity to make such practices explicit (Bess, 2000).

Specific provisions of the Fisheries Act 1996 link Māori principles to management of the fisheries. During consultation, the Act requires the Minister of Fisheries to have "particular regard to *Kaitiakitanga*" [5], or the tradition of guardianship of natural resources. The Act declares its object with regard to the customary fishing provisions as "better provision for the recognition of rangatiratanga and of the right secured in relation to fisheries by Article II of the Treaty of Waitangi" [6]. The Act establishes the Crown's authority to "make regulations recognizing and providing for customary food gathering by Māori and the special relationship between tangata whenua and places of importance for customary food gathering" [7]. However, the Act limits harvests "to the extent that such food gathering is neither commercial in any way nor for pecuniary gain or trade" [8]. Regulations authorised by the Act (promulgated later) allow designation of exclusive customary fishing areas known as *mataitai* reserves, managed by local iwi through *marae* (meeting house) committees or *katiaki* (guardians) of the tangata whenua [9]. Another provision in the Act establishes the Crown's authority to declare a coastal or estuarine area as a "*taiapure*-local fishery," *taiapure* is a legal term coined for this management tool combining *tai* (sea) and *āpure* (patch) to describe its territorial nature [10]. Taiapure are declared upon consideration of a proposal stating "why the area ... has customarily been of special significance to an iwi or *hapu* [sub-tribe] either ... [a]s a source of food; or ... [f]or spiritual or cultural reasons" [11]. A committee appointed from nominations of representatives of the local Māori community manage the taiapure-local fishery (the appointees themselves need not be Māori) [12]. Such management committees have the power of recommending fishery regulations "for the conservation and management of the fish, aquatic life, or seaweed in the taiapure-local fishery," to be promulgated by the Minister of Fisheries [13]. Other traditional marine management measures include *tapu* (spiritual restriction) and *rahui* (reserve); the Fisheries Act partially recognizes these instruments [14], but with time limitations that do not reflect traditional practice (Wickliffe, 1995). While inclusion of customary fishing practices in legislation helps to recognise and define Māori traditions, the requirement of Ministerial approval for designation of taiapure, mataitai and rahui limits the recognition of rangatiritanga (Memon, *et al.*, 2003). Essentially, legislative acknowledgement of these practices amounts to a declaration of co-management aspirations with details left to be worked out through adaptive management (Id.)

During the same time these fisheries laws and regulations were sorted on a national level, Ngāi Tahu brought a claim for grievances under the Treaty including claims over marine resources (Waitangi Tribunal, 1992). These claims were settled with the Crown by legislation in the Ngāi Tahu Claims Settlement Act of 1998. Ngāi Tahu agreed to the settlement terms in a spirit of reconciliation (Ngāi Tahu Negotiating Group, 1997). In the Act, the Crown explicitly acknowledged the iwi's "cultural, spiritual, historic, and traditional association" to "Te Mimi o Tu Te Rakiwhanoa (Fiordland Coastal Marine Area)" [15], allowing Ngāi Tahu the opportunity to reassert its rangatiratanga by meaningful participation in management of the area (Id.). Also amongst its provisions, the Crown acknowledged the association of Ngāi Tahu with "taonga fish species," and made explicit Ngāi Tahu's role in their future management, including a role in developing regulations concerning mataitai and taiapure [16]. Customary fishing regulations were determined by Ngāi Tahu in coordination with other South Island iwi working closely with the Ministry of Fisheries, spelled out in Fisheries (South Island Customary Fishing) Regulations of 1999 (Bess, 2000).

The development of the Fiordland marine guardians

The Fiordland area was designated a national park in 1952, but boundaries of the park stop at the mean high-water mark and do not include the marine area (Department of Conservation, 2003). In 1990 the Fiordland area was inscribed on the UNESCO World Heritage List upon application by the Department of Conservation (DoC), the Royal Forest and Bird Society (a conservation NGO), and the Ngāi Tahu Māori Trust Board, but again specifically did not include the marine area . The agency that oversees national parks in New Zealand, DoC, wanted to include the marine area (UNESCO, 1986). But local stakeholders in the marine area including Ngāi Tahu, commercial fishermen, recreational interests and community groups, resisted designation of the marine area for fear of increased visitation pressures upon the marine environment and the possibility of well-intentioned yet poorly-designed restrictions to counter such pressure (Carey, 2004). Nonetheless, these stakeholders recognised the need for conservation measures, and joined together in 1995 to form the Guardians of Fiordland's Fisheries (Tautiaki Ika O Atawhenua) to "manage and conserve Fiordland's fisheries resources for the use and enjoyment of future generations" (Guardians of Fiordland's Fisheries, 1999).

To start, the Guardians' documented the nature and extent of both traditional and current fishery interests, gathering the collective knowledge of the members of the group. The Guardians ensured participation of independent scientists to incorporate the biological and physical data needed for context. In 1999, the group published the results of its investigation to inform discussions going forward. Previously, no single document gathered all of the data relevant to the management of the resources together with the communities' interaction with the resources (Id.).

The Guardians' agreed to seven principles guiding its work towards recommending an integrated strategy for the Fiordland marine area:

- To ensure sustainable utilization of resources;
- To support existing fisheries management framework;
- To ensure the interests of tangata whenua and other stakeholders are identified, recognized and involved in management decisions;
- To ensure equity of access amongst stakeholders;
- To prevent uncontrolled expansion of effort by any sector;
- To identify research or information needs;
- To adopt a cautious approach to any proposed new use of the resources.

(Guardians of Fiordland's Fisheries, 1999)

The group expanded to include interests of environmental NGOs, and was able to sort out issues through a novel intra-group bargaining process of mutual respect and shared interests dubbed "gifts and gains" for its approach to promoting tradeoffs between members (Guardians of Fiordland's Fisheries & Marine Environment, 2002). From this basis, the Guardians developed a proposed integrated management strategy for the Fiordland marine area, released in October 2002 for public comment (Id.). Subsequently, the Guardians published the final strategy in September 2003 (Guardians of Fiordland's Fisheries & Marine Environment, 2003).

Originally, the group considered the fisheries management tool of taiapure as suitable for accomplishing its consensus strategy; but after review, it became clear that while taiapure may be useful for customary fishery management, it does not work well at coordinating the non-fisheries aspects of an integrated strategy such as increased visitations, bioinvasion risks, and pollution concerns (Id.). Thus, implementation required a broader overarching mechanism - legislation. The strategy was submitted to the Ministers of Fisheries and the Environment with a request to honour the package as a whole; Ministers supported the approach and passed the Fiordland Marine Management Act on 13 April, 2005.

This Act established a new management body patterned on the prior stakeholder group to advise on actions taken in the Fiordland area - the Fiordland Marine Guardians (FMG) [17]. The Act is important to Ngāi Tahu because it explicitly requires Ngāi Tahu representation on the FMG, ensuring future katiaki involvement in management of marine resources [18]. The Act established eight new marine reserves in representative areas the stakeholder group chose, as well as an eight-year moratorium on additional marine reserve applications to give time to monitor effects of the new designations [19]. The Act places additional restrictions on each user group, but importantly, these were restrictions agreed to from the "gifts and gains" process, rather than imposed by central government from afar [20].

The process of establishing the Fiordland Marine Guardians may provide a useful template for co-management of marine resources elsewhere. For jurisdictions that may wish to follow this model, the author makes the following recommendations:

- *Identify* all members of the community: Traditions of Ngāi Tahu exist complimentary to other users of the resource – recognizing this allowed traditions to be honoured and respected in management processes, a comprehensive solution would not be possible without inclusion of all interests.
- *Encourage* discussion between all users: Open and honest communication between stakeholders as to where their interests lay provided the insight of common values and opportunity to work together.
- *Seek* additional information: The Guardians invited scientists to join their effort, adding more information to support the validity of its recommendations to legislative decision makers, bolstering the group's management authority.
- *Listen* to areas of agreement between parties: Positions developed from informed consensus of stakeholders only work if officials recognize the balance of tradeoffs within the communities of interest represented.
- *Act* on recommendations: Respect the time and effort of stakeholders and incorporate the communities' recommendations within a reasonable period to encourage continuing engagement of affected communities.
- *Follow-through:* Once a plan is in place for incorporating community and traditional values, make certain that such values continue to be respected in practice. The continuing role of Ngāi Tahu on the Fiordland Marine Guardians assures the indigenous community will continue to have input on management of its traditional marine resources into the future.

References

Bess, R. 2000. New Zealand Māori claims to fisheries resources. Paper read at *IIFET 2000: Microbehavior and Macroresults*, 10-14 July, at Corvallis, Oregon, USA, ISBN: 0-9763432-2-3.

Boast, R P. 1989. "The Treaty of Waitangi: a framework for resource management law." *Victoria University of Wellington Law Review*, ISSN: 1171-042X, 19 (Monograph 1):1-67.

Carey, P. 2004. "Guardian Angels." *North & South*, July, 70-78.

Dawson, R. M. 1992. "The genesis of the New Zealand Parliament's Treaty of Waitangi (Fisheries Claims) Settlement Act, 1992." *The International Journal of Water*, ISSN (Online): 1741-5322, 1 (1):80-101.

Department of Conservation. 2003. *Fiordland National Park: General Information* [brochure]. Available from http://www.doc.govt.nz/pdfs/southland/FNP-pdfs/FNP-General-Information-Brochure.pdf (last accessed 28 July 2006).

Department of Conservation. 2005. *Fiordland (Te Moana o Atawhenua) Marine Area* [pamphlet]. Wellington, New Zealand: Department of Conservation.

Department of Conservation, and Ministry for the Environment. 2000. *New Zealand Biodiversity Strategy: Our Chance to Turn the Tide.* ISBN 0-478-21919-9.

Goodall, M., ed. 1997. *Te Whakatau Kaupapa O Murihiku: Ngāi Tahu Resource Management Strategy for the Southland Region.* Wellington, New Zealand, and Christchurch, New Zealand: Aoraki Press. ISBN 0-908925-08-5.

Guardians of Fiordland's Fisheries. 1999. *A Characterisation of Fiordland's Fisheries: Beneath the Reflections.* Dunedin, New Zealand: Ministry of Fisheries.

Guardians of Fiordland's Fisheries & Marine Environment. 2002. *Draft Integrated Management Strategy for Fiordland's Fisheries & Marine Environment.* Wellington, New Zealand: Ministry for the Environment. ISBN 0-478-07675-4.

Guardians of Fiordland's Fisheries & Marine Environment. 2003. *Fiordland Marine Conservation Strategy.* Wellington, New Zealand: Ministry for the Environment. ISBN 0-478-11904-6.

Memon, P. A., Sheeran, B., and Ririnui, T. 2003. "Strategies for rebuilding closer links between local indigenous communities and their customary fisheries in Aotearoa/New Zealand." *Local Environment,* ISSN 1354-9839, 8 (2):205-219.

Munro, J. 1994. "The Treaty of Waitangi and the Sealord deal." *Victoria University of Wellington Law Review,* ISSN: 1171-042X, 24:389-430.

Ngāi Tahu Negotiating Group. 1997. "Five generations of Ngāi Tahu men and women have grown old in the shadow of our tall trees." In *Te Karaka Special Edition: Crown Settlement Offer Consultation Document,* ISSN: 1173-6011. Christchurch, New Zealand: Ngāi Tahu Publications Ltd.

Roberts, C. M., and Hawkins, J. P. 2000. *Fully-protected marine reserves: a guide.* Washington, D.C. USA and York, UK: WWF Endangered Seas Campaign and Environment Department, University of York. ISBN: 2-88085-239-0.

Robinson, G. M. 2002. "Treaty and tribunal: redressing longstanding grievances in Aotearoa/New Zealand." *The Round Table,* ISSN (Online): 1465-4008, 367:613-624.

Sobel, J., and Dahlgren, C. 2004. *Marine reserves: a guide to science, design, and use.* Washington, D.C.: Island Press. ISBN: 1-55963-841-9.

UNESCO. 1986. *Report of the rapporteur: Tenth Session of the World Heritage Committee* (Unesco Headquarters, 24-28 November 1986): United Nations Educational Scientific and Cultural Committee.

Waitangi Tribunal. 1992. *The Ngāi Tahu sea fisheries report.* ISBN 0-86472-103-X.

Warne, K. 2000. "A Special Place: Fiordland: New Zealand's Southern Sanctuary." *National* Geographic, ISSN: 0027-9358, December, 68-85.

Wickliffe, C. 1995. "The co-management of living resources and Māori customary fishing rights." Paper read at The *Way Forward: Collaboration and Cooperation 'In Country'; Indigenous Land Use Agreements Conference,* 26-29 September, at Darwin, Northern Territory, Australia. ISBN 0-644-36100-X.

Endnotes

[1] United Nations Convention on Biological Diversity Art. 8(j).

[2] Translations from Māori derive from Goodall (1997).

[3] Fisheries Act 1996 § 12(1).

[4] *Id.*
[5] *Id.*
[6] Fisheries Act 1996 § 174.
[7] Fisheries Act 1996 § 186(1).
[8] *Id.*
[9] Fisheries Act 1996 § 186(2).
[10] Fisheries Act 1996 § 175.
[11] Fisheries Act 1996 § 177.
[12] Fisheries Act 1996 § 184.
[13] *Id.*
[14] Fisheries Act 1996 §§ 186A, 186B.
[15] Ngai Tahu Claims Settlement Act of 1998, Sched. 102.
[16] Ngai Tahu Claims Settlement Act of 1998 §§ 298-311.
[17] Fiordland (Te Moana o Atawhenua) Marine Management Act 2005 §§ 12-29.
[18] Fiordland (Te Moana o Atawhenua) Marine Management Act 2005 § 15.
[19] Fiordland (Te Moana o Atawhenua) Marine Management Act 2005 §§ 7, 10.
[20] Fiordland (Te Moana o Atawhenua) Marine Management Act 2005, Preamble (6).

Acknowledgements

The author acknowledges the initial assistance of Peter Adds, Tumuaki/Head of School of the Victoria University of Wellington Māori Studies Department, and Paul Irving of the Ministry for the Environment; any errors, however, are solely attributable to the author.

James Mize currently attends Victoria University of Wellington in its Master of Development Studies programme (MDevSt, expected 2007), under the sponsorship of a full grant from the US Fulbright Graduate Student program administered by Fulbright New Zealand. His thesis investigation centres on New Zealand's experience with implementing marine reserves, the Fiordland area comprises one case study in his research. He is currently on leave from the law and management postgraduate degree program at the University of California, Los Angeles (JD/MBA, expected 2008) where he focuses on environmental law and sustainable management. He has published two law review articles on marine reserve implementation processes in California in the Channel Islands and in prospective State processes. He received his Bachelors degree in Social Sciences at the University of Washington in Seattle. Prior to these academic pursuits, he had a lengthy career in the commercial fisheries in Alaska where he achieved his Captain's license. Among other accomplishments, he promoted native interests through joint ventures in the State's Community Development Quota program, and represented commercial interests in legislative processes. Although he does not go to sea anymore, he remains an avid fan of all things marine, as his continuing research interests demonstrate.

The Beauty of the Place

Carol Davies[1], Karen Paroissien[1], Jasmine Payget[2] and Jacinta Tobin[3]
[1] Department of Environment and Conservation NSW, AUS
[2] Blue Mountains City Council, AUS
[3] Private researcher

Using action learning we are searching for common ground. The beauty of the place is our starting point for working together and also a starting point for communities. We are working with three different groups of volunteers for the environment in the Blue Mountains, Blacktown and along the Cooks River to understand more about their motivations. All of these communities are on Darug land.

The Australian Heritage Commission (1999) have identified that there are cultural heritage values in landscapes, that the landscape itself holds the importance. How is the landscape of the Blue Mountains important? What is significant about the Cooks River and Blacktown Community Gardens that draws people to care for these places? Where is the place where people can change their perspectives, becoming more aware of the changes needed to sustain both natural and social life into the future?

The authors have been working together to explore these and other questions. In dialogue we explore the complexities of Aboriginal and non-Aboriginal people caring for the land together. Aboriginal perspectives of the landscape challenge us all to connect with the land in a deeper way.

Introduction

What makes one person oblivious to environmental degradation, to the scars left by today's society on the land, in the air, the water and the effects on plants, animals and ourselves? And yet at the same time others not only appreciate the world around them but respect and care for the place where they live, including volunteering countless hours of their time to clean-up, land care, water monitoring, interpretive tours, community gardens and wildlife rescue.

The authors have come together to explore connection to place. We wanted to collaborate, to share our experiences and learn more about what motivates people to want to make a difference. Is it related to a sense of connection? What gives us this connection? We have all worked with communities and for our own communities. In our current roles we work for state and local government and a community based organization. Work on this paper has provided us with the opportunity to reflect on our own work as well as our own connection to place.

We began with a meeting in a little café in Springwood, in the lower Blue Mountains. One of us lives in the Blue Mountains, two on the plains, in the Western suburbs of Sydney and one lives on the coast, on Botany Bay. We started by speculating to what degree experiences of beauty in nature help motivate people to protect and care for the environment. Conversation soon turned to water, an issue that has received a great deal of media coverage in recent times.

Each of us are concerned about water in different ways. In Nth Katoomba, the Yosemite Creek has been significantly affected by erosion and sedimentation in the catchment. In 1972, the plunge pool at the base of MiniHaHa Falls was more than 25 metres deep. It is now ankle deep. The western suburbs of Sydney are expanding to the base of the Blue Mountains. Thousands of homes are being built in an area where the amount of rainfall is noticeably less in the past couple of years. Last month the State Government extended restrictions on the domestic use of groundwater from the Botany Sands Aquifer. Ground water use in many Sydney suburbs is now banned. Jacinta explained her interest in Aboriginal perspectives on ground water.

Method

We have chosen three groups of volunteers for the environment:

- RiverLifers
- Blacktown Community Gardens
- North Katoomba – bushcarers, landcarers and community gardens volunteers

Each of these communities are connected by being on Darug land.

We asked to meet with them to ask about their connection to their local creek or river and what has made them respect and care for the environment. The questions are based on previous research conducted in relation to motivations of environmental volunteers (Kelly, et al., 2006) We then expanded the questions in order to explore connection to place and water in particular, in more depth. Reflection from Jacinta provides dialogue on Aboriginal and non-Aboriginal perspectives on deeper caring for the land.

At the time of writing we are able to report on the responses of the first focus group, held with the Riverlifers in late August 2006. The call for papers for the conference was the catalyst for our coming together but our collaboration will continue and we have learnt to expect the unexpected.

The Riverlifers

One of the authors has been working with a group of volunteers who guide tours of the Cooks River in Sydney. The Riverlife Interpretive Tour Program explores the relationship between the urban environment and the Cooks River through community led activities. The tours are free and cover a variety of topics depending upon the passions of the tour guide. The program is supported by Canterbury, Marrickville and Strathfield Councils. Most recently a grant from the NSW government program, *Our Environment*

It's A Living Thing has been received to support sustainability training of the Riverlife tour guides to assist in their role as community educators.

Seven of the Riverlife interpretive tour guides volunteered their time to come and talk with us and share their inspiration. We began by asking them to consider a drop of water. Where did it come from? Where is it going to? What was their connection to it? We asked them to draw this, to express it visually and then explain what they had drawn to the rest of the group. The Riverlifers' drawings expressed diverse connections with water. "That's my open door and a pathway to the River," explained one. Other comments were: "I connect with water daily ...We need it [water] to operate in the world and to relax" and "It's part of life, DNA, my story". Some drew and referred to the water cycle: [the] cycle keeps going around, scoop a bucket of water from the river, drink, sewer, small creek, the river [again] round and round."

Respecting and caring for the environment

Common themes emerged in the Riverlifers' stories and explanations about what made them respect the environment and want to care for it. Some reflected on being young and playing outside: "I grew up in an estuarine area. It's where I learned to play and I haven't stopped doing that." Other memories were about being taken away from the city as a child: "Once I left the city, thought I had gone to heaven."

For others it was about being taught about respect for the environment. One spoke about being in the cubs and scouts. Another talked about an Aunt and Uncle who took him away camping and fishing when he was a child. Another told a story about his grandmother who was one of the people who started the National Trust.

"From an early age I was dragged around to natural areas they were trying to save. I saw things being lost and things being saved".

The participants spoke of this respect continuing into adulthood, of the need for places for recreation and for 'safe havens'. "In Sydney all the rivers run into the harbour – it's like a jewel."

Yet another them\e to emerge was feeling a sense of responsibility.

"It was the visible pollution that got me first. I was curious to see what people were doing. How they were managing environmental problems in an urban area. I was impatient to find out why it doesn't look good."

They emphasized the need for their contribution. It was evident that the Riverlifers valued the Cooks River as being part of their place on earth right now. They were aware of how the River made them feel. One participant told us:

"I love the River. Rushing along the river sometimes I'm half way along before I look up and realize where I am – and I stop and say 'sorry - hello river'. My day's completely different after that"

Two others commented that they have to do something because its in their backyard and if they don't do something, who will?

Wanting to make a difference

We asked the group if they would be a Riverlifer on any river. Most said they cared for the Cooks River because of the terrible shape it was in. Only one said they would care for any river in the same way. This is different to the research conducted by the CSIRO which reported that:

"...participants generally felt that the activity was more important than the place and they would be able to substitute one place for another to do their environmental volunteering." (Kelly et al., 2006: 12)

Most commented that the Cooks River needs more care because of what it has suffered. "If it was pristine I would be motivated to interact with it but maybe not to do something." The state of the river, the visible pollution had been part of their motivation to take action. They did however also mention the importance of the social interaction, the physical activity of walking and the pleasure of leading tours, often taking people down to the river for the first time, hoping they will also learn to care for the river.

Strengthening their connection

It was evident from the group discussion that their connection with the Cooks River has strengthened as a result of caring for it. "This [being a Riverlifer] is another extension of who I am." They spoke of the need for people to know the river, to care about it and like it for them to want to take responsibility for it. The Riverlifers noted changes to the Cooks River over the years:

"There used to be a floating sea of debris glistening in the sun. Now, sometimes you get shots of pristine water." and

"The smell of the mud - it smells like a River."

And finally a detail from one of the drawings:

"[The] drop equals a tear because of what humans have done and the neglect. What is beautiful could be more beautiful."

One water for all

"In the very beginning, the Earth was formless and empty of life. The Creator Spirit, in the form of the Rainbow Spirit, shaped the land, it's mountains, seas, rivers and trees...

These images in our stories are essentially the same as those depicted in Genesis 1:2 and 2:4. In the beginning the earth is specifically designated as formless and empty

(tohu wabohu) but covered with water. Likewise, in Genesis 2:4, the earth is depicted as a vast desert without life or form." (Archie & Corowa, 1997: 29)

The Cooks River focus group talked about their sense of responsibility for their environment and water ways. About how they believed that they can make a difference but mostly how their connection to place, has been strengthened through the process of caring for their environment. How their fond childhood memories of nature and the interaction with nature has stayed with them throughout their lives. Have these wonderful volunteers touched their unconscious connection to all that is?

As Aboriginal people, some of us feel the Earth in many ways. We are connected to crystal, earth, plant, insect, fish, reptile, bird, animal and of course each other. Some understand this though our Kinship system, which embraces our connection to all life here on Earth. We have a responsibility for looking after our family or Kin, which was a way of life for our people in the past, in the future and in the present.

In their book *Sing the land, signing the land,* Waston, & Chambers talk about the Yolngu people and the moieties or kinship systems *Yirritja* or *Dhuwa.* They say:

"The most important function of the system is the orderliness it imposes on the relations of individuals and groups to each other, to the land and to all things in the Ylongu world... the system determines potential marriage partners; it designates the precise parcels of land which individuals and groups 'own' or for which they are responsible; it provides a coherent spiritual account of the genesis of life on earth: and when combined with 'messages' delivered by living agents within the system, it provides practical information about which plants and animals are available for harvesting. It designates the various sets of people responsible during a burial ceremony." (Waston & Chambers, 1989:37)

This connection goes much further a field than just Australia and the Earth itself, it spreads across our solar system and the known universe. This connection is not just for Aboriginal people alone it is part of every life force on this beautiful planet. We as Aboriginals in our past have had time to stop and listen to our environment. We have watched it change, looked for the signs and communicated freely without a barrier of superiority but with a feeling of equality and responsibility.

It was said by one of the volunteers that she stops when passing the Cooks River and says "Hello". This made my heart: sing to hear that she has honoured the river by giving it an acknowledgement of being a living entity.

Dr Masaru Emoto talks of an experiment that he and this team performed at Fujiwara Dam in central Japan with a Shinto priest and the photos produced by capturing the frozen water crystals.

"The priest stood on the edge of the lake and performed the incantation for about an hour, all of which I videotaped. After the incantation, the priest and I conversed; within fifteen minutes, my crew called me over.

"This is incredible! The water is getting clearer right in front of our eyes," someone said. And it was absolutely true. You could clearly see that the water was becoming more and more transparent as we looked at it. We were even able to make out the foliage at the bottom of the lake, which had been hidden by the cloudy water.

We next took photographs of crystals. The crystals made with water from before the incantation were distorted, and looked like the face of some in great pain. But the crystals from water taken after the incantation were complete and grand." (Emoto, 2004: 90)

We as humans are made up of 70% water, which is mirrored and magnified by our planet Earth, which is made up of 70% water itself. Now through the great advancement in our science we can see the effects our actions, thoughts and words have on the micro and the macro environment we live in and are a part of. As Aboriginal we have those old ones who can still sing up the rain and have a deep understanding of our song lines. We can recreate our Garden of Eden, which has always been here and which as Aboriginal people we never left. The more we have volunteers working with the Australian environment, the closer we can come together as living being on this Earth and connect with all that is.

How to continue this work

As community educators, we frequently review our work and evaluate our time and effort against both outcomes determined by our workplaces and community committees. We also listen to the still small voice inside our hearts to check our journey. Are we listening well enough? If we are in partnership between community organizations (especially with Aboriginal community organizations) and government agencies, is this based on respect and two way learning & acknowledgement? Where is the place where people can change their perspectives, becoming more aware of the changes needed to sustain both natural and social life into the future? How do we as co-learners in this process, take these lessons to the organizations and communities we work within?

We know that one of the great motivators can be a significant transformative experience/s, such as that of John Seed (2003):

"I was gripped with an intense, profound realization of the depth of the bonds that connect us to the earth, how deep are our feelings for these connections. I was no longer acting on behalf of myself or my human ideas, but on behalf of the earth... on behalf of my large self, that I was literally part of the rainforest defending myself."

These sorts of experiential learning experiences are nurtured in Earth Education courses and through Earth Journeys schools programs in the Blue Mountains region. For adults, the structured educational opportunities are rarer. Bushcare provides a structured learning process but requires persistence to break through to new perspectives. Community gardens vary greatly and provide a different set of learning to value the local environment. Other motivators can be the challenges in learning about how to share what you know with others. Some Riverlifers found the training "full-on" but have found their new skills useful in helping the "grow my community – to get them to care about it."

Blue Mountains City Council works on a priority catchment process. Sub-catchments are chosen for infrastructure work with integration between council and government agency projects. Community engagement is sought and actively welcomed and fostered. Understanding more about motivations for community members will help Blue Mountains City Council design more appropriate programs for working effectively with its range of communities.

All of the authors will test these and other learnings with other communities with whom we are working in partnership. We will engage in a more active process of asking what motivates people in the new communities; testing whether there are different motivations. The educational programs will more consciously incorporate community perspectives and indigenous perspectives as we continue our collaboration.

References

Archie, N & Corowa, D. (1997) *Rainbow Spirit Theology: Towards an Australian Aboriginal Theology by the Rainbow Spirit Elders Australia*: Harper Collins Religious.

Australian Heritage Commission (1999), *Natural Heritage Places Handbook, Applying the Australian natural heritage charter to conserve places of natural significance,* AGPS.

Emoto, Dr M., (2004) *The Hidden Message in Water*, Hillsboro, Beyond Words Publishing, Inc.

Kelly, G., Hosking, K., Beaty, M. and Measham, T. (2006), *Environmental volunteers... Insights into their motivations,* Canberra ACT: CSIRO Sustainable Ecosystems, www.cse.csiro.au

Seed, J. (2003) "The path through the rainforest" in Ed. J. Cameron, *Changing places – re-imaging Australia*, Longueville Books.

Waston, H., & Chambers, D., with the Yolngu Community at Yirrkala (1989) *Sing The Land, Signing The Land*, Geelong, Victoria: Deakin University.

Carol Davies is studying for a Masters of Environmental Management. She works as a project officer in the Community Education Section of the Department of Environment and Conservation. She has worked with diverse groups to develop community projects and is interested in trialing action research methodologies to learn more about what brings communities together. She's lived in Australia for the last 15 years and loves the bush and the rocks and the water here, but lived for the first part of her life on the west coast of Canada, beside a roaring creek with enormous boulders and evergreen trees. This experience of having strong connections to two distinctly different environments triggered her interest in learning more about what connects people to place.

Karen Paroissien has a BA, a Bachelor of Social Studies and a Master of Education (Adult Education). She is currently enrolled in a Master of Sustainable Heritage Development at ANU. Karen has worked in community development, education and training. For the past 2 years she has worked as a Senior Education Officer with the Department of Environment and Conservation NSW, designing and managing community education projects. She has lived in Sydney most of her life and has always been close to the beach. The roar of the waves and the smell of the sea always makes her feel like she's on holidays. And a place is also its people. There are happy people at the beach. She is interested in the connection between how we are feeling and where we are - if we feel good, we learn, we're healthy, we share and we care for the environment.

Jasmine Payget has a Bachelor of Science and a MA (Environmental Studies) and works as the Environmental Education Officer at Blue Mountains City Council. Previous publications and areas of research interest are Equal employment opportunity, encouraging women into science and other non-traditional work areas, education, Sustainability, Community Participation and Open Government and Community Engagement. 'I grew up in southern Ontario in Canada next to a creek where beavers lived. Across the road were three huge maple trees whose leaves gave us lots of pleasure and whose sap gave maple syrup. Yum, a form of bush tucker?! In Australia I lived near the Cooks River and then 'discovered' the Flinders Ranges. My understanding of semi-arid landscapes changed when I had to do an assignment on the Western Myall, a fabulous acacia species. In the Blue Mountains the angophoras will be blossoming when the conference is being held. I hope to help foster a connection for conference participants to the place where the conference is held.'

Jacinta Tobin has a Master of Social Ecology and has worked in the Aboriginal Community for the last 10 years. Her passion runs deep when it comes to the unknown and spiritual matters. She has worked to raise the issues of the Child Protection; environmental destruction and the Darug people and her family's history. She is currently the Chairperson of the Darug Tribal Aboriginal Corporation. 'From the age of 5 I had lived in very extreme climates: from the bitter cold of Ballarat, Victoria where I was born, to the stark heat and humidity of Nhulunbuy on the top right hand peak of the Northern Territory overlooking the Gulf of Carpentaria. Then at the age of 4 my family moved to Darwin. It was in 1974 that I gained a healthy respect for Mother Nature, when Cyclone Tracey had come to visit on Christmas Day. We were flown down to be with my Mother's family in Sydney, NSW where we stayed and where I grew up. I went to Darwin in my twenties for a short time then returned back home to live in the Blue Mountains and now I am back living on the plains. Through understanding my Aboriginality I have come to have a stronger connection to my traditional lands of the Sydney region and take comfort when I can in the Blue Mountains.

The Material Component of the Aboriginal Cultural Landscape: Mapping Country Through Predictive Modelling

Malcolm Ridges
GIS Research & Development Group, Department of Environment and Conservation (NSW), AUS

This paper reports on work being conducted in the Greater Blue Mountains World Heritage Area (GBMWHA) to map the distribution of Aboriginal archaeological features using predictive modelling. The project combines with other studies being conducted in the region under the banner of 'Mapping Country'. The 'Mapping Country' program was initiated by the Blue Mountains World Heritage Institute (BMWHI), in partnership with the Hawkesbury-Napean CMA, Blue Mountains City Council and the Department of Environment and Conservation, NSW. Knowledge of the distribution of Aboriginal features throughout the landscape is providing a better understanding of the Aboriginal Cultural Landscape. The objective of the Mapping Country program is to generate information that will assist Aboriginal communities and planners develop a more productive working relationship around the development of regional conservation strategies.

Introduction

Aboriginal people experience a very strong sense of connection to the country they and their ancestors are descended from. For them, this attachment to country extends well beyond a sense of belonging, and involves a custodial responsibility to care for country, the successful implementation of which is central to their sense of well being (Waters, 2006). For this reason, it is important for Aboriginal people to be fully integrated into land management. For if they are not, there is a very real sense that for them they are letting down their country, their community, and themselves.

However, consideration of cultural heritage as an integral part of land management is only a relatively recent practice in Australia. And indeed, as Sullivan (1992, 172) has argued, the view that the Australian landscape is also an Aboriginal cultural landscape is only a recent realization for many Australians. Even after 30 years of giving greater emphasis to the importance of Aboriginal cultural heritage (see Kijas 2005), there still remain conceptual conflicts such as with the notion of 'wilderness' and the idea of a traditional Aboriginal cultural landscape (Hooper, 2006a; Mowaljarlai 1992). State and Federal Governments are now in the process of developing policy towards productive partnerships with Aboriginal communities focused on the social and cultural benefits of full engagement in land management (e.g., DEC, 2006).

However, the issue still to be clearly resolved in setting up such partnerships, is what should be focus of the relationship between Aboriginal communities and government bodies. It has been highlighted for example that the notion of simply 'involving' Aboriginal communities in conservation programs, is in itself reinforcing politically that cultural values are still peripheral to core conservation objectives (Birkhead & Smith, 1992). A possibly more productive step is to recognise the conservation value of cultural landscapes in their own right, as has recently been done by the Council of Europe (Déjeant-Pons, 2002). If this is done, then it shifts the agenda towards focusing on conservation efforts that ensure the persistence of a functioning Aboriginal Cultural Landscape into the future. This is not unlike the corollary objective of ecology, which is to ensure the persistence of biodiversity into the future (Cowling, *et al.*, 1999).

Increasingly, it is being realized that such a position involves more than just conserving evidence of cultural features (in a museum sense), and instead requires actively continuing cultural practices that maintain a cultural landscape. This follows the UNESCO view, which is that for a cultural landscape to have integrity, conservation programs must ensure that the behavior that constitutes that landscape is maintained according to the level of authenticity recognized in that community (see ICOMOS, 1994). For example, maintaining historic agricultural landscapes in Europe has meant practising historic farming practises despite the cost this imposes economically (Fairclough, 2002). The idea that maintaining a cultural landscape requires cultural practice, comes much closer to the way many Aboriginal people view the importance of cultural practice itself being an approach to conserving cultural heritage (Mowaljarlai, 1992).

One of the difficulties with adopting such a framework however, is the difficulty some Aboriginal communities have in articulating their own cultural landscape. This is particularly the case in places like the Blue Mountains, where there has been a long history of Aboriginal people being removed (forcibly) from their land, and has subsequently led to a considerable loss of traditional knowledge and detachment of from country. Projects like 'Mapping Country' (Hooper, 2006b), are attempting to help communities organise and synthesise the traditional knowledge they still hold, as well as recognizing the social value of the experiences of Aboriginal people during the contact period (see Byrne, *et al.*, 2001). These efforts are important for helping Aboriginal communities restore their cultural landscape.

It is also here that anthropology and archaeology can play a supporting role to help re-establish information about how Aboriginal people traditionally utilized the landscape. Archaeology can be useful to Aboriginal communities in providing explanations of long term temporal change, as well as detailed examination of the activities that took place in the past at particular places. A further utility of archaeology lies in the systematic reporting of the distribution of the material remains of Aboriginal behavior in the past. It is this last application of archaeology that is the focus of this paper. Distribution patterns of Aboriginal features, for instance rock art sites or places with stone tools, are important evidence for reconstructing how Aboriginal people utilized a landscape in the past. Understanding where people situated different activities, how they moved through a landscape and relationships between places with cultural value are important building blocks for reconstructing what the Aboriginal cultural landscape may have been like in the past. The link between a traditional cultural landscape, and its reinterpretation in the present, is important for Aboriginal people trying to reconnect to their country.

Taking the step from interpreting individual sites to extrapolating regional settlement and land-use patterns has been done with mixed success by archaeologists in the past. The majority of these attempts have focused on deriving descriptive accounts of the way Aboriginal people utilized a region (eg Stockton, 1993). However, actually mapping the pattern of land-use and activity has generally proved to be a much more difficult task (Gaffney & van Leusen, 1995), especially when trying to untangle these patterns from issues of archaeological detectability and temporal preservation (Hall & Lomax, 1996). For instance, it has tended to be done with landscape units applied over a region, to which qualitative statements about typical behavior are assigned (e.g., Purcell, 2002).

Otherwise, particularly in a regulatory framework with the protection of registered Aboriginal sites, the emphasis is placed upon the site itself, where it is represented spatially as an isolated point. This has the unfortunate consequence of giving the perception that there is an absence of cultural value between these points, when in fact the density of points really only reflects a combination of preservation potential and archaeological survey effort. The reality of course is that Aboriginal people perceive value in the entire landscape. Consequently the representation and management of this aspect of their cultural heritage, as a set of isolated points, lies in stark contrast to their perception of it as a cultural landscape. One of the challenges for conserving and describing indigenous values is to explore ways of representing a spatially discrete and biased archaeological sample as a cultural landscape.

One way that the cultural landscape (at least its material component) can be described in a spatially continuous way is to model the distribution of archaeological material. Predictive modelling is a technique that, when applied to spatial archaeological data, attempts to describe the likelihood of finding archaeological features occurring at any given location in the landscape (Kvamme, 1988, 325). In this way a picture of how the likelihood of archaeological features occurring varies throughout the landscape is built up by calculating the predictions on a grid cell by grid cell basis within a GIS. The result is an image illustrating the spatial pattern of archaeological material throughout the whole region.

Figure 1 illustrates the results of performing the predictive modelling for the GBMWHA using all the available archaeological information recorded in the region. This involved just over 5000 archaeological sites, along with an equivalent number of random points distributed in varying density to reflect variation in archaeological survey intensity across the region (for method see Ridges 2006). The model itself was formed using Generalized Additive Modelling with a binomial family within S-Plus using a custom algorithm called GRASP (Lehmann, *et al.*, 2002). The following variables were used to form the model, and build on well-established approaches to modelling hunter-gatherer archaeological sites (Warren & Asch, 2000):

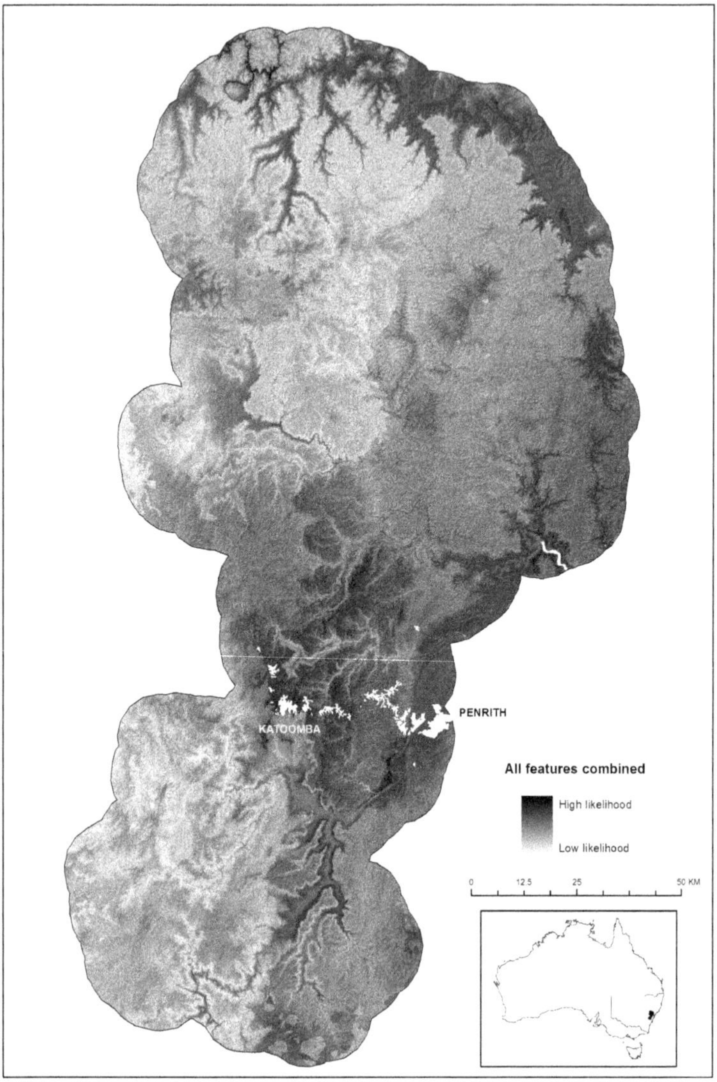

Figure 1 *Predictive model for the GBMWHA. The darker the shading indicates a greater likelihood of finding archaeological features. White areas indicate urban.*

- aspect;
- elevation;
- average annual rainfall;
- average annual temperature;
- geology (simplified into key geological units and rock types);
- accumulated visibility (derived by summing how frequently each grid cell is visible from within 25km of 17,000 observation points distributed in a grid pattern over the region);

- three dimensions of landscape similarity (as modelled using multi-dimensional scaling). This was achieved by building a matrix compiled from the amount of each of Keith's (2004) vegetation types occurring within the landscapes mapped by Mitchell (2002) that occur within the region;
- A weighted proximity to stream lines where the weight was stream order calculated using the Strahler method. The proximity was measured using cost distance where the cost was slope modified for walking speed using Tobler's formulae (see Ridges 2005, 128).

Although the map shown in Figure 1 indicates the variation in spatial pattern of finding archaeological evidence of any kind throughout the region, what it doesn't show is the variation in different types of behavior. Different types of Aboriginal behavior varied throughout the landscape in response to a complex array of resource patterns, social factors, and settlement patterns. Consequently, the pattern seen in Figure 1 needs to be dissected so it is possible to see how that pattern is further composed. This is an issue which is rarely explored in hunter-gatherer predictive models (Kvamme, 2005), despite how amenable the data and method of modelling are to doing so (see Ridges, 2003). The focus of this study is to pursue this avenue specifically, for it goes to the heart of describing the complexities of past indigenous cultural landscapes.

Figure 2 illustrates the variation that exists throughout the region when different kinds of archaeological features are modelled individually. For these models, the input data, method of modelling and variables were all the same, except that only those archaeological sites that contained the feature of interest were used for input into the model. For the present study, models for stone tools, rock art sites, grinding grooves, stone arrangements, and scarred trees have been derived. The choice of features was driven primarily by the frequency with which these features have been recorded in the region.

There are a variety of other feature-types that also occur within the region. Importantly however, the features that have been modelled reflect a variety of different behaviors. For instance, stone tools reflect primarily economic/subsistence behavior associated with the places where Aboriginal people undertook activities in the landscape. Art sites reveal aspects of the social/ideological landscape. Grinding grooves again reflect economic behavior, but quite a different aspect of it- stone axe sharpening. Stone arrangements are variously associated with ceremonial activity and/or Aboriginal burials. And scarred trees are usually associated with economic activity such as canoe or coolamon manufacture, or if they were carved decoratively, with burials. The five models therefore seen in Figure 2 start to show some of the complexity of different kinds of Aboriginal behavior in the region, not merely variation in the distribution of archaeological features.

The complexity of the material component of the Aboriginal cultural landscape really emerges when these models are used to explore the similarity in combination of features that is predicted to occur in each grid cell. This is illustrated in Figure 3, where similarity in color reflects similarity in the combination and likelihood of the features occurring. This map was generated using

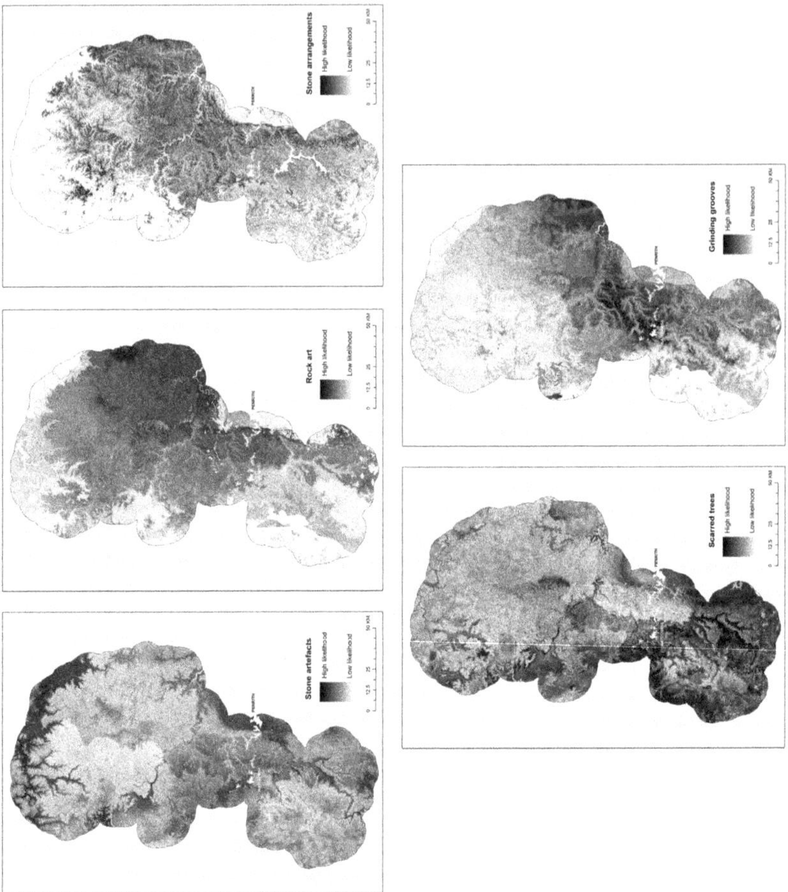

Figure 2 Predictive models for individual feature types. White areas indicate urban

the models in Figure 2 as input for a principal components analysis, calculated on a grid cell by grid cell basis. The first three components of the result were used to assign a color to each grid cell. Figure 3 begins to display what might be the beginnings of the spatial complexity in the way Aboriginal people utilized the landscape in this region. It also begins to illustrate some of the spatial pattern in the cultural landscape in a behavioral sense. Importantly, this pattern is continuous spatially, and fits much closer to the way Aboriginal people perceive their cultural landscape, and is a positive step away from simply viewing cultural heritage as a set of isolated sites.

Information that describes some of the spatial pattern in the traditional Aboriginal landscape, like that in Figure 3, permits some important possibilities in planning and decision making for cultural heritage values. For instance, using the patterns seen in Figure 3, it would be possible to construct a conservation strategy that sought to maximise the representation of all colors (or feature combinations). The idea being that such a representative sample would enable a community to connect with all aspects of their cultural landscape, in so far

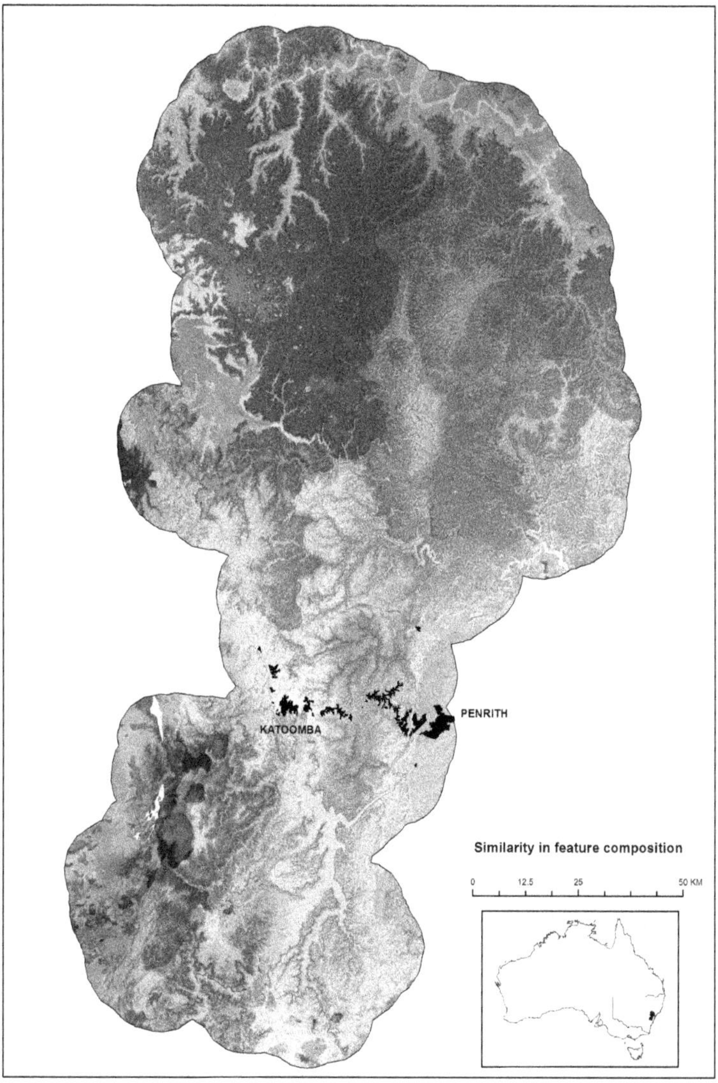

Figure 3 *Models of individual features types combined into a single color (greyscale) figure. Similarity in color (or shade) indicates similarity in the combination of features predicted to occur. Black areas indicate urban*

as it is represented in archaeological sites. This would require much more than just looking at a representative sample of 'sites', and instead require look at planning for an entire landscape using systematic planning principles (Margules & Pressey, 2000).

For Aboriginal communities, maps like that in Figure 3, articulate, spatially, some of what they recognise as the way their ancestors utilized and existed in the landscape. Importantly, armed with such information to represent their heritage spatially, Aboriginal communities can potentially use it to engage in a variety of planning contexts. These contexts include:

- connecting to country projects and cultural revival programs;
- site management issues;
- directing archaeological surveys, both on-reserve and within the regulatory environment;
- within the local government environment;
- with the management of reserves;
- with a variety of state and federal land-management bodies;
- decision making and conservation priority setting.

However, of most importance, is that these principles can be applied under the control and direction of Aboriginal communities themselves. Potentially this is where empowerment of Aboriginal communities can be realized, by ensuring their full control of this information and enabling them to use it to help obtain the outcomes they desire. For instance, the planning domain become maps of site potential and condition, without the need to reveal sensitive information such as the location of recorded sites. In doing so, it potentially offers to foster a much more productive engagement between Aboriginal communities and land-use planners. Additionally, it also goes someway to reversing the trend of viewing the indigenous landscape as a relic landscape instead of a living cultural landscape.

References.

Birkhead, J. & L. Smith. 1992. Introduction: conservation and country- a reassessment. In *Aboriginal involvement in parks and protected areas* (eds) J. Birckhead, T. De Lacy & L. Smith. Canberra: Australian Institute of Aboriginal and Torres Strait Islander Studies. ISBN: 085575236X, pp1-5.

Byrne, D., H. Brayshaw & T. Ireland. 2001. *Social significance. A discussion paper.* Hurstville: NSW National Parks and Wildlife Service. ISBN: 0731363655.

Cowling, R.M., R.L. Pressey, A.T. Lombard, P.G. Desmet & A.G. AEllis. 1999. From representation to persistence: requirements for a sustainable system of conservation areas in the species-rich mediterranean-climate desert of southern Africa. *Diversity and distributions* **5**, 51-71. ISSN: 1366-9516.

DEC. 2006. Aboriginal people and cultural life. Retrieved 7 September 2006 from: http://www.nationalparks.nsw.gov.au/npws.nsf/Content/Aboriginal+people+and +cultural+life.

Déjeant-Pons, M. 2002. The European Landscape Convention, Florence. In *Europe's cultural landscape: archaeologists and the management of change.* (eds) G. Fairclough & S. Rippon. Brussels, Belgium: Europae Archaaeologiae Consilium. ISBN: 907697502X, pp. 13-24.

Fairclough, G. 2002. Europe's landscape: archaeology, sustainability and agriculture. In *Europe's cultural landscape: archaeologists and the management of change.* (eds) G. Fairclough & S. Rippon. Brussels, Belgium: Europae Archaaeologiae Consilium. ISBN: 907697502X, pp. 1-12.

Gaffney, V. & M. van Leusen. 1995. Postscript- GIS, environmental determinism and archaeology: a parallel text. In *Archaeology and geographic information systems: a European perspective.* (eds) G. Lock & Z. Stancic. London: Taylor and Francis. ISBN: 074840208X

Hall, R. & K. Lomax. 1996. A regional landscape approach to the management of stone artefact sites in forested uplands in Eastern Australia. *Australian Archaeology* **42**, 35-38. ISSN: 0312-2417.
Hooper, S. 2006a. Decolonising country. Paper presented to the Finding common ground. A day of dialogue about Wilderness areas in the Greater Blue Mountains, Sunday May 7, 2006, Blackheath., Blackheath, Blue Mountains, NSW, 2006a.
Hooper, S. 2006b. *Mapping country report: archaeology in the Blue Mountains.* Blue Mountains World Heritage Institute. ISBN: 0-9802894-2-4.
ICOMOS. 1994. The Nara Document on Authenticity. Retrieved 7 September 2006: http://www.international.icomos.org/naradoc_eng.htm.
Keith, D.A. 2004. *Ocean shores to desert dunes : the native vegetation of New South Wales and the ACT.* Sydney: Dept. of Environment and Conservation. ISBN: 0731367804.
Kijas, J. 2005. *Revival, renewal and return. Ray Kelly and the NSW sites of significance survey.* Sydney: Department of Environment and Conservation (NSW). ISBN: 1741371023.
Kvamme, K. 2005. There and back again: revisiting archaeological location modelling. In *GIS and Archaeological Predictive Modelling.* (eds) K. Westcott & M. Mehre. Boca Raton, Florida: Taylor and Francis. ISBN: 0415315484, pp. 3-38
Kvamme, K.L. 1988. Development and testing of quantitative models. In *Quantifying the present and predicting the past: theory, method, and application of archaeological predictive modeling* (eds) W.J. Judge & L. Sebastian. Denver, Colarado: U.S. Department of the Interior. ISBN: 88600552, pp. 325-428.
Lehmann, A., J.R. Leathwick & J.M. Overton. 2002. GRASP: Generalized regression analysis and spatial predictions. *Ecological Modelling* **157**, 189-207. ISSN: 0304-3800.
Margules, C.R. & R.L. Pressey. 2000. Systematic conservation planning. *Nature* **405**, 243-253. ISSN: 0028-0836.
Mitchell, P. 2002. *NSW Ecosystems: background and methodology.* Unpublished report. Dept. of Environment and Conservation.
Mowaljarlai, D. 1992. Wayrrull- Aboriginal traditional responsibility in cultural resource management in the Northwest Kimberleys of Western Australia. In *Aboriginal involvement in parks and protected areas* (eds) J. Birckhead, T. De Lacy & L. Smith. Canberra: Australian Institute of Aboriginal and Torres Strait Islander Studies. ISBN: 085575236X, pp. 179-189.
Purcell, P. 2002. *Aboriginal cultural heritage assessment. NSW Western regional assessments. Final report September 2002.* Resource and Conservation Assessment Council. ISBN: 1740291786
Ridges, M. 2003. Numerous indications. The archaeology of regional hunter-gatherer behaviour in northwest central Queensland, Australia. Unpublished PhD: University of New England.
Ridges, M. 2005. Regional dynamics of hunting and gathering. An Australian case study using archaeological predictive modelling. In *GIS and Archaeological Predictive Modelling.* (eds) K. Westcott & M. Mehre. Boca Raton, Florida: Taylor and Francis. ISBN: 0415315484, pp. 123-143
Ridges, M. 2006. *ASDST. Aboriginal sites decision support tool. Derivation of state-wide models for NSW.* Unpublished report. Department of Environment and Conservation (NSW).
Stockton, E. 1993. *Blue Mountains Dreaming: the Aboriginal heritage.* Winmalee, NSW: Three Sisters Production. ISBN: 0646148843.
Sullivan, S. 1992. Aboriginal site management in national parks and protected areas. In *Aboriginal involvement in parks and protected areas* (eds) J. Birckhead, T. De Lacy & L. Smith. Canberra: Australian Institute of Aboriginal and Torres Strait Islander Studies.

ISBN: 085575236X, pp. 169-178.
Warren, R.E. & D.L. Asch. 2000. A predictive model of archaeological site location in the Eastern Prairie Peninsula. In *Practical applications of GIS for archaeologists. A predictive modelling kit.* (eds) K.L. Wescott & J.J. Brandon. Philadelphia: Taylor and Francis. ISBN: 0748408304, pp. 5-32.
Waters, K. 2006. *Report on co-management and wellbeing.* Waters Consultancy. History and Cultural heritage A report to the Cultural Heritage Division, DEC. Unpublished report. Department of Environment and Conservation (NSW).

Malcolm Ridges's primary research interest is in the application of systematic conservation planning to the conservation of Aboriginal cultural landscapes. My work in this field has involved archaeological predictive modelling of Australian Aboriginal features and using such information as the basis of mapping conservation priorities. My related interests lie in spatial data analysis, regional archaeological studies, GIS and site recording. After studying for a PhD in archaeology, I have worked for the past five years in the field of ecological conservation planning, but have recently moved back into a cultural heritage focus.

THEME THREE

Creativity, Organization and Complexity

Kay Fielden, Unitec

In the creativity, organization and complexity session ten papers were presented that covered a diversity of issues relating to the main them area. These issues arose across the board: in education and in industry; natural resources concerning water and primary industry; and ranged from complexity within and across business organizations and higher education, particularly looking to the future. Creativity was called for in challenging traditional leadership. Complex issues pertaining to environmental sustainability, environmental issues facing fisheries, mediation, mentoring and action research as a viable and sustainable tool for managing major organizational change were also presented.

It was apparent that there is a need to question basic assumptions in looking for creative ways to solve the 'wicked' problems facing us environmentally as we look into a future that is struggling with global dilemmas. Local solutions, seeing repeated patterns no matter where systems thinking is applied, creating meaning for all stakeholders, exploring the dynamics that arise from conflicting paradigms and questioning imposed commencement and conclusions to 'project' solutions were all described and debated. Perhaps the most important outcome of this theme session is that these dilemmas have been aired and local solutions proposed. If we are true to our beliefs that we are true systems thinking then a tolerance for confusion, multiple views and many paths to agreeable action is more important.

Sustainable Workplace: A Study of Mobile Knowledge Work - Practice Motivating Theory Development

Duska Rosenberg
School of Management, Royal Holloway, University of London, UK

1. Introduction

This paper draws on the results of several research projects focused on studies of interaction in the real-life workplace. The projects share one key characteristics – they all involved strong collaboration with industry, notably the construction industry and the related disciplines of architecture and workplace design. The focus of this paper is on the input of empirical workplace studies to the development of a human environment framework that captures key features of both face-to-face and computer-mediated communication. Computer-mediation of day-to-day cooperative activities is regarded as an integral part of the 'hybrid' workplace, that is, the workplace that integrates physical and computerised resources to facilitate communication and collaboration across spatial, organizational and social divides. From a theoretical point of view, the main aim of this work is to outline an analytical approach that takes into account the role of broad non-linguistic context in shaping co-action, with particular emphasis on spatial and organizational features. From a practical point of view, such a framework is needed to inform the design of collaborative work environment of the future where the traditional workplace resources are enhanced with facilities provided by information and communications technologies.

The requirements of real-life workplace for support of day-to-day communicative and cooperative activities were initially analyzed within the CICC project[1]. The research was user-driven and interdisciplinary, focused on the development and use of interactive technology in large-scale construction projects. It was motivated by the needs of people working in construction and manufacturing industries, where poor communication causes serious problems in day to day activities which require continuous cooperation and coordination (Fruchter, 1998).

New technologies such as multimedia, mobile telephones, wearable computers and video were introduced into a large-scale construction project[2] to address the communication requirements of construction teams. The technology support was designed to improve communication within and between teams characterized by professional and cultural diversity, and to offer a richer information environment for the repair of breakdowns and misunderstandings. Such a facility is particularly useful in agile project teams, where a stable or-

1 European research project funded under Framework IV ACTs programme: CICC (Collaborative Integrated Communications for Construction, ACTS No. 017), September 1995-March 1999.
2 Bluewater, a £20M project, building a shopping mall in Kent, with 120 subcontractors. Workplace studies were carried out over a period of three years.

ganizational form is absent as a team is created for the purposes of a particular project and is dissolved once the project is completed. The work on the project itself represents a period during which cooperation is vital. Problems of lack of shared culture have to be addressed and this is usually done with a series of induction meetings and seminars at the start of the project. This is a vital yet time-consuming team building stage of the project, and is undermined by teams joining projects at later stages with consequent integration problems (D14, CICC report, Rosenberg, 1998).

To support natural interactions in this context, the working assumption is that team members together build a shared environment where they cooperate to solve their work problem. They coordinate their actions and focus on common artefacts (for example documents, drawings or a computer screen) in the process of negotiating the meanings of words or images presented there (Robinson, 1993). They thus create the common ground – "a sine qua non for everything we do with others… the sum of [the participants'] mutual, common or joint knowledge, beliefs, and suppositions" (Clark 1996: 92).

Within the boundaries of the common ground the participants can identify the objects referred to, come to understand each other's goals and purposes, cooperate and coordinate their actions. Indeed, common ground is regarded as fundamental to all coordination activities and to collaboration (Clark & Brennan, 1991). In this context, one of the key research questions was 'How do people create the common ground in situations where the contact between them is influenced or mediated by technology?' (Rogers, 1993; Heath & Luff 1991).

Usable interactive technology is expected to facilitate the processes that shape human cognition and communication in the significant social and cultural contexts, thus fitting in with normal human activities in the workplace. The technology is thus regarded as an integral part of the entire information environment of the 'hybrid workplace' created by the interaction of people, organizations and artefacts where information is generated, exchanged, stored, processed, internalized and externalized.

2. User and workplace requirements

From the perspective of day-to-day activities in the workplace, individuals work in any number of places, their homes, other people's offices, as well as various public places such as libraries, parks, and cafés. They also work when traveling – in cars, trains, planes, and in lounges, waiting rooms and lobbies. They maintain contact both with their homes and their offices, enjoying flexible working conditions because information and communications technologies enable them to communicate and work together, albeit 'at-arm's-length'.

The key issues for workplace design concern the nature of such a distributed workplace and the resources such a workplace has to offer. Such resources include not only work-space and furniture but also technology-enabled connectivity to other work-spaces and the representations of people who are not physically co-present.

Moreover, being connected is a pre-requisite for re-creating the essential features of an effective workplace in a mediated environment. A 'hybrid' work environment, that integrates physical and mediated spaces for communication and collaboration, must provide access to different kinds of work space as required by its occupants, as well as allowing them to develop normal social and cultural relationships based on trust, friendship and organizational identity and belonging. This is a challenge for distributed teams, since communities have traditionally relied on proximity to bind them together into a 'local culture' where members share the available resources, speak the common language, dress and behave in ways that are appropriate for the customs, values and beliefs of their community.

If the visions of the workplace of the future can be expressed in few words, the most appropriate would be 'distributed networked enterprises'. In the world of business, the current ideal form has shifted from large multi-national corporations rich in property, equipment and workforce, towards smaller, but more flexible autonomous units that are more adaptable and responsive to changes in the business environment. An appropriate metaphor is that of an elephant giving way to a gazelle. Consequently, the sustainable workplace of the future will be distributed because its workforce will be increasingly more mobile and because large hierarchical organizational structures will be replaced by smaller and less hierarchical, but highly networked organizations in consortia and partnerships with similar enterprises working together as agile teams.

The research into various aspects of the workplace of the future was primarily carried out in two EU-funded, large-scale projects CICC (c.f. above) and SANE[3], but it was complemented by a number of smaller projects focusing on communication in physical and media environments in collaboration with the designers and developers of Virtual Reality applications and context-aware technologies[4], the use of drama techniques in the design of virtual and augmented reality interfaces[5], as well as the work oriented towards the role of language technologies in the context of computer-mediated communication in the workplace[6]. The application of this research falls into three main domains – introducing information and communication technologies into a distributed organization in construction and manufacturing[7], education[8], architecture and workplace design[9]. Special emphasis is given to understanding the impact of the technology-based innovation on the established patterns of collaboration and

3 EU Framework 5 IST project Sustainable Accommodation for the New Economy, IST 2000-25-257.

4 BT University Research Fellowship at the BT Research Laboratories during 1998-99.

5 LIVE (Language in Virtual Environments) 1998-2000, sponsored by ISCE (Institute for the Study of Coherence and Emergence)

6 SCALE (Internet-based intelligent tool to support collaborative argumentation-based learning in secondary schools), EU F5 2001-2003, and Leverhulme Visiting Professorship 2003 – 04.

7 Sponsored by STENT Foundations Plc and BICC (British Insulated Cable Company), both part of Balfour Beatty Construction.

8 More specifically, using Virtual Reality in education, funded by the Nuffield Foundation scholarship.

9 PhD scholarship funded by the Academy of Science, Finland.

communication in these domains.

The key outcome of these projects is our increased understanding of the information environment in the workplace that takes the centrality of interactive technology a step further. The mobile workplace is location-independent as it enables people to work anywhere, anytime since it extends the resources from a physical space to a 'virtual' or 'augmented' space mediated by technology.

The study of computer-mediated communication in a location-independent workplace is focused on addressing several research questions concerning the key features of context of coordinated action and communication (co-action) in such a workplace. Firstly, we aimed to establish the extent to which space, people, work and technology, are interrelated, and the extent to which these relationships influence co-action either directly or indirectly. Secondly, we developed a framework to identify the extent to which the interrelated settings of space, people, work and technology can be configured so that they improve facilities for co-action. Finally, we examined the theoretical and methodological issues involved in our attempt to understand and design the workplace of the future in order to enhance its proximity, permanence and productivity.

3. Space, people, work and technology

As the ultimate aim of this research is to inform the design of the hybrid workplace and of the information and communication technology as an integral part of it, we pay special attention to the nature of the broad-based human environment in the real-life workplace. We do this by examining primarily the roles, responsibilities and contributions of the individuals to team work, whose patterns of interaction involving the interplay of individual and join action are the main focus of our empirical study.

One of the most interesting features of the real-life workplace concerns different kinds and different degrees of responsibility that individuals have in particular communications in the work context. These responsibilities may be assigned, that is, they are a part of the formal duties and positions, such as Project Director, Project Manager, Assistant Project Manager, depending on the nature of the work tasks. They are also assumed, that is, in any particular instance, an individual is expected to make a contribution in accordance with her expertise, experience or preference, such as advising colleagues on IT matters or researching a particular topic. These are not 'given' but adopted depending on the particular circumstances in which a particular team engages in a particular task.

The assumed roles and the resulting contributions are only partially determined by the status of an individual in the organization, reflecting instead the requirements of day-to-day collaborative activities and individual engagement with others. It is in this context that the interaction between individual and joint action provides the basis for creating and maintaining the common ground.

The outcomes from the empirical studies of day-to-day collaboration in the real-life workplace suggest that individuals take on roles that are formally assigned from an organizational aspect and informally assumed from a communicative or conversational aspect. As they negotiate and define boundaries in collaborative environments, the ways in which different roles are assumed informally, varied from situation to situation, and frequently included the use of space. This seemed to be quite a powerful way for individuals to pursue their own intentions. For example, if someone walked to the front of the meeting, it was clearly the way to express that s/he wanted (and had the right to expect) the attention of the whole audience to focus on him/her, compared with someone else, who just talked from his place or perhaps stood up to say his/ her opinion. Someone else could 'take the floor' in a more indirect and yet powerful way, such as walking to the white board and pulling things together by writing a few lines, without needing to say much or anything.

This kind of individual action and its recognition by others is of critical importance in negotiating the nature and the degree of involvement, participation and contribution that an individual makes to the joint activity. In the physical work setting, the interplay between the individual and joint action occurs spontaneously and most of the time, un-problematically, but in mediated settings it has not been possible to represent the context of the situation to sufficient degree of richness and subtlety.

The initial phase of our empirical study was thus directed towards increasing our understanding of the broad non-linguistic context of a traditional workplace so that it can guide the design of the hybrid workplace of the future. Our initial hypothesis is therefore that we need to understand how organizational and conversational roles of individuals in teamwork shape the interplay between individual and joint action and thus indirectly influence the manner in which the common ground is created and maintained.

A further hypothesis is that the roles and responsibilities individuals hold by virtue of their status in the team (organizational roles) and are able to assume in the course of their work with the team (conversational roles) influence how they define their work tasks and how they organize their workspace. These in turn influence their requirements for workplace resources (including technology) and their use of the resources to interact with their co-workers. From this we attempt to distinguish how work space, work tasks, and technology function together as key elements of the work context in both traditional and hybrid workplace.

However, the clear distinction we make in our analytical framework between organizational and conversational roles is not entirely consistent with how these roles, in practice, influence the key features of the work context; the work space, work tasks and technology. Our empirical results show that there is a systematic ambiguity between 'assigned' organizational roles and 'assumed' conversational roles. For this reason we assume that the key features of the work context are also constrained and/or defined by conflicting influences such as organizational culture, power relationships, timescales and other workplace features. In particular, we focus on the ways these constraints manifest themselves as legitimate requirements for access to people and information on the one hand, and privacy and confidentiality to protect from outside intrusion on the other.

Moreover, it is these constraints that lead to patterns of variation in the work context – which is precisely what our approach needs to explicate if it is to inform the design of a 'hybrid' workplace where computer-mediated setting provides the same degree of proximity, permanence and productivity as an effective physical workplace is known to do. Empirical studies in both CICC and SANE projects show that mobile workers need *permanent spaces of interaction*: spaces that are not purely informational but provide a sense of permanency (or belonging) to the team and the organization. In addition, informants demonstrated that mobile workers need *proximate spaces of interaction*: spaces that can sustain appropriate degrees of closeness between workers and foster interaction between colleagues working at different times and in different locations, whilst maintaining access and privacy that individuals and groups are entitled to by virtue of their roles and responsibilities. The studies in the SANE project also identify that mobile workers require *productive spaces of interaction*: spaces that allow access to shared resources from any number of locations and which provide mobile workers with the flexibility to organise the space and time of their activities.

For example, if an individual, in her 'assigned' role as a consultant has the option to work from home, but needs to draw upon the expertise of others she might 'assume' the role of a researcher and commute to the office in the expectation of meeting up with colleagues in her own team and from other projects. The contextual variations offered by commuting to the office whilst limiting her activities in some respects give her the opportunity to explore others. Therefore, an individual will choose a particular work setting depending on the particular role she has assumed to bring her closer to completing the work task. An effective workplace (and this holds true of a location-independent one as well) thus needs to provide her with work settings that support a range of 'possible' communicative activities and opportunities for effective co-action with her colleagues.

However, when the individual arrives at the office she might be unexpectedly constrained not only by the availability of her colleagues, but her tasks and activities might also be redefined by the distractions of phone calls from clients, or perhaps by her manager who decides to capitalise on her unexpected presence in the office by calling a spontaneous meeting. The individual, although still able to meet up with some of her colleagues, is obliged to reconcile the ambiguity between her organizational and conversational roles through alternating her assigned (consultant) and assumed (researcher) roles respectively. Moreover she is obliged to form her collaborative network by adapting the 'potential' to the 'actual' communicative activities in her work context.

From interviews and focus groups with the informants (knowledge workers who spend substantial amounts of time working away from their office) we concluded that it is the 'actual' communicative activities that determine how individuals and teams create the common ground, in their words – 'how they manage knowledge and information in their own systems and in their own networks'.

Organizations assign roles but the critical factor is the assumed roles, because these are the roles individuals can, when working together, generally interchange in order to adapt to constraints on communicative activity. It is these constraints that produce variations in the work context, but the drivers of these variations are the roles and responsibilities that individuals assume in relation to the communicative activities they undertake. For example, an individual might assume a role that allows them to contribute their skills to the dynamic relationships that are formed in a new collaborative event. Such an event might necessitate a response to not only a prior communicative situation, but also the unexpected, day-to-day, series of related events and circumstances. The individual, therefore, must assume a role that allows him or her to adapt to the constraints imposed in a previous communicative situation - one that obliges them to dynamically reconstitute, and expand those roles and responsibilities in ever more complex variations. Hence roles become the drivers of the variations in the work context that determine the appropriateness of its resources for co-action and communication.

4. Work settings for co-action and communication

As communication is essentially in service of performing common tasks, there is a wide variety of conversational modes that a theoretical framework has to account for. Interaction sometimes takes the form of face-to-face conversations in meetings, but it just as often involves "communication at-arm's-length". This is conducted by means of shared representations of information in, for example, documents e-mailed to interested individuals, in project and bulletin boards which constitute public information spaces, and, more recently, through technology-mediated channels, such as media and virtual spaces, web archives, chat logs and others.

One of the implications of the real-world perspective for the theoretical framework is that we need to take into account the ways people engaged in conversations make use of information obtained from other channels and other resources. These channels and resources provide information that is as important for creating the common ground as the information obtained from the conversation itself (Rogers, op.cit). In technology-mediated conversations ways must be found to make such information accessible to all those who share the mediated workplace. For example, in the virtual workplace it will be necessary to design representations to show not only speakers/listeners and their talk, but also the background information in the setting, and in documents, drawings and objects they use as shared artefacts to focus on and organise their talk.

Another implication concerns different degrees of individual involvement in conversations or meetings. As Clark (op.cit, p. 197) points out, "listeners who participate in a conversational interaction go about understanding very differently from those who are excluded from it". In the design of hybrid work environments we need to account for different degrees of commitment and responsibility that are required of individuals, as well as identifying the appropriateness conditions for changes in their involvement. This is to account for the fact that in an open-plan office, for example, people frequently join conversations as and when they perceive the topic to be relevant to their immediate concerns and drop out of conversations to get on with more urgent tasks. Furthermore, people often left 'traces' of their activities in the office that informed others of their whereabouts. For example, in the management hut at a construction site, engineers frequently left recent documents open on their desks as deliberate traces of their actions to be 'picked up' by colleagues in their absence, or placed their boots and hardhats in visible positions near the desk to indicate that they were away from the office, but not on the construction site.

These observations of the strategies people used to coordinate actions of mobile team members lead us to distinguish three types of conversational roles. First, we defined the participants whose aim is to establish reciprocal relationships that guide the creation of the common ground and who actively control the process. Second, we characterise over-hearers who observe, and indeed overhear, conversations whilst relying on their shared knowledge of the work context in order to interpret the participants' actions and behavior. They thus infer the significance of the participants' words and actions, implicitly sharing in their reciprocal relationships. Third, we characterise the trackers who do not have access to the communicative activity itself, but are able to rely on their mutual knowledge of the work context in order to reconstruct what was said and what was decided.

There seem to be significant differences in the nature of informational resources that are required by "participants" who will share the contributions and responsibilities for taking up each other's projected actions, "over-hearers" who will pick up information about the conversation through observing and interpreting the actions of active participants, and "trackers" who will find out about it through accessing traces of participants' actions in records or other shared artefacts.

Much research has been done in the creation of the common ground in synchronous co-located communication (Clark, op.cit), but the empirical studies of real-life workplace show that the context of communication there presents additional challenges, as illustrated in Figure 1.

First of all, the importance of overhearing in an open-plan office, for example, is confirmed by the informants, who as a rule, need to be aware of a number of conversations taking place around them at the same time. When they share the work space with other members of their team, they also need to be in a position to join conversations and leave as appropriate for the effective performance of their work tasks. Thus far this is only possible in the traditional, co-located workplace where people, informational resources and other shared artefacts are integrated into the work settings, whilst the facilities for computer-mediated communication support mainly the active participants, not overhearers or trackers.

Secondly, in situations where individuals can choose whether to be in the office or not, we need to explore under what circumstances they make those choices, and in particular, how the assumed roles and responsibilities influence the choices. As one of our informants, a project manager and, within his team, an acknowledged expert in information and communication technologies, said: '... I then go to the office so that everyone can have a piece of me....'

The research questions following from these empirical observations relevant to the development of an analytical framework concern the key fea-

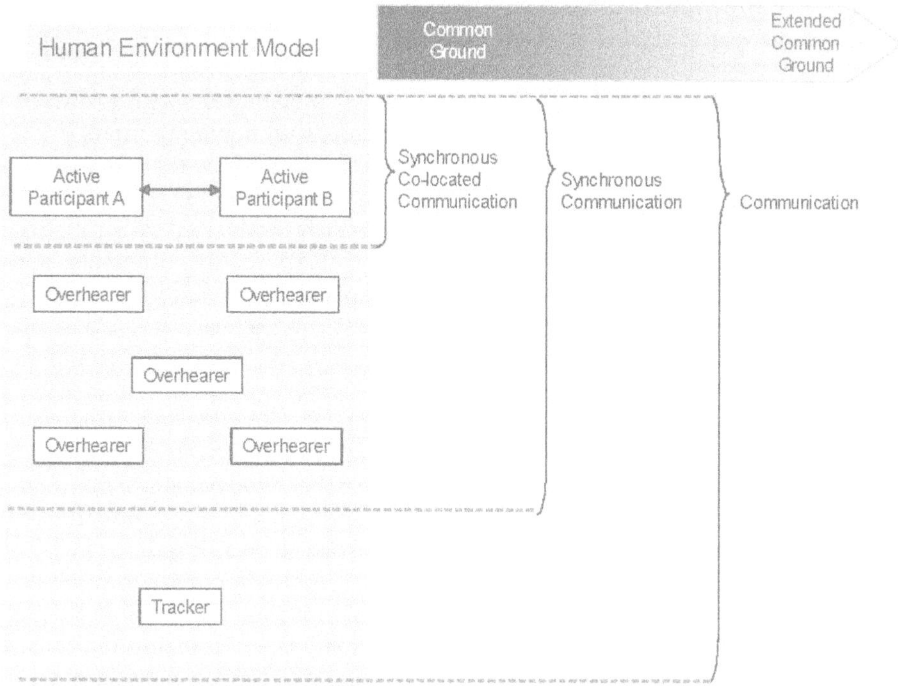

Figure 1

tures of the work context for co-action in the real-life workplace and, from the design point of view, the range of the communication channels that need to be provided in order to facilitate and constrain co-action in such a workplace.

In particular, our current research is focused on providing answers to the questions: 'What are the resources that a physical workplace provides, not only for active participants, but also for overhearers and trackers? 'What determines how these resources are used to support their choices of assumed roles and responsibilities?' Finally, 'how is the resulting shared knowledge retained in the common ground of participants, overhearers and trackers?'

In the case of participants, over-hearers and trackers in non-collocated settings, communication is possible only through the functions of interfaces and gatekeepers that provide informational links and boundaries respectively. These concern technology-based resources available in synchronous and asynchronous communication in co-located and non-co-located settings, as shown in the diagram above. Again, the key research question is how is the resulting shared knowledge retained in the common ground of participants, overhearers and trackers, and, to paraphrase our informant, why can't everyone have a piece of him when he is out of the office, but connected with them all in some form of a mediated setting?

In mediated settings, we need to understand in what ways and to what extent the technology constrains how the information from variety of sources gets integrated into the common ground of participants, over-hearers and trackers, that is, how are various communication channels coordinated? Furthermore, what function do individual channels have in this context and how they relate to the use of shared artefacts, and the functions of interfaces and gatekeepers?

These questions have been addressed in collaboration with architects and urban planners, as well as technology specialists who were partners in the SANE project. As mentioned above, our empirical results show that the distinction between 'assigned' organizational roles and 'assumed' conversational roles is rarely clear cut in practice. Their interdependence is moderated by conflicting influences such as organizational culture, power relationships, timescales and other workplace features. In the post-empirical phase of our research, we focus primarily on the ways these constraints manifest themselves as requirements for access to people and information on the one hand, and privacy and confidentiality on the other.

Key issues were addressed with reference to a spatial model that classifies workspace into three categories. The metaphors for these categories are cloister, club and café. "Cloister" is a private space that does not allow interruption from outsiders and is suitable for handling and exchanging confidential information, by individuals alone or by active participants of a closed team where overhearing is barred and tracking significantly limited to the chosen few. "Club" is a space where access is restricted to "members only" but allows spontaneous as well as formal exchanges among them. Participation, overhearing and tracking are all acceptable modes of co-action provided the members have legitimate assigned and assumed roles. "Café" is an open-access space where anybody can join and contribute[10].

From the point of view of space we can recognise three main categories – private space that individuals withdraw to avoid contact with others (a metaphoric example is that of a cloister in a monastery), privileged space where only certain categories of people are allowed in (such as a club for directors, doctors, naval officers, etc.), and public where everyone can enter (for example, a café). Different types of space require different boundaries. In the physical environment walls and doors provide the strongest boundaries between public spaces on the one hand, and private and privileged on the other, although, in an open plan office, there may be partitions to symbolise a visual boundary, or a person wearing headphones is recognized as having established an audio boundary.

Similarly, groups of people may indicate their boundaries through configurations in space, a circle for example, which does not automatically allow outsiders to join in (c.f. Kendon, 1990). In a public space, there are no boundaries, but such a space may be occupied by groups or individuals who have created their own restricted or private space by virtue of their behavior. We therefore recognise that the three 'pure' categories of space are in reality a series of 'interlocking' spaces, where an individual may be simultaneously in a private space

10 The metaphors were originally developed at DEGW, one of the SANE partners specializing in workplace design, and were subsequently formalized in the context of the uniform framework that takes into account the place, people and process factors in workplace design. For the purpose of developing the Human Environment Framework, we aligned the spatial metaphors with appropriate conversational roles and responsibilities.

inside a public one, or where a group may be in an interlocking privileged/public space.

These distinctions are important for the study of people's behavior in such spaces, and in particular, for the study of communication where the resources available in such spaces influence individual and joint actions, and the interplay between them. From the perspective of co-action and communication, these three main types of space correspond to personal, social and open zones where different norms of appropriate behavior apply. In particular, people in the personal zone are expected to show openness towards each other, intimacy and trust, where misconduct may be perceived as rejection and result in personal offence. In the social zone, the rules established by a particular community determine who is in and who is out, as well as how appropriate behavior should be regulated. In the open zone, there are no membership restrictions, but sanctions against misbehavior may result in police intervention, fines or any other form of legal or political control. These norms of appropriate behavior regulate the social dimension of private, privileged and public spaces and influence the way people use these spaces.

Thus the initial framework for analyzing interactions in the three kinds of space used the metaphors of cloister, club and café as the starting point for describing key contextual features of the workplace. Special attention was given to the requirements for privacy of people, confidentiality of information, as well as presence and sharing in the work context.

Subsequent analysis of communicative activities in these spaces is aimed at capturing the dimensions of the work context that characterise the 'interaction spaces' people create in the process of establishing the common ground. The methodology relies on observable actions, changes in shared artefacts and the communicative effect of these "implicit communications" on participants, over-hearers and trackers[11].

5. Collaborative work environments: Theoretical and methodological issues

As discussed above, we developed the CWE approach to cover a broad range of communication phenomena that have been observed as part of the empirical workplace studies. Initial empirical observations provided the basis for articulation of hypotheses and research questions that were iteratively refined through investigations focused on co-action in a 'hybrid' workplace. Such an eclectic approach to the empirical study was motivated by the requirement that the empirical findings inform the development of an analytical framework where the organizational, spatial and social perspectives on co-action are accounted for at the same level of analysis. This framework can then guide the design of the workplace of the future since it has preserved the full complexity of the human environment in such a workplace.

11 An earlier version of this work has been published in the proceedings of the JSAI-Synsophy workshop on Social Intelligence Design, Japan, 2001, D Rosenberg: "Communicative Aspects of Social Intelligence".

5.1 Boundaries in CWE

Our working assumption that people create an interaction space which establishes boundaries in terms of space and time around particular communicative events lead us to focus on the nature of these boundaries and their representations in the hybrid workplace. Within those boundaries people can accomplish mutuality of perspectives regarding the other participants, their goals and purposes, the resources available to achieve their work and communicative aims. In other words, they share the same environment in which they can see the same people and objects, recognizing their availability in the interaction space. They also have pre-conditions for establishing reciprocity of perspectives, that is, they have expectations of individual contributions to the joint activity and achieve a degree of awareness and responsiveness that help them recognise appropriate and meaningful action in the social 'space' that they jointly create through interaction.

To capture the complexity of these relationships, and the need to cover a broader range of communication phenomena than those covered by other studies of human communication (c.f. Clark op.cit. Rogers op.cit., Hindmarsh op.cit. for example), we have focused our work on different degrees of distance between people in an interaction space. In our context the distance is not only spatial, but also organizational and social. The aim was to take account of those key themes and issues raised by informants in the empirical studies that related to location independent computing and ubiquitous networking on the one hand, and organizational identity and the creation of communities of practice, on the other.

Guided by these considerations, our analysis is focused on co-action and the common ground whilst taking into account the 'place' perspective, that is, shared resources, shared spaces and shared objects in a particular work space. The 'process' perspective is also needed to capture the nature of the work tasks, goals and objectives as it shapes the use of space and the resources it provides. The 'people' perspective emphasizes the degree of participation and involvement in joint activities and tasks. It is grounded in the place and process aspects which moderate the form that co-action and communication take in a particular situation. These perspectives when integrated in the Human Environment Framework and Model capture the variations in the work context that constrain and/or facilitate the range of assumed organizational and conversational roles and thus shape day-to-day activities in the real-life workplace.

In mediated settings, these perspectives provide the guidance for describing the characteristics of the workspace in terms of its ability to provide effective gatekeepers that control access to people and are constrained by privacy requirements, access to information constrained by confidentiality requirements and interfaces that facilitate varying degrees of participation and sharing which taken altogether distinguish private from privileged from public workspaces.

We thus approach the analysis of empirical data with the hypothesis that the cloister, club and café types of workplace can be described in terms of the above contextual features. The possible configurations of these contextual features that define the three basic types of space also help to define the hybrid spaces that are most common in real life, the 'interlocking' spaces, such as, club in café that can be seen when a small group of people holds an informal meeting in an open-plan office, or cloister in café when an individual puts headphones on in order to distance herself from the surrounding activities.

In this way we expect to improve our understanding of the regularities and uniformities in joint activities in the workplace. Furthermore, based on this understanding we expect to inform workplace design by identifying the key features of human environment that must be preserved in mediated and hybrid cloisters, clubs and cafés. Working within this framework we aim to inform the design of the mobile workplace, when the symbolic nature of people's behavior is dependent on the information displayed on the computer screen, so that the shared representations of privacy, access, presence and sharing may be incorporated into an effective hybrid workplace.

5.2 Distances in physical and social space

As discussed above, our key analytical stance is that communication and collaboration is best viewed as the interaction between individual and joint actions in the context of shared work tasks and collaborative activities and that action, interaction and context can best be explored through a uniform concept of a shared interaction space.

The main variations in the shared interaction space are illustrated in Figure 2 that shows how, in general terms, the three types of work space, namely, private, privileged and public differ from one another in terms of the opportunities for interaction and in terms of mutual visibility of the individuals

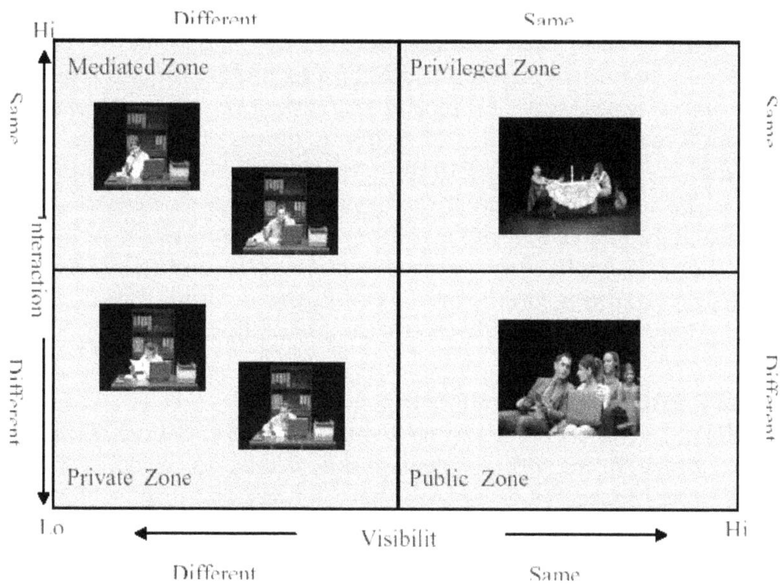

Figure 2 *Interaction zone matrix for hybrid workplaces*

occupying the work space. It also shows how a well-designed mediated space enhances these opportunities by making new configurations possible.

Furthermore, the interaction zone matrix can be adapted to capture varying degrees of privacy in the work context, so that we can examine the implications of variations in the work context. A simplified example can illustrate how spatial, organizational and social distances may guide appropriate behavior – if members of the same team (social aspect) share the same work space (spatial aspect) and work on the same task (organizational aspect), then the degree of access to people and information is high and the requirement for privacy and confidentiality is low. Thus the presence and the sharing required for effective co-action and communication are also high. At the other extreme, if members of different teams share the same work space but work on different tasks, then they need to create boundaries within the shared work space that will help them to accomplish privacy and confidentiality, reducing the feeling of presence and sharing to the minimum.

Between those extremes, there are variations that seem particularly interesting for the design of a hybrid environment. For example, when members of the same team work on the same task but do not share the same space, which is frequently the context of computer-mediated communication, the requirements for access, presence and sharing are high and the demand is for the mediated setting to provide rich and flexible communication channels that resemble, as closely as technology advances allow, the traditional, co-located workplace.

In contrast, if members of different teams share the same work space and work on the same task, the degree of access to people and information is restricted to the part of the work context that does not conflict with team loyalties. In such a case, mediated settings have the advantage, since boundary controls are easier to establish and maintain when some channels of communication may be switched on and off 'on demand' as it were. It is in this sense that we can explore the advantages of mediated settings over the co-located ones.

Thus the initial empirical study of user and workplace requirements leads through a series of iterations and refinement to the formulation of hypotheses that may be validated empirically by means of informants' feedback, or experimentally, when the variations in the context of mediated work settings have to be specified in advance of the design and implementation of the 'hybrid', technology-enabled workplace. The practical requirements of workplace and technology design have encouraged us to cross the methodological boundaries of distinct traditions in the studies of human communication, and this is the ongoing concern discussed in the section that follows.

6. Theory and practice working together

For the purposes of this study, we focus on issues of boundary control that is necessary for individuals to adapt their actions to the collective norms. In terms of space, we distinguish between private, privileged and public spaces for co-action and communication. Such spaces regulate access to people, bearing in mind their need for varying degrees of privacy, and access to information bearing in mind different degrees of confidentiality, where access to particular items is often restricted to a certain category of employee. Access to people and information regulated by constraints imposed by requirements for privacy and confidentiality thus characterizes the nature and the extent of connectivity in the workplace.

However, in order to create and maintain conditions for individual and joint co-action, it is not enough to design connectivity into the workplace. Such, profoundly social dependencies, demand that the workplace provides appropriate conditions for interactivity, that is, to design spaces that allow proximity of its occupants, that offer them a sense of permanence and that support their productivity. It is only through proximity, permanence and productivity of the workplace that people are empowered to create trust, friendships and partnerships based on trust. In addition, as has been recognized in the multi-disciplinary literature concerned with work culture (cf. Foley, 2004), organizational identity of a company's workforce is a pre-condition for them having the sense of belonging to their organization, community and culture.

Thus the key design issues for the workplace of the future can be summarized as connectivity and interactivity. Connectivity refers to access to people and information, moderated by facilities for boundary control that regulate different degrees of privacy and confidentiality. Interactivity refers to the affordances in the workplace that enable people to regulate the nature and the extent of sharing of its resources, so that they can create trust and partnership through natural forms of interaction.

The key research issues in this context focus on connectivity and interactivity of a work space as the means for establishing communication channels that are flexible and adaptable to the requirements of particular, situated, interactions. Sufficient visibility between participants is required and also sufficient awareness on the part of the participants of their environment, physical or mediated. Essentially, research issues concern the relationship between individual and joint action that underpins the way people work together, that is, how co-action is accomplished, through interaction, to solve shared problems, how relationships based on trust and personal contact are created and how people make use of the affordances in the workplace to make the integration of individual and joint actions more effective.

To address these research issues we developed a repertoire of methods and techniques that allowed us, in the first instance, to discover key features of physical, traditional, workplace that supports effective communication and collaboration. Since such methods have been developed to capture the full complexity of observable patterns of interaction and behavior, they were by necessity limited to the investigation of the existing workplace in the organizations of our partners. However, we need to extend our 'toolkit' to be able to project our understanding of the workplace of today to shed light on the connectivity and interactivity in the workplace of the future where information and communication technologies are an integral part.

Empirical studies of introduction of technologies into the real-life workplace bring to light the profound changes in the structure of our societies in a very short time. Indeed, there was a warning against the unqualified acceptance of rapid technological change before a knowledge base had developed that captures experiences of individuals and fosters information and knowledge sharing between and within the community. Our interpretative analysis of office-based employees in a number of organisations suggesting that inappropriate implementation of technology disorientates, invades, isolates and ultimately separates individuals from the conditions of their work. At the same time, the analysis also suggests that people at work are establishing a prevailing knowledge base that provides a range of new opportunities for not only co-action and communication, but also increasing mobility within and without the workplace, ultimately raising employment prospects, and nurturing a greater sense of self-sufficiency, accountability and responsibility. Novel forms of interaction space, that is, physical space enhanced with communications technologies, make it possible for the boundaries that determine organisational and social co-presence to become invisible.

Workplace of the future will incorporate technologies as an integrated component that ensures connectivity and interactivity of the mobile workplace. It should be emphasized at this point that usability studies of relevant technologies suggest that the technology should not be regarded as replacing essential features of the physical workplace, but enhancing the workplace by providing additional facilities for mediated collaboration and communication (c.f. Rosenberg, *et al.*, 2004). This means that we investigate natural interaction in the workplace with the view to augmenting the facilities and ultimately enhancing the experience of collaborative work.

Taking into account the interplay between the physical and the social space where people communicate and collaborate, the Interaction Space Approach aims to explain how communication serves the purpose of making work performance more effective and how conversations are organised to facilitate effective teamwork.

In our approach, we have followed an empirical method for the collection and analysis of the data which involves an examination of the interview text and its context, in other words - what the informant says explicitly, and what he or she means. It also considers discourse and values, in other words how the informant engages with others and why they do what we observe them to be doing. Moreover, in this kind of approach it is important to interpret their words in relation to the researcher's observations.

The main advantage of using an empirical study of roles and responsibilities in real settings is that it offers a range of methods and techniques to describe what people say and do in particular situations. These methods are especially suitable for bringing to light the implicit constraints that underpin their talk and behaviour. The constraints can be related to their knowledge of the language and the way it is used, the subject matter of a particular conversation, the social norms and patterns of behaviour. In sum, these constraints are about "what members of a community need to know in order to behave appropriately in culturally significant settings" (Saville-Troike, 1972). Within the research projects described above, we use these methods to bring to light the implicit constraints and preferences that underpin and drive the work context. In addition, our aim has been to illustrate how these constraints lead to variations in the work context, which ultimately influence how individuals create and maintain conditions for co-action and communication in a workplace that enables permanence, proximity and productivity of shared work.

Last, but not least, we conclude that the requirements of the hybrid work environments that will characterise the workplace of the future for technology-based resources are not to mimic the contextual features of the traditional co-located workplace, but to enhance it. In other words, mediated work settings are, and should be different, as the main benefit they offer is to increase the range of possibilities for effective co-action and communication. To accomplish this, we may need to establish closer collaboration with 'practitioners' concerned with the design of the workplace and the technology that will enhance it, and we hope that this work is a step in the right direction.

References

Clark H., (1996). Using Language, Cambridge University Press.

Clark H., and Brennan S., (1991). 'Grounding in communication', 127-149 in Resnick L.B., Levine J.M., and Teasley S.D., (eds) Perspectives on socially shared cognition, American Psychological Association.

D14, CICC project report: http://www.rhul.ac.uk/~uhtm059

Dourish P., and Bellotti V., 1992, 'Awareness and coordination in shared workspaces', *Proceedings of CSCW'92,* ACM, New York

Fruchter, R. (1998) Roles of computing in P5BL: problem-, project-, product-, process-, and people based learning. *Artificial Intelligence for Engineering Design and Manufacturing, 12,* p.65-67

Heath, C. & Luff, P. (1991) Collaborative activity and technological design: task coordination in London Underground control rooms. In *Proceedings of the 2nd European Conference on Computer Supported Cooperative Work* p. 65-80. Eds. Bannon, Robinson & Schmidt., Amsterdam, The Netherlands. September 25-27.

Hindmarsh J., Fraser M., Heath C., Benford S. & Greenhalgh C., 1998, 'Fragmented Interaction: Establishing mutual orientation in virtual environments', *in ACM98 Proceedings of CSCW98*, Seattle, November 14-18, pp 217-226

Hughes, J.A., Randall, D. & Shapiro, D. (1992) Faltering from ethnography to design. *CSCW 92: sharing perspectives. Proceedings of the Conference on Computer Supported Cooperative Work*, October 31 to November 4, Toronto, Canada. Turner & Kraut (Eds.). N.Y.: ACM Press. p. 115-122.

Kendon, A. Conducting Interaction: Patterns of Behavior in Focused Encounters . Cambridge: Cambridge University Press, 1990. A volume in the series Studies in Interactional Sociolinguistics, John Gumperz, series editor

Perry M, Fruchter R, Rosenberg D (1999): "Co-ordinating Distributed Knowledge: a Study into the Use of an Organisational Memory", in Cognition, Technology and Work , An International Journal for the Analysis, Design and Use of Joint Cognitive Systems, Springer-Verlag 1:142-152.

Robinson, M., (1993). 'Design for unanticipated use' in Proceedings of the third European Conference on Computer-Supported Cooperative Work. Eds, De Michaelis, Simone & Schmidt. September 13-17, Milan, Italy, p. 187-202. Kluwer: Netherlands

Rogers, Y. (1993) Coordinating computer-mediated work. *Computer Supported Cooperative Work, 1,* 295-315

Rosenberg, D (1998): 'Communication in Real and Virtual Environments', unpublished report of the British Telecom University Research Fellowship (available from the

author).
Roseerg, D (2001): "Communicative Aspects of Social Intelligence", JSAI-Synsophy workshop on Social Intelligence Design, Japan, June 2001
Rosenberg, D., Foley, S. & Kammas, S.(2005): "Language Technologies for the Future", in AI&Society special issue on Social Intelligence Design for Mediated Communication
Rosenberg D, Kammas S & Foley, S (2003): "Interaction Space Theory: A Linguistic Approach To Computer-Mediated Communication"in "Linguistics and the New Professions: from Linguistic Research to Linguistic Enterprises", Materiali Linguistici Series, Franco Angeli- Milan , A.Giacalone & E. Rigotti (eds.)
Saville – Trojke (1972), Ethnology of Communication. Blackwell, London
van Maanen, J. (1979) The fact of fiction in organisational ethnography. *Administrative Science Quarterly, 24,* 539-550

Acknowledgements

Special recognition is due to my colleagues who have worked with me on the research projects which provided the motivation for the theoretical aspects of the Human Environment Framework, namely Jean Cornillon, Simon Foley, Anzela Huq, Stavros Kammas, Mirja Lievonen and Mark Perry.

Systems Thinking for Sound Management in Hostile and Other Complex Adaptive Environments

Graeme Couch, Denis Cooke, Stuart Cumming, Richard Doyle, Barry O'Donnell, Alec O'Halloran & Thane Scott
Couch & Associates Pty Ltd, AUS

Operating "human endeavours" successfully in today's turbulent environments presents a special challenge for managers. While, the essential principles of management are simple, the difficulties encountered frequently lead to managers neglecting these principles, exposing them to unnecessary risk in their drive for success. This paper proposes a fresh, practical view of how strategy, governance, projects and operations, resources and decision-making need to be integrated (whether in commercial, government, non-profit or mixed endeavours) to maximize the prospects of success in today's complex environments. Complexity and turbulence are here to stay, and dictate the need to have in place responsive management arrangements. Any endeavour's chances for surviving and prospering in hostile circumstances, are determined by its own ability to adapt in an intentional and cohesive, but agile manner. The paper presents views derived from practical, wide-ranging experience on:

- **The essential function management and the simplicity of sound management (while accepting its profound difficulty);**
- **Decision-making structures, processes and deployment necessary for sound governance;**
- **Complex adaptive systems behavior and its implications for decision-making; and**
- **Learning and intelligence processes as key supports for decision-making.**

Introduction

Increasingly today, discourse among leaders engaged in the governance of "human endeavours" reflects concern over the turbulence or "complexity" of the environment in which they are operating.

The authors are associated in a "loose-tight", "virtual business", "Couch & Associates Pty Ltd" (C&APL). The associates are independent freelance managers and consultants, whose work has been largely shaped by the turbulent circumstances faced by their clients' organizations and the need to manage them. The associates' backgrounds cover a wide range of disciplines (e.g., marketing, engineering, financial, IT & HR development, supply chain) and experience (across Government, Commercial and Non-Profit sectors, and including manufacturing, utilities, regulators and service providers in Australia and in Asia), working jointly on assignments as circumstances require, and independently otherwise.

This paper is part of an exercise commenced to draw general but important conclusions from this diverse work base and from observations in management. The experience base is particular in that it reflects observations of a small group of professionals. However, it is somewhat wider than what is experienced typically by individual managers, even in larger endeavours, where the experience base is not used as an effective, collective learning tool. Our aims are to share, corroborate, where appropriate modify, and apply the conclusions more openly and systematically. In this paper, selected conclusions are presented as a series of propositions, to invite comment, rather than as a document presenting final research findings. Further analysis is to be developed and made available on www.capl.biz .

A result of this exercise has been the development and exploration of an alternative perspective on what is required fundamentally for managing survival and success in the more turbulent environments facing human endeavours. The alternative perspective has been learned and developed (often informally as much learning happens in practice), and applied (often implicitly) where appropriate in the course of operating and resourcing the business over the past 15 years.

The paper outlines early views emerging from this exercise, supplementing the outline with brief notes describing fundamental features of:

- Human endeavours from a complex adaptive systems (CAS) viewpoint;
- Complexity as encountered in human endeavours, and;
- Essentials for managing under complexity.

These descriptions are presented at an "elemental" or irreducible level that is common across the different contexts that the authors have encountered both in the C&APL business and in their independent businesses. The encounters canvass commercial, government, non-profit, informal and institutional contexts, covering both distinct entities, and divisions, teams and individuals within entities. While description at an elemental level can appear at first to be overly theoretical, understanding is important for underpinning sound management under complexity in all of the contexts.

Background

It is not the purpose of this paper to critique or debate the considerable literature about systems thinking (or "soft" systems practice) in strategic management. The authors of this paper have found their views - e.g. in concepts of meta-resources (potential and actual), decision-making (process and structure), decision support (intelligence and learning) – to be complementary to works by notable authors encountered in these fields. Kim Warren, 2002 in "Competitive Strategy Dynamics", addresses firm performance in terms of resources and capabilities, and writes compellingly of the "hard face of soft factors"; intangible resources "have a powerful impact on business performance over time, by affecting strongly the rates at which tangible resources are won or lost". While "hard" systems engineering has made major contributions to humankind's capacity to implement major physical projects, its scope has had to expand to accommodate increased complexity, particularly of human requirements (Checkland, 1984). Reflecting these human requirements, recurring themes in modern strategic management concern political feasibility, participation, stakeholder management, alternative futures, organizational change and analytical support (Eden & Ackerman, 1998).

In embarking on this paper the reader is referred to two key concepts (Couch, 2006), concerning "the nature of human endeavour", and "complexity". The term, "human endeavour", used to cover any type of enterprise established and operated for a particular purpose, is defined by its objectives, scope of activity, and its meta-resources (everything that is available, actually or potentially, to be deployed wholly or in part, for the purposes of the endeavour, including, intangible assets, people and value-generating relationships, both internally and externally). A complex adaptive systems view of the endeavour highlights (a) its open nature, having no inherent or fixed boundaries, and (b) its agent-based nature with "agents" having uncertain or changing roles and relationships, and having degrees of autonomy in the manner and effectiveness of their deployment and actions in (or contrary to) the interests of the endeavour.

"Complexity" is an often used but fuzzy and frequently misused term. It is helpful to think of complexity manifesting itself at three levels or orders, (a) size or scale, (b) inter-relatedness, and (c) adaptation. In the CAS sense it is not **size** or **scale** (e.g. of a project), including variability, per-se that generate complexity. Rather, it is the extent or density of **"inter-relatedness"** of issues and interests; where the resolution of one issue, which may affect any single interest group, depends on and affects the resolution of other issues and impacts on other interest groups, all demanding satisfaction. Finally, at the third level, **adaptation** describes the willingness of interest groups (or "agents" having particular interests) to combine and to "escalate" the issues through opportunistically seeking to further their private interests, often at the endeavour's expense.

Overview of results

Key to the alternative perspective foreshadowed earlier is a new understanding of "management", and particularly its focus on: (a) the essential tasks inherent in the management function; and (b) the way that management functions are structured and deployed throughout an endeavour.

To summarize:

1. Many surveys highlight new and severe challenges facing managers trying to organise their endeavours and cope with information overload in the turbulent environments where they operate. (Future Monitor Survey, 2006 is one example) Such challenges include competition and regulatory inter-

vention, utility reforms, globalization and work across cultures, perverse behavior and corrupt practices, terrorism risk and security, employment and intellectual property issues, technology development and the increasing rate of transactions that new technology enables. While they are all different, and examples only, it is when such matters converge disruptively, that conventional approaches to management fail.

2. In a systems theory sense, these turbulent environments are behaving as Complex Adaptive Systems (CASs). For an endeavour, to survive and prosper within a complex adaptive environment, it must itself develop attributes of a CAS (as in Ashby's Law - Stafford Beer, 1995). Endeavours that remain captured by rigidities (e.g. in organization, information handling, conducting communications, and in hiring, training and developing staff), and facing complexity (critical issues arising with little notice, through activist, opportunistic, hostile behavior) are risking their survival and prosperity by not establishing the new levels of agility needed to adapt. This agility:
    - Needs to be initiated from the governing board;
    - Needs to extend to all who are engaged in management and operations;
    - Requires that the endeavour's people have access to, and use, material information, and are learning continuously while making decisions and performing their work;
    - Should emphasise simplicity and integration, rather than impose additional hurdles of management complication; and
    - Needs to foster openness and innovation.

3. For this purpose, endeavours need to foster a new, dynamic understanding of management. The authors are proposing an understanding in which traditional static views of management tasks (whether focused on management activities, such as planning, organizing, staffing, implementation, and control, or focused on distinct functional layers, such as strategic, tactical and operational layers) are extended by a dynamic view of the management function. In this dynamic view, the elemental building block of management is decision-making.
    - Decision-making is conducted with the effect of deploying resources (or meta-resources) available to the endeavour in order to pursue the strategic directions and aims that will make for its success.
    - When distributed throughout the endeavour, streamlined but rigorous, and conducted in a cohesive, iterative, recursive, multi-level, framework, decision-making is able to transform a conventional organization into a CAS.
    - Decision-making needs to be efficiently distributed at all levels where management decisions are made, structured transparently, aligned through clear accountability relationships to the strategic directions and aims of the endeavour, and enhanced by intelligence

to which decision-makers (whether recommenders or approvers have ready access), and a commitment to continual learning.

4. This view represents a progression in management thinking, (a) moving beyond Taylorism, with its focus on outcomes and objectives, and Deming Quality, with its focus on technical process and systems (Dalrymple, 2000), (b) extending the idea of "knowledge workers" as anticipated by Drucker, 1993, and (c) resolving conflicts between the traditionally conflicting "hard" or systems schools and the "soft" or human relationships school of management thought (O'Connor, 1999).

This perspective has application to defence, where CAS thinking is being actively researched, as well as in business, the public sector, and non-profit enterprise. Increasingly in the authors' view, explicit CAS thinking will be widely and explicitly used for strategy development and implementation in all sectors.

Decision-making: The essential role of management

While making sound decisions can be difficult, the essential principles are simple; and all value-generating employees can and should know them (or should learn and continuously improve their ability to apply them). Everyone, i.e., everyone who has custody of resources, and needs to make judgements about exercising that responsibility is in some sense a manager. While different workers operate at different functional levels, this covers everyone in the endeavour whose work cannot be automated. At the more operational levels, larger numbers of workers may be making many more, incrementally smaller decisions, compared with senior managers working at high functional levels making fewer larger decisions. At all levels, however, value is still being captured (or destroyed). When anyone in an endeavour is not making sound decisions that are in the best interests of the endeavour, then value (or the prospect for success in the endeavour) is diminished.

Sound decision-making requires:

- Sound **decision-making process** (that include gathering material information [from intelligence, consultation & research], identifying objectives and issues, generating and evaluating alternatives, planning for implementation, recommending for approval, implementing, and reviewing and adjusting – see Table 1 for key stages to be included in a sound decision-making process);
- Sound **decision-making structures** (that provide for (a) transparency and independence as far as practicable between essential stages, and (b) assignment of single point accountability for implementation, including risk and value management, organization, staffing and leadership, operation, milestones and performance, controls and reporting, securing of resources and meeting of costs); and

- Sound **decision-making content** (that covers (a) selection of driving strategies for the endeavour [or part of the endeavour for which the manager has responsibility], and (b) deployment (including delegation) throughout the endeavour of decision-making functions, over and above the other usual matters to which managers give attention.

All supported by:

- **Intelligence** (covering what is happening [environment scanning] or anticipated, and how things work and connect [modeling] in the endeavour and its environment [internally and externally], and what determines success (or failure) [fitness]; and
- **Action learning** (providing, where and when required, the essential skills and understanding of principles for decision-making, sharing of experience [of both success and failure], continuous improvement).

Space limitations prevent drilling down in this paper to the next level of detail on these principles. Table 2 lists key features that emerge from their application. It is unfortunately common to observe: claims by decision-makers that they have no alternatives; unnecessary or partial decisions that will fail to generate end-benefits; inefficient report and submission writing procedures that simply waste resources rearranging others' work; decisions that are not assessed for their feasibility, their alignment with the rest of the overall endeavour, or their impact on other interests; and no provision for implementation, control, costs, or progressive review and adjustment. The key to formalizing decision-making lies in learning, understanding and tailoring to the endeavour rather than in fostering a "tick-the-box" approach.

As a matter of practice, decisions made in the course of managing an endeavour are aggregated into sets covering all aspects of its operation, as illustrated in Figure 1. For example key operational decisions in a business will be aggregated to address such functions as, production, quality, maintenance and operations, service, sales, procurement, inventories, capital, HR, IT, logistics, projects, and their management.

The essential elements of the decision-making process, properly learned, understood and implemented, promote integration (as outlined below) between the endeavour's strategy and governance at the highest functional level, and the management of its projects and operations.

Integrating strategy, governance and operations

The key challenges for governance of any human endeavour are:

- **Strategy** - Developing, adopting, articulating and communicating the strategic directions and aims by which the endeavour is to conduct its projects, operations and management; and
- **Implementation** - Ensuring that the strategic directions and aims are implemented with integrity.

Stage	Comment
Intelligence	Characterizing the current situation and projections in the endeavour and its environment. This require information or intelligence (as far as is known and is material) concerning the endeavour and its environment.
Consultation & research	Consultation with (or equivalent research on) stakeholders impacted by the decision, or if not practicable, those who represent them or understand their perspective
Objectives and issues	• Describing gaps to be closed (social, economic, business and interest group gaps) covering waste to be reduced, opportunities to be seized, and stakeholders to be satisfied; • Articulating broad aims and desired directions, and success / failure criteria; • Identifying issues arising and criteria for their resolution.
Alternatives	Generating then evaluating, against the objectives, alternatives for producing priority benefits, and selecting alternatives to be recommended based on the evaluation
Plan	Preparing the plan of action for implementing the recommendations, and the management arrangements
Recommendation	Framing "implementable" recommendations for approval (or amendment)
Approval	Giving effect to the recommendation (as recommended or amended)
Implementation	Implementing the plan as approved or amended.
Review	Monitoring, reviewing and as necessary adjusting the decision as commitments are made and results emerge

Table 1 *Key stages in the decision-making process*

Sound governance is determined by the way that decision-making is conducted, structured and distributed, since it is the underlying mechanism that will determine an endeavour's likelihood of success.

To combine agility and governance, decision-making needs to be:

- Efficiently deployed throughout the endeavour to its projects and operations;
- Rigorous, streamlined and accessible to all managers; and
- Aligned consistently to the overall strategic directions and aims of the endeavour.

Decision making with these characteristics must be part of an integrated chain of accountability that traverses the endeavour's functional levels. This is secured through:

Feature	Comment
Intentional choice between alternatives	Without alternatives a decision is not needed - the decision compares and selects the best
Necessity / sufficiency for creating value	The decision should be what is required and sufficient to give effect to a course of action for generating and capturing benefits (or establishing options)
Streamlined process	Avoidance of long reports and re-writing that does not add value
Intelligence support	Access to and use of relevant information for environment scanning, business modeling and aligning of interests
Learning support	Sharing, promotion and improvement of decision making skills and judgement
Feasibility	Assessment that implementation will work and benefits will be delivered, allowing for risks in functionality, quality, timeliness, and demand
Strategic alignment	Consistency with the endeavour's strategic directions
Governance	How it will be known that the decision was sound and its implementation is working
Implementation	Single point accountability for implementation (including management of value and risks) and control
Consultation / research	Account taken of complex interests, and reactions anticipated through consultation and research
Costing	Provision made for costs and deployment of resources required for implementation (allowing for risks), including timing or delay costs
Review	Planning for monitoring and adjustment at critical points

Table 2 *Essential features of decision-making*

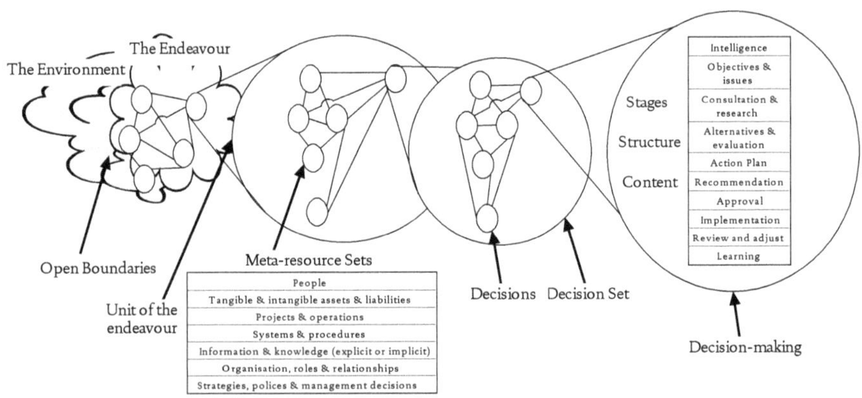

Figure 1 *Decision-making context: The Endeavour in its complex adaptive environment*

- The provision of intelligence required to inform decision-making for a business unit is "two-way", but includes information from the higher functional levels within the endeavour, i.e. (a) the strategies for directing the endeavour, (b) what makes for success of the endeavour, and (d) how things work and connect.
- Explanatory business models (information on "how things work and connect" within the endeavour and its environment) address accountability relationships between functional levels, as well as service relationships between units in and beyond the endeavour.
- Individual links in the chain of accountability relationships are formed explicitly, e.g. by transparently distinguishing recommendations and approvals for decisions, and by explicit monitoring and controls.
- Action planning (with single point accountability for implementation tasks), for a decision, together with monitoring and reporting ingredients in the decision-making process provide the basis for performance measurement.

A decision that concerns the endeavour's operations will take account of the agents impacting materially on those operations (and those agents' principal relationships, agendas and interests), such as:

- Suppliers and users of the endeavour's finished or intermediate products and services;
- Customers who are paying for them;
- Personnel of the endeavour;
- Sponsors (supervisors, managers, owners, boards, directors and controllers, and investors) in the accountability chain who are bearing the risks of performance;
- Regulators who can impose obligations on the endeavour and enforce compliance;
- Interest groups that can emerge (in the CAS sense) or organise to escalate issues and use their influence to do so - customer interest groups, shareholder activists, unions, environmental and social responsibility interests;
- Others who can impact, directly or indirectly on the success of the endeavour, including competitors of the endeavour and key stakeholders;

Conclusions

Application of the principles, described briefly in this paper as part of the C&APL project, has been found in practice to be effective for supporting key decisions and their contribution to strategy, governance and operational management in a range of endeavours. They do not make decision-making easy, but understanding and applying the principles assists directors of endeavours to manage decision-making and governance, and to have confidence in the decisions of their less-senior managers. By removing the commonly ad-hoc nature of decision making the process can be streamlined and made rigorous. Articulating, promoting and learning the simple principles, allows decision-making to be more efficiently distributed through the endeavour, giving it an agility not possessed in rigidly directed organizations.

The approach outlined in this paper views management not as a set of procedures, but more fundamentally in terms of distributed, integrated and aligned, accountable decision-making. The resort to fundamentals is oriented to the needs of endeavours to develop their agility for succeeding in emerging complex adaptive environments. Attention to basic principles will provide mechanisms that are based on: (a) value in service relationships (e.g., customers and service users); (b) Competitive performance and cost management that provides confidence to managers and investors / sponsors; and (c) Positive reputation and preference for the endeavour in its dealings with all stakeholders.

A key requirement is that management must be simple, accessible, and useable by all who have custody of resources in the endeavour, but also rigorous in achieving practical and sound results in which investors and other stakeholders can have confidence. Decision-making that neglects these simple principles, exposes the endeavour to unnecessary strategic and performance risks.

There is a sense in which the approach outlined in this paper may be described as a "return to basics". While that is a result, it was not the original intent in forming the proposals. Historically, it is a common initial reaction to new theoretical frameworks that they merely restate principles that are already well understood, and the authors recognize the hazards of trying to funnel observations from a wide experience base into a few general propositions. This paper marks an early stage in this exploration and research project. The principles outlined are being articulated in light of the authors' interests, experience and study. Interested readers can contact the authors by email gcouch@capl.com.au or through www.capl.biz to be kept informed, and more importantly to interact.

References

Checkland, P. B. (1984). *Systems Thinking, Systems Practice,* Chichester, England. John Wiley & Sons, ISBN 0471279110, pp. 125ff.

Couch, G. H. &ors (2006). "Integrated Decision-Making for Leading Good Governance and Efficient Operations in Turbulent Environments", Institution of Engineers, Australia Engineering Leadership Conference 2006, paper accessible at http://www.engineeringleadershipconf.org.au/userfiles/File/Graeme%20CouchPaper.pdf.

Dalrymple, J. F. (2000). "From F Winslow Taylor to W Edwards Deming: Over a Century of Progress?" RMIT University, Centre for Management Quality Research, unpublished, accessible at http://www.cmqr.rmit.edu.au/publications/jdtaylordeming.pdf.

Drucker, P. F. (reprint ed. 1993). *Management: Tasks, Responsibilities, Practices*, Collins, ISBN 0887306152, p.76.

Eden, C. & Ackerman, F (1998). *Making Strategy*, London, England. Sage Publications, ISBN 076195224, p. 1, and generally.

Future Monitor Survey, February 2006, a venture of the Financial Times and Future Monitor, accessible at http://www.futuremonitor.com/.

March, J. G. (reprint ed.1989). *Decisions and Organizations*, Blackwell, ISBN

0631168567.
O'Connor, E. S. (1999). "The Politics of Management Thought: A Case Study of the Harvard Business School and the Human Relations School", *Academy of Management Review*, ISSN 0363-7425, 24(Jan.): 117-131.
Stafford Beer (reprint ed. 1995). *Decision and Control*, New York, NY: John Wiley & Sons, ISBN 0471948381, p. 279ff.
Warren, K. (2002). *Competitive Strategy Dynamics*, John Wiley & Sons, ISBN 0471899496.

From Industry Practitioner to Educational Professional: Evaluating the Learning Institute as a Learning Organization

James Cannan
Unitec New Zealand, NZ

The purpose for this research was to evaluate the effectiveness of the Applied Technology Institute as a learning organization particularly in the area of formal/informal training of new staff, and to identify the effectiveness of the Institution in the context as a learning organization. New Staff to the teaching profession at the Institute predominately originate from industry where they have usually been practitioners involved in engineering or other technical backgrounds. To be successful as lecturers, formal or informal training in teaching and learning is necessary, and this research was aimed at the Institute to establish how well it met the challenge as a learning organization when its normal function is to offer education and learning for students. From the survey and discussion groups the consensus was that the Institute did strive to be a learning organization and it was clear that staff were willing to learn and take up the challenges and move forward and indeed improve the notion as a learning organization for the institution. The future learning needs analysis identified with this research for improving learning opportunities has clearly shown that for an organization to truly become a learning organization, it must then have a conscious learning strategy and plan in place to allow for the continued effort of its staff to develop and continually learn.

Introduction

Workplace Learning

Research into learning at the workplace has brought about a great deal of learning perspectives and different points of view from various researchers. Rapid changes in technology and increased competition between national and international organizations has led to a renewed interest by organizations to re-evaluate the training and educational needs of the workforce.

There is a general agreement amongst stakeholders that workplace learning is becoming an important feature and that employees skills and vocational knowledge has to be improved through positive workplace training.

Workplace learning has been defined in several ways and Boud and Garrick (1999: 2) said there should not be just one definition ... "many perspectives are needed not only because of the diversity of work and the differences which exist even within a single organization but because learning in the workplace is so multifaceted".

Workplace learning or On the Job learning as it is also known has attracted a great deal of interest and substantial research has taken place around learning opportunities in the workplace and the workplace is viewed as having a number of roles in learning. Marsick (1987) gave a broad definition of workplace learning as:

"The way in which individuals or groups acquire, interpret, reorganize, change or assimilate a related cluster of information, skills and feeling. It is also primary to the way in which people construct meaning in their personal and shared organizational lives." (Marsick, 198: 4)

In later work, Marsick identified three primary categories as learning opportunities which are listed as formal, informal, and incidental.

The formal learning opportunities are normally provided in a structural way in a classroom-based activity suitable for learning, that involve a teacher and a group of learners usually known as a course, seminar or conference either in-house or at another venue located away from the workplace.

Informal learning is not deemed to be an ad hoc process; rather, there is an interrelationship between informal learning and work. The informal learning refers to the learning experiences that occur naturally as part of the work and learning can be incorporated within the work by, for example, by providing employees with a variety of tasks, or by arranging the work in a manner which maximizes learning opportunities. Informal learning may include on the job learning, performance planning, coaching and self-study by reading journals and manuals, and potentially all structured activities that lead to learning. (Marsick & Watkins, 1990, 1999).

Incidental learning as defined by, Marsick and Watkins (1990) identified 'incidental learning' as a subset of informal learning where learning opportunities occur, have not been planned but occur spontaneously as a by-product of opportunities to work with new or challenging projects with expert professionals or with knowledgeable peers. Marsick and Watkins (1990) also suggest that incidental learning includes learning from mistakes, learning by doing, and learning through a series of interpersonal experiments.

Ross-Gordon and Dowling (1995: 315) provide their definition of incidental learning as "spontaneous action or transaction, the intention of which is task accomplishment, but which serendipitously increases particular knowledge, skills, or understanding. Incidental learning is unintentional and unexamined. It is not based on reflection; thus the learning is embedded in the learner's actions".

When incidental learning occurs, it is an unplanned activity and can be the product of some other activity and the learner discovers something while in the process of doing something else. To act upon and pursue that discovery, the learner must pull away from the primary or planned task and examine the discovery clearly before he/she can learn from it. Awareness of opportunities and the value of such learning may be brought to learners' attention by emphasizing the outcomes they might anticipate through incidental learning Mealman (1993).

Other writers have challenged the earlier work of Marsick and Watkins of informal learning as being intentional and incidental learning as not, and Garrick (1999: 218-219) suggests that incidental is incorporated within the term informal, and 'separating informal and incidental learning seems to represents a false dichotomy... existing definitions do not problematize the phenomena adequately'

Rylatt, (1994: 10) describes workplace learning as: 'A sustained and high leverage development of employees in line with organizational business outcomes'.

The research undertaken over the past two decades have established that people learn in a variety of ways and individuals have their own known preferred styles of learning that lead to improved competence and performance. (Cervero, 1988, 1992; Houle, 1980).

As stated by Dryden and Vos (2001) "we learn visually, auditory, tactile/kinaesthetically, through print –orientated material, or group interaction".

Research into how people learn show that most job related proficiencies are learned outside the formal classroom, and suggest that where the learning occurs and how it's used is closely associated to how adults learn. (Greenagel & Lagay 2003)

Concepts of the learning organization

Much has been researched, written and discussed about the notion of learning organizations, and Senge (1990) was one of the early researchers to develop some key theories of a learning organization. With his work The Fifth Discipline, Senge (1990) developed five disciplines that determined the architecture of a learning organization and when used collectively facilitates organizational learning. With his work he encapsulated the essential integration process of the individual, the organization and the work environment. He states that the five vital disciplines include personal mastery, mental models, building shared vision, team learning and systems thinking. These disciplines have evolved from the premise that an organization should focus its attention on conditions that motivate people to do great things for themselves and their organization.

The definition of a learning organization given by Senge (1990) is that: "where people continually expand their capacity to create the results they truly desire, where new and expansive patterns of thinking are nurtured, where collective aspiration is set free, and where people are continually learning how to learn together".

An earlier definition of a learning organization stated by Beck (1989: 22) argued that "a learning organization is one which facilitates learning and personal development of all its employees, whilst continually transforming itself".

Watkins and Marsick (1992: 298) offered an alternative detailed definition of a learning organization who argued that: "a learning organization is one in which learning is a continuous, strategically used process, - integrated with, and running parallel to, work - that may yield changes in individual and collectively held perceptions, thinking, behaviors, attitudes, values, beliefs, mental models, systems, strategies, policies and procedures".

Institutions as learning organizations

There is a substantial amount of literature on the subject of workplace learning and on the concept of organizations becoming learning organizations. However on the subject of universities and tertiary institutions there appears to be a limited amount of literature focussing specifically on lecturers as learners in there own workplaces and institutions as learning organizations. Also there appears to be very little in the way of studies for non academic personnel making the transition from one form of workplace as an industry oriented organization to that of academic position as a lecturer or a tutor in a tertiary organization.

Arising out of the literature and research findings Patterson (1998) discussed the idea as to whether universities which are dedicated to learning are in fact learning organizations. Due to universities adopting strategies of alliance and pursuing new forms of partnerships and collaboration across tertiary sector boundaries, the question was raised as to whether maintaining a distinction between the university and non-university sectors are still appropriate. Patterson (1999) addresses the many questions that pertain to the concept of learning organizations. The term "learning organization" has variously been defined and the important characteristics captured and Garvin (1993) list these as that the organization learns through creating, acquiring, and transferring new ideas and knowledge, and changes its behavior to reflect these.

Patterson argues that many universities have responded to the challenge of significant and rapid changes in the technological, economic, and social-cultural environment of the post industrial society, and they have applied new knowledge and changed the way of behaving and have indeed become learning organizations.

Previous work carried out by Ollin (1996) indicated that demands were being placed onto Further Education (FE) institutions and on academic staff to perform to high quality standards through rigorous quality assurance procedures, external inspections and self assessment and published information on performance.

These conditions forced onto them by external forces have necessitated the need by FE institutions to adopt the accoutrements of the business culture and have adapted a more "clearly defined purpose structure, structure and accountability then ever before" Ollin (1996: 2).

Research carried out by Matthews (1999) investigates the real meaning of workplace learning, and goes on to suggest that for workplace learning to work in any situation, certain opportunities, conditions and features need to be in evidence.

Matthews (1999) suggests also that an examination of the concepts and the elements involved in workplace learning is necessary, as workplace learning is difficult to define and is considered to cover many different activities. Matthews (1999) argues that a workplace could be somewhere else other than located in a physical location and gives an example of academic staff within the university system carrying considerable amounts of work such as marking and preparation work at home.

Some of the barriers for universities to become a true learning organization are found directly within the universities and White and Weathersby

(2005) state that as lecturers they teach the students about learning organizations and conscientiously design courses to include elements of learning organizations, as the ideal for work organizations, including opportunities for innovation, self-managed work teams, experiential learning, double and triple-loop learning, and learning lessons about how things work discovered through "mistakes." but "we work in institutions that rarely practice the even the simplest tenets found in the theories of learning organizations".

Research carried out by Retna (2004) focuses on the increasing interest in the concept of the Learning Organization. The research investigates the role of tutors in a university in New Zealand and discusses the idea of universities becoming learning organizations along the lines suggested by Peter Senge (1990). She argues that tutors play a crucial role in this process, and that student learning is affected by the competency of tutors, and training in turn plays an important part in developing effective teaching.

Method

The approach used with this research was based on a combination of qualitative and quantitative process, with the main emphasis towards the qualitative methodology as this research was based on a case study on institutional organizational learning, and from the results it was hoped to be able to determine and describe the phenomenon of formal, informal and incidental learning compared to existing theoretical perceptions and writings on this topic as discussed in the literature review.

Attributes which occurred from the literature as characteristics of learning organizations were identified and compared with qualities of factors listed by various writers outlined in the literature review.

The survey questionnaire was based on prior research and knowledge of workplace learning, and the aim was to initially establish and gain information as to the available learning opportunities that new staff had participated in around the area of formal and informal training opportunities at the Applied Technology Institute organization.

Following the survey questionnaire semi-structured interviews with focus group discussions were carried out to explore certain issues regarding the institute as a learning organization, and for new staff how effective the institution is for them as a learning organization. Responses to the questions were recorded manually and then the staff's perceptions and opinions recorded electronically.

Throughout the programmes that are offered over the ten main disciplines within the Applied Technology Institute (Electrotechnology, Automotive Engineering, Marine, Joinery, Carpentry, Interior Décor, Plumbing, Welding) an approximate total of 60 staff are involved in teaching and learning. A sample of 22 staff who were either new or relatively new lecturers were selected to be part of the case study as this is recognized to be a common sample size of the population for a case study of this nature, according to Gay & Airasian, (2000).

Before any data was collected ethical approval was sought from the appropriate Research Ethics Committee and from there participants gave their permission on the consent form and all information given would be kept confidential and that in no way would identities be revealed in reports from the case study. All participants received an information sheet that explained the nature of the research with clear instructions on how to respond to the survey questionnaire.

The survey questionnaire used was of a descriptive design based on Gay & Airasian, (2000) idea that survey instruments contain four main elements to allow for the respondents to answer in a range of modes, thus not favouring any one particular mode. Findings from the surveys and interviews were examined using a combination of descriptive analysis.

Survey questionnaire

The survey questionnaire for the new staff consisted of 23 questions that covered 5 areas pertaining to learning organization concepts and learning opportunities within the organization.

Personal questions for staff were focused on questions such as:
Length of time worked as a lecturer at the Applied Technology Institute? Industry experience? Main reason for starting work as a lecturer? Positive or Negative aspects in the job?

Questions regarding learning on formal learning such as:
Learning strategies in place for new staff? Sufficient formal training in teaching prior to taking classes? (This included questions on the following methods, Learning technologies, People capabilities, Induction training, Professional training courses at UATI, Other academic course, GDHE courses) List of possible formal learning opportunities at the institute (includes: aware of courses, unaware of courses, actively participated, not had opportunity)

Questions regarding personal professional development plan such as:
Formal professional development plan with your supervisor?
Saticfaction with professional development plan over the last year? Satisfaction with learning and developing professionally in your current position? Additional support in role as a lecturer?

Questions regarding learning on informal learning and incidental learning such as:
Identification of possible informal learning activities at Unitec. (list includes: Coaching by Team Leader, mentoring by Team Leader, mentoring by peers, performance planning, peer evaluation, student evaluation, self directed learning)
Identification of possible incidental learning activities at Unitec. (list includes: Observation of other lecturers' when teaching, Observation from students when teaching, Observation of other lecturers' when assessing, Observation of other lecturers undertaking research, Observation of best practice by others, Identify-

ing best practice by other organizations, Joint lecturing activities, Peer tutoring, Working on projects, Problem solving with peers or supervisor, Reviewing problems or errors with peers/students

Questions regarding level of support from peers and supervisor such as:
Working in a team situation? Main advantages/disadvantages when working in a team environment? Level of support from your team members in present job? Level of support from your manager in your present job? Level of support from the staff development unit in your present job?

Questions regarding the organizational structure such as:
Rate its organizational structure? Conscious learning strategy in place for new staff?

Questions regarding on the job learning and off the job learning such as:
(list includes: I have learnt more about my job by observing my peers, I have received advice from others, I have discussed quality issues with my team, I have positive work satisfaction, I have learnt more about my job through experiential learning, I have learnt more about my job through interaction with students).
Agree or disagree with the following: My off job learning is interesting,
The professional training course tutors are effective, the training courses allow me to be innovative and try out new ideas, I develop valuable skills and knowledge from the formal training courses, My professional development training programme fits in well with my normal work, I have completed my education professional development plan in the expected time frame

The survey also invited further comments on improving the organization.

Focus groups interview questions
Separate semi structured staff discussion sessions were carried out to explore certain issues regarding the institute as a learning organization, and to determine how effective the Applied Institute was for them as new staff. Prior to the focus group discussions pre-circulated questions were distributed to the participants to indicate the nature of the topics to be included in the following discussions.

Questions for the focus groups included reasons and motivation for joining an institution and changing professions from an industry perspective to an educational one. Other questions centred on the transition from an industry oriented profession to an educational career and how could the institute help in the transition. Questions were asked about the understanding of organizational learning and further learning opportunities with regard to formal/informal and other types of learning and training. As well questions relating to the culture of the organization and whether there were any perceived barriers in accessing formal or informal learning at the Applied Technology Institute.

Quality education is an important issue in any teaching organization and in the focus groups staff were asked as to whether they felt that quality was featured in the organization.

On job learning was an area that was addressed also and what it meant for the staff now as educators within an institution when formally on job learning was more familiar as an industry focused experience.

Results

Analyses and discussion

The main aim of this research project was centred across two viewpoints: one, to focus on new staff to the organization and to gain an insight as to how effective from their perspective the Institute is in the area of formal/informal and incidental learning; and two, evaluate the effectiveness of the Applied Technology Institute as a learning organization.

In order to meet this aim, four objectives were set and these will be considered in turn.

- Evaluate the effectiveness of the Applied Technology Institute as a learning organization.

From the results gained from the questionnaire survey and focus groups the data sampling gathered from the participants indicated that the majority had indeed originated from an industry background and were experienced at some form of technical occupation prior to joining the teaching profession. The experience that had been gained in an industry background had been valuable as in the main the staff were familiar with the concept of a learning organization. When analyzing the results as to the reasons for changing careers it was evident that one of the main reasons for doing so was because of the prospect of challenging work and secondly opportunities for further professional work was a factor with pay not featuring very high at all. Dealing with the question of the institute functioning as a learning organization a high proportion of respondents felt that the organization had a conscious learning strategy in place and tried hard to foster learning within the organization. What was startling in the results in the domain of formal training for new staff was that a very high proportion of staff had either not been given the opportunity to carry out formal training for teaching skills or were unable to do so due to time constraints or other reasons. There was a mixed response on the structure of the organization where the majority believed the structure to be semi-flexible and the minority deciding the structure to be either bureaucratic or too rigid. From the aspect given earlier by Ortenblad (2004) "a learning structure is a flexible organization ... and is preferable to one that is bureaucratic ... the organization needs continual learning to occur in order for it to remain flexible".

As indicated in the focus groups it was highlighted that the organization was conscious of a learning culture and was very positive in continuing with this strategy. However there were some very good suggestions made as to how the cultural awareness could be improved and reiterating some of these included focussing on the vision of the organization and enabling effective methods and providing support to allow for staff to develop professionally. Allowing applied research to build on relationships between the organization, staff and industry and providing adequate meeting time for staff to share their experiences and best practice.

The effectiveness of the Applied Institute as a learning organization was discussed and the consensus from the focus groups and survey was that it was a learning organization but still had lots of work to do to ensure that new and existing staff continues to learn and develop new knowledge. The respondents were confident that the organization believed in its staff and did focus on quality and that the Institute places its students first and strives for continuous improvement in teaching and learning.

Teamwork in the organization was shown to be very strong amongst the staff and the comments made were very positive and reinforced the learning organization structure by: "sharing ideas, sharing knowledge and support, spreading the workload across the team, peer course development and assessment, synergy in research and development, speed to get tasks done, consistency, developing more solutions to problems, constructive discussion for more successful outcomes, team outcomes tend to be of better quality, easier working in a group situation, development of expertise in a particular area".

- Determine the learning opportunities that may be available for new and existing staff.

Marsick and Watkins, (1990, 1999) indicated that for workplace learning to occur, three primary categories are required as learning opportunities which are listed as formal, informal, and incidental through informal learning, formal learning, and incidental learning.

This research project attempted to determine the extent of learning opportunities that was available for new and existing staff to the institution in each of the learning opportunities listed. The results of the questionnaire and focus groups revealed that formal learning in the institution was available and the majority of the staff were aware of the formal learning activities but only 25% had actually participated in these activities.

From the results taken regarding induction training only a small amount of staff had participated in this form of training showing a concern in the Induction training programme. It was clear that the induction process required some modification to ensure that all staff starting work in the organization in the future experienced a positive start in their employment.

Also as gathered from the focus groups and survey questionnaire issues surrounding the management of performance development plans needs to be addressed to improve agreed outcomes.

Suggestions made by the staff for further formal learning activities included professional development coaching and professional individual growth programme and staff development training forums on department policy and procedures and operational systems.

The area of informal learning has shown that the staff have either not been aware or unfamiliar with some of the types of activities that can contribute to the idea of a learning organization and creating new knowledge in the identified activities that may contribute to informal learning. The very positive section in this category was in the response for student feedback where the majority of staff had participated in student evaluation and had involvement in the area of self directed learning that leads towards learning.

Further ideas on informal training did arise from the results including providing enhanced staff collaboration to combine project and research work to share experiences and best practices.

Incidental learning identified by Marsick and Watkins, (1990, 1999) is a sub-set of informal learning where learning opportunities occur, the questionnaire survey provided results again in the low awareness or lack of opportunity to participate bracket, apart from observation from students and reviewing problems.

The responses relating to questions on on-job learning proved to be extremely positive regarding the on-job learning experiences as a lecturer in quality and interaction with the students.

Other issues concerning On-Job training related to attachment with specific leading industries to improve technical capabilities as a suggested strategy to aid in delivering up to date technical knowledge.

The off-job learning experiences for the staff that had been involved in some form of training or learning development programme gave a reasonably high satisfaction rate of 66% across all staff surveyed which was encouraging.

- Identify the areas of formal or informal needs analysis in order to make recommendations for improved training initiatives.

The areas of formal learning that have been identified with in this study that require further attention are:

* induction training;
* the management of performance development plans;
* staff awareness of identified training programmes;
* professional development coaching;
* staff development training forums on department policy and procedures and operational systems;
* technical training in the areas of new technologies;
* Professional Development and in tandem facilitate research projects.

The areas of informal learning that have been identified with in this study that require further attention are:

* staff awareness regarding identified informal and incidental learning activities with low participation levels;

* enhanced staff collaboration to combine project and research work;
* applied research opportunities;
* attachment with specific leading industries to improve technical capabilities.

- Identify what other research may be necessary for institutions/universities of Technology to adopt to embrace the notion as learning organizations.

This research project was limited in the sense that the full representation of the Applied Technology Institute as a learning organization was not covered as this research project pertained only to new staff that had been employed for a relatively short period of time as educators. The main aim and focus was to gain an insight as to how effective from their perspective the Institute is as a learning organization. This aim has been achieved with this project as there have been valuable results gained from this study. However perhaps a wider research project could be attempted to bring in longer serving staff and a wider cross section of the organization with not only academics, but administration staff and management staff to gain an integrated research outcome of all staff engaged in the organization in order to evaluate and view the full extent of the institute as a learning organization.

Summary

This section has detailed the main results obtained from the data collection described previously. Results from the staff survey questionnaire and the focus groups were compiled and these results have been interpreted and analysed in terms of the objectives for the project and have provided a basis for discussion.

Conclusion

This research project attempted to gain an insight as to how effective the Applied Technology Institute is as a learning organization from the perspective of new academic staff, and to evaluate the possible learning opportunities that exist through the context of formal, informal and incidental learning.

Academic staff that are new to the teaching profession in the applied technologies field predominately originate from industry where they have usually been practitioners involved in engineering or some other technical background. To be successful as lecturers in their new profession, all new staff need some form of formal or informal training in teaching and learning, and this research was directed at the Institute to determine how it measured up to the challenge as a learning organization when its normal function is to offer education and learning to its students.

The effectiveness of the Applied Technology Institute as a learning organization in the training of new and current staff was an important question and analyzing the results from this research, the findings have shown that the institute has many identified strengths when it comes to the notion as a learning organization.

The consensus from the focus groups and survey was that the Applied Technology Institute at Unitec did strive to be a learning organization following along the lines of Garvin (1993) who lists these as "the organization learns through creating, acquiring, and transferring new ideas and knowledge, and changes its behavior to reflect these". However lots of work is still required to ensure that new and existing staff continue to learn and develop new knowledge for the Applied Technology Institute to aspire to becoming a learning organization in the true sense bringing together all four aspects of a learning organization such as one based on an integrated model developed by Ortenblad (2004).

This research followed along the lines of prominent researchers in the field of learning organizations and a key focus was on the differing perspectives in the area of formal and informal learning opportunities that the new staff experienced.

Some very good initiatives and suggestions have resulted from the outcomes and these suggestions for improvement will be recommended to enhance the formal and informal learning activities in the institute.

Critical factors identified in learning needs for new staff from an industry background in the transition from engineers to educators included the institute developing effective induction and professional development plans and providing adequate support mechanisms that encouraged new and existing staff to develop teamwork and sharing of knowledge with each other.

The future learning needs analysis that has been identified for improving learning opportunities has clearly shown that for an organization to become a learning organization the organization must have a conscious learning strategy and plan in place to allow for the continued effort of its staff to develop and continually learn whether it be through a formal plan or through on the job learning. An organization such as an applied technology institution has clearly a role to play as a learning organization now and in the future.

References

Beck, M. (1989), "Learning organizations: how to create them", *Industrial and Commercial Training*, Vol. 21 pp.21-8.

Garrick D. &, J. (1999), Understandings of workplace learning. In *Understanding learning at work*. eds. D Boud & J. Garrick. Routledge, London.

Garrick, J. 1999, *The dominant discourses of learning at work*. In Understanding learning at work. Eds. D. Boud & J. Routledge, London Garrick.

Garvin, D.A. (1993), "Building a learning organization", *Harvard Business Review*, Vol. 71 No.4, pp.78-91.

Gay L.R, & Airasian, P. (2000), Educational research: *Competencies for analysis and Application* (Sixth ed.). Upper Saddle River, New Jersey: Prentice Hall.

Greenagel, F & Lagay, (2003) Lead *Balloons, Stone Canoes, and Learning Styles* viewed (online) at www.learningcircuits.com

Harris R, Simons M, M & Bone J. (2000) *More than meets the eye? Rethinking the Role of the Workplace Trainer* NCVER

Matthews, P. (1999), Workplace learning: developing an holistic model, *The Learning Organization*. Volume 6 Number 1 1999 pp. 18-29.

Marsick, V. ed (1987), *Learning in the workplace.* Croom Helm, London.

Marsick, V.J. and Watkins, (1990), *Informal and incidental Learning in the Workplace,* Routledge, London and New York, NY.

Marsick, V.J. and Watkins, (1999), *Facilitating Learning Organizations; Making Learning count,* Gower, Aldershot.

Mealman, C.A. (1993), *Incidental Learning by Adults in a Non-traditional Degree Program*: a Case Study. Presented at the Midwest Research-to-Practice Conference in Adult, Continuing, and Community Education, Columbus, Ohio , October 13-15, 1993. http://www.nl.edu/academics/cas/ace/facultypapers/CraigMealman_Incidental.cfm (accessed October 2005).

Olin, R. (1996) *Learning from industry: human resource development and the quality of lecturing staff in further education.* Quality Assurance in Education Volume 4 Number 4 1996 pp. 29-36.

Örtenblad, A. (2002), "A typology of the idea of learning organization", *Management Learning,* Vol. 33 No.2, pp.213-30.

Ortenblad, A. (2004), "The learning organization: towards an integrated model", *The Learning Organization*, Vol. 11 No.2/3, pp.129.

Patterson, G. (1999) The learning university. *The Learning Organization,* Volume 6 Number 1 1999pp.9-17.

Retna, K (2004) *Universities as learning organizations*: Putting tutors in the picture, Herdsa 2005.

Ross-Gordon, J. M., and Dowling, W.D. (1995) "Adult Learning in the Context of African-American Woman's Voluntary Organizations." *International Journal of Lifelong Education 14*, no. 4 (July-August1995):306-319.

Rylatt, A. (1994), *Learning Unlimited: Practical Strategies and Techniques for Transforming Learning in the Workplace*, Business and Professional Publishing, Sydney

Senge, P.M. (1990) *The fifth discipline: The Art and Practice of the learning Organization*, New York: Doubleday.

Watkins, K.E., Marsick, V.J. (1992), "Towards a theory of informal and incidental learning in organizations", *International Journal of Lifelong Education*, Vol. 11 No.4, pp.287-300.

White, J, and Weathersby R (2005) Can universities become true learning organizations, *The Learning Organization*, Volume 12 Number 3 2005 pp. 292-298.

Here Be Dragons: Mapping Future Possibilities Using an Emergent Framework

Lynn Allen & Trudi Lang
Curtin University of Technology, AUS

"Here Be Dragons"[1] is a phrase believed to have been inscribed by mediaeval mapmakers to warn of the dangers beyond the boundaries of their defined world. Dragons were symbols of the unknown. The history of cartography is a history of representing reality. From sextants to satellite photography, new tools have led to new types of maps. Mapmakers have much in common with systems thinkers. They select boundaries and perspectives, making conscious choices on where to place their attention. Over time, maps of the same space change as do conventions for drawing, and rules for construction and interpretation. As new knowledge changes previous maps become but transitional objects in the evolution of new perspectives, perspectives that drive the dragons further away. What dragons do today's organizational leaders expect to discover at the edge of their worlds? In particular, what approaches could they adopt to develop shared meaning and action in uncharted futures?

Today's Dragons

Today our scary creatures go by the names of Uncertainty, Ambiguity, Complexity and Conflict. They constitute our dragons of unknowing and bewilderment. While some leaders push these dragons to the edge of their universe others prefer, like St. George, to tackle them head on.

Uncertainty: The Dragon of Confusion

Uncertainty comes from a lack of information so the major task in reducing it is to gain more information. However, with the blurring of the roles of public, corporate and community organizations and the trans-national influences drowning us in contradictory signals, the search for the information to reduce this uncertainty can prove difficult.

Ambiguity: The Dragon of Meaningful Contradictions

Ambiguity occurs when no matter how much more information is obtained, the situation is still no clearer. McCaskey (as cited in Weick, 1995: 93) also suggests ambiguous situations have multiple and conflicting interpretations as well as goals[2]. Ambiguity is a key aspect of 'wicked' problems (Rittel & Webber, 1973) which are characterized by being unclear, definable from varying perspectives with no agreement on the problem situation itself, and no simple cause and effect relationships whose resolution would lead to a single solution.

Complexity: The Dragon of Ubiquitous Connections

Complexity arises from too much information. Everything seems connected to everything else. Individuals have access to more and more information but it is not necessarily reliable or authoritative and its value depends on the questions that are asked. Nor is such information context or value free. This poses unique challenges in organizations as Griffin (2002:218) notes:

"There is staggering complexity in the interdependency of people in a large organization in the movement of the living present. This is compounded by similar complexity in the many other organizations it interacts with. It is astounding that we continue to hold fantasies that single persons or small cliques of persons can steer such complexity to achieve targets they have set in advance."

Conflict: The Dragon of Coercion

The concept of Power is a notion much studied, from Sun Tzu's *The Art of War* to Machiavelli's *The Prince.* It seems each culture and age has its own treatise on wielding and maintaining power. In our times we have *The 40 Laws of Power* (Greene, 2005).

Power is wielded in organizations in a myriad of ways: positional, informational, technological, financial, familial, nepotism etc. Many power struggles are over identity or territory. In both cases they are debates about boundaries. As a result, Jackson (1991: 164) says the social world is:

"created by people who have conflicting aims and intentions and bring different resources to bear when the social construction is taking place. It follows that the social world escapes the understanding and control of any one person or group of people. It takes on the form of a highly complex and structured external reality that exercises constraint on the individuals who make it up."

In organizations where people are seeking to make sense of the world and develop shared meaning about purpose, direction and success, what frameworks can we provide that will deal with these four dragons?

The difficulties of exploring new territories with old maps

For more than fifty years there has been one 'solution' after another. Re-engineer; restructure; outsource; downsize; balance the scorecard; manage for total quality; benchmark and so on. Sometimes they do not bring the promised results[3].

These approaches share an assumption that the world can be commanded by the manager. The machine metaphor is a foundational premise, a metaphor that embodies the scientific method in management practice where it is expected that observation, measurement and micro-process analysis will yield control.

If our view of the future is not to be an extrapolation of the past, we need processes that allow the quiet, creative and even subversive voices to be heard. In other words, agreement on preferred futures emerges from dialogic processes, not from a latter day Moses descending the corporate mountain with the latest '10 ways to...' As Vickers (1987:44) suggests:

"The conceptual world has a life of its own. Like the life forms in the physical world, ideas are born, grow, clash, excite, modify or destroy each other or coexist by reaching strange compromises with their rivals. ... We need an ecology of the conceptual world..."

Given this, how many of the tools and methods in today's leader's toolbox will usefully assist them to develop their teams' shared future space?

Standing on the shoulders of giants

As Stacey (2002) says, 'There are voices from the fringes of organizational theory, complexity sciences, psychology and sociology that are defining a newer, participative perspective. With this inter-subjective voice people speak as subjects interacting with others in the coevolution of a jointly constructed reality.'

There is a burgeoning literature in areas that foreground human intelligence: organizational learning; knowledge management; storytelling; creativity and innovation; and emotional intelligence. Some writers have moved from the machine metaphor to the biological suggesting the interrelatedness and connectedness of all things. But is this to replace one nonhuman metaphor for another? What would happen if we brought innately human attributes to the forefront of our thinking?

Coming together from different backgrounds and knowledge domains, the authors of this paper have been on a journey of exploration on this topic. The journey included two Deep Conversations with Professors Peter Checkland and Kees van der Heijden whose intellectual heritage and inspirational support we acknowledge[4].

The result of this journey thus far is the *ariadne* methodology which shares its essential logic with three of what Checkland calls 'the four key thoughts' of the Soft Systems Methodology (SSM) (Checkland, 1999: A7) relating to modeling human activity systems from a declared *Weltanschauungen* as a learning system.

The remainder of the paper describes the current state of *ariadne*, a framework that foregrounds human attributes for Applying Research and Innovation to Advance the Development of Networked Enterprises (*ariadne*).

Introducing *ariadne*

Ariadne was a Cretan princess who assisted the Greek Theseus to kill the Minotaur in the Labyrinth by providing him with a magic thread and a sword. She persuaded the designer of the labyrinth, Daedalus, to provide her with the map. This ancient piece of lateral thinking removed the monster and stopped the annual slaughter of Greeks sacrificed to the terrifying half man, half bull - a Greek dragon.

Ariadne, the methodology, is an alternative approach to strategy work where more creative futures are needed. *Ariadne* provides a way for groups to arrive at new shared conceptual spaces in a process Inayatullah (2002:109) calls 'coevolutionary anticipatory action learning.' The flow is iterative, moving at differing speeds and in differing directions, partnering the known with the unknown, interweaving the mundane and the mysterious.

The basic shape of each phase

Before presenting the methodology, it is useful to reflect on its underlying logic which is outlined in Figure 1.

Keeping in mind the intention to be creative, conversational, collaborative, and coevolutionary, the essential activities are:

a. For any topic of study there exists knowledge which humans can access, determine its relevance and contest it.
b. Consideration of this knowledge inspires the creation of transitional objects from many perspectives. These objects may be conceptual models and rich pictures (as per SSM); other modelling techniques, eg Bryson et al (2004); stories; scenarios as in van der Heijden (2005).

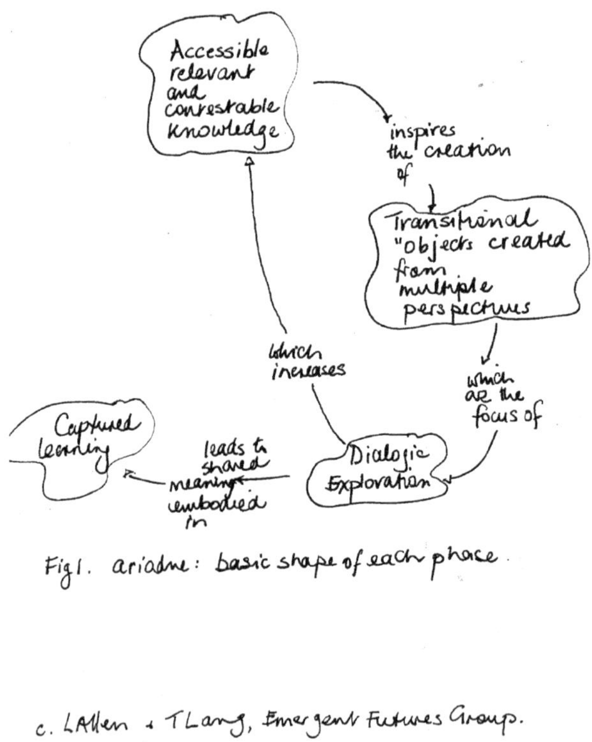

Figure 1 *The basic shape of each phase of the methodology*

c. These objects are the focus of dialogic exploration the purpose of which is learning. Here we use conversation, decision support tools, psychodrama, shared story-telling, drawing, etc.
d. The process of developing shared meaning leads (hopefully) to the synchronicity and accommodation that are then captured in learning objects of some sort. These could be a discussion paper, a project plan, new models, stories, or a strategy document.

This structure informs both the methodology as a whole and each of its activities.

Ariadne: An emergent futures methodology

Let us now turn to the full methodology (Figure 2) to explain how this holistic approach facilitates thinking about the future in innovative ways.

The representation is consistent with an SSM conceptual model. As Checkland (1999: A49) reminds us, the world is not made of systems that we can engineer but the inquiry process can be designed as a system. As we are using a methodology and not a step-by-step tool there are some general principles to remember. Each 'phase' is in essence a sub-system so *ariadne*, can be holographically applied to design each activity. For each phase, decisions are made about methods and outcomes. Those outcomes, essentially the transformation of the sub-system, feed into the next phase of the process but can also be trans-

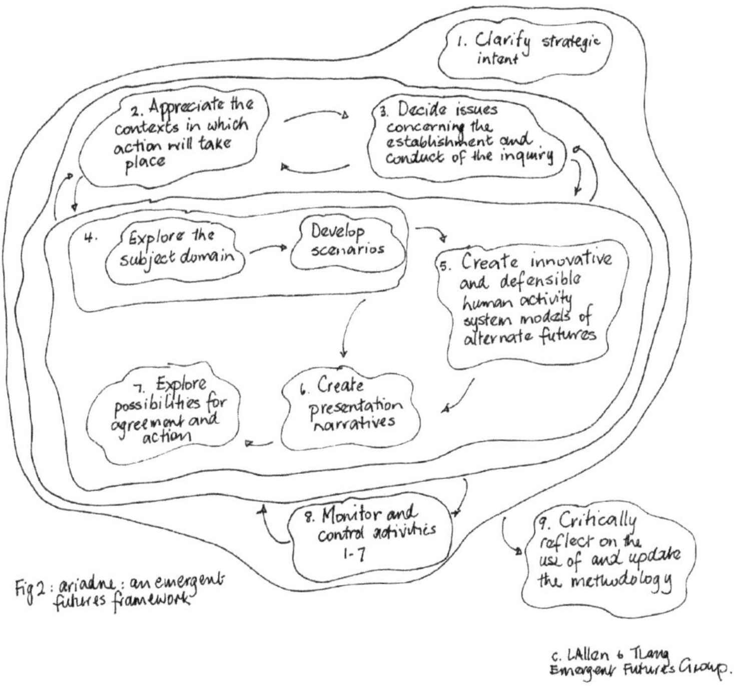

Figure 2 *The ariadne methodology*

lated into 'deliverables' if required by clients or participants.

In addition, each phase (as well as the overall project) is monitored against performance criteria that may be agreed in advance and/or modified in progress. We will say something about critical reflection later. Outlined below is a short introduction to *ariadne*'s activities. While each activity relies on the logic outlined in Figure 1, particular knowledge domains and capabilities are emphasized here.

1. Clarify strategic intent

This activity sits outside the whole framework because it is the foundational driver and checking point, the beginning and the end. Here we gather the questions to be addressed but we phrase them in open, aspirational and non-outcome focused language, thus avoiding a single 'problem' statement. Given that we are undertaking what Midgely (2002) calls a 'systemic intervention' we also take a fluid approach to the boundaries of our intentions. This activity influences and is influenced by all other activities.

2. Appreciate the contexts in which action will take place

Here we seek to gain an understanding of the environment in which action will be taken to address the strategic issues. We explore what social, political and other factors inform the design and conduct of the inquiry.

This phase is grounded in the perceptions of those associated with the enterprise. The aim is to gather as much richness as possible. Decisions about how to handle power are also made here.

3. Decide issues concerning establishment and conduct of the inquiry

Here we design the Inquiry System, giving attention to factors such as roles (e.g., who is the client, the decision taker, the inquirer, etc), the transformation sought from the inquiry, available time and resources, the appropriate methods to be used, determining how success will be judged, and how milestones, etc will be communicated.

An important decision made here is the choice of boundary for the Inquiry System. As Flood (1999: 92) points out:

"mental constructs determine what is in view and might be taken into account at the moment and what is out of view and thus excluded from consideration. The boundary judgment will therefore determine at this point in time the client, issues and dilemmas of concern, and purposes to pursue."

Activities 2 and 3 feed into and are fed by activities 4 – 7.

4. Explore the subject domain and develop scenarios

Here we move beyond the immediate context of the strategic issues and look at developments in the relevant subject domain(s). We do this for two reasons: to generate new insights about innovations that we can incorporate into our strategic directions; and to gather perceptions on key trends and drivers of change.

This helps ensure that potential choices are aligned with the major changes that could occur in the future. Dator (2002) refers to this as 'surfing the tsunamis of change.'

A subject domain is a branch of knowledge or field of study chosen in relation to the strategic intent. For example, if the inquiry concerned the future of business education at Curtin University of Technology the subject domain would be business education in universities.

This knowledge feeds into the scenarios that are written using a compelling storyline, which are named and checked for plausibility and internal consistency. Good scenarios are relevant, effective, mutually exclusive, and sustainable (Schoemaker 1995).

5. Create innovative and defensible human activity systems models of alternate futures

Here we create models of a range of futures as alternate directions. Ideas are 'mined' from previous activities, distilled from scanning the subject domain (s) and the scenarios as well as conversations about the social, cultural and political contexts. Using SSM principles for conceptual model building, we produce not representations of the "real" world but devices to encourage discussion and learning, "transitional objects" that exist solely for the purpose of the process, not as some future reality.

Then we pause to consider whether our models would be deemed innovative by checking on our strategic intent. We are not considering whether they would be acceptable to stakeholders or clients. Rather we are still seeking richness and imaginative alternatives.

We then select defensible models to take forward usually to a larger group than that which has developed the models. Here we bring the innovative ideas closer to imagined implementation, not for rejection, but for testing their viability and likely acceptance and workability. We sweep again through the subject domain material for missed ideas or new perspectives. In addition, material is organized and documented ready for story building and later defending.

6. Create presentation narratives

Here we take the selected models and other information and transform them into narratives in formats and with content that will most suit the situation. This stimulates a wider discussion without going through the whole process again. We use narrative methods to achieve this. The stories' purpose is to stimulate discussion by sharing perceptions; seeking accommodations; learning more widely and deeply; and combining the logic of the models with emotions.

7. Explore possibilities for agreement and action

Now we hope emergent agreement will happen. There are no guarantees or easy ways to achieve this. This is an intention. It may not work. One can be responsible for the process but not the outcome, and whatever the outcome, it is the "right" one for the enterprise.

Many skills are needed to lead groups to accommodation where every individual feels they have been heard and understood and not only have they found some common agreement that they can they 'live with' but they are committed to them intellectually and emotionally.

What is hoped for in this phase, is an agreement on the way forward that can be accommodated by key stakeholders as well as an agreement on what the change is going to look like.

Issues of power will more than likely surface here with a vengeance. We may even get some surprises. Often when it comes to the actual implementation, even those who were previously supportive may baulk at their responsibilities in bringing about the change in which case we loop back to earlier phases because it means the work is not complete.

Conclusion

A*riadne* has been developed in response to the characteristics of the world in which we now find ourselves. It is not a map of the world. More like a trusted companion that reassures us that whatever paths we discover will create wonderful opportunities and whatever dragons we meet along the way, we will flourish. The dragons of uncertainty, ambiguity, complexity, and conflict will be seen for what they are: creatures of our imagination that learning and new shared meaning will transform into creatures of wonder, innovation and perhaps still a little mystery. For our perceptions of the world are our individual stories. We hope that *ariadne* is a contribution to the development of processes for thinking and acting for the future in this world.

References

Bryson, J.M., Ackermann, F., Eden, C. and Finn, C.B. (2004). *Visible Thinking: unlocking causal mapping for practical business results*, Chichester, John Wiley & Sons.

Checkland, P. (1999). *Soft Systems Methodology: a 30-year perspective*, Chichester: John Wiley & Sons.

Dator, J. E. 2002, *Advancing Futures: futures studies in higher education*, Westport, Connecticut: Praeger.

Flood, R.L. (1999). *Rethinking the Fifth Discipline: learning within the unknowable*, London, UK: Routledge.

Greene, R. (2005) *The 40 Laws of Power*, Sydney: Hodder & Stoughton.

Griffin. D. (2002) *The Emergence of Leadership: Linking self-organization and ethics* London: Routledge.

Inayatullah, S. (2002) *Questioning the Future: Futures studies, action learning and organizational transformation*, Taiwan: Tamkang University.

Jackson, M. (1991) *Systems Methodology for the Management Sciences*, New York: Plenum Press.

Midgley, G. (2000) *Systemic Intervention: philosophy, methodology and practice*, New York: Kluwer Academic.

Rittel H. & Webber M. (1973) 'Dilemmas in a General Theory of Planning', *Policy Sciences*, 4 (pp. 155-169).

Schoemaker, P (1995) 'Scenario planning: a tool for strategic thinking', *Sloan Management Review*, vol. 36, No. 2, pp. 25 - 40.

Stacey, R (2002) 'Series preface to Complexity and Emergence in Organizations' printed

in Griffin, D. *The Emergence of Leadership: linking self-organization and ethics*, London, UK: Routledge.
van der Heijden, K (2005) *Scenarios: the art of strategic conversation*, Chichester, UK: John Wiley & Sons Ltd.
Vickers, G. (1987) 'Norms and meanings.' in *Policymaking, Communication, and Social Learning: Essays of Sir Geoffrey Vickers*; Adams, G, Forester, J and Catron, B (Eds) pp. 39-49 ,New Brunswick, NJ: Transaction Books.
Weick, K (1995) *Sensemaking in Organizations*, Thousand Oaks, US: SAGE Publications.

Notes

[1] 'Hic sunt dracones' http://www.maphist.nl argues there is only one occurrence, on the Lenox Globe in the New York Public Library
[2] Other characteristics include:

- Nature of problem is itself in question
- Information (amount and reliability) is problematical
- Multiple, conflicting interpretations
- Different value orientations, political/emotional clashes
- Goals are unclear, or multiple and conflicting
- Time, money, or attention are lacking
- Contradictions and paradoxes appear
- Roles are vague, responsibilities are unclear
- Success measures are lacking
- Poor understanding of cause-effect relationships
- Symbols and metaphors are used
- Participation in decision-making is fluid

[3] For example, "American corporations are expected to spend some $US 50 billion per year during the latter part of the 1990s on reengineering projects, with 80% of that amount going into information systems. More than two thirds of those efforts are likely to end in failure, according to ... Michael Hammer. According to a survey by Arthur D.Little Inc, only 16% of executives say they are fully satisfied with the results of their reengineering efforts while 39% say they are totally dissatisfied. Finally, according to a survey of 400 Canadian and American firms by Deloitte & Touche the main reasons for reeingineering failures seem to be the significant resistance to change and the lack of consensus and commitment among senior executives."
Boyer,M., Robert, J. and Santerre, H. (2001) 'Industrial Restructuring in the Knowledge-based Economy' in Doing Business in the Knowledge Based Economy: facts and policy challenges.' ed. Lefebvre,L., Lefebvre, E. & Mohnan, P. (Boston: Kluwer Academic Publishers) p.398
[4] For a record of the first Deep Conversation with Professors Checkland and van der Heijden see http://www.jcipp.curtin.edu.au/local/docs/DeepConversation.pdf

Lynn Allen is a Professorial Fellow at Curtin University of Technology and co-founder of the Emergent Futures Group. This Group centres its work on developing and teaching methodologies that enable organizations to create holistic futures by taking advantage of the creativity of their staff, stakeholders and communities. She is a long-term user of the Soft Systems Methodology (SSM) in planning and development and teaches this at Masters level. Her senior management experience extends over the university, private and public sectors. She completed a twelve-year term as State Librarian and CEO of the Library and Information Service of Western Australia (LISWA) in March 2001. During that time she contributed widely to policy and strategy development across government. Lynn was a member of the Council of Curtin University of Technology for nine years, the last two (1998-2000) as Pro Chancellor. She has been a member of the Science, Industry and Technology Council of WA, the WA Information Policy Council and the public sector's CEO Consultative Group. In recognition of these contributions, in February 2001 Lynn was awarded an honorary Doctor of Letters for her service to Curtin University of Technology and the world of information technology innovation.

Trudi Lang is a Senior Research Fellow with the John Curtin Institute of Public Policy (JCIPP) on secondment from the Curtin Business School (CBS). She is also a co-founder of Curtin University of Technology's Emergent Futures Group. Previously, she was the manager of the Scenario Planning and Research Unit at Curtin's Graduate School of Business which was established in 1996 with support from Motorola and Compaq (Australia) and worked with over 50 organizations in Australia and Asia. Trudi has worked for number of years as a strategic foresight consultant, researcher and educator in Australia, Asia and Europe both in the private, government and university sectors. She has consulted to numerous private and public sector organizations in Australia and has taught in the International Management Programme at *Reims Management School* in France and presented scenario planning masterclasses in Singapore and Malaysia. Trudi is also cocreator and co-facilitator of *Navigating the Maze*, a strategic development programme for senior executives building innovative and adaptive organizations.

The Story in Leadership

David Green
David J Green & Associates Pty Ltd, AUS

Heavily influenced by the traditional scientific paradigm, modern organizations are often obsessed with approaches and methodologies which strive to be rational, objective and logical. Yet, all organizations are human systems in which their people struggle on a daily basis to find connection and meaning in their worklives – and these central dimensions are scantly serviced by a scientific approach, comprising as they do elements which are best seen as subjective, non-rational, and personal. Effective leadership is concerned with working productively with human dynamics, as distinct from management, which focuses on resource control. As such, it is qualitative, and personal, creating enabling conditions, as opposed to problem-solving - for open, trusting and robust relationships, and the continual and ever-changing creation of meaning. It is here argued that the tools of traditional science are largely ineffective in working with human forces, that we must look outside the relatively recent preoccupation with rationality for the human vehicles we need to do a human job. It is here that the ability to create, fashion and render story and metaphor emerge as a central capability for the exercise of leadership. It is story which throughout history has been a primary vehicle for the creation and dissemination of meaning – in the creation of relationships and trust, in building a sense of shared purpose and value, in modelling behavior, in giving inspiration, in communicating and guiding, and in providing a thread of meaning through times of change. Organizations which seek to provide a creative, ethical, stimulating and humanly satisfying work environment will increasingly be those in which leadership in these terms is encouraged, where the real human dynamics are productively engaged, rather than repressed – and where their employees constantly challenge and regenerate their story.

The story in leadership

We shape our self
to fit this world
and by the world
are shaped again.
The visible
and the invisible
working together
in common cause,
to produce
the miraculous.
I am thinking of the way

the intangible air
passed at speed
round a shaped wing
easily
holds our weight.
So may we, in this life
trust
to those elements
we have yet to see
or imagine,
and look for the true
shape of our own self,
by forming it well
to the great
intangibles about us.

"Working Together" by David Whyte

We are our stories. As individuals, the narrative representations which we construct and rehearse and relate are both the mythology we enact and the process by which we make sense of ourselves in our worlds. In organizations, and their sub-cultures and groups, there are analogous, though often disparate, narratives in formation and representation, which serve to create and render meaning in our worklives. In this setting, storytelling is very much alive, but generally disowned and discounted, associated with rumors, gossip, or 'spin'. We often fail to see how integrally connected story is to meaning-making, and just how much we employ it to do just that for us. Ask experienced managers to tell a story to a purpose, and all too quickly they play-act teachers talking to children...or worse, tell a story, or analogy, then explain the point, just in case it wasn't clear.

As Morgan (1986) pointed out twenty years ago, the dominant metaphor for our companies and governmental agencies is that of the machine. Even when managers refrain from phrases such as 'running like clockwork' or 'well-oiled machine', the implicit model for conceptualizing their organizational systems is mechanical – or mechanistic. The only 'knowing' which can be trusted, or at least justified, is that based on measurement. Business books, with titles such as "The 7 Habits of Highly Effective People", "The 21 Irrefutable Laws of Leadership", or "How Full is Your Bucket?", are the surrogate technical manuals required by managers who see their job as identifying and solving problems, and fixing things - linear, sequential, operational and exhortational. They offer the promise of simplification in an uncertain, uncontrollable and complex world. And they are the companions to the 'tools' required to diagnose and re-engineer the organization – from pareto diagrams to decision analysis to accountability mapping.

It is hard, after all, as a senior manager, to be held accountable for leading a system which you can't actually control, though the mantra of 'shareholder value' does offer the possibility of immediate, superficial change, tied to short-term performance measures, linked in turn to often excessive remuneration benefits – a kind of 'chain letter' passed from one CEO to the next. In a messy world, leading by analysis and decree offers the promise of control, or at least of some sense of order and momentary stability.

This inflation with rationality and analysis is, of course, rooted in traditional science thinking, the paradigm of technical, academic, and business fields; in organizational life, it has habitually become the paradigm for business management and leadership, and linked to a widely-endorsed managerial fantasy of control. It 'makes sense' to use the traditional science paradigm to underpin the business of technical, financial, strategic, administrative, and operational work. Customers, after all, expect consistent high quality and reliability in products and services, and rationality, logic and measurement are essential to that end.

It is no surprise to us, then, that whenever managers seek to better their organizations, they will approach the task within a mindset which has dominated their formal education, their technical and management development, their worklives, and the public discourse of the workplace. As a consequence, their work is defined through analysis, planning, strategizing, process-driven design, frameworks, modelling, systems, measurement, technology, and processes. Performance management and rigid meeting agendas at the team level; Six Sigma and Du Pont safety training at the corporate. A change of senior manager has become equated with an organizational restructure, process redesign, or new frameworks for regulating employee behavior.

But in the process, modern organizations have lost their connection to the non-rational, essentially human nature of their own being. Thirty years of globalization and agglomeration, of the corporatization of language, of the justification of far-reaching decisions in notions such as 'wealth creation' and 'shareholder value', and galloping uncertainty and disorder – a fragmentation possibly greater than at any time for 500 years (Kelly, 2006) – has if anything intensified organizational desperation for order and the reassurance of numbers. And yet, whether we harness its potential or not, story remains probably the most potent means of shaping and representing meaning in our direction, purpose, relationships, and engagement with our lives – including here, particularly, our worklives. As individuals or collectives, our lives reflect the continuous iteration of our being and our narratives. Our identities are always in construction through this interaction between experience and sense-making. And, as we 'make sense' of our experiences, our stories become the myths which our lives express. In our organizations, our shared identity is constantly moderated and maintained in our shared stories.

The central fact to contend with in the leadership of organizations is that they are human, or social, systems (Cohen & Prusak, 2001). They are best understood through a non-rational lens, as complexes of subjective realities, formed and expressed through patterns of personal and collective meaning creation and dissemination. Despite our best efforts to control our behavior and thought, to make defensible decisions, and to make ourselves (or at least our staff) predictable and orderly, we are in fact emotional, subjective, and distinctly non-rational in how we engage with ourselves and each other. When Wheatley (1992) began to explore the concepts and metaphors of 'new science' for their application to human systems, she initiated a process of legitimizing a more fitting perspective for the understanding and engagement of human forces in organizations.

This approach to leadership is characterized by an acceptance that meaning-making is central to being human, and is a process which is internal, emotionally integrated, personally and socially constructed, and essentially uncontrollable (that is, no-one can legislate for others what they will believe, what they will know, or who they will trust). When we add a systemic frame, and the effects of inter-relationship, the business of 'leadership', as opposed to 'management', may be seen as the capacity to work effectively with non-rational dynamics, at various system levels – intrapersonal through to organizational. Looked at from another perspective, leadership is the expression of Gardner's (2000) interpersonal and intrapersonal intelligences. And, as Wheatley (1996) highlighted, the foci for leadership energy will be on those aspects which make the greatest difference to system effectiveness – 'relationship' (or connectivity), 'identity' (or the socially-constructed ideations that create coherence and cohesion), and 'information' (or freely-available data and interpretation necessary for action). Arguably, none of these aspects can be relationally transacted without story.

Many related notions emerge from the carriage of leadership in these terms. Problem- solving is replaced by identifying and modifying patterns (no cause and effect relationship is assumed); the modelling (in both senses – behaving and mentally mapping) of desired patterns replaces the modelling of problems and dysfunctionality; we look for the patterns of the patterns (fractals); we accept that unconscious (i.e., out-of-awareness) process is at least as powerful a force as conscious behavior; we accept and work with multiple realities, rather than seek to impose one reality; and the focus moves from feedback (or control) loops to amplification (or self-fulfilling) loops, as key dynamics.

The incessant efforts of managers to use the tools and methods of rationality, conscious design and control to work with essentially non-rational, unconsciously patterned and uncontrollable dynamics, are doomed to mediocre results. The high failure rate of acquisitions and mergers to realize the financial and market gains which are predicted at the planning stage, is testimony to the relatively rare attention given to working with human dynamics.

What comprises the exercise of leadership? If it's not about rationality, analysis, measurement or control, then what does it entail? For those of us involved in leadership development, what capacities and dynamics might we aim to help build in client organizations, or with individuals?

When leadership is seen as embedded in a context of non-rational dynamics, and as an essentially human activity, we can see that effective – and ethical – leaders demonstrate several distinct capacities. They welcome the act of engagement with others and are skillful and open in doing so. They turn their face towards others, and use the opportunity to explore rather than debate, learn rather than judge, elicit rather than pronounce. In so doing, they create an imperative for openness. They show a capability for self-understanding, combined with a willingness for the discomfort of personal learning and exploration; they are prepared to make mistakes, and are personally secure – when these aspects are manifest, they are experienced by others as congruent (which in turn increases trust).

Effective leaders demonstrate that they understand what it means to be a participant in the human system to which they belong; they know that they cannot *not* participate – that the leadership question is about the nature of that participation, not the fact of it; and in so doing, they draw others into the act of belonging, rather than fragmentation. They also show that they understand that the fact that no person can legislate, or control, another's beliefs, relationships, or knowledge, far from making them impotent, helps them to understand the power of modelling the qualities and actions they espouse – otherwise, why would others follow them down any particular path? Rather than dictate behavior, leaders 'do it'; then, rather than command, they invite others to suspend their disbelief and test whether they, too, might take the risk of new directions, new behaviors, and new identities. They know that the higher the level of interpersonal safety and trust, the more likely it is that others will accept the invitation.

Rather than seeking to 'change the culture' through electronic media, powerpoint slideshows, corporate roadshows, and 'events' at five-star hotels – all of which reinforce disconnection, formality, fabricated order, and hierarchy – effective leaders engage with others by using human means to do a human job. They are skillful in using, and co-constructing, symbols, symbolic actions, rituals, and metaphor and story.

'Leadership' may be seen as a term to describe a certain quality of relationship, in which, at that point in time, a particular person, or people, make both an offer and an invitation, with which others choose to engage. As such, the giving and receiving of leadership involves an exchange which is value-laden and rich in personal meaning. And while we might expect rational judgement to be better exercised as managers progress through the hierarchy, it is clear that leadership may be found – and increasingly needs to be found – at all levels in an organization.

Wherever leadership is brought to bear, story is an integral element in its effectiveness. This includes the 'why' of work – in the form of shared purpose and aspiration, which can only ever be captured and shared through metaphor. It also includes the 'where' – an organization's vision, direction, desired future or outcomes, best expressed in a way which connects to emotion and creates clear mental images. There is the 'how' – in the form of shared values and of the ethics of engagement, both internally and externally; it is through examples and illustrations that we build a shared knowledge of what these abstractions might mean to each other, which are then brought to life by constantly challenging each others', and our own, actions, against those espoused values. And there is the 'what' – the subjectively-laden content of our discourse, which, as Gargiullo (2005) has argued, sees knowledge and information constructed and shared primarily through story.

Leadership also works to create a shared sense of the 'who' – as a focus on identity, it can be to draw out the story and re-present it in such a way as to amplify the organization's sense of self. But it is also socially conceived – we are defined by who we relate to, and the quality of those relationships. Thus, a 'shared story' is not an artefact of data collection but of relatedness unfolding, adjusting, over time. The important thing about narrative is the process of its construction and maintenance between the story-holders. The shared story we tell and enact is embedded in our relationships, and is a living thing. Every time we have that story (or metaphor) printed or published, we risk the loss of its living character, by ossifying it in print. Stories with life must be told, re-told, personalized, and evolutionary. And, behind the shared stories, there is the dynamic, modelled by leaders, of the willingness to share our individual stories. It can be argued that relationships and trust are the product of perhaps just two things: the assurance of consistent and compassionate behavior over time; and the disclosure of and listening to each others' stories – both from the narrative itself and the way in which it is told, which take us past stereotypes to a deeper knowing of the other.

The relational skills of leadership evidence a continual infusion of metaphor and story, not just in the opening of the self to others, but in transactions which are clearer, richer and more compelling than by using technical or corporate language. Metaphor, in the form of story, anecdote, illustration and instance, evidences rapport, and is the vehicle for communicating ideas and information. It enables us as teachers, learners and coaches. And it brings the future to life in the present.

The quality of leadership is also about the depth and insightfulness of our self-understanding. Our tales about ourselves – whether shared with others or held within, others' tales of themselves, and others' tales of us, all stimulate us to reflection and the possibility of re-writing our personal mythologies with greater awareness.

A continuous challenge for leadership lies in effecting successful change and transition, and this is an area in which ritual, symbolic behavior, and story are vital. The preoccupation of managers with rational approaches leads them too often to the re-design of structure, processes, systems, and required behavior, as if the human aspects can be folded in later – or perhaps during the 'implementation' phase. But also engaging with the transitional dynamics – the human dimension – is critical to success, or else the corporate culture, the non-rational dynamics, the power structure, the emotions, and the changing identity and relationship issues are all essentially ignored. If only the corporate culture could be simply dictated by senior managers! But, as in all significant relationships, the telling and re-telling of our personal accounts and responses are an essential part of real engagement as complete human beings; and the joint pursuit of the emerging shared story for the 'new order' can only be achieved through the face-to-face co-creation of the new narrative. This contrasts strongly with the corporate pronouncements, slogans, jargon, 'spin', and 'corporate-speak' which is dry and lifeless, and usually handed down from the 'top team', or worse, by communications consultants – outsiders.

The conventional patterns of organizational life lie in disconnection, so-called 'objectivity', and detachment, and it is in this manner that much of what passes for leadership is framed. In addition, there is often a pervasive collusion to 'keep things nice'. But Whyte (2002) has argued over many years for engagement with the *whole* experience of organizational life – not just the light and the tangible, but that which is "the dark and invisible" side of that life, since it is the "very chasm from which our personal destiny emerges" (p. 5). At no time previously in the history of the modern organization has this been more important than now, at a time of immense uncertainty and global upheaval.

The common response to uncertainty by managers is to do more of what already fails, that is, to seek more control. To put a name, a human face, a feeling, or a genuine 'warts and all' engagement into organizational life is to reveal to ourselves the frightening edge of chaos and conflict, and to have to confront that place where our practice is weighed against universal human values (or our espoused ones). This is the place in daily organizational life where we are truly accountable to each other – seen every day in those areas we dare not name and face together, such as excessive and manipulated CEO remuneration levels, or the human realities of constant retrenchments ('rationalizations'), or the falsifications of corporate language and 'spin doctors'. And organizational development practitioners too often play their part in perpetuating rational and disconnected methodologies as the means to working with the uniquely human forces within our organizations, and in so doing, help to ensure that violence, aggression, power, and unjustness – as well as compassion, fairness, happiness, and grace – cannot be openly engaged, known and honoured. It is in relationships sustained by the stories which are true to us that we are led towards what Orton calls our "workable truth".

The post-globalization context for business, described by Saul (2005) and by Kelly (2006), is truly a world of turmoil with multiple emerging forces competing to drive our global futures. Effective leadership in both society and in organizations will seek to chart a course which seeks the best for their people, while honouring their right to share in the responsibility. To voice our true stories and work together to build a genuine shared narrative to represent and shape our destinies takes courage, but the leader within us has no better alternative for creating a coherent, relational and congruent course.

References

Allan, J., Fairtlough, G. and Heinzen, B. (2001). *The Power of the Tale: Using Narratives for Organizational Success*, Chichester: John Wiley.

Atkinson, Robert. (1995). *The Gift of Stories*, Westport, CT: Bergin and Garvey.

Berman, Michael and Brown, David. (2001). *The Power of Metaphor*, Carmarthen, UK: Crown House.

Bolman, Lee G. and Deal, Terrence E. (1997). *Reframing Organizations: Artistry, Choice and Leadership (2nd Ed,* San Francisco: Jossey-Bass.

Brown, J.S., Denning, St., Groh, K. and Prusak, L. (2005). *Storytelling in Organizations: Why Storytelling Is Transforming 21st Century Organizations and Management,* Burlington: Butterworth-Heinemann.

Bruner, Jerome. (2002). *Making Stories: Law, Literature, Life*, Cambridge, MA: Harvard

University Press.
Capra, F. (1996). *The Web of Life*, New York: Anchor Books.
Cohen, Don and Prusak, Laurence. (2001). *In Good Company : How Social Capital Makes Organizations Work,* Boston, MA: Harvard Business School Press.
Conger, Jay A. & Assoc. (1994). *Spirit at Work: Discovering the Spirituality in Leadership,* San Francisco: Jossey-Bass.
Deal, T. & Kennedy, A. (1982). *Corporate Cultures: The Rites and Rituals of Corporate* Life, London: Addison-Wesley.
Denning, Stephen. (2005). *The Leader's Guide to Storytelling: Mastering the Art and Discipline of Business Narrative*, San Francisco: Jossey-Bass.
Fulford, Robert. (1999). *The Triumph of Narrative: Storytelling in the Age of Mass Culture*, New York: Broadway Books.
Gardner, Howard. (2000). *Intelligence Reframed: Multiple Intelligences for the 21st Century*, New York: Basic Books.
Gardner, Howard. (1995). *Leading Minds: An Anatomy of Leadership*, New York: Basic Books.
Gargiulo, Terrence L. (2005). *The Strategic Use of Stories in Organizational Communication and Learning*, New York: M.E. Sharpe Inc.
Kearney, Richard. (2006). *On Stories,* London: Routledge.
Kelly, Eamonn. (2006). *Powerful Times: Rising to the Challenge of our Uncertain World*, New Jersey: Wharton School Publishing.
Kennedy, Allan A. (2000). *The End of Shareholder Value: Corporations at the Crossroads*, Cambridge, MA: Perseus Publishing.
Kurtz, Ernest and Ketcham, Katherine. (1992). *The Spirituality of Imperfection: Storytelling and the Search for Meaning*, New York: Bantam.
McAdams, Dan P. (1993). *Stories We Live By: Personal Myths and the Making of the Self*, New York: Morrow.
Morgan, Gareth. (1986). *Images of Organization*, Bristol: Sage.
Neuhauser, Peg C. (1993). *Corporate Legends and Lore: The Power of Storytelling as a Management Tool*, Austin, TX: Self-published.
Orton, Torrey. 'Facing the Facts' Seminar, Melbourne, 02 Dec 2005.
Parkin, Margaret. (2004). *Tales for Change: Using Storytelling to Develop People and Organizations*, London: Kogan Page.
Sanders, T. Irene. (1998). *Strategic Thinking and the New Science*, New York: Free Press.
Saul, John Ralston. (2005). *The Collapse of Globalism and the Reinvention of the World*, Melbourne: Penguin Viking.
Senge, Peter et al. (2005). *Presence: Exploring Profound Change in People, Organizations, and Society*, New York: Currency Doubleday.
Watson, Don. (2003). *Death Sentence: The Decay of Public Language*, Sydney: Knopf/ Random House.
Wheatley, Margaret and Kellner-Rogers, Myron. 'The Irresistible Future of Organizing', Seminar, Sundance, UT: 25-28 Sept 1996.
Wheatley, Margaret. (1992). *Leadership and the New Science*, San Francisco: Berrett-Koehler.
Whyte, David. (2004). *Crossing the Unknown Sea: Work as a Pilgrimage of Identity*, New York: Riverhead Books.
Whyte, David. (2002). *The Heart Aroused: Poetry and the Preservation of the Soul in Corporate America*, New York: Currency Doubleday.

David Green is an independent consultant in organizational development, focussing mainly on leadership development, organizational change and dynamics, and relational systems approaches. His background includes working in the public sector, in a major utility, and in an IT company, before setting up his consultancy in 1994. Since then, he has worked with a wide range of public and private client organizations, including TNT Logistics, BHP, Queensland Department of Main Roads, Ernst and Young, Linfox, the Department of Innovation IT Economy Sport and Recreation Queensland, Simon Engineering, and Sustainable Industries in the Queensland EPA. His approach draws on human systems and organizational dynamics ideas, and always involves experiential, or living, learning environments. In recent years, he has initiated the use of narrative approaches, including storytelling, in all his projects, in a wide range of ways, and in three organizations has conducted residential programs for managers in the craft of story-making and telling as a leadership art.

Every Journey Starts with a Single Step: Yarra Valley Water's Journey Towards Environmental Sustainability

Francis Pamminger[1] & Julian Crawford[2]
[1]Yarra Valley Water, AUS
[2]EcoSTEPS, AUS

This Paper discusses the innovative and pioneering activities undertaken over the last three years by Yarra Valley Water (YVW) to make it a more environmentally sustainable organization. YVW recognizes that both the well-being of our community and the strength of our business depend on the health of our environment. This means we must provide our water and sewerage services in an environmentally sustainable way 'within the carrying capacity of nature'. Three years after starting, the organization asked the question: "What do we do now to progress further?" EcoSTEPS were engaged to help and did this by delivering a training program to all employees in April 2006. This paper focuses on the key learnings from the process and is applicable to all organizations and communities. Topics addressed include:

- **Behavioral and cultural change processes and impacts;**
- **Challenges we are now working with;**
- **Innovative Education for Sustainability techniques employed.**

Introduction

Yarra Valley Water is committed to environmental sustainability. We recognise that the well being of our community and the strength of our economy are dependent on the health of our environment and accordingly strive to provide water and sewerage services within the carrying capacity of nature.

This paper has been written to summarize our progress to date and present the challenges we are now dealing with. We are proud of our achievements to date and equally recognise that we are not yet sustainable. We also recognise that our goal to be environmentally sustainable is arguably fruitless on our own. For example, will we experience any environmental benefit if we achieve our objective to have no net greenhouse gas emissions if no one else also does anything? We accordingly share our learning equally for others to learn, and in the hope that there may also be people who can help us.

Behavioral and cultural changes processes and impacts

We are proud of what we have achieved to date. We have a clear and bold commitment to achieve environmental sustainability. It is articulated in a business policy, endorsed by our Board, it is integrated into day to day business activities, and monitored monthly by our Executive through a Balanced Scorecard reporting process. We have done this by clearly articulating what environmental sustainability means in the following way. To provide our services within the carrying capacity of nature we will:

- Reduce greenhouse gas emissions to below 10,500 tons CO2-e by 2008, which is 50% of our 1990 total.
- Reduce average water consumption to below 306 liters per person per day by 2010, which is 80% of our 1990's total.
- Reduce the amount of waste to landfill to below 10% of the total waste by 2010.
- Establish quantifiable targets for nutrients discharged from our sewage treatment plants by 2006.

To get to this stage however was in itself quite a journey. Reflecting on the major steps the organization had to face and solve, four are identified. Organizational *buy-in* was required, a *definition* of sustainability was required, *awareness* to the challenges was required, and a means of moving from the *philosophical to pragmatic* was required.

Buy-in

The seed for change was first recognized at a business planning workshop towards the end of 2002, which was attended by the businesses management team. They had gathered to initiate the planning process for the coming financial year. To gain clarity of focus a facilitator using a 'fish-bowl' process asked three questions of three different groups. The first group was asked to identify the purpose of the business ten years ago. The second group was asked to identify present drivers, and the third group was asked to identify what future drivers should be.

Reflecting back on the answers, the first group was very clear on the past drivers. They quickly came to agreement and could not be challenged by the facilitator. The second group struggled. Not out of effort, but rather that a lot of work was being done for a lot of different drivers. The experience left people tired and arguably reflecting where the organization was at. While the third group, similarly to the first, quickly agreed that an opportunity existed to improve our environmental performance and that is where we should point the company.

Not only was this a very powerful process to gain *buy-in,* it was further expanded through the company. Coinciding with the engagement of a new Managing Director in 2003, a number of 'world café's' were conducted throughout the business including all employees. The objective was to explore what individuals perceived the business should do to become an environmental leader. The collective outcome of this work was captured on a mosaic that was produced and hung around the business in prominent positions.

We monitor employee opinion of our commitment to environmental sustainability through four questions included in our employee opinion survey

Figure 1 *Mosaic compiled from individual thoughts and desires for environmental improvement*

conducted every two years. This records an average 25% increase over the last survey period, hence a high level of employee 'buy-in' to the change process.

Definition

Having established a collective business agreement that we needed to do something, the next step was to determine what this should be. We chose The Natural Step process to help us with this. This was chosen because of the perceived fit that this had with the business's predominant paradigm, which is one built on the reductionist scientific principles. The predominant professions employed at Yarra Valley Water are engineers and accountants. We liked the scientific robust definition it provided for sustainability.

The Natural Step defines sustainability in the following way. In the sustainable society, nature is not subject to systematically increasing:

1. concentrations of substances extracted from the earth's crust.
2. concentrations of substances produced by society.
3. degradation by physical means,and, in that society
4. human needs are met worldwide.

Awareness

All employees had exposure one way or another to the Natural Step principles and in this way we created a common business definition for sustainability, which is to provide our services within the carrying capacity of nature. It started first with the Management team, who through an environmental audit against the Natural Step principles, explored our existing performance. This consisted of three half day workshops. Through to those with less involvement in sustainability receiving only a one hour training workshop where we only explained the general concepts.

A major component of the training was raising everyone's awareness of the need to change. The most significant contributing element to this was the Ecological Footprint concept developed by Mathis Wackernagel and Bill Rees, as described in Wackernagel *et al.* (2000). This was later reinforced with a visit by Mathis to Yarra Valley Water in February 2004. The Ecological Footprint taught us all that if everyone lived the lives we did we would need three planet earths to support us. Mathis humorously explains, given that he has delivered a scientific paper that was peer reviewed and approved, that proves that we do not have this available to us.

Philosophical to pragmatic

Having buy-in, a clear definition, and awareness, only brought us so far in our journey. There was no denying that many employees were enthusiastically engaged, however, having reached this, there was still uncertainty as to 'what we should do on Monday', so to speak.

We now had before us many different concepts, for example, The Natural Step, Ecological Footprint, Environmental Management Systems, the waste hierarchy, and we were confused as to how they all fitted together. The paper by Robert *et al.* (undated) provided the theoretical framework for us to understand this, and we consequently incorporated this thinking into the development of our environmental strategy.

It is very important to bridge the gap between the philosophical and the pragmatic as this is a major stumbling block for some people. So much so that it impedes their engagement with any action. Another way that we have addressed this is to use a metaphorical description.

It is true that sustainability by definition is unachievable. Taken to its absolute definition, and maybe philosophical, the plant earth only has a life of about 4.5 billion years. Arguably also, we as individuals also are not sustainable; we also have a limited life. But against this, there are still actions we can take 'on Monday', to move from the philosophical to pragmatic. Commonly cited and easily available indicators for human health are blood pressure, cholesterol, and weight. And equally, when we think of the environment, we must also identify the key variables that we need to focus on. It is those that we need to target 'next Monday'.

Challenges we are now working with

Three and a half years after starting on our journey to become environmentally sustainable we asked ourselves: "What do we now have to do to further improve?" Surprisingly, for everything we had done, employees were of the opinion that we were: 'talking the talk but not walking the walk.' Faced with that, we set about designing a training program to take the next steps on our journey. It is this program that EcoSTEPS has helped us with.

It was our primary goal to move from the self talk 'that expected the business to do something' to achieving individual ownership where employees could say 'this is what we are doing to make Yarra Valley Water more sustainable'. Fundamental to this was the recognition that the decisions we make are based on the knowledge we have and the opinions we have.

When we look at how decisions have changed with regard to drink driving, sun exposure, seat belts, hosing driveways and recycling water and so on over the last twenty years, this becomes more transparent. As knowledge has changed, we have changed our responses. Equally, we now need to look forward and ask how we incorporate knowledge and opinions on environmental sustainability.

Innovation education for sustainability techniques employed

It is against that background that we specifically asked EcoSTEPS to design our environmental training program for all employees of the business. Given that we find it easier to change the opinions of others rather than change our own, the training program had to find a way to address this issue. We chose an experiential process where people could see how others made decisions, and arguably how an individual may use a different process when considering decisions either at home, in their work group, or how the business makes decisions may highlight the processes. Alternatively, they would see it how others made decisions, and then hopefully extrapolate this into how they also made decisions.

The Overall Aims of the Training Program were articulated at three levels, namely:

1. Personally - What does Environmental Sustainability mean to ME (and my family) personally?
2. Organizationally -What does Environmental Sustainability mean to US collectively as an organization?
3. My Job - In terms of my role/job/team at YVW, what does Environmental Sustainability mean?

The Outline Contents of each training session were:

- Environmental Sustainability – What is it?
- What are the big issues for us as individuals?
- Sustainability Approaches
- What's YVW been doing as an organization?
- What more can we do? Case Studies
- Progressing Sustainability
- Reflections & Next steps...

Over 300 staff and contractors attended the training program. There was a formal evaluation process based on the Dept of Environment & Conservation (NSW)'s 'Does your Project Make a Difference – Guide to Evaluating Environmental Education Projects and Programs'. The Overall Aims of the Program were largely achieved and set an platform for the next steps on our journey.

Conclusions

Yarra Valley Water is proud of its achievements to date with regard to achieving environmental sustainability. However, as we are not yet there yet, we can say that our progress is like a journey of many steps. We start by taking one step at a time. Each step represents a challenge and consequently requires a solution before we can progress to the next. Our steps to date have included achieving buy-in from employees that we needed to do something. Then we had to get collective agreement on a common definition of sustainability. This was followed by instilling an awareness that 'sustainability' is important and that we all need to change. Then we had to develop a framework to allow us to move from the philosophical to the pragmatic facilitating a move to action. Most recently we have addressed how to address the challenge that the values that individuals have may be the next barrier for us to progress to the next stage. EcoSTEPS were used to develop a specific training program to help with this.

References

Dept of Environment & Conservation (NSW) (2004). *Does your Project Make a Difference – Guide to Evaluating Environmental Education Projects and Programs*.
EcoSTEPS website, www.ecosteps.com.au (Accessed 1 October2006).
Mathis Wackernagel; Chambers, Nicky; Simmons, Craig (2000). *Sharing Natures Interest. Ecological Footprints as an indicator of sustainability.* Earthscan. London.
Robert, Karl-Henrik; Schmidt-Blek, Bio; Aloisi de Larderel, Jacqueline; Basile, George; Jansen, J. Leo; Kuehr, Ruediger; Price Thomas, Peter; Suzuki, Motoyuki; Wakernagel, Mathis: *Strategic sustainable development – Selection, design and synergies of applied tools.* Journal of Cleaner Production 10 (2002) 197–214.
Yarra Valley Water (undated). *Towards environmental sustainability - Our journey.* www.yvw.com.au/NR/rdonlyres/BA195021-D98F-4B19-BEC9-F94DB275D2CF/0/OurJourney.pdf (Accessed 1 October 2006).

Mentoring Three Local Association Staff: Our Experiences With Being Integrative and Sometimes Innovative

Richard Monypenny
James Cook University, AUS

We report our experiences with an association of local service providers as we under took mentoring, capability development and cultural transformation (mentoring for short) of their three senior staff in their project facilitation skills, on top of their existing very good policy, management and project management skills. The aim of our intervention was to extend the driving and the success of their business strategy. Most, if not all, of the separate component parts that we use to be integrative in how we do our work in mentoring the three staff are well proven (not new). But we do claim that at least one way in which we combine the separate component parts, is in fact innovative. Specifically, the way we combine separate parts to make the link between inputs and outcomes achieved. The experiences reported are those relevant to the Systems Conference. Experiences related to other aspects of our work, are reported elsewhere.

Introduction

We have experiences regarding our work with an association of local service providers as they make concrete steps towards closer cooperation with some of their main local stakeholders so that they, in fact, become partners.

We see these concrete steps towards closer cooperation as one of the early stages in the medium term process to successfully develop a local industrial cluster.

The association's core business since 1994 is implementing and managing projects that are largely about teaching technical skills. The project officers have very good project management skills.

In this paper we report on our experiences from our work with three senior full-time association staff. The aim of our intervention was to extend the driving and the success of their business strategy by improving their project facilitation skills, on top of their existing very good policy, management and project management skills.

We see the project facilitation skills first, as the main building block and second, as the general framework for closer cooperation and increased value adding onto existing organizational tasks and activities.
Most, if not all, of the separate component parts that we use to be integrative in how we do our work in mentoring the three staff are well proven (not new).

But we do claim that at least one way in which we combine the separate component parts, is in fact innovative. Specifically, the way we combine separate parts to make the link between inputs and outcomes achieved. We usually call this link the value adding onto existing organizational tasks and activities.

The experiences that we report in this paper are those relevant to the Systems Conference. Experiences related to other aspects of our work, like for example aspects directly related to improvements that could help increase local service provider wellbeing, are reported elsewhere.

The paper has two parts:

- A very brief outline of the context for our work with the three association staff.
- An outline of a few aspects of our experiences that are most relevant to the participants at the Systems Conference:

    1. Being Integrative. We use the project facilitation skills to integrate all aspects of our intervention.
    2. Being Innovative. We combine known separate parts in an innovative way to make the link between inputs and outcomes achieved. We usually name this link as value adding.
    3. Active listening as an all-pervasive process of interaction between the four of us (The three Division staff and my self).
    4. Active listening as a vehicle for improving project facilitation skills between the four of us.
    5. Value adding as the outcome from improving project facilitation skills.

The context for our work with the three association staff

In this section we outline the context of our work with the three staff. The aim of this section is to give the background and the context for our experiences in the next section. The main take home message from this context section is that the context within which we were interacting with the three staff is, and always has been, a very good working environment with high staff morale, in which all staff are strongly encouraged to be good project managers. In practice this environment can be divided into two parts:

- That part that is relatively stable, like for example salaries.
- Second, that part that is very dynamic, with significant opportunity to be proactive to unexpected opportunities, like for example to be offered extra funding from an unexpected source.

The main aspects of the context are:

1. The association of local service providers reported in this paper is the Townsville Division of General Practice. Its members are most of the General Practitioners who practice in Townsville. The Division has been in

operation successfully since 1994. Its core activity to date has been largely about teaching technical skills to General Practitioners and to staff that work with the General Practitioners.

2. We now want to place the Division in such a position so that it will have an increasing role, along with a few other stakeholders, in being the main drivers of the success of our future industrial cluster. To date, most of the successful industrial clusters, both in Australia and overseas, have been in manufacturing industry. One of the next challenges for industrial clustering proponents, practitioners and researchers is the development of industrial clusters of organizations that are mainly service providers, rather than mainly manufacturers.
3. We want our Townsville cooperation between the Townsville Division of General Practice, the State Development Centre Townsville (which is part of the Department of State Development Trade and Innovation, Queensland) and the School of Business at James Cook University, to be at the forefront of these developments. When we stared our work with the Division in mid 2005 our challenge was how do we help the Division to awaken itself to the realization of the extraordinary opportunity that it had within its grasp? And to do this without triggering the negative side effects that often come associated with such undertakings.
4. Some of the important aspects of why we felt that "the Division had an extraordinary opportunity within its grasp" are:
    a. A well-established reputation of being a reliable organization in the delivery of its core business.
    b. A well structured and experienced governance structure that includes a Board of Directors, a CEO and an Operations Manager.
    c. A recently established partnership arrangement with four other Regional Divisions of General Practice covering all of North Queensland to pursue projects of mutual interest.
    d. A well-established and experienced local Regional Economic Development agency, the State Development Centre Townsville.
    e. A well-established and experienced local Regional University, James Cook University.
5. From the start we acknowledged that our challenge to successfully develop a local industrial cluster was unlikely to succeed if we used a reductionist approach. Rather we started from our evidence from our observed reality that the combination of health and regional economic development was likely to be about the most complex system that we could have picked to work on. But, we were very lucky:
    a. To have access to well developed human capital in the form of the appropriate combination of staff skills and staff experience and
    b. To be operating in a work environment with the appropriate level of organizational flexibility, executive decision making and senior

staff buy-in that got us to where we are now.

6. What really makes the work that we do in the Townsville Division of General Practice, different is that the Division's relationships are maintained irrespective of the 'Grant or Project' funding, and that the relationships help us move towards solutions as a team, providing mutual support as required, and picking up grant and project funding as it becomes available, to progress already identified and agreed objectives.
7. The literature related to the various aspects of our work in the Townsville Division of General Practice is relatively extensive. However, we are not aware of any literature that reports on projects similar to ours. This extensive literature is covered in 11 papers presented recently (August 2006) to the SEGRA 2006 Conference. The Conference link is http://www.managementsolutions.net.au/segra/index.html. For details on the papers please see the attached Technical Appendix.
8. The literature related to the various systems thinking aspects of our experience in the Townsville Division of General Practice is relatively extensive. However, we have largely followed the well know authors, like for example: For example Checkland and Scholes (1990), Flood (1990), Senge (1990) Maani and Cavana (2000). For details please see Monypenny (2001).
9. So given that the conference is designed so that papers can be mentored and can evolve I have included a few references that have the literature rather than to include the literature directly here. These are: Monypenny (2001); Monypenny and Mellor (2005). The full papers are available by request from Richard.Monypenny@jcu.edu.au

To close this context section, we would like to reinforce the idea of a very dynamic and invigorating context within which we were interacting with the three staff. The next section looks at a few aspects of our experiences.

A few aspects of our experiences

In this section we outline our main experiences. The main take home message from this experiences section is that any valid evaluation of our intervention needs to include the real dynamics of the system that underlie the normal activities of The Townsville Division of General Practice. We acknowledge that it is probably very difficult to create such a dynamic system from scratch but it is certainly very invigorating to have the opportunity to actively participate in such an intervention.

The main aspects of our experiences are:

1. Being Integrative. We use the project facilitation skills to integrate all aspects of our intervention.
    a. First, we use project facilitation skills as the main building block for creating closer cooperation and increased value adding for local service providers.

    b. Second, we use project facilitation skills as the framework with which the association integrates its activities in order to focus effort on creating closer cooperation and increased value adding for the association.
    c. We chose project facilitation skills as the integrative concept because:
        i. The Division is largely a project based organization and all project officers have good project management skills.
        ii. The concept of project facilitation is an easy extension from the already well understood concept of project management.
2. Being Innovative. We are especially proud when we look at the way we combine different well known components to make the link (or the relationship) between inputs used and the outcomes achieved. We usually name this link as value adding. What our innovation has really done is:
    a. To give an overall reason for pulling together many apparently disjointed skills and tools. Like for example active listening, value adding, individual and personal coaching, walk the talk (or other similar), insiders vs. outsiders (or similar), etc.
    b. To deliver the use and application of this innovation in real time, in real Division decision-making, in real resource demanding situations, in real trade-off situations.
    c. To use our innovation by choice and proactively rather than reactively.
3. Active listening as an all-pervasive process of interaction between the four of us (The three Division staff and my self).
    a. Yes, active listening is often seen as what some people pretend to do when they do not really want to listen. What we did was much harder to do. It is like when you do a Google search for something simple like ... and three hours later you are still finding interesting aspects that are related and relevant to where you started but with many more new opportunities to follow up.
    b. Small group skills support active listening very well, especially as the group goes through the stages of development: forming, storming, norming and performing (or similar).
4. Active listening as a vehicle for improving project facilitation skills between the four of us.
    a. The driver of this experience was the staff's interest to hear something different. That is, driven by yes it is working but there must be something else?
    b. Supported by my interest to understand where they were at individually and as a group of three.

        i. To have successful communication and growth we needed to have a shared starting point, a common interest. It is difficult for two to share if they have little in common.
        ii. My strategy is to learn where they are at, not that they need to learn where I am at.
5. Value adding as the outcome from improving our project facilitation skills.
    a. Get in with what they want then work on what they really need.
    b. Systems thinking (action research, experiential learning or similar) that supports the reality that every thing is related to every thing.
    c. That we usually underestimate the complexity of the systems that we study.
    d. That learning is important, that is once we have finished a given intervention, if we had to do it again we would do it differently.
6. The outcome or conclusion from our mentoring, capability development and cultural transformation (mentoring for short) of the three senior staff in their project facilitation skills, has been extremely successful. This success can be indicated in that:
    a. The regular organizational meetings of the three senior staff are now automatically run using their improved project facilitation skills. Rather than the normal meeting format used in the past.
    b. The value adding onto existing organizational tasks and activities is now a regular consideration in decision making in allocating resources to organizational tasks and activities.
7. In summary, the take home messages from our experiences are:
    a. Dynamic and invigorating. Given that the Division is such a complex, dynamic system and it is very invigorating to work in, the main outcomes from our experiences are very different from the main outcomes that we would expect from our experiences from a less dynamic and invigorating work place.
    b. Congruent management. Given that the Division is a complex, dynamic and invigorating work place, the management and leadership style and all intervention activities need to be congruent. This is especially important for internal staff in relation to external consultants that are brought into the Division to do a specific task.
    c. Learning from our experience. Given that the Division is such a complex, dynamic and invigorating work place, we are successful largely because we are part of a well functioning system. We did not have to create the well functioning system we just had to help it to be more successful.
    d. Our advice to others. Given that the Division is such a complex, dynamic and invigorating work place, from our experiences, we acknowledge that success comes not just from our own skills and

efforts but from the summation of all aspects of the system. Our contribution is usually, but a small part. In the medium term it is unwise to let success go to our head.

Acknowledgements

This paper was developed as part of collaboration between the Townsville Division of General Practice, the State Development Centre Townsville in the Department of State Development Trade and Innovation (Queensland) and the School of Business at James Cook University in Townsville.

References

Checkland, P. and Scholes, J. (1990) Soft Systems Methodology in Action, John Wiley, Chichester, UK.

Flood, R.L. (1990) Liberating Systems Theory, Plenum, New York, and USA.

Maani, K.E. and Cavana, R.Y. (2000) Systems Thinking and Modelling: Understanding change and complexity, Prentice-Hall, Auckland, New Zealand.

Monypenny, R., (2001) Teaching project management: A systems challenge. In: Systems in Management 7th Annual ANZSYS Conference 2001 31 - 39 (eds W Hutchinson & M Warren). Edith Cowan University, Perth, WA

Monypenny, R and Mellor, P (2005, September) Building Regional Development Practitioner Skills: A Collaborative Strategy. Proceedings of the 2005 SEGRA Conference, Yeppoon.

Monypenny, R. (2001, November) Teaching Project Management: A systems challenge, Proceedings Systems in Management 2001 7th Annual ANZSYS Conference, Perth, pp 31-39.

Senge P.M., (1990), The Fifth Discipline : The art and practice of the learning organization, Doubleday, New York, USA.

Technical appendix: A short review of some of the relevant literature

The literature related to various aspects of our work in the Townsville Division of General Practice is relatively extensive. However, we are not aware of any literature that reports on projects similar to ours.

One significant spillover of our work in the Townsville Division of General Practice is the writing of a number of papers related to the current interests of the Division. These papers can be divided into two groups.

- First those that review some of the aspects of the extensive literature that is more relevant to the Division's current interests. These are papers 1-5 below.
- Second, those that report various aspects of the Division's activities. These are papers 6-10 below.

Currently (October 2006) 10 of these papers are in various stages of being DEST refereed by SEGRA . The Conference link is http://www.managementsolutions.net.au/segra/index.html. The full papers are available by request from Richard.Monypenny@jcu.edu.au A brief abstract of each of these papers is also included.

1. International Trade and Local Regional Development (Bano and Monypenny)

Our research interest is the intersection of the following three different but related areas of study:

a. The provision of local primary heath care services.
b. The understanding of those aspects of the regional economic development literature that are relevant to the international trade in primary health care services.
c. The understanding of those aspects of the international trade literature that are relevant to the international trade in primary health care services.

This paper reports the outcome of a project that reviews the literature on the following aspects of international trade:

a. Intra and inter-industry trade.
b. Intra-industry trade indices.
c. The demand similarity theory.
d. Economies of scale, product differentiation and market structure models.
e. Research and development expenditure, human capital, and the technological and the product life cycle models.
f. Trade diversion and trade creation.

2. Comparative Advantage (Castorina and Monypenny)

A good understanding of the literature on comparative advantage is seen as a precursor to increasing cooperation between stakeholders, especially in industries linked to international trade. Increased cooperation between stakeholders is seen as one of the early stages in the development process that might lead in the medium term to the establishment of a formal cooperation agreement or to an industrial cluster.

It is widely believed by economists that comparative advantage is a key determinant of international production and trade patterns. Gaining a deeper understanding of the concept of comparative advantage enables regions or countries to increase their welfare by exporting in those goods in which they know they have a comparative advantage.

In Section II we review the evolution of trade theory from Classical to neo-classical and then to Modern theory. Section III is a practical application of determining comparative advantage.

3. The Knowledge Economy and Local Mainly Service Industries (Jenkinson and Monypenny)

This paper reports on the outcome of a project to review the literature on cluster development as related to the knowledge economy. The paper has three parts:

a. A brief review of the literature on both Porter's Cluster Theory and on the evolution that has taken place post Porter.
b. An outline of the differences and the similarities between the application of the above literature to mainly manufacturing industries and to mainly service industries. We place a strong focus on what can be relatively easily transferred from manufacturing to service industries and what is likely to be far more difficult to transfer to service industries.
c. A brief discussion of the main advantages and disadvantages of using the above literature to support an application to local primary health care.
    i. Good theoretical concept and capture of cross-disciplinary concepts.
    ii. Problems with applications.

4. Has the Diamond Lost its Sparkle: An analysis of Porter's Cluster Theory (Jenkinson and Monypenny)

This paper reports on the outcome of a project to review the literature on Porter's Cluster Theory. The paper considers the development of a key aspect of Michael E Porter's competitiveness theory, that is, cluster theory. The paper has four parts:

a. Development of Porter's Cluster Theory and the consideration of economic theories leading up to 1990 Cluster Theory by Porter.
b. Porter's theory – Diamond Model & Porter's definition.
c. Post Porter Cluster Evolution:
d. Advantages / Disadvantages

5. Industrial Clustering (Lines and Monypenny)

The performance of the national economy is a composite of the performance of regions within the nation. It is now relatively well understood that the performance of the region is largely influenced by the strength and performance of local industry clusters.

Industrial clustering and 'cluster theory' has in recent times emerged as a leading economic policy widely accepted for increasing industrial efficiency.

This paper is the outcome of a project to review the literature on both the positives and the negatives from industrial clustering. A good understanding of the literature on industrial clustering is seen as a precursor to increasing cooperation between stakeholders. Increased cooperation between stakeholders is seen as one of the early stages in the development process that might lead in the medium term (say 8-12 years) to the establishment of a formal cooperation agreement or to an industrial cluster.

6. The Business of Health (Mellor and Monypenny)

This paper reports the outcome of a project to better understand the economic drivers of the local industry of the Business of Health. The paper is divided into the following three sections:

a. We review the significance of the Health Sector from a business perspective and we address a range of relevant issues.
b. We briefly outline the main differences and similarities between the Business of Health and the other significant local business sectors.
c. We briefly outline a few of the main reasons in support of the role of the Business of Health as a contributor to the forces of local regional development.

We conclude, that to date significantly more emphasis has been placed on other sectors of similar overall contribution like for example minerals processing and that this paper could be a starting point so that one of the next sectors to receive increased focus of the forces for regional development, could be the Business of Health.

7. A Baseline for Progress (Arlett and Monypenny)

Progress since the establishment of the Townsville Division of General Practice in 1994 has been significant. But it has been difficult to establish just how significant the improvements have been because a Baseline was not established at that time.

This paper reports the outcome of a project to now establish such a Baseline. Our aim in determining our local progress is to identify what has worked well locally and what still needs our aggregate attention so that we can continue to improve our whole of community outcomes. The paper is divided into the three sections:

a. The sources and attributes of both primary population health data and data on wider measures of wellbeing.
b. A brief review of the literature of the evaluation of progress from baseline data in primary health care.
c. A brief review of the perceived priorities of local stakeholders about the variables best used to measure progress.

8. Sustaining Competitive Advantage in the Primary Health Care by Working Together (Randall and Monypenny)

This paper reports on a project to review of literature on the characteristics of regional clusters and the involvement required by individuals, organizations and the region to transform the challenges of a rapidly changing competitive environment into a sustainable, competitive advantage, creating value to cultivate a vibrant, healthy regional economy.

Based on the findings from the review of the literature, the paper looks at the Townsville/Thuringowa primary health care sector for examples of the characteristics of regional clusters.

Finally, we conclude and we make recommendations for further research that might help cultivate a vibrant, healthy regional economy. We focus specially on assisting practitioners in the primary health sector. A significant first step is a regional needs analysis.

9. The Combination of Private and Social Capital in the Primary Care Sector (Scheuber and Monypenny)

We report the outcome of a project to highlight issues that present when time limited taxpayer funds are provided to the private health sector in attempts to drive the reform agenda of the Australian Health System. We see that these issues have implications for local regional development.

We focus on the Primary Care Sector which is particularly influenced by the Australian Government's Health Service Delivery policies and implementation strategies. These policies and implementation strategies are designed to meet health service delivery needs at the broadest of levels.

The example that we use to present the issues is the work that we have undertaken in the Townsville Division of General Practice to establish a regional After Hours Primary Medical Care Service.

In concluding, the paper states that developing a local working partnership and complementarity between the private and social contribution is easier when there is an effective mechanism for expressing and arriving at appropriate policy decisions. In our case one of the local organizations that can continue to facilitate these ongoing relationships in the Townsville Division of General Practice.

10. Managing Frequent Flyers: Developing and Maintaining Improved Relationships in Local Service Delivery (Scheuber and Monypenny)

We report the outcome of a project to better understand the difficulty of developing and maintaining improved relationships in local service delivery. We see these relationships as important building blocks in the early stages of the development process towards increased cooperation used for regional development. We have three positive take home messages:

a. We accept that positive relationships take time to build;
b. We accept that there are significant differences in terminology used between organizations;
c. We focus on whole of community outcomes to ensure that all opportunity costs and benefits are considered in the planning.

To close this Technical Appendix, on a short review of some of the relevant literature, we see that a significant spillover of our work with the Townsville Division of General Practice is being able to make available these papers to other interested parties, like the participants at this Systems Conference.

Richard Monypenny, BIng. Agr., MS, PhD, is Associate Professor in the School of Business, James Cook University, on the Townsville Campus. His teaching responsibilities are in the two third year subjects EC3420:03 Regional Economic Development and EC3401:03 International Trade. His research supervision responsibilities are for Honours theses and for Ph D theses. He is the Postgraduate Liaison Officer in the School of Business. In this role he spends considerable time and energy in mentoring students and growing them into their Honours and Ph D programs. He is always interested in helping individuals who are looking for academic mentoring and supervision. Richard's areas of research interest include: Local Primary Healthcare in collaboration with the Townsville Division of General Practice; Regional Economic Development in collaboration with The State Development Centre, Townsville (Queensland State Government), and; International Trade and the Local Implications from International Trade.

Meditating in the Margins

Kay Fielden
Unitec Institute of Technology, NZ

In this paper the concept of mediation as an institutional tool to render organizational conflict invisible is discussed. In an organization's human activity system there is a marginal space between the dominant controlled organizational system and its environment in which mediation takes place. These concepts are illustrated with a hypothetical case that demonstrates the complexity of this marginal system.

Introduction

Alternate dispute resolution (ADR) within organizations has become common practice in meeting employment relations legislation (Report, No date) and providing a healthy work environment. It appears that ADR has been relegated to the margins of organizational activities as institutional issues become individual problems (Herr, 2005). In a hypothetical case informal mediation as an ADR is hidden from the official organizational system and confined to a border system (Figure 2). All stakeholders in the informal mediation process are considered. This case demonstrates: how a border system is linked to an official system; and the implications of hiding ADR.

Mediation defined

"The process by which the participants together with the assistance of neutral person/s, systematically isolate disputed issues in order to develop options, consider alternatives and reach agreement that will accommodate their needs." (Leadr, 2002)

In the Leadr (2000) model (Figure 1), the mediator maintains a detached, neutral and unbiased position. Initially, a joint meeting of parties occurs. Subsequently, separate sessions are held. The mediator helps parties to move towards own solutions. Both parties are interviewed separately and then brought together to negotiate a solution. Characteristics of a mediator include: listening, empathy and neutrality. Informal mediation in this hypothetical case does not necessarily involve bringing both parties together. Both parties must agree for this to happen.

Mediation and organizational structures

Organizational mediation tends to be bargaining rather than therapeutic (Warters, 2000). Both parties are encouraged to reach a mediator-facilitated solution.

It is helpful in considering organizational ADR to consider a border system in which complex activities may occur (Figure 2). The inner border-edge provides a boundary with the organizational system (OS) whilst the outer

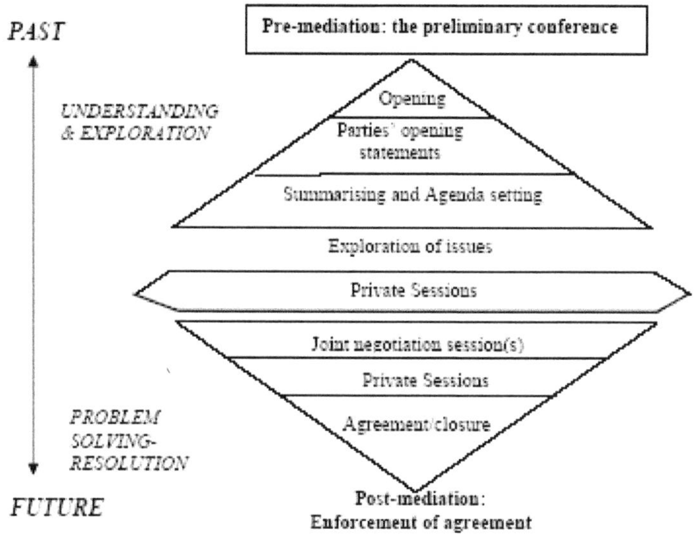

Figure 1 *Mediation Model (Leadr, 2002)*

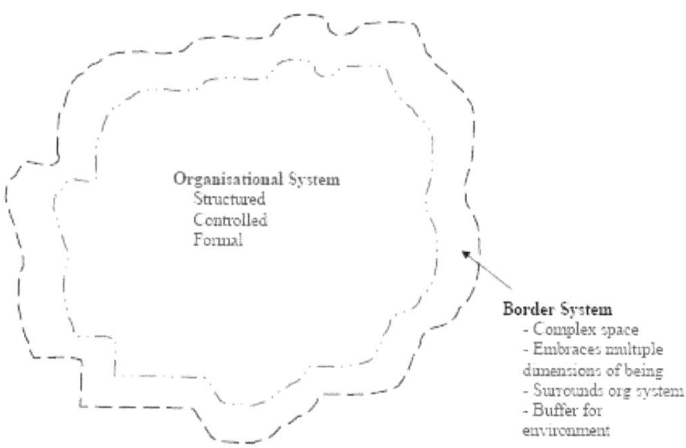

Figure 2 *Border system defined*

edge separates the OS from its environment. The outer edge is defined by the external view provided by public websites, advertising, annual reports and media statements. The inner edge lies within the OS and provides a buffer between private and public information and processes. Confidential meetings, hidden structures and roles within an organization inevitably reside within a border. In a healthy organization the inner border-edge becomes porous allowing the free flow of information as well as protecting required confidentiality. Unofficial transformational processes either within formal personnel structures or as formal institutional procedure exists within the border. Informal mediations are transformational processes that occur within the border. In this case, control mechanisms within border systems may become an organizational issue. Communication within the border may also pose organizational issues as hidden activities may not have clearly defined communication lines. It seems that conflict that arises within the official OS is mandated as requiring some form of ADR which is then conducted within the border system.

A hypothetical case

The ADR within institution X is conducted by voluntary internal service positions. 15 contact harassment people (CHPs) and an informal mediator (IM) perform these roles voluntarily as well as their normal employment contract.

The contact harassment list

The organizational conflict resolution procedures (Figure 3) require a list of CHPs as a first contact point.

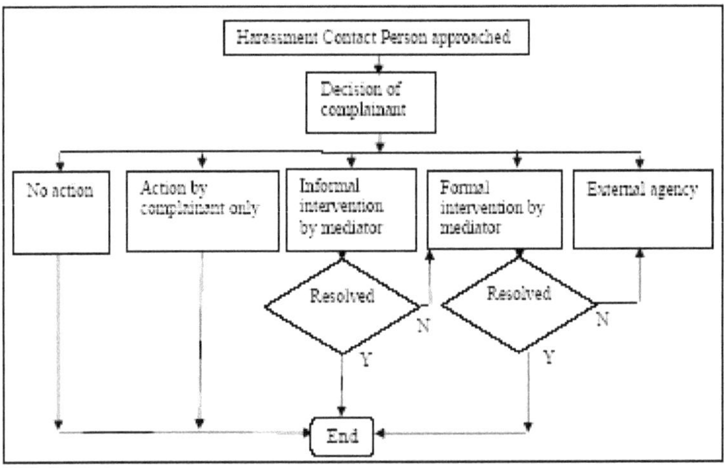

Figure 3 *Alternate dispute resolution procedure*

These people are organizational volunteers from a range of positions. Logistically, CHPs are required to cover academic discipline, student support, allied staff, onsite contractors and student residences. Volunteers are requested at the end of each year to undergo selection and training.

The informal mediator (IM)

Informal mediation is required as the third step in problem resolution (Figure 3) in which the IM is a neutral facilitator (Figure 1). Mediation skills and knowledge include: organizational knowledge, classroom practice, teaching and learning skills, interpersonal understanding and an ability to empathise. Erickson(2000) suggests that a "designated insider" is often better placed to facilitate ADR.

The conciliator

The conciliator has an official salaried position within Institution X and acts as the final internal arbiter before outside resolution is sought.

The human resource manager (HRM)

One HRM is responsible for the selection, training and operational management of any internal ADR. Both the HRM and conciliator report independently to the registrar who is a member of senior management team (Figure 4).

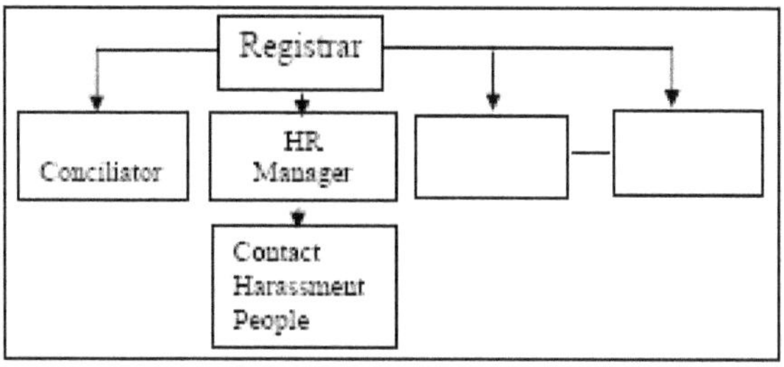

Figure 4 *Organizational structure for dispute resolution*

The IM has no official position within the organizational structure (Figure 4) and therefore is invisible within the institution. Erickson(2000) suggests that the most effective 'designated insider' in ADR is one that is visible, and that credibility for such a mediator needs to be reinforced from within the organizational structure. Also, in the official ADR brochure the only referral to informal mediation is in the procedural flowchart (Figure 3) in which the IM is not named. Invisibility therefore is reinforced across the institution as well as within the official organizational reporting lines.

The case

After a student complaint was raised with a CHP the IM was briefed by the HRM and separate appointments were made with both complainant and respondent. The respondent fulfilled her meeting obligations promptly. The complainant did not. It was only after intervention of another family member that the complainant returned the IM's call. When the respondent met with the IM she was angry at having a complaint laid against her. In her employment as a debt collector she was familiar with legal ADR. The reason for the complaint made against her was stated as two occasions when the respondent had spoken inappropriately to the complainant. (Any incident must be of a repeated nature. Do two occasions constitute repetition?) The respondent was well-prepared for the informal mediation meeting with a written statement and a letter to the complainant withdrawing her remarks (No apology was made either to the IM or in her written statement).

In the complainant's telephone conversation the IM was informed that the complainant was unable to meet because she worked. The IM was also told that the complainant did not want to carry the matter any further and that this had been conveyed to the person with whom she had previously spoken. (There is some confusion here about which the 'other woman' was – the CHP or the HRM. The IM's assumption was that it was the CHP, because explanations about rights and process originate with the CHP. This particular CHP was very sensitive to what constitutes sexual harassment.)

The complainant and respondent agreed about what was said on two occasions in which offence had been taken. However, there was not agreement about the intent behind these two utterances. The respondent believed she was being friendly with the group of younger students. The complainant, on the other hand reacted to the words as if they were a demeaning. (The complainant did not label the two occasions as harassment – just that she and her friends did not like what was said.) It was only after the complainant met with the CHP that the harassment label was placed on the situation.

Agreement was reached that the complainant would stay away from the respondent on campus. The complainant also stated that she would complain again if anything else was said against her. The respondent's main wish was to finish the semester and pass the paper.

This should have completed the informal mediation, with one more phone conversation to the respondent to relay the agreed terms. However, it seemed to the IM that there were a number of issues that needed attention. Questions posed about this mediation were: the role of the CHP; the direction the mediation process took; the lines of communication in the established ADR procedures; the HRM's role; what constituted a complaint; and the invisibility of the IM in the ADR process.

Organizational border systems

Figure 2 shows an OS that is highly structured, controlled by vision, statutory requirements, policy, operational procedures that operate with a formal hierarchical structure in which there was limited communication flows. Surrounding the formal OS is not just a boundary, but a border system that encompasses a complex organizational 'shadow space' in which multiple dimensions of being are possible. This border system provides a buffer from the outside environment, particularly when conflicts arise. It is within the border system that ADR occurs as the organization relegates disputes from public to confidential spaces. Mediated processes are confidential – public disputes are therefore rendered private and often individualized (Herr, 2005). These mediated processes happen within the border system, away from the formal OS and hence become invisible. All stakeholders involved in any dispute will view the border system and its links within the formal system differently.

The registrar's view

The registrar has ultimate responsibility for ADR and as a senior manager holds the institutional view that: conflicts should be resolved within the organization; formalized procedures and policies protect the institution; and these should operate in the most cost-effective way possible (hence the voluntary services of CHPs and IM). The unspoken view of the registrar was that ADR should happen away from the public and media attention is to be avoided. In placing ADR within the border system (Figure 2) any mediated solution is rendered invisible both within the formal OS and the outside environment. The confidential nature of mediation also isolates particular disputes from one another within the border system, particularly if these disputes are facilitated by different organizational volunteers.

The conciliator's view

The conciliator only becomes involved when ADR goes beyond informal mediation. The last internal resource for ADR is offered by the conciliator. Because the conciliator's position is officially recognized with a direct link to the registrar (Figure 4), this provides an important link between formal and border systems. The close contact between the IM and conciliator: provides a supervisory service once a dispute has been resolved; and more importantly sets up a two-way link between formal and border systems. However, this link is weakened by invisibility and lack of acknowledgement of informal mediation. Further weakening the link is the different nature of energetic spaces that exist between formal and border systems. The border system is one in which discussions are often emotional - trust needs to be established for ADR to move forward. A mediated solution is just as likely to be found with emotional rather than rational agreements. Within the formal system discussions normally take place in a rational framework.

The informal mediator's view

The IM operates completely within the border system. The role is invisible to the formal hierarchy (Figure 4), the IM is not named (the ADR process only refers to 'informal mediation' (Figure 3) as a step in the ADR process. The IM's indirect links to the formal system are: with the conciliator who reports to the registrar on a regular basis; and outcome reporting to the HRM.

The HRM

The HRM's role is to ensure that organizational policies and procedures are followed for ADR in a timely fashion and that outcomes are reported back to her from the IM. Experienced HRMs in ADR bring to the role considerable tacit knowledge that help to ameliorate decisions, and to monitor and expedite mediated solutions.

The CHP

The CHP's role is to listen to: the complainant only; advise on individual rights; assess whether there is a complaint to uphold; make sure that the complainant is familiar with organizational policies and procedures; and advise the complainant on subsequent ADR steps. The CHP does not have direct contact with the IM or the conciliator. The CHP's reporting line is to the HRM. CHP's may call on each other for debriefing and support, as extra voluntary activities.

The complainant

Often a complaint arises from a sense of unease about a situation –"something not being quite right", a lack of understanding about organizational policies and procedures, and no knowledge about who he/she should be discussing the complaint with in the first instant. The complainant is reliant on the CHP's interpretation of the situation to dictate what ADR steps need to be followed.

The respondent

The respondent is often unaware that her/his behavior lies outside of organizational and/or individual norms. As in the case described, the respondent's reaction to being notified of a complaint laid against her was one of anger and defensiveness.

Discussion

This case highlights some important issues for organizational ADR. In line with a major tenet of complexity systems(Stacey,1996), it is evident in the case described that the role of the CHP is vital in setting initial conditions. Indeed, it appears that selection and training of CHPs is an important precursor for setting an appropriate direction for ADR. Because CHPs are voluntary, this poses a dilemma for the institution which is: how can performance be monitored, evaluated and improved when the service provided is voluntary. If a CHP holds an extreme view about harassment, and ADR, this is likely to affect the direction taken. In the case described, a radical view appears to have escalated a general complaint about in-class behavior to mandated ADR.

Expanding this organizational boundary to a border system provides a space in which ADR activities may be hidden. ADR policies and procedures are set and initiated within the OS, and conducted within the border system after communication from the HRM to the IM. This communication provides the link across the internal edge of the border system from the OS.

Findings

When we consider what might happen if ADR is moved from invisibility within the border system into the official OS we need to ask what would be the fallout. Whilst CHPs are recognized and named within the OS, the IM is not. CHPs operate within the OS within a voluntary capacity. The IM operates within the border system. By not naming or paying for ADR services, the organization relegates this important activity to the border where there is no quality control, performance management, monitoring or evaluation. It seems logical and necessary that the mediator be formally recognized so that a feedback loop can be established back into the formal system and policy and procedures reviewed.

A result from this hypothetical case should be a debriefing session with the HRM and the registrar to review existing policies and procedures to ensure that the official system is providing appropriate support for ADR. If indeed, the border system is the appropriate place for mediation to occur then perhaps a more porous inner edge is required from the border system.

The CHP's role in ADR is pivotal in setting the direction for ADR to take. Whilst the CHP's role remains voluntary the organization is left with a major dilemma on quality control for ADR.

Conclusion

An ADR hypothetical case has been the vehicle to explore the presence of an organizational border system. Multiple dilemmas have emerged in this organization. Financial pressure, voluntary service and a culture that renders ADR invisible all contribute to challenges in maintaining a healthy organizational culture.

Exploring the links across the inner edge of the border system and transformational processes that occur within the border shed new light on ADR in organizational settings.

References

Erickson, M. (2000). The Insider's Edge.*CMHE Report,1*(2).http://www.campus-adr.org/chher/reportarticles/Edition1_4/Variations1_4.html accessed 17 May,2006

Herr, K. (2005). Administrators mandating mediation: tools of institutional violence cloaked in the discourse of reconciliation.*International Journal of Leadership in Education,8*(1),21-33.

Leadr.(2000). *Mediation Model.*Auckland.Workshop notes.

Report, G. (No date). *Employment Relations Law.* http://www.executive.govt.nz/era/erl_over.htm#9 accessed 18 May,2006

Stacey, R. (1996). *Strategic Management and Organizational Dynamics*(2 ed.).Great Britain:Pitman.

Warters, B. (2000). Thinking About Variations in Campus Mediator Style.*CMHE Report,1*(4).http://www.campus-adr.org/chher/reportarticles/Edition1_4/Variations1_4.html accessed 17 May,2006.

Kay Fielden is an Associate Professor in the School of Computing & Information Technology, Unitec New Zealand. She is the school's Academic Leader for Information Management and Research Coordinator. Kay has been an active researcher in ICT for many years and initiated this particular project. She teaches and supervises at the postgraduate level, and mentors staff research within the school. Kay has been an academic for more than 20 years and has developed wide interests in researching, teaching and consulting.

Creativity, Complexity and Organization in the DPI&F Ecology

J. Varghese & G. Ching
Queensland Department of Primary Industries and Fisheries (DPI&F)

The Queensland Department of Primary Industries and Fisheries (DPI&F) exits within an agricultural ecology. CEO, Jim Varghese will discuss how the Department actively uses generative dialogue to enhance inquiry, discovery and collaboration to build social capital and deliver its strategic priorities. The presentation will illustrate how social capital is grounded in the innovative three frame methodology of achieving excellent performance through strategic relationships and alignment to deliver public value. Case applications referred to in the presentation will include:

- **The Gladstone Oil Spill;**
- **Review of the shark safety program:**
- **Community development and leadership at Lockhart River.**

These will highlight the collaboration to develop commitments of multiple stakeholders and leverage insight to support public values, solve problems of collective action and indeed social capital.

Introduction

The Queensland Department of Primary Industries and Fisheries (DPI&F) exits within an agricultural ecology. With ecology studying the interactions between living organisms and their environment, agricultural ecology examines the ecology of agricultural systems and the natural resources required to sustain them. Within this context, I will discuss how the Department actively uses generative dialogue to enhance inquiry, discovery and collaboration to build social capital and deliver its strategic priorities as existing within an agricultural ecology. The presentation will illustrate how social capital is grounded in the innovative three frame methodology of achieving excellent performance through strategic relationships and alignment. I will present three distinct stories to illustrate the DPI&F approach and successes to sustaining social and natural capital. These are:

- The Gladstone Oil Spill;
- Review of the shark safety program;
- Community development and leadership at Lockhart River.

These stories will highlight the collaboration to develop commitments of multiple stakeholders and leverage insight to support social capital, solve problems through collective action and build sustainable, productive relationships.

Harvard University Professor, Mark Moore provides the idea of a strategic triangle with the intersection of legitimacy and support, public value and organizational capabilities as essential in strategic management in Government.

He challenges public managers to imagine and articulate a vision of public value that can command legitimacy and support and is operationally doable in the domain for which they have responsibility.

This framework helps us as a public manager to connect what we believe is valuable and requires public resources, with improved ways of understanding our public value and how we connect with them.

In all of these cases, I have employed the principals of The Three Frames approach as a "human living system" to support social and natural capital development and to maximise public value. They illustrate the norms, networks and relationships that enable:

- Information flows
- Norms of reciprocity (mutual aid) that connect people who are diverse
- Collective action
- Broader identities and solidarity that help translate an "I" perspective into a "we" perspective.

So what is the three frames?

The three frames

Consistent with the ideas of social capital, this approach combines the three interacting learning frames:

- The relationship frame aims to build and sustain a safe, trusting, non-threatening environment in which people feel able to share their thoughts, feelings and values honestly with others;
- The performance frame looks at what we want to achieve and provides a clear and measurable picture of what we want and need to achieve to meet our goals; and
- The alignment frame looks at the relationship between or within an organization and its members and identifies any blockages that are stopping them from achieving the goals as set out in the performance frame.

In applying the three frames, I use dialogue-based learning circles to gather together people with diverse interests to share information, ideas and perspective, build constructive relationships on behalf of achieving high performance results. These "critical friends" may include representatives from industry, local, Commonwealth and State Government, the private sector and the tertiary sector etc. and staff from my Department.

There are sometimes complementaries and tensions within these dialogues but they provide the space for questioning and reflection to help reach some answers together.

The ground rules for engagement in these dialogues, which support creating a safe and trusting environment include:

- We welcome multiple viewpoints and maximum interaction
- Our behaviors encourage cooperation with and acceptance of others
- We inquire, explore and learn about shared issues.

I support the idea of inquiry through dialogue and the work of pioneering leadership thinker Meg Wheatley who suggests:

"Invite in everyone who care to work on what's possible. Know that creative solutions come from new connections. Be intrigued by the difference you hear and expect to be surprised."

Wheatley (2001:1) introduces the notion of human living systems to create successful organizational change. She suggests that it is valuable to start with the assumption that people, like all life, are creative and good at change. Machines on the contrary cannot tolerate great variance and are established to run in certain environments. They have no flexibility or resiliency to deal with extraordinary levels of change. She argues that if you think of an organization as a living system, then it has the capacity for great flexibility and resiliency and the ability to adapt, to change and to grow. She argues that once we stop treating organizations and people as machines, we move into the paradigm of living systems. They do not require to be engineered, tuned and repaired at a technical level without reference to its human and social context.

Stories

Story 1: The Gladstone Oil Spill

The first story I'd like to share to demonstrate these ideas at work is that of the Gladstone oil spill.

Seafood is considered an icon in Queensland, with the industry worth a total of $xx on the domestic and export markets. When a tug boat collided with a bulk carrier during berthing operations at Gladstone Harbour just before midnight on 24 January, spilling about 25 tons of heavy fuel oil (bunker plus kerosene or diesel), DPI&F and other government partners took quick action to re-open markets to seafood from Gladstone Harbour to limit the immediate environmental impact on sensitive seagrasses and mangroves as well as local dolphins and dugong populations and the social implications for the local community.

The Minister for Primary Industries and Fisheries, myself and the department's Deputy Director-General Fisheries travelled to Gladstone to meet with local stakeholders in a Dialogue for Action forum to discuss an appropriate strategy to re-assure consumers and the community. Participants, from which Harvard University Mark Moore describes as the authorizing environment to give legitimacy and support, included the Member for Gladstone, the Mayor of the City of Gladstone, chairman of Gladstone Shire, a representative from the Shire of Calliope and the Gladstone Port Authority.

After an insightful dialogue, the participants agreed to three key issues in developing an action plan:

- testing of seafood samples to ensure all fisheries products caught in Gladstone Harbour were safe to eat;
- an efficient and effective claims process for all affected businesses;
- a communication campaign to reassure the public that Queensland seafood was safe to eat.

While Queensland Transport managed the actual oil spill clean-up – about 4kmn wide and 20km long – the DPI&F-led Queensland Government Food Incident Response Group (FIRG) coordinated and managed a stringent sampling and laboratory-testing program for all commercial catches of prawns, crabs, finfish and scallops coming into Gladstone Harbour, as well as a targeted sampling program in the areas hardest hit by the spill. Live trout fishing operators were also targeted.

The logistics of coordinating the sampling program proved to be challenging. To ensure minimal impact on the industry, samples were collected from commercial operators in Gladstone almost as soon as they arrived in port. These samples then had to be appropriately packaged, tagged, and immediately sent to Brisbane for laboratory testing for tainting and oil contaminants to ensure they were suitable for sale. Commercial and recreational fishers were asked to avoid the area until tests confirmed all seafood was safe.

In total, laboratory tests on more than 100 samples during the first three weeks following the oil spill confirmed there was no evidence of contamination or tainting and that the seafood was safe to eat. There were also no confirmed reports of dead fish or birds.

In addition, the department, through FIRG, quickly introduced a communication strategy to inform the industry about the sampling program and its progress. An advertising campaign alerted the local community of the steps being taken and to reassure them of the quality of the area's seafood, while a wider communication campaign targeting key industry stakeholders both within and outside Queensland focussed on the quality of the State's seafood.

Thanks to the efficiency of the response from the commercial fishing industry and other stakeholders in the local area, and government agencies, all commercial fishers were fully operating as normal within xx days of the spill.

Story 2: Review of the shark safety program

The second story also relates to an issue of the natural ecology.

For the past 44 years, Queensland has had a Shark Safety Program in place to minimise the risk to beach swimmers from the threat of shark attack.

The effectiveness of the program was publicly questioned after the death of a 21-year-old Queensland girl at Amity Point on North Stradbroke Island, just off the Brisbane coastline, on 7 January 2006 – the first fatality from a shark attack on a protected beach in Queensland in more than four decades.

The Queensland Government and DPI&F, which manages the Shark Safety Program, responded quickly to the tragedy and initiated a state-wide review of the program.

The department first conducted a Dialogue for Action forum with key stakeholders in March to discuss the program and issues impacting on it. Stakeholders representing six government agencies, environmental groups, Gold Coast City Council and Redlands Shire Council, CSIRO (for its maritime research knowledge), tourism bodies and surf lifesaving associations participated in the forum. Interstate counterpart, the New South Wales Department of Primary Industries also took part.

After the three hours of dialogue, all xx participants agreed that the forum was highly valuable in gathering collective and divergent views and appreciated the opportunity to come together as a group. They found a common ground and understanding of the challenges and what was possible to maintain an effective Shark Safety Program.

The review report found the Shark Safety Program was effective and offered swimmers the best available protection from shark attack. The report said while the netting and drum-line program had a minor impact on whales and other vulnerable species, the Government must make swimmer safety its priority.

To ensure swimmers were better educated about sharks and understood the practical steps for swimmer safety these initiatives were put in place:

- presenting swimmer safety information in such a way that it is suitable for use in schools;
- working even more closely with Surf Lifesaving Queensland in its education activities and with local government and tourism bodies;
- increasing advertising and community service announcements for swimmer safety, especially during holiday periods.

Story 3: Community development and leadership at Lockhart River

The last story I will share is one that I am particularly encouraged by and brings me great personal satisfaction. Unlike the two previous stories, it refers more to the human ecology and living systems of a community. It relates to my responsibility as the Government Champion for the Lockhart River community located more than 780 km north of Cairns with 800 residents. The Lockhart River community presents its own set of unique challenges, diversity and opportunities for creating social capital as a remote indigenous Queensland community. People here experience significant economic and social disadvantage.

As the Government champion, I knew that to be able to add value I needed to gain legitimacy and support with the community.

I achieved this through building key relationships and using dialogue-based Learning Circles for improved ways of shared understanding, learning and taking action together.

I worked on building trust and collaboration within the community as a foundation for action.

Since 2002, the Lockhart River Council, the community, representatives from the State and Federal governments have participated in regular Learning Circles.

From the very beginning, I was conscious of not following a trend of flying in, conducting a meeting, and flying out the same day.

Each time I visit, and I have visited the community on 12 different occasions, I stay a couple of days either camping at an outstation, such as Doti's Camp at Chilli Beach, or staying with the community. In essence the Learning Circles provide an opportunity for up to 80 people to sit together to discuss areas of concern, whether they be health, education, housing or justice.

This gives me time to share meals and engage in other daily activities that enable the community to get to know me as a person and allow me the opportunity to find out what they are thinking through numerous informal conversations.

These occur every three or four months. Importantly, they are not just events. Work continues in between the Learning Circles to ensure the conversations continue and agreed actions are taken. They have produced positive, tangible results in the community including:

- The development of the Lockhart River Community Plan;
- The creation of the Lockhart River Fishing company known as "Puchiwu", in a community where there is no history of small business;
- Providing local employment, for the first time at Lockhart River, seven new community houses are being built using local labour;
- The production of a CD called "Music of Lockhart" which is a mix of traditional, contemporary and church music.

While, I led these Learning Circles, the community with increased social cohesion now own them and insist that this is the way they want to engage with us. My colleague, Denise Hagan observes that there is renewed enthusiasm in the community for dealing with challenges as well as a reduction in conflict and criticism. It would suggest that the social capital created by the dialogues enhances the community in terms of its wellbeing, access to resources and ability to cope with the issues that face them.

Reflection

Each of these stories had a distinct complexity and challenge. They provide the practices and routines that engage people in learning and to find the courage to speak their truths. Dialogue through learning circles proved to be versatile, applicable at many levels and in all kinds of issues. The use of creative approaches grounded in dialogue and social and natural capital provided a way forward and productive, sustainable outcomes. These outcomes became the property of the groups and its communities rather than owned by individuals.

Beyond these examples, some 60? learning circles have been held in DPI&F as a primary method for how we do business. In fact, the Minister for Primary Industries and Fisheries, Tim Mulherrin, himself facilitates industry dialogues with examples including Cyclone Larry, the Fishing Industry and Sugar Cane SMUT.

They are also used for dialogues on individual staff and the Department's achievements as a review mechanism. Similarly, representatives of industry, across Government, academia are invited to participate in the dialogues to add to the richness of conversation and perspectives. This is unique and innovative approach to the traditional performance management systems.

Conclusion

In conclusion, the words of the Lockhart River community, I think, sum up the power of social and natural capital and the use of dialogue to nurture engagement and achieve results:

"You-me working together, not standing alone.
You-me it's up to us, we have got to help one another.
You-me I know we can do it."

References

Flower, J. (1993). "The power of chaos: Excerpts from a conversation with Meg Wheatley," *The Healthcare Forum Journal,* 36, http://www.well.com/user/bbear/wheatley.html.

Hagan, D. (internet). "The learning circles of Lockhart River: Building productive partnerships," http://www.engagingcommunities2005.org/abstracts/Hagan-Denise-final.pdf.

Moore, M. H. (1995). Creating Public Value: Strategic Management in Government, Cambridge, MA: Harvard University Press, ISBN 0674175581.

Wheatley, M. (2001). "Innovation means relying on everyone's creativity," *Leader to Leader*, Berkana Institute.

A Reflection on the Use of Action Research in a Major Change Project

John Molineux
Australian Public Service, AUS

The problematic use of action research in the context of a large organizational change project is explored. The paper uses a systemic approach to identify and discuss issues around participation, reflection and the role of an insider as the action researcher. Recommendations are given in relation to future projects.

Introduction

This paper reflects on my use of action research as a change methodology at a large Australian government agency. The research included my direct participation in the design and implementation of a major cultural change project.

A number of issues arose out of a systemic reflection on the assumptions, processes and outcomes of the project. One of these was around the difficulty of using action research in a large organizational change project. This is particularly true due to its emphasis on participation, which is inevitably time consuming, and therefore makes the methodology problematic in the context of change in a large organization. In addition, there were some significant issues related to the reflection processes of action research in understanding the whole system that the project was working within. Finally, the role of an insider working as an action researcher raised further issues.

Arising out of the systemic understanding of this project, the paper includes a number of recommendations about the use of action research for future large change projects.

Action research in business

The nature of action research (AR) in business is represented in texts as an adaptation of the original form of action research to organization development, where the major emphasis is on planned change.

The common use of action research as a participatory approach to change was used by many organizational development practitioners in the 1970s and 1980s. French and Bell (1990: 108) even claim that the "organization development process is basically an action research program in an organization designed to improve the functioning of that organization". They also outline many successful action research interventions over several decades. Also, Burnes (2000: 270) notes that "action research has enjoyed a large following over the years". Since then, organizations have continued to use participatory cyclical processes, but have not necessarily labelled them as 'action research'.

An action research cycle is represented in Figure 1 (Coghlan and Brannick, 2001). In this model, four phases of action research are noted, and continue in a repetitive cycle of learning and improvement, which Waddell et al (2004: 34) note "may lead to rediagnosis and new action". Coghlan and Brannick (2001: 19) see the cycle continuing in a series or "spiral of action research cycles".

Many organizations use an action research approach to change management without labelling it as action research. This is because they find that a methodology that links research, action and participation in a cyclical approach as useful. For example, the large action research change program in which I was involved (Molineux, 2005) was never labelled 'action research' within the organization. The methodology, and my intent in using it, was only known by a small group involved in managing the implementation of the change project. The intent of this implementation group was to use participative techniques in a systemic understanding of change using strategic human resource management. Action research is the methodology that fits in well with not only participation, but also with an understanding of the complex systems involved in implementing a major organizational change project. Chisholm (2001: 324) notes that action research "attempts to generate knowledge of a system, while, at the same time, trying to change or develop it" and "attempts to contribute to general knowledge about systems and the dynamics of changing them".

Its usefulness in a business context such as this is because of its application of research to practical situations. This practicality is recognized as a problem-solving ability by Gaventa and Cornwall (2001: 75), who state that action research "focuses first on problem-solving, and more secondarily on the knowledge generated from the process", generating "knowledge which will lead to improvement, usually...organizational improvement or for the solution of practical problems". Also, Dick (2002: 162) notes it is beneficial in situations "with highly uncertain outcomes" as action research allows time to "build enough understanding to decide which methodology and methods best suited" the research situation and questions. He adds that "action research offers substantial flexibility and responsiveness to a complex situation", so is "particularly useful...for others who need responsiveness to complex situations – people such as managers or professionals". Avison *et al.* (2001: 44) note that "no other

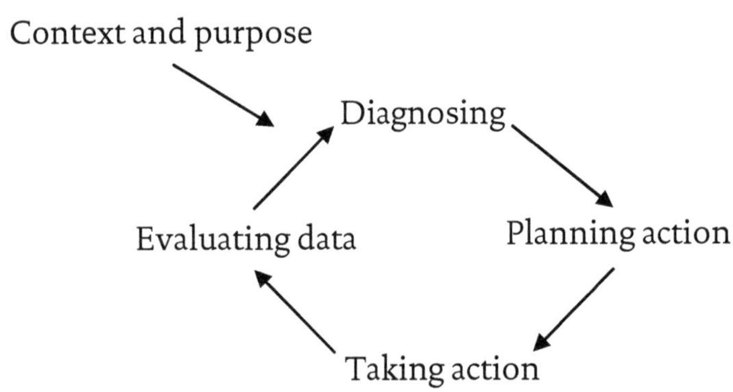

Figure 1 *The action research cycle (Coghlan and Brannick, 2001: 17)*

research approach has the power to add to the body of knowledge and deal with the practical concerns of people in such a positive manner".

This leads to my first recommendation, which is that action research is an excellent approach to use by organizations that desire to implement change in a complex and dynamic environment.

The importance of reflective cycles in action research

Reflection is a critical component of action research and is an integral part of the action research cycles. Greenwood and Levin (1998: 115) state that "doing AR means engaging in a process of mutual action and reflection". They add that action research practitioners "must continually reflect on experiences from the field, seeking what is necessary to keep a change process moving and tracking what is being learned". Its importance is outlined by Coghlan and Brannick (2001: 19), who note that:

"It is the dynamic of this reflection cycle that incorporates the learning process of the action research cycle and enables action research to be more than everyday problem solving."

Regular, systematic and critical reflection, according to Dick (2002: 168) "enables you to develop the understanding that accrues to become your contribution to knowledge and, directly or indirectly, to action" and so is "an essential, iterative part of action research".

Martin (2001: 202) also notes that reflection leads to "the uncovering of new interpretations and perspectives", which may "challenge prior beliefs and understandings and reframe what they know". Rose and Haynes (1999: 214) stress the importance of reflection, particularly in complex organizations and suggest that "mature reflection has become crucial to progress, for change is unlikely to be simple to achieve, and the way forward is unlikely to be obvious to everyone concerned".

A reason for the reflection periods is noted by Greenwood and Levin (1998:97), who suggest that in action research "there are always more possible futures than appear at first to be open" and therefore effort needs to be put to "reanalyze the past", and to consider "what other, possibly more desirable, futures may be available".

Two processes of reflection are noted by Schön (1983, 1987). One of these is reflection-in-action, which is the ability to reflect on the process while engaging in the action itself. The other is reflection-on-action, which means working through experiences gained from actions after the fact. Both of these processes were evident in my research.

Reflection in my research was an interweaving process that arose as a deliberate part of the action research cycles, but also at other times triggered by events, ideas or quiet times. In this research, Checkland's (1985) FMA model was used as a tool for deep reflection and rethinking about the implications of unexpected action that arose during the conduct of the action research. This occurred between Phases 1 and 2, and between Phases 2 and 3 of my research.

The reflection cycles in the process of this research and my own reflexivity led to new thinking about strategic human resource management and about the relevance of the theory of punctuated equilibrium to the research problem. In my research, reflection was essential and resulted in new understandings of the application of theory in the field, and on the methodologies used.

Checkland and Holwell (1998a: 13) state that "any research in any mode may be thought of as entailing the elements shown [in the FMA model]". This is represented in Figure 2. Checkland (1985:758) notes that the FMA model "is a very general model of the organized use of rational thought". Checkland and Holwell (1998a: 13) describe it as:

"Particular linked ideas F are used in a methodology M to investigate an area of interest A. Using the methodology may then teach us not only about A but also about the adequacy of F and M."

The importance of this in action research, Checkland and Holwell (1998a: 13) note, is that a "change to or modification of F, M, and even A has to be expected in action research".

They believe that the "susceptibility to change F, M, and A in research in which the researcher becomes involved in the flux of real-world social situations" (p.13) implies an essential need to "declare in advance the elements F M A" (p.14), that is "the intellectual structure which will lead to findings and research lessons being recognized as such." Changing situations, then, are likely to test "the adequacy of F and M and the appropriateness of A" (p.14).

In action research, Checkland and Holwell (1998a: 13) state the declared framework is in the sense of "research themes within which lessons can be sought", and the researcher "enters the 'social practice' of a real-world situation in which the themes are relevant and becomes involved as both participant and researcher". Reflection of experience on the declared F and M "may require some rethinking of earlier phases—and again, it is the declared intellectual framework of F and M which allows this to be done coherently" (p.14). Checkland and Holwell (1998b: 25) note that the reflection "can yield findings of various kinds, such as learning about F, M, A, or about the research themes; or new themes may be defined as a result of the experience".

I found that this approach to reflection was holistic and systemic and yielded significant benefit in my research. In the project, the framework of ideas was stated in a research proposition. However, my reflection on evidence from the applied action area indicated that the framework of ideas required some adjustment. At the end of Phase 1 of the action research, noted in Figure 3, the cultural change implementation projects were abolished and I needed to reflect on why such important and successful change projects were terminated after senior management of the organization had so strongly supported them. I was forced to rethink the original framework of ideas and methodology, as the area of action was bringing up new and unexpected information. It was during this time that I discovered that the organization was the subject of macro-cycles, in accordance with the theory of punctuated equilibrium. Subsequently, I re-

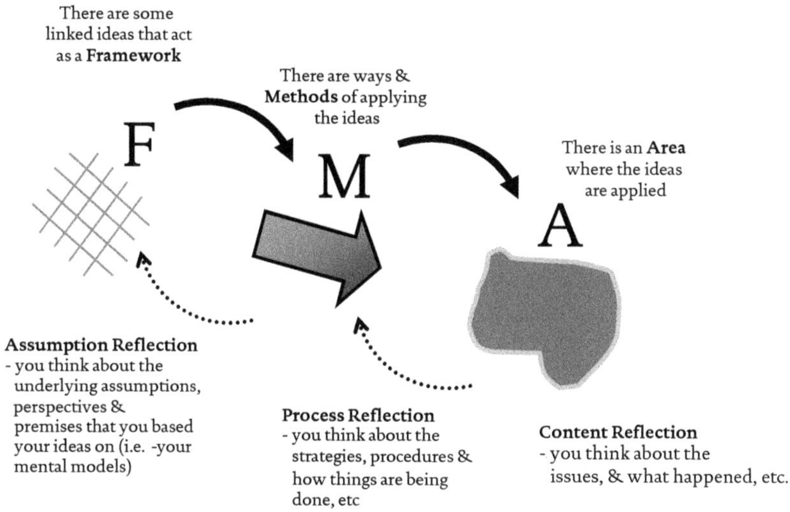

Figure 2 *Checkland-Mezirow template (Sarah, et al., 2002).*

thought the framework and the methodology of the research project and generated new insights.

Included in the model at Figure 2 are Mezirow's (1991) reflection types. Mezirow notes that transformative learning occurs when fundamental mental frameworks are questioned and revised, and in action research such reflection leads to the uncovering of new interpretations and perspectives. Three types of reflection are noted: *Content reflection* examines the issues and events in the area of action (*the what*); *Process reflection* examines the methods and processes used (*the how*); and *assumption reflection* examines the underlying ideas and mental models (*the why*). I used these three types of reflection in reflecting on the action research at the conclusion of Phase 1.

In my research, there were three major phases of the action research and a total of ten cycles, which are represented in Figure 3. The cyclical model of action research used consisted of the four elements: *Evaluate*; *Reflect*; *Design Action*; and *Take Action*. In Phase One, the cycle was repeated six times, with each cycle taking around three or four months to complete, although the cycles tended to overlap at times.

Phase One extended for 25 months, which was the bulk of the time involved. It commenced with the Agency Executive's decision to approve the change project, so implementation action could commence. I was a key member of both the design team for the change project (for eight months prior to implementation) and the implementation team. A large number of projects were set up to implement it, and I was involved in coordinating and guiding these projects.

The change projects commenced late in 1999 and operated for 18 months before being terminated in June 2001. Many of these projects were fully or partly implemented by June 2001, and the outcomes from the projects contributed to a shift in the organization's culture. The reasons for the termina-

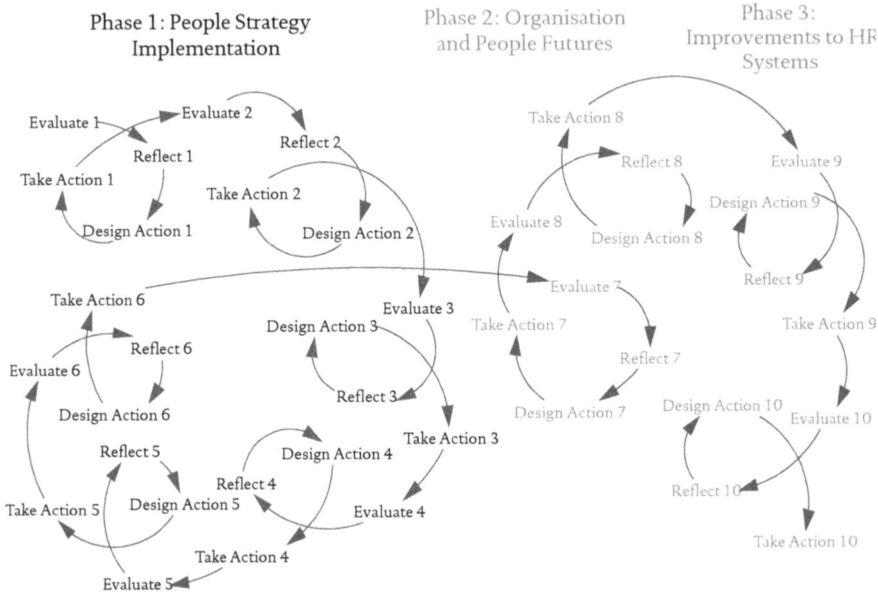

Figure 3 *A representation of action research phases and cycles in a major change project (Molineux, 2005).*

tion of the projects, along with discoveries about the macro-cycles of government influencing the decision to terminate, were noted by Molineux and Haslett (2002). The discoveries relating to the macro-cycles were a major finding arising from the reflection process used through this research.

The second phase lasted a much shorter period, for nine months. In this phase, I was a key member of the Organization and People Futures team. This team was involved in designing new HR strategies for the Agency, as well as rethinking some of the original ideas of the change project, based on extended research in HR.

Phase Three lasted ten months. During the third phase, I was involved in two different projects that were implementing improvements to the HR systems of the Agency. These activities arose from the original work of the cultural change team, and data collection ceased twelve months after completion of the third phase of the project.

I was able to transfer the action research to other HR improvement projects, following the abolition of the Organization and People Futures team. The first of these projects resulted in the ninth cycle in this research, and involved work originally commenced in the cultural change project. The second project in Phase Three involved the redesign of the Agency's Performance system using Soft Systems Methodology.

I was able to continue forms of action research following the abolition of the projects. Through undertaking these systemic reflection processes, I understood that I was able to continue research in different forms, facing setbacks and dealing with real world problems, whilst working in a system where his control was limited. This leads to a key learning and recommendation arising from this work in using systemic reflection processes and in trusting the systemic view for the discovery of answers to problems.

The problematic nature of participation

Participation is a fundamental principle of action research. For example, West and Stansfield (2001: 266) state that "Attention to the notion of collaboration is vital in action research". Also, Reason and Bradbury (2001: 2) note that action research:

"...is only possible with, for and by persons and communities, ideally involving all stakeholders both in the questioning and sensemaking that informs the research, and in the action which is its focus."

Greenwood and Levin (1998) discuss the need to achieve a balance in action research of three elements: research, participation and action. The process of generating knowledge through participation is reinforced by Brulin (2001: 441), who notes that knowledge "that aspires to...be of practical use has to be developed jointly with the users of the knowledge", so "researchers have to form knowledge in interactive relationships with practitioners". In another commentary, Dick (2002: 164) states that "there is a spectrum of participation from which one can choose" in action research, which "may be done with the style and level of participation that suits you [the action researcher] and your participants".

Whilst it is noted that participation is an essential part of action research, its use in large organizational change projects is somewhat problematic. This is due to the sheer numbers of people involved or impacted by major change in organizations. In an effort to become more participative, organizations have used techniques such as industrial democracy and representative boards to enhance participation in management. Change management, however, can be much more of a problem as it can be very emotive, and participative management techniques can invoke resistance rather than collaboration.

Systems theory offers some alternative approaches for participation in action research projects. I used one of these, Soft Systems Methodology (Checkland, 1981), in the course of the first and third phases of this research. I found that projects that used a codesign approach using SSM had much greater acceptance and success than projects that did not use SSM. This finding was confirmed through a number of interviews with project managers and the implementation team.

Also, I noted that there is a range of possibilities in relation to participation. In a large organization, it is not possible to involve everyone as a collaborator in codesign of change programs. It is more likely that action researchers may involve people in different ways and at different times of the project.

A possible spectrum of action research participation is included as Table 1. The spectrum ranges from a researcher conducting a research project with one client or manager, to a fully collaborative action research project involving all relevant members of the organization. It should be noted that one action

Type of action research participation	Examples of participants involved
Client-oriented action research intervention	Researcher and one client or manager
Manager-sponsored action research	Researcher and management team
Project-based action research	Researcher, managers and/or project team
Representative action research	Researcher, managers and representatives from parts of the organization
Participative action research	Researcher, managers and relevant organizational members participating
Collaborative action research	Researcher, managers and relevant organization members collaborating
Collaborative action research with co-researchers	Researchers and clients working jointly as co-researchers

Table 1 *Spectrum of action research participation in a large organization*

research intervention may cut across several different forms of participation in different parts of its cycle. For example, an action researcher may be engaged by the CEO in the first instance, and in the first phase or cycle may be working only with a small team. Later, during subsequent cycles, the researcher may be designing interventions with representative groups, and later be implementing the change in a collaborative way with organizational members.

In my research, the level of participation varied depending on the cycle and the situation context. At times it involved a small group, at other times it involved larger groups of participants, clients and others. The smaller group of participants consisted of a core group and several project leaders. These participants had a much higher level of participation and involvement than workshop participants and other stakeholders. In relation to the categories suggested in Table 1, the design stage of the change project would be categorized as project-based action research, as I worked with managers and a project team. The implementation stage of the project would be categorized as representative action research, where the implementation team worked with representatives from other parts of the organization. The third phase of the research involved codesign with the collaboration of relevant organization members, so this part of the research would be categorized as collaborative action research.

A recommendation from this reflection is that participation may be used in a variety of ways at different times in the life of an action research project.

The role of an insider in an action research project

In relation to the study of organizational culture, Schein (2000: xxvi) outlines three routes for a researcher:

"(a) infiltration, in which the participant observer becomes a true insider; (b) a formal research role agreed to by the insiders; and (c) a formal clinical role, in which the insiders ask the outsider to come into the organization as a helper/ consultant."

In my case, I would be considered as a 'true insider', but not in terms of 'infiltration', as I was already a member of the organization. In another case, this type of insider role allowed De Guerre (2002: 332) "insight and opportunity that could not occur for outsiders". He found that from the inside, he "had the privilege of seeing first hand the nature of new local, contextual, and immediately applied social science iteratively, leading to the generation of new knowledge". The insider role, however, has the potential to create a dilemma between organizational commitment and academic achievement and obligations. Coghlan and Brannick (2001: 49) note that the organizational role "may demand total involvement and active commitment", whereas the research role may require "a more detached, more theoretic, objective and neutral observer position."

There are also specific issues for managers who undertake insider action research, such as in my research. Some of these issues are outlined by Coghlan (2001: 50). He believes that "managers who undertake action research projects may be located anywhere in their organization's hierarchy", but their location in the hierarchy "has undoubted implications for what may be researched and how". For example, higher-level executives "may have more access" but may by excluded "from access to informal and grapevine networks". Middle- or lower-level managers "may find upward access difficult and be confined to their function or division". Other issues for the researcher that he notes include maintaining a "desired career path" (p.50) and "pre-understanding" (p.51) of knowledge, insights and experience of the organization and various participants.

An advantage of being an insider is that the researcher is "already immersed in the organization" (Coghlan, 2001 :51), so has built up an extensive understanding of how it works. He notes that this can also be a disadvantage, in that the researcher "may assume too much and not probe as deeply as if they were outsiders or ignorant of the situation". Other problems include role conflict and impacts on organizational relationships.

One of the most difficult problems Coghlan (2001: 52) sees is the possibility of the research being considered political and "might even be considered subversive". It is therefore critical that the researcher carefully negotiate their access and use of organizational information. Coghlan (2001: 53) therefore sees a need for manager-researchers "to be prepared to work the political system" and to "maintain their credibility as an effective driver of change and as an astute political player".

Many of these aspects outlined above were issues during my research. For example, I was able to access information that would not generally be available to an outsider. I also understood the context and history of the organization in undertaking change, and had built relationships with key stakeholders.

As I wished to maintain a career in the organization, playing the political system was an important consideration. For myself, the intensity of the level of involvement in the project over three years and the subsequent abolition of the implementation projects did have an impact. An outsider could be more dispassionate. However, arising from this reflection, I recommend that it is important for the insider to become resilient and to work through setbacks and pitfalls to discover new understanding. I also recommend that action researchers carefully position themselves within the organization.

Recommendations

I have noted a number of recommendations arising from the systemic use of reflection from my research:

1. Action research is suitable in complex and dynamic environments

Action research should be considered as a highly suitable methodology in undertaking large organizational change projects where the issues are complex and the environment is dynamic. This is because action research is naturally systemic, and can readily adapt to these environments.

2. Action researchers need to understand the systemic view

The action researcher should always remember to view the system at which they are working on a number of levels. The researcher should not just consider the project they are working on, but its ongoing context in the macro-system of the organization and its interaction with its environment. Understanding the impact and relationships at this level will help the researcher to implement a more effective approach to change management. It is recommended that the researcher actively advocate the systemic approach, particularly in meetings with organizational management, such as during the design and implementation of a project.

3. Action researchers can choose from a spectrum of participation

There are a range of possibilities in involving people in change programs. The action researcher needs to consider the stage of the project, the relationships with key stakeholders, and the appropriateness of involvement of organizational members at these project stages. The researcher can then work out which form of participation would be the most effective for each of these project stages. It is recommended that researchers consider the types of participation noted in the spectrum at Table 1.

4. Action researchers need to have reflexivity and flexibility

It is important for an action researcher to critically review their own action and impact on the organization. The action researcher needs to consider that their intent and the outcome of the project may be quite different, so they need to be able to be responsive and flexible to deal with changing requirements on the run. This not only involves interaction with others, but a critical reflection and re-evaluation in the researcher's own approach and goals. FMA is recommended as an approach for reflection in this context.

5. Action researchers need to position themselves within the organization

The action researcher needs to place themselves in a position within the organization where they can most benefit themselves and bring about the desired change. The researcher must first understand their role and the political and social position they have within the organization. To achieve a place of influence that would benefit the project, the researcher needs to build close relationships with key clients, and gain sufficient political and positional support to enable effective contribution to the outcomes of the project. It is recommended that the researcher obtain organizational written agreement about their role and reporting relationships.

References

Avison, D., Baskerville, R. and Myers, M. (2001). Controlling action research projects. *Information Technology & People*, 14 (1): 28-45. ISSN 09593845.

Brulin, G. (2001). The Third Task of Universities or How to Get Universities to Serve their Communities! In P. Reason and H. Bradbury (Eds.), *Handbook of Action Research: Participative Inquiry and Practice*: 440-446. Sage, London. ISBN 076196645.

Burnes, B. (2000). *Managing Change: A Strategic Approach to Organizational Dynamics*. Pearson Education, Harlow, UK. ISBN 027303376X.

Checkland, P.B. (1981). *Systems thinking, systems practice*. Wiley, Chichester, UK. ISBN 0471986062.

Checkland, P.B. (1985). From optimizing to learning: A development of systems thinking for the 1990s. *Journal of the Operational Research Society*, 36 (9): 757-767. ISSN 01605682.

Checkland, P.B. and Holwell, S. (1998a). Action Research: Its Nature and Validity. *Systemic Practice and Action Research*, 11 (1): 9-21.

Checkland, P.B. and Holwell, S. (1998b). *Information, Systems and Information Systems*. John Wiley & Sons, Chichester. ISBN 0471958204.

Chisholm, R.F. (2001). Action Research to Develop an Interorganisational Network. In P. Reason and H. Bradbury (Eds.), *Handbook of Action Research: Participative Inquiry and Practice*: 324-332. Sage, London. ISBN 076196645.

Coghlan, D. (2001). Insider Action Research Projects. *Management Learning*, 32 (1): 49-60. ISSN 13505076.

Coghlan, D. and Brannick, T. (2001). *Doing Action Research in Your Own Organization*. Sage, London. ISBN 1412902479.

De Guerre, D.W. (2002). Doing Action Research in One's Own Organization: An Ongoing Conversation Over Time. *Systemic Practice and Action Research*, 15 (4): 331-349. ISSN 1094429X.

Dick, B. (2002). Postgraduate programs using action research. *The Learning Organization*, 9 (4): 159-170. ISSN 09696474.

French, W.L. and Bell, C.H. (1990). *Organization Development: Behavioural Science Interventions for Organization Improvement*, 4th Ed. Prentice Hall, Englewood Cliffs, NJ. ISBN 013242231X.

Gaventa, J. and Cornwall, A. (2001). Power and Knowledge. In P. Reason and H. Bradbury (Eds.), *Handbook of Action Research: Participative Inquiry and Practice*: 70-80. Sage, London. ISBN 076196645.

Greenwood, D.J. and Levin, M. (1998). *Introduction to Action Research*. Sage, Thousand Oaks, CA. ISBN 1412925975.

Martin, A.W. (2001). Large-group Processes as Action Research. In P. Reason and H. Bradbury (Eds.), *Handbook of Action Research: Participative Inquiry and Practice*:

200-208. Sage, London. ISBN 076196645.
Mezirow, J. (1991). *Transformative Dimensions of Adult Learning.* Jossey-Bass, San Francisco, CA. ISBN 1555423396.
Molineux, J. (2005). *Systemic strategic human resource management enables organizational cultural change: An action research study.* Unpublished PhD thesis, Monash University, Australia.
Molineux, J. and Haslett, T. (2002). Working within organizational cycles – a better way to implement action research projects in large organizations. *Systems Practice and Action Research,* 15 (6): 465-484.
Reason, P. and Bradbury, H. (2001). *Handbook of Action Research: Participative Inquiry and Practice.* Sage, London. ISBN 076196645.
Rose, J. and Haynes, M. (1999). A soft systems approach to the evaluation of complex interventions in the public sector. *Journal of Applied Management Studies,* 8 (2): 199-216. ISSN 13600796.
Sarah, R., Haslett, T., Molineux, J., Olsen, J., Stephens, J., Tepe, S. and Walker, B. (2002). Business Action Research in Practice—A Strategic Conversation About Conducting Action Research in Business Organizations. *Systemic Practice and Action Research,* 15 (6): 535-546. ISSN 1094429X.
Schein, E.H. (2000). Sense and Nonsense about Culture and Climate. In N.M. Ashkanasy, C.P.M. Wilderom and M.F. Peterson (Eds.), *Handbook of Organizational Culture and Climate*: xxiii-xxx. Sage, Thousand Oaks, CA. ISBN 0761916024.
Schön, D.A. (1983). *The Reflective Practitioner: How Professionals Think in Action.* Basic Books, New York. ISBN 1857423194.
Schön, D.A. (1987). *Educating the Reflective Practitioner: Toward a New Design for Teaching and Learning in the Professions.* Jossey-Bass, San Francisco, CA. ISBN 1555422209.
Waddell, D.M., Cummings, T.G. and Worley, C.G. (2004). *Organization Development and Change,* 2nd Ed. Thomson, Melbourne, Vic. ISBN 0170111350.
West, D. and Stansfield, M.H. (2001). Structuring Action and Reflection in Information: Systems Action Research Studies Using Checkland's FMA Model. *Systemic Practice and Action Research,* 14 (1): 251-281. ISSN 1094429X.

Before undertaking his doctoral studies, **John Molineux** had completed a Bachelor of Business, a Graduate Diploma in Management, and a Master of Management. John has worked for over three decades in the field of HR and is currently a human resource manager for a large Australian government agency. He completed his PhD in Management at Monash University. This involved applying his ideas around systemic human resource management in a major organizational cultural change program, using an action research methodology.

THEME FOUR
Systems Thinking and Dynamics

John Barton, Barton Consulting
Bob Cavana, Victoria University of Wellington

The nine papers presented in this theme raise questions ranging from the nitty-gritty of data collection and the practice of systems intervention and problem framing, to various aspects of the philosophy of systems thinking and ethics. If this selection of papers is indicative of what is currently occupying researchers' minds (and the authors represent an interesting mix of backgrounds and international affiliations), then it strongly suggests a search for stronger fundamentals, particularly relating to the nexus between systems thinking and knowledge management.

The scene is set by the trio of papers presented by Sheffield and Guo. These papers provide an invaluable exploration into the fundamentals of knowledge management, starting with an attempt to sort out the rather confusing way the terms data, information, and knowledge are used. They provide new insights into current approaches to knowledge management by viewing the area through the lens of Habermasian knowledge interests (technical, practical and emancipatory) and knowledge worlds (technical, interpersonal and personal). They conclude with a proposal for a normative theoretical framework incorporating knowledge perspectives, sources of boundary judgements, system roles and knowledge dynamics, leading to a process of ethical inquiry.

As testament to the usefulness of Sheffield and Guo's contribution, their framework can be used to position the other contributions. For example, Ison and Schlindwein identification of the objectivist versus interpretivist epistemological positions that appear to have dominated systems thinking until recent times are directly related and can be compared with critical pluralism as explained by Sheffield and Guo. And at an even simpler level, if we recognise that knowledge is what allows us to interpret data to provide information that may lead to action, and hence more data, then we start to realise the importance of systemic framing of knowledge. Consequently, Tepe, Haslett, and Osborne raise the question of how frames are chosen when identifying appropriate decision rules in Agent Based Modelling. Dhawan, O'Connor and Borman contribute to a growing body of evidence concerning the possible benefits or otherwise that come from formal education in framing dynamic problems using the stock-flow-feedback approach of System Dynamics. Coady, Cavana, and Pearson explore a conceptual framework, again based on System Dynamics framing, for better understanding the tricky area of public health policy. Monypenny confronts the vexing problem of the often disparate and dysfunctional way professionals work at cross purposes because of an inability to develop a common framework within which to conduct dialogue. Vélez-Castiblanco draws on the philosophy of Wittgenstein, language pragmatics, and methodologies as language games to make sense of systems frames, particularly relating to importance of boundary setting in systemic inquiry.

Proceedings of the 12th ANZSYS conference - Sustaining our social and natural capital
Katoomba, NSW Australia, 3rd-6th December, 2006

Contribution of System Tools to Dynamic Decision-Making

Rajat Dhawan, Marcus O'Connor & Mark Borman
School of Business, The University of Sydney, AUS

Systems tools (both for conceptual and computer modelling) are widely used for the study and managing complex systems. However, the effect of these techniques on a person's dynamic decision-making abilities is yet not fully known. To explore this relationship, a controlled experimental study was conducted on graduate students. Results show that most participants initially had poor understanding of basic dynamic situations. However, the completeness and accuracy of their mental models improved considerably with system interventions. Specifically participants' ability to discern between stocks and flows, identify causal relationships and feedback improved by around 27% after a conceptual systems tools intervention. These abilities further increased by around 4% after participants underwent a computer modelling intervention. This study is significant to the field as it will contribute to the long-standing debate on the links between system tools and dynamic decision-making, particularly the relative contribution of conceptual tools and computer modelling tools.

Introduction

Systems tools (both for conceptual and computer modelling) are widely used for the study and managing complex systems. Their focus is to study the relationships between components in the system, especially feedback loops and the patterns of behavior generated by them.

Though both these system methodologies are widely accepted as aids for making decisions in complex systems, surprisingly the relationship between these and dynamic decision-making is yet to be fully explored (Cavaleri & Sterman, 1997; Huz *et al.*, 1996; Doyle, 1997). There have been some studies conducted in the past to explore participants' understanding of dynamic tasks like those by Sweeny and Sterman (2000) and Kainz and Ossimitz (2002).

This study seeks to explore the relationship between decision making in dynamic tasks and systems methodologies in a pre-test/post-test design. The aim of this experiment is to test participants' ability to make decisions in dynamic situations: i) using conceptual system modelling tools alone; ii) using a combination of conceptual and computer modelling tools. Conceptual systems tools evaluated in this study include causal-loop diagrams, feedback loops, system archetypes and behavior over time graphs. These tools collectively also comprise of qualitative systems thinking. Computer modelling tools considered in this study involved computer modelling using the concepts of stocks and flows and feedback. These concepts collective comprise of quantitative systems thinking (sometimes also referred to as system dynamics). For the remaining part of the paper conceptual systems modelling tools are referred to as systems thinking and computer modelling tools are referred to as system dynamics.

With this aim the following research questions were examined:

1. Is the understanding of a complex system enhanced when participants use systems thinking as a decision aid as compared to their understanding before being taught about systems thinking?
2. Is the understanding of a complex system enhanced when participants use a combination of systems thinking and system dynamics as a decision aid as compared to their understanding using only systems thinking and controls?

The paper consists of five sections including this introduction. The second section discusses the method that was followed to conduct the experiment. The subsequent section is devoted to results and discussion. The paper concludes with the major finding of this research, its limitations and avenues for further research.

Method

Subjects

Thirty-one graduate students enrolled in the School of Business at the University of Sydney, participated in the study. All students were either in their penultimate or final semester at the University and some had prior work experience. Students had to undergo the study in order to fulfil course requirements and were graded. Five out of the thirty-one students failed to satisfy the criteria of the experiment as they missed either one of the lectures or one of the tests. These were excluded from the data set.

Design

The experiment is structured around a scenario of a firm that provides consulting and IT services. The scenario describes in sufficient detail the operations of the firm. It then describes a problem of periodic oscillations in revenue over time, the measures taken by the top management and other parties involved, as well as the response of the employees towards those corrective policies. The task for the participants was to analyze the situation and assess the causes of the periodic oscillations.

All participants were administered three tests (one pre-test and two post-tests termed post-test 1 and post-test 2). Students were given the case study a week prior to the pre-test and had ample time to read and re-read the scenario. In the following week they were administered the pre-test. In between pre-test and post-test1, the participants were taught systems thinking. Participants were taught system dynamics modelling between post-test1 and post-test2. All teaching was done in a lecture/tutorial format and participants were required to complete tasks as homework.

The tests on all three occasions were similar. However, the questions were not provided to the participants after the any test, nor were the students aware that similar questions would be repeated in subsequent tests. Many questions were re-phrased and the data and context was changed for quantitative questions. Keeping in mind that there is a possibility of learning of questions from one test to the other, future experiments have been designed to address this issue.

Procedure

The experiment was spread over 5 weeks and took place either in a lecture hall or a computer lab. All tests were individual and participants were seated at sufficient distance from each other.
The detailed schedule of the experiment is given below:
Week 1: Participants were provided with the case study at the end of the lecture. They were asked to read the case during the week and they were also told that they would be quizzed on the scenario in the next week's lecture.

Week 2: All participants underwent the pre-test in the first hour. After the test, students underwent a short course on systems thinking.

Week 3: The systems thinking lecture continued from the previous week and concepts were revised. After a short break, all participants underwent the post-test1. For the remaining period they were introduced to the concept of system dynamics modelling. Participants were asked to familiarize themselves with popular system dynamics software (Powersim) during the week.

Week 4: The lecture on system dynamics modelling was continued from the previous week. All participants devoted significant time to model in Powersim, analyze the outputs and conduct sensitivity tests. At the end of the session, participants were given questions to practice at home.

Week 5: The first two hours was a revision class where all the concepts from week 2 to 4 were reinforced. During the last hour, participants were administered the post test 2.

The total time spent by students on systems thinking and system dynamics was 10 hours each. This included time spent in the class as well as time taken to complete related tasks at home. The lectures, related handouts and exercises were drawn from standard systems thinking/system dynamics texts and covered majority of the concepts of these two methodologies.

Data analysis

The data collected from participants was both qualitative and quantitative. The method of their analysis is described below.

Quantitative analysis

Quantitative data included participants' response to questions relating to discerning between stocks and flows, calculating the values of certain variables (based on the data provided) and measuring the change in confidence after and before the intervention. Part of this analysis (discerning between stocks and flows) was based on the framework utilized by Kainz and Ossimitz (2002).

Qualitative analysis

Most of the data collected from participants was in the form of narratives. Analysing these narratives to measure change in participants' mental models required a coding criteria. The criteria utilized in this study are based on two previous studies (Doyle et.al. (1998) and Maani and Maharaj (2004)). An expert model of the problem was constructed before hand. Each answer of the participant was then matched to this model and the percentage of correctness was calculated by identifying correct causal relationships. This helped in measuring the completeness of the participants' mental model pre and post intervention. Additionally the completeness of the mental models was also measured by measuring the type and number of relationships identified by participants.

Results and discussion

This study attempted to explore the links between two system interventions (systems thinking and system dynamics) and dynamic decision making. Figure 1 below presents the results of this study. The graph displays the performance of participants in three tests – pre-test, ST test and SD test. It provides the percentage of participants who got correct answers in each of the four questions.

During the course of analysis of the results, various things came up. As found by previous studies (Sweeny and Sterman (2000), Kainz and Ossimitz (2002)) the dynamic decision-making abilities of participants were surprisingly low. The authors were surprised to see such poor results from graduate students, all of whom had a mathematical background. Our results were very similar to those conducted by Kainz and Ossimitz (2002). However, results improved to a great extent by a short course on systems thinking and explanation of stocks and flows. Results further improved (though not as much as before) once participants were given further exposure to stock and flow modelling in the SD lectures.

The results of the question "what in your view is causing the periodic oscillations in revenue?" were then analysed. Without any systems intervention, more than 50% of the participants could only identify one-to-one relationships. Most of these were quite clearly mentioned in the case study. Furthermore participants were provided a week's time to read the scenario. Participant responses were mostly superficial and lacked any in-depth analysis of the relationship between key variables. Few recognised the "big picture" by identifying the relationship between the decision, its short-term effect and the long-term unintended consequences. However, post-intervention most participants improved in their ability to recognize the key relationship. They could trace relationships as well. Many did an in-depth analysis of the problem and some even attempted to compare the situation correctly with well-known system archetypes. One participant wrote "from the short-period, more project in hand,

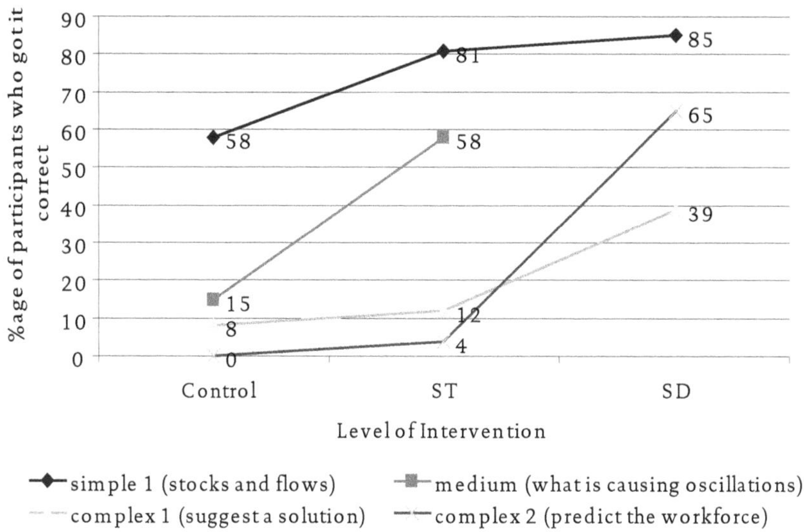

Figure 1 *Results*

more new employees hired, and relieve the work pressure, make revenue increase. However, from long-period, this lead to increase in quit-rate, experienced employee decreased, that makes the project fail, this make the revenue fall down. This is a repeatable situation". This participant not only identified most relationships, but traces variables with their effects on each other, identifies the short-term versus long-term effects and also that this is a cyclic process. Another participant wrote "since employee pressure increase, company will try to increase hiring rate to decrease employee pressure, but there is delay in their process, and other unintended consequences happen faster than this delay, so employee pressure is accumulated, which cause performance to slow down = revenue slow down". The intervention, made a significant impact on the ability of participants to think causally, focus on the delays of the system and analyze the situation by tracing causal links. Many students were able to identify the feedback loop that was the cause of the oscillations.

The third question "advise a long-term solution to the problem" required participants to utilize the knowledge that they had gained from their analysis of the problem and suggest measures that would help in alleviating the problem. Without any intervention, most (58%) participants focussed their suggestions on 'planning' for the project before hand. After undergoing the systems thinking intervention, unlike previous questions, response hardly improved. Participants who thought that reducing the delay would be helpful, increased by 19%. However there was no change whatsoever in the number of participants who suggested retrenchment to reduce. Surprisingly there was a reduction in the number of participants who indicated that recruitment be increased (by 4%) and that there was a need for planning of the project (by 12%). After participants underwent a system dynamics modelling session, most could model the scenario in Powersim. The suggestions that they provided subsequently told a different story. A large number of participants suggested that a long term solution was that recruitment should increase. 8% more participants thought that reducing the hiring delay was one of the strategies to alleviate the problem and 15% more thought that impeding retrenchment was a good strategy. 7.7% more participants thought that improved planning was required. The improved perception of the underlying cause of the systemic problem might be expected since participants interacted much more with a formal model and simulated it. This indicates that although systems thinking alone does help in understanding the relationship between key variables and delays and in discerning between stocks and flows, it is not very useful in situations where the scenario is very complex. In this case, computer modelling, simulation and sensitivity tests give a much deeper insight into the existing problem.

Conclusion

This study was conducted to explore the links between system interventions and dynamic decision making. The findings suggest that both systems thinking and system dynamics are useful methodologies in understanding complex systems. Specifically, participants' ability to take decisions about simple tasks such as discerning between stocks and flows, and identifying causal relationships and feedback improved by around 27% after a systems thinking intervention was introduced. These abilities further increased by around 4% after participants underwent a system dynamics modelling intervention. Interestingly, in a more complex task that involved numerical data to be processed, required an in-depth analysis and subsequently involved suggesting a solution to alleviate a systemic problem, systems thinking hardly had any positive effect on the participant's decision-making. However, for the same situation, participants' mental models improved significantly after system dynamics modelling intervention. The results of this study confirm the results of some of the previous studies done in this area (Sweeny & Sterman, 2000; Kainz & Ossimitz, 2002; Gary & Wood, 2005) and give a deeper insight on the impact of system interventions on dynamic decision-making.

This study has its own limitations. Given that the study was conducted in a University setting and not with real-world practitioners, these results cannot be generalized to a great extent. Also, both systems thinking and system dynamics provided to the participants were short in duration due to the structure of the course. The authors are involved in further experiments to explore these relationships in a more rigorous experimental setting.

References

Cavaleri, S. and Sterman, J.D. (1997). "Towards evaluation of systems thinking interventions: a case study" System Dynamics Review Vol. 13 (2): 171–186.

Doyle, J.K. (1997) "The cognitive psychology of systems thinking" System Dynamics Review Vol. 13 (3): 253–265.

Doyle, J. K., Radzicki, M. J., Trees, W. S. "Measuring Change in Mental Models of Dynamic Systems: An Exploratory Study" Report No. 14 May 26, 1998 accessed from https://www.wpi.edu/Academics/Depts/SSPS/Research/Papers/14.pdf on 1st March 1, 2006.

Doyle, J. K. and Ford, D.N. (1999). "Mental models concepts revisited: some clarifications and a reply to Lane" System Dynamics Review Vol. 15 (4): 411-415.

Gary, M.S. and Wood, R. (2005) Mental Models, Decision Making and Performance in Complex Tasks. Proceedings of the 2005 International System Dynamics Conference. System Dynamics Society: Albany, NY.

Huz, S, Andersen, D.F., Richardson, G.P., and Boothroyd R. (1996) "A framework for evaluating systems thinking interventions: an experimental approach to mental health system change".

Kainz D. and Ossimitz G. (2002). Can students learn stock–flow thinking? an empirical investigation. Proceedings of the 2002 International System Dynamics Conference. System Dynamics Society: Albany, NY.

Maani, K. and Maharaj, V (2004) "Links between systems thinking and complex decision making" System Dynamics Review Vol. 20 (1): 21–48.

Sterman, J.D. (1994). Learning in and about complex systems. System Dynamics Review 10(2/3), 291-330.

Sweeny, L.B. and Sterman, J.D. (2000) " Bathtub dynamics: initial results of a systems thinking inventory" System Dynamics Review Vol. 16 (4): 249–286.

Rajat Dhawan, BE (CompSc) VTU MInfTech, is a PhD student at the School of Business, University of Sydney. He holds a Bachelor of Engineering in Computer Science from India and a Masters degree in Information Technology from the University of Sydney. Prior to commencing serious research, he has been tutoring in the schools of Business, Information Technologies, Electrical and Information Engineering at the University of Sydney. His academic interests are in the fields of decision making and system dynamics. He has briefly worked with on a systems development project with the Indian Air Force Selection Board. Research interests: System Dynamics, Systems Thinking, Decision-Making

Marcus O'Connor, BCom (Hons) MCom (Hons) PhD UNSW, started at the University of Sydney in 2002. He has a distinguished record of intellectual and administrative leadership. His scholarly expertise covers forecasting, decision making and computer based decision support systems. His research program has made outstanding and innovative contributions to the DSS literature. His research interest are: judgemental forecasting, using IT in decision making and judgement, information sharing. His work is widely published in top international journals and has been supported extensively by competitive ARC Grants. He is the Associate Editor of the International Journal of Forecasting. Research expertise: Business information systems, Improving resource allocation decisions, Information sharing, Judgemental forecasting, Supporting decision making, Using IT in decision making and judgement

Mark Borman, BSc (Hons) UMIST MSc York PhD Strath. Prior to joining the University of Sydney Mark worked for a number of years in senior consulting and executive roles in the UK, USA and Australia; most recently as Head eBusiness Strategy at the Australian Broadcasting Corporation. He has considerable experience in market analysis, market research, strategy development, IT implementation, project management, risk management and procurement. Mark has a PhD in Management Science and lectured in Information Systems for six years at the University of Strathclyde in Glasgow during the early 1990s. His current research interest is in understanding the role of information technology in inter-firm collaborations. Research Expertise - Business information systems, Strategic innovation and implementation of business information systems.

Agent Based Modelling of Real Issues: Acquisition of Data to Support AB Models of Social Phenomena

Susanne Tepe[1], Tim Haslett[2] & Charles Osborne[2]
[1]School of Applied Science, RMIT University, AUS
[2]Department of Management, Monash University, AUS

Agent Based Models (ABMs) are powerful models for describing the interactions between and among individual agents within a dynamic situation, such as in a community. In contrast to System Dynamics models, ABMs are based on rules describing agent behavior. If modelled correctly, the ABM can mimic the behavior observed in a community and new behaviors can emerge as a result of agents' interactions. But where does the modeler acquire the data to underpin and define the rules which can then be modelled? This paper describes and evaluates the advantages and short comings of various data acquisition techniques, with particular reference to social modelling. Gathering data via documented historical information, personal experience, group model building, qualitative analysis of public comment data, quantitative methods such as discrete choice questionnaires, mixed qualitative / quantitative methods such as the Cynefin technique and the use of ABM archetypes are examined. Recommendations are made for the use of appropriate techniques for different situations.

Introduction

Agent Based Modelling (ABM) can be a powerful tool for systems modellers. As opposed to modelling the stocks and flows as in Systems Dynamics (SD) models, ABM focuses on groups of agents and their interactions. Whereas SD models define the system 'in focus' from a big picture or top down point of view, ABM allows the system to emerge from the interaction of the agents, a more bottom up point of view. While both SD and AB models rely on the language of mathematics to represent their actions, SD generally uses mathematics to describe flows through the system based on quantitative data, while AB uses equations to define the rules of agent interaction. Both modelling types are capable of describing non-linear dynamics, with SD focusing on interacting balancing and reinforcing loops and ABM describing its rules for agent interaction with non-linear equations and IF / THEN statements. Essential to the function of SD models are feedback loops and delays in response. While these can happen in ABM, more focus is placed on agents undergoing 'state changes' (eg from a passive 'state' to an active 'state') as a result of parameters that influence them. While state changes and flows from stock to stock can be similarly defined by the dynamics of the model in both SD and ABM, ABM allows these dynamics to change as a result of recent history (Anderson & Johnson 1997; AnyLogic, 2005; Axelrod, 1997; Axelrod & Tesfatsion, 2005).

While SD and ABM are both useful techniques for the systems' modeler, they must be used to address appropriate issues. While applicable in other areas, SD works better for complicated interacting systems involving quantitative data while ABM is more appropriate for examination of issues where individuals interact, such as social phenomena. Table 1 summarizes these observations.

Model formation

Despite these differences, all systems modelling require input from the environment in the form of data to formulate and validate the model. The data from the environment influence the model formation in different ways and at different times during the model formation. In order to describe this process more thoroughly, it is appropriate to describe the assumptions that underpin my description of model formation. Figure 1 shows a summary, very generic diagram of the model formation process.

As simplistically depicted in Figure 1, in establishing a systems model, the modeler starts with internal pictures of the 'world' being modelled, which Senge, *et al.* (1994) refer to as a Mental Model. Because at this point the Mental Model belongs to the modeler, Figure 1 refers to it as a Personal Mental Model (PMM). As the modeler interacts with and gathers data from the environment and reflects on this experience, their PMM evolves in clarity and the modeler's depth of understanding about this 'world' improves. The data drawn from the environment may take the form of discussions or interviews with other people concerned with the 'world' being examined. The modeler may acquire input from written documents or historical accounts concerning this 'world'. In addition, the modeler may discuss this 'world' with a group who in turn share their own versions of their mental models with each other and as a group they evolve a deeper understanding of this 'world'. As the modeler acquires this additional input, the PMM evolves into a richer understanding of this 'world', which Figure 1 refers to as an Evolved Mental Model.

At this point, the modeler may choose to cease keeping this model in their head and choose to record the model. While this can be done using any of the multitudes of systems modelling tools, in Figure 1 these models are differentiated by whether the data used to form the model is qualitative or quantitative. Generally, the modeler will first attempt to develop a qualitative model, such as a causal loop diagram (CLD). In the development of the CLD it is likely that PMM will continue to evolve, as indicated in Figure 1 by the feedback loop between the Qualitative Model and the Evolved PMM. When the modeler has a satisfactory CLD and if they have numerical data, they may choose to develop a quantitative model. Again, feedback from the model as it evolves will influence the PMM of the modeler.

Data acquisition

Given the above assumptions about the way that a systems model evolves, it is clear that data from the environment enters the model at various stages. Data from interviews with stakeholders, written documents

	Systems Dynamics Modelling	**Agent Based Modelling**
Aim	To describe complex multifaceted interacting systems	To evolve system behavior from the interaction of individual agents
Premise	System is defined by the modeler	Agents interact to define the system
Key components	Stocks and flows	Individual agents
Viewpoint	Top down / big picture / broad	Bottom up / individual
Outcomes	Behaviour over time	Emergent behavior due to agent interaction
Role of mathematics	Equations describe flows	Equations describe rules of interaction
Role of history	Not influenced by history (each model run is usually a fresh run)	Agents influenced by their history (agents change to new states as a result of experience)
Non-linear dynamics	Due to balancing and reinforcing loops	Due to rules of interaction
Essential elements	• Feedback loops • Delays	• Agents undergo State changes • Groups of agents have a distribution of responses
Best areas of application	Complex systems with available quantitative data	Social issues where individual actors make decisions

Table 1 *A simplistic comparison between systems dynamics modelling and agent based modelling*

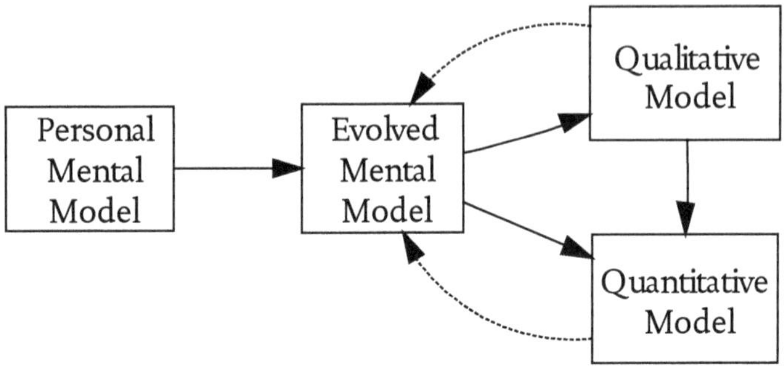

Figure 1 *Process of model development*

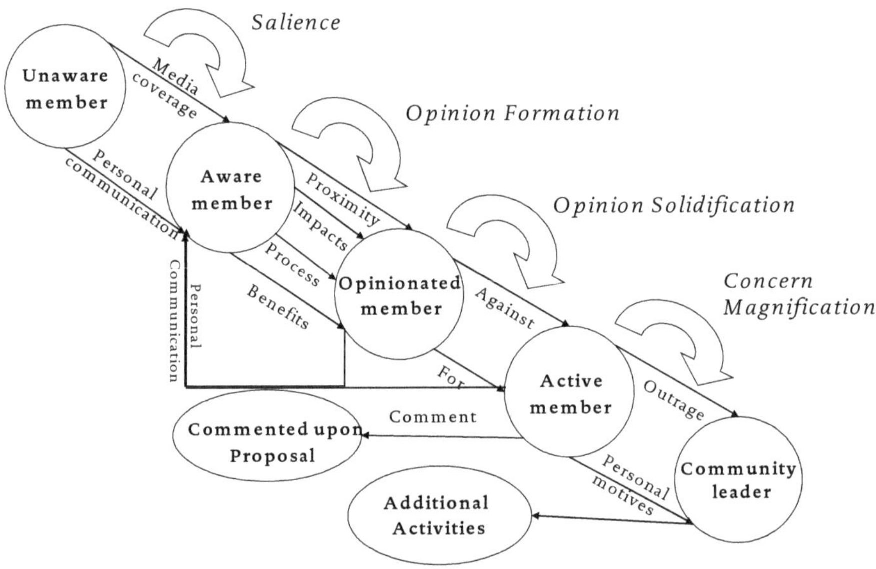

Figure 2 *State changes of community member agents*

and group discussions enrich the personal mental model of the modeler, which in turn can be recorded as a qualitative model. Numerical data can be used to convert the qualitative model into a quantitative model, such as a Systems Dynamics or an Agent Based model. But where does this information come from and how can it be used in Agent Based Modelling of a social system?

Depending on the issue at hand, the modeler may or may not have personal experience with the issue or the 'system in focus.' As the modeler attempts to define and understand the issue, they will form a mental model of their idea of the system in focus and what influences it, who the stakeholders or actors are, what environment the system operates in, what system boundaries need to be defined, etc. Generally the modeler recognises that, as an individual, this is a shallow one-sided view of the issue and will desire to enhance their understanding by acquiring other viewpoints. At this point, the modeler may or may not choose to record their model as a qualitative model. Always at the back of the ABM modeler's mind is the question: how does this information relate to the rules of individual interaction? Who are the actors, what influences them, how do they interact with other actors and how can I represent this as mathematically expressed rules?

The modeler may decide to enhance their understanding of the issue through acquisition of other input. They may choose to read documents concerning the issue. In the case of these authors' research on community acceptance of hazardous waste facilities, there is a book written that describes a community who became outraged about a plan to establish a hazardous waste facility and their successful blocking of the proposed plan (Strangio, 2001). This book was useful for providing additional insights to supplement the authors' experience, particularly concerning which groups should be construed as 'agents' and what parameters influence the agents at which stages of the process. For example, it became clear that an 'agent' community member could be represented in the variable states of 'unaware' of the issue, 'aware' of the issue, 'active' in their approach to the issue or 'community leader'. The parameters that influenced them included: the credibility of other community members who contacted the individual agent through word of mouth, the salience of media reporting and the potential impact of the development. Based on the mathematical relation between the parameters, the agents could undergo a state change from one state to the next. Figure 2 represents this set of state changes in a diagrammatical form.

In addition to deriving information from personal experience or from written material, a modeler can acquire information from the stakeholders in the issue. Implicit to this comment is that the modeler must recognise who the stakeholders are; this is usually as a result of previous personal experience or from case study input. Interviews with stakeholders can inform the personal mental model of the modeler in similar ways to reading a case study, but in a much more interactive way. The interviewee can correct any misunderstandings by the modeler and can specifically address topics of interest to the modeler. Interviews provide a broader range of viewpoints on the issue to supplement the modeler's own view. Because the stakeholder groups effectively become the 'agents' in the model, it would be best to have interview data from several individual stakeholders in each group to ensure that diversity of opinion is represented. The focus of the interviews should be about the parameters that influence the stakeholder / agent in changing from one state to the next. Consistency of view about the parameters is desired within the one agent group, while the range of opinion about the strength of influence of this parameter can be used to define the shape of the distribution of response. For example, all participants interviewed about what makes a person pay attention to a news article concerning the potential development of a hazardous waste facility (e.g. the news article's salience), indicated that geographical proximity of the proposed location to their own life was an important factor. In modelling terms, this data was interpreted as a relatively strong factor (so a numerically bigger value) with a narrow distribution of response.

Similar to interviews with individual stakeholders, a modeler can interview several member of one stakeholder group at one time, usually in a focus group type arrangement. This helps with ensuring that the breadth of stakeholder views are canvassed while also helping the stakeholders reflect on comments made by others in the group. This is not dissimilar to the group model building process used by SD modelers. Again, for ABM, the focus of the discussion should be based on the parameters that cause the stakeholder agents to change from one state to another, but obviously in a language that is consistent and sensitive to the issue at hand. In the example at hand, the government regulators were interviewed as a group, to get a consistent description of the process of review for development proposals.

All of the above methods for gathering information are qualitative in nature. They are very useful in forming and enhancing the mental model of the modeler, which in turn can be converted into CLD or ABM, but they do not give any numerical data about the mathematical relationships, the frequency of any events or the relative value of one influencing parameter compared to another.

At this point the modelers can test their mental model by using an ABM archetype. If the modelers recognise similarities between their issue and one already developed by a soft ware package, they can modify the existing software to reflect their own case. In the case of the authors' research, the AnyLogic software had a model of the Bass Diffusion Model, where prospective purchasers became actual purchasers due to influence of advertising and word of mouth (AnyLogic, 2005). This model was able to be modified to reflect how community members could be influenced to become 'aware' of the issue and to become 'active' based on media reports and on contact with other community members. This archetype was then embellished to include more state changes and to incorporate feedback and reinforcing loops (see Figure 2). While it is possible to get the model to run and for community member agents to change state, the mathematical relationships and the values attributed to each parameter are contrived; there is no basis in reality for why a given parameter has a given value, only that those values make the model 'work'.

Numerical data for ABM modelling is more challenging to acquire than data for SD models. SD models frequently use existing data concerning levels of stocks and rates of flows from existing situations. However, for the authors' research, it has been very challenging to acquire data from existing sources. One opportunity has been to examine the public comment submissions required by the Environment Protection Authority; in determining if the hazardous facility is to go ahead, the EPA is legally required to accept public comment on the plans. The comments were analyzed to determine the types or themes of issues of concern to the public. In addition, some numerical data were converted to ratios, eg the number of comments on a given theme compared with the total number of comments. While the prospect of gaining data in this manner is quite attractive, it has its own limitations. It is known to the EPA that public comment statements are heavily biased according to who submits comment, with people in favour of a development much less likely to comment than those opposed to it. In addition, simply counting the number of comments does not provide any insight into who made the comment; intuitively a comment made on behalf of a group of people should have more 'weight' than a comment made by an individual.

If existing numerical data do not prove to be useful, de novo data can be acquired through the use of questionnaires. There are many, many forms that a questionnaire can take, all of which are influenced by the mental model of the modeler. A questionnaire that will produce usable data for ABM modelling must derive information about the variables that the modeler believes influence state changes in the agents while producing frequency and relativity data. Choice based questionnaires, such as Discrete Choice or Contingent Choice methods where the respondent must state a preference for a given choice within a hypothetical scenario or as a real choice are very useful for this purpose (Train 2002). In addition, methods that generate Bayesian distributions and probability statistics can also be used in ABM (AnyLogic, 2005).

An interesting method with possible application to ABM is the Cynefin technique (Snowden, 2000) of narrative analysis. In this technique, qualitative data are gathered by asking respondents questions that generate stories. The respondents are asked to index their own stories against the characters (agents) and themes (parameters) which appear in their story. While this technique has not been used by the authors, it appears to provide an interesting blend of quantitative data which can be enriched by the narrative analysis. It allows for model development as well as detection of 'surprises' from respondents which can add to the mental model of the modeler.

Discussion

Agent Based Modelling can be a very powerful technique for modelling interactions between groups of agents in order to observe if relatively simple rules can cause unexpected behaviors to arise. But if ABM is to be more than just entertainment, it must be able to reflect real life experiences, and thus must be able to take data from experience and incorporate it into the model.

Qualitative inputs are essential in clarifying the mental model that drives the ABM, however quantitative data must be acquired eventually if the model is to meaningfully reflect the life experience. Qualitative data are probably easier to acquire; virtually every stakeholder has an opinion that can contribute to the development of the modeler's mental model. Quantitative data are more difficult to acquire, primarily because it is unlikely that existing data will be in a format to be usable to drive the rules of agent interaction and the data will need to be acquired de novo.

The use of ABM archetypes is a feasible stepping stone between qualitative and quantitative models; however, the software suppliers will need to extend the availability of archetypes and make the process for modification more transparent.

References

Anderson, V & Johnson, L 1997, *Systems Thinking Basics: From Concepts to Causal Loops*, Pegasus Communications, Inc., Waltham, MA. ISBN 1-883823-12-9

AnyLogic 2005, *Standard Training Handout Materials*, XJ Technologies. www.xjtek.com

Axelrod, R 1997, 'Advancing the Art of Simulation in the Social Sciences', in ReH Conte, Rainer (ed.); Terna, Pietro (ed.) (ed.), *Simulating social phenomena.*, Springer, Berlin, pp. 21-40. ISBN: 3-540-63329-4

Axelrod, R & Tesfatsion, L 2005, 'A Guide for Newcomers to Agent_Based Modeling in the Social Sciences', in LTaKL Judd (ed.), *Handbook of Computational Economics, Vol. 2: Agent-Based Computational Economics*, North-Holland, Amsterdam, the Netherlands, vol. 2. ISBN 0444898573

Senge, P, Roberts, C, Ross, R, Smith, B & Kleiner, A 1994, *The Fifth Discipline Fieldbook: Strategies and Tools for Building a Learning Organization*, Nicholas Brealey Publishing, London. ISBN 1-85788-060-9

Snowden, DJ 2000, 'The Art and Science of Story or "Are you sitting uncomfortably?"', *Business Information Review*, vol. 17, no. 3, pp. 147-56. ISSN: 0266-3821

Strangio, P 2001, *No Toxic Dump! A triumph for grassroots democracy and environmental justice*, Pluto Press, Annadale, NSW ISBN 1 86403 179 4.

Train, K 2002, *Discrete Choice Methods with Simulation: Cambridge University Press*, Cambridge University Press. ISBN: 0-521-81696-3.

A Paradigmatic and Methodological Survey of the Knowledge Management Literature: 2000 to 2004

James Sheffield[1] & Zining Guo[2]
[1] University of Auckland, NZ
[2] Xelocity Limited, NZ

Knowledge management research lacks a cumulative tradition. This critical review of the literature employs a Habermasian inquiring system and McGrath's classification of research methods to identify gaps and recommend how to fill them. All articles from eleven first-tier journals for the period 200-2004 are classified by research paradigm, research methodology, and research interest (technical, practical, emancipatory). The key finding is that inquiry in KM is starkly unbalanced. Overuse of the positivist paradigm and its dominant research method (sample survey) prevents the exploitation of the highly relevant insights available via the interpretivist and critical pluralist paradigms and the field study method.

1. Introduction

The knowledge management (KM) literature is expanding rapidly. However, surveys of the literature have shown that KM research lacks conceptual integration and cumulativeness. Knowledge management research is in an emergent, theory-building stage. (Croasdell, *et al.*, 2003) An overarching framework is needed to unify research. (Rubenstein-Montano, *et al.*, 2001)

This lack of integration in KM research motivates the need for the practical use of philosophy. An inquiring system - a philosophically grounded, universally pragmatic framework for inquiry - is required to provide a coherent framework. (Courtney, 2001; Courtney, *et al.*, 1998; Richardson & Courtney, 2004; Richardson, *et al.*, 2001; Sheffield, 2004, 2005)

This paper reports on a survey of the research methods employed in KM research, and analyzes findings in conjunction with a Habermasian inquiring system previously developed by the authors. (Guo & Sheffield, 2006) All articles relevant to knowledge management published in leading journals in 2000 through 2004 are classified. The results are analyzed for trends and gaps. Recommendations are made.

This paper is organized in the following way. Section 2 reviews the Habermasian inquiring system and provides conceptual necessary definitions of concepts. Section 3 explains the survey methodology. Section 4 presents the findings. Section 5 discusses key findings. Section 6 identifies gaps in the literature and recommends new directions for KM research.

2. Research objectives

2.1 Research interests

Habermasian inquiring systems draw from both the systems thinking and critical thinking traditions. The key architectural element is Habermas' knowledge interests (viz. technical, practical, and emancipatory). Guo and Sheffield's (2006) Habermasian Inquiring System consists of describing how four other design elements (Habermas' three rationalities, Churchman's roles, knowledge dynamics, and research paradigms) are positioned within this structure (Table 1).

Knowledge Interest	Technical	Practical	Emancipatory
Habermas' Rationalities	Technical	Organizational	Personal
Churchman's Roles	Measure of Performance	Decision Maker	Client
Knowledge Dynamics	Knowledge Application	Knowledge Normalization	Knowledge Creation
Research Paradigm	Positivism	Interpretivism	Critical Pluralism

Table 1 *Habermasian Inquiring System and research paradigms*

The following provides a brief definition of *research interests.* Habermas recognizes three knowledge-constitutive interests, viz. technical, practical, and emancipatory. The *technical* interest is concern for human "work", which encompasses any interactions with the physical world. The *practical* interest is concerned with interpretation of language and intersubjective communication. The *emancipatory* interest is concern for emancipation from colonization of the life-world. It provides a dialectical synthesis of, and a self-reflection on, both the technical and practical approaches. (Habermas, 1968) Knowledge interests, as instantiated in research activities, constitute research interests.

2.2 Research paradigms

Research paradigm generally refers to "the progress of scientific practice based on people's philosophies and assumptions about the world and the nature of knowledge". (Collis & Hussey, 2003) There are various categorizations of research paradigms. (Burrell and Morgan, 1979; Deetz, 1996; Morgan and Smirchich, 1980). Positivism, interpretivism, and critical inquiry are frequently identified as the main paradigms for business research. (Cavana, *et al.*, 2001; Crotty, 1998; Ulrich, 2001)

Table 2 is adapted from (Cavana, *et al.*, 2001; Sheffield, 2004). It provides a detailed conceptual definition of the positivist, interpretivist, and critical pluralist research paradigms.

Research Paradigm	Positivism	Interpretivism	Critical Pluralism
Assumptions	Objective world which science can measure and "mirror" with expert, privileged knowledge	Inter-subjective world which science can represent with concepts and indicators; social construction of reality	Material world of structured contradictions and/or exploitation which can be objectively known only by removing tacit ideological biases
Aim	To discover universal laws that can be used to predict human activity	To uncover socially constructed meaning of reality as understood by an individual or group	To uncover surface illusions so that people will be empowered to change their world
Stance of researcher	Stands aloof and apart from stakeholders and subject matter so that decisions can be made objectively	Becomes fully involved with stakeholders and subject matter to achieve a full understanding of the stakeholders' world	Involved with stakeholders so that surface illusions can be identified, but urges subjects to change their world
Values	Value free; their influence is denied	Values included and made explicit	Values included and made explicit
Types of reasoning	Deductive	Inductive	Deductive and inductive
Research plan	Rigorous, linear and rigid, based on research hypothesis	Flexible, and follows the information provided by the research stakeholders	The imperative for change guides the actions of the researcher
Typical research methods and type(s) of analysis	Experiments; questionnaires; secondary data analysis; quantitatively coded documents; statistical analysis	Ethnography; participant observation; interviews; focus groups; conversational analysis; case studies	Field research, historical analysis, dialectical analysis, feminist studies, case studies of personal experience and injustice
Goodness or quality criteria	Conventional benchmarks of "rigor"; internal and external validity; reliability and objectivity; technical excellence validated by objective truth	Trustworthiness and authenticity; Fit with social norms; interpersonal consensus validated by rightness of advocacy (speech acts) and actions	Historical situatedness; erosion of ignorance and misapprehensions; sincerity of beliefs; action stimulus; personal commitment validated by truthfulness

Table 2 *Research paradigms and research methods*

2.3 McGrath's classification of research methods

Research methods are "generic classes of research settings for gaining knowledge about a research problem". There are various schemes to classify research methods. By comparing different classification schemes, for instance, McGrath (1982), Alavi & Carlson (1992), and Pervan (1998), it is decided that McGrath's classification scheme would be used, with adaptations if necessary.

McGrath classifies eight research methods in a circumplex. (Figure 1) The eight methods are positioned as pie segments located within a triangle of measures of quality (validity). Three desiderata (viz. generalizability, precision,

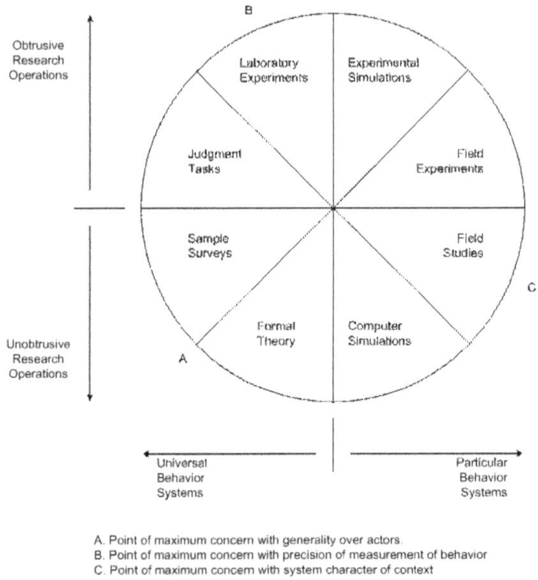

Figure 1 *A classification scheme for research methods (McGrath, 1982 :73)*

and realism of context) are mapped on the circumplex. McGrath argues that reliance on a single method will inevitably produce research that is "flawed" in that it cannot perform well against all three desiderata. (McGrath, 1982)

The strength of McGrath's classification is that it links research methods to research paradigms. Positivist methods provide greater internal and external validity in the measure of universal constructs. Interpretive and critical pluralist paradigms are required to appreciate the existential reality that a particular social setting has for those who in-habit (i.e. en-act) it.

3. Survey methodology

3.1 Modified classification scheme

Key aspects of the classification scheme are shown in Figure 2. To better reflect KM research activities, "system development" was added as a methodology. (Adams & Courtney, 2004; Nunamaker, *et al.*, 1990-91)

Each article included in the literature survey will be classified by the following five concepts.

A. *Demographics*: The demographics of journal name, year of publication, author and country affiliation are recorded. (Croasdell, *et al.*, 2003; Pervan, 1998)

B. *Empirical or Non-Empirical*: An article is judged to be *empirical* if the researcher acquires original data as his/her primary source of evidence. (Alavi & Carlson, 1992; Pervan, 1998) Otherwise, it is judged to be *non-empirical.* A non-empirical article is further classified into *computer simulation*, *theory-building*, or *literature review.*

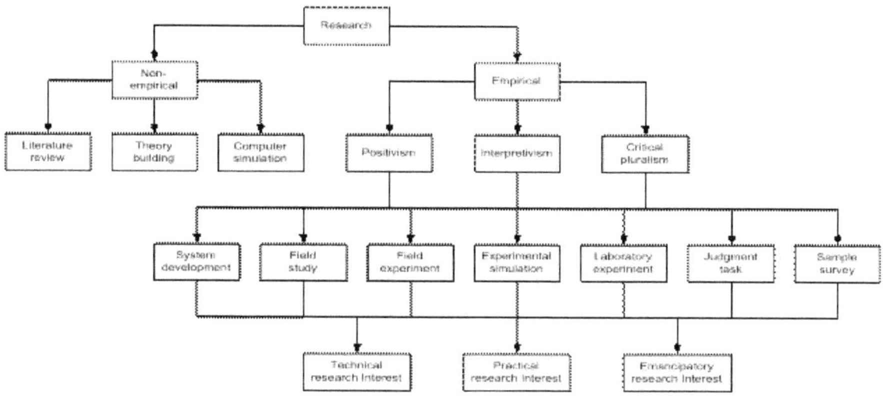

Figure 2 *The modified classification scheme used in the present study*

C. *Research Paradigms*: The research paradigm of an empirical article is judged to be *positivist, interpretivist, or critical pluralist.* Careful attention is given to research aims, stance of the researcher, types of reasoning, research plan, research methods, and the writing style. (Cavana, *et al.*, 2001)

D. *Research Methods*: The research method of an empirical article is judged as *system development, field study, field experiment, experimental simulation, laboratory experiment, judgment task, or sample survey.* Particular attention was paid to the way the researcher designs and undertakes the research, i.e. its generality, precision, and realism of context; universal vs. particular behavior systems; and obtrusive vs. unobtrusive research operations.

E. *Research Interests*: The research interest of an empirical article is classified holistically as *technical*, *practical*, or *emancipatory*. Attention is given to surfacing the researcher's motivation, domain of the research problems, research foci, and the overall "payoff" of the research.

3.2 Framing of the sample

Journal rankings are reviewed and 11 top-tier journals in information systems and management science are chosen. The journals are listed below. Abbreviations are used to identify each journal. (Table 3)

In summary, the sample frame is all articles published in the 11 journals over the five-year period 2000-2004 whose primary focus is knowledge management, organizational learning, or organizational memory.

3.3 Sampling procedure

The sampling procedure is summarized as follows. All copies of these journals for the five-year period 2000-2004 are searched for knowledge management research articles. All articles in all volumes in the sample frame were located and the abstracts read to confirm relevance.

Two non-author reviewers were trained in a formal face-to-face one-hour class. One of the non-author reviewers was an "extra" whose services proved not to be required. A total of three reviewers were expected to participate in this literature survey. One of the authors and a non-author served as pri-

Journals in Information Systems	Journals in Management Science
MIS Quarterly (MISQ)	Management Science (MS)
Information Systems Research (ISR)	Organization Science (OS)
Communications of the ACM (ACM)	Academy of Management Journal (AMJ)
Journal of Management Information Systems (JMIS)	Administrative Science Quarterly (ASQ)
Decision Sciences (DS)	
Decision Support Systems (DSS)	
European Journal of Information Systems (EJIS)	

Table 3 *List of journals surveyed*

mary reviewers. The second author served as a potential third reviewer.

The two primary reviewers classified the articles separately, and then met to compare ratings. When disagreement occurred, the two reviewers attempted to resolve the disagreement through communication. If disagreement persisted, the decision was to be adjudicated by the third reviewer.

It transpired that disagreements were invariably settled by the primary reviewers via one or more iterations of discussion. This collaborative process produced a history of judgment calls that was codified into a formal protocol. The development of a formal protocol enabled the two primary raters to achieve 100% inter-rater concordance or reliability.

4. Survey findings

4.1 Annual output of KM journal articles

The purpose of this subsection is to establish the relevance. The ABI/Global database was searched for those scholarly papers that had "knowledge management" in the citation or abstract fields. The count of articles for each of the years 1995-2004 is reported in Figure 3.

4.2 Distribution of articles

The issues of the 11 journals between 01 January 2000 and 31 December 2004 contained 174 research articles relevant to knowledge management. The counts are reported in Figure 4.

4.3 Country affiliation

The country affiliations of the authors are categorized into three regions, namely, North America, Europe, and Asia-Pacific. Since some articles are co-authored by researchers from different countries, the number of country affiliations (194) exceeds the number of articles (174).

It is found that North America is by far the biggest contributor with 140 articles. Europe contributes 31 articles. Asia-Pacific contributes 23 articles.

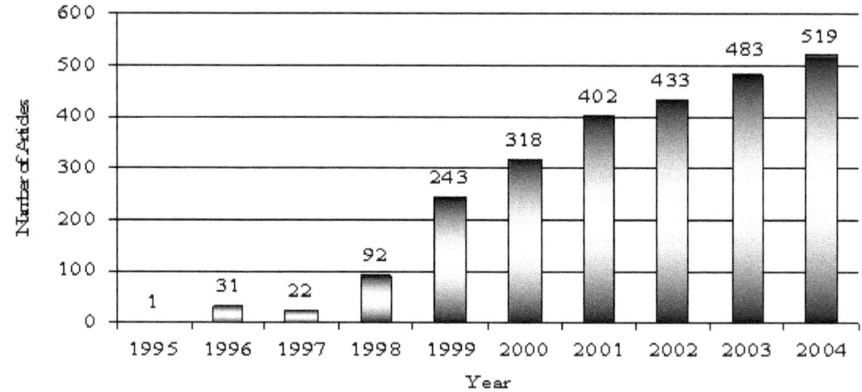

Figure 3 *Trend in the number of articles for the ten-year period 1995-2004*

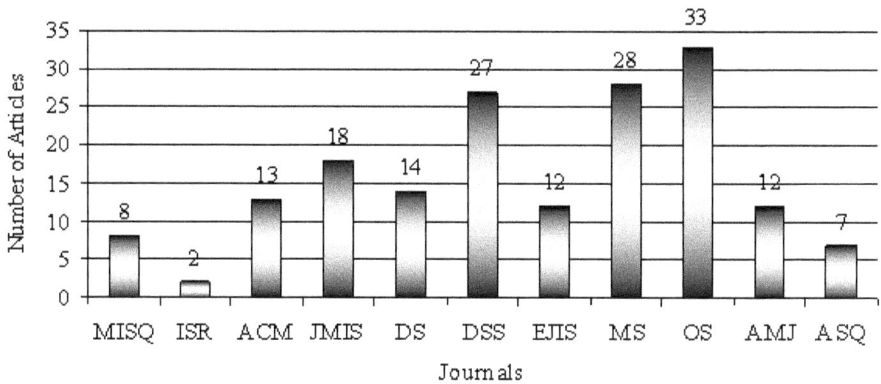

Figure 4 *Distribution of articles for the survey period 2000-2004*

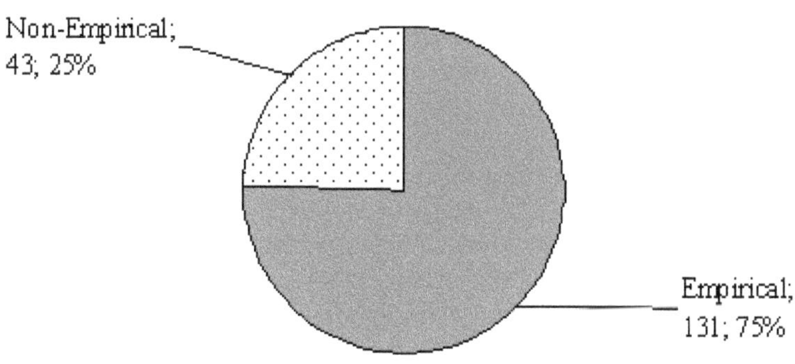

Figure 5 *Distribution of empirical and non-empirical articles*

4.4 Non-empirical or empirical

Among the 174 articles, 25% (43 articles) represent non-empirical research whilst 75% (131 articles) represent empirical research. (Figure 5)

4.5 Types of non-empirical research

Among the 43 non-empirical articles, 60% represent theory-building research; 35% represent literature review research; 5% represent computer simulation research. (Figure 6)

4.6 Research paradigms

Among the 131 empirical articles, 76% are positivist; 22% are interpretivist; and only 2% are critical pluralist. One article reported both an interpretive field study and a positivist sample survey, making the total 132 rather than 131. (Figure 7)

4.7. Research methods

Among the 131 empirical articles, sample survey is found to be the most widely adopted research method - 48% were sample surveys; 31% were field studies. The counts are reported below. One article reported both an interpretive field study and a positivist sample survey, making the total 132 rather than 131. (Figure 8)

Research methods adopted to support positivist, interpretivist, and critical pluralist articles are analyzed separately. The results are presented in Figures 9, 10, and 11.

4.8 Research interests

Among the 131 articles of empirical research, 78% of the total represents technical research interest, 20% represent practical research interest, and 2% represent emancipatory research interest. One article reported both an interpretive field study and a positivist sample survey, making the total 132 rather than 131. The distribution of research interests is illustrated in Figure 12.

5. Discussion of key findings

Firstly, the results show that *positivism is the dominant research paradigm.* This confirms findings of other studies (Arnott & Pervan, 2005; Chen & Hirschheim, 2004; Hopp, 2004; Schultze & Leidner, 2002). The results also show that interpretivism (22%) has taken its foothold in the KM research community. However, it is notable that critical pluralist research (2%) remains marginalized.

Secondly, the results show that *sample survey is the most widely used method of acquiring evidence.* Field study is the second most used method. It is possible that the significant volume of field studies may reflect the increasing acceptance of interpretivism, and the need for deep involvement of the researcher in the research "site". (Cavana, *et al.*, 2001) It may also confirm the immature, theory-building stage of KM research. (Croasdell, *et al.*, 2003)

Thirdly, the results show that *technical research interest dominates the research community.* Practical research interests motivate 20% of the empirical research surveyed. The emancipatory research interest is rare and marginalized.

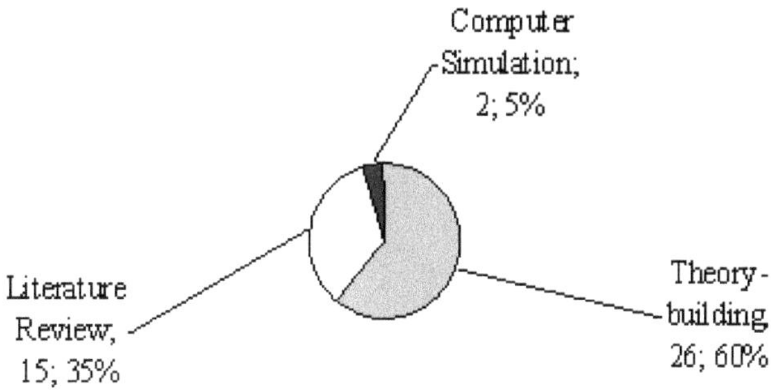

Figure 6 *Distribution of types of non-empirical articles*

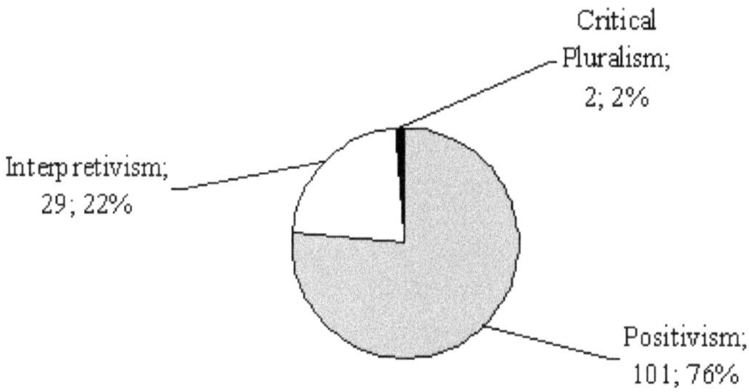

Figure 7 *Research paradigms of empirical research*

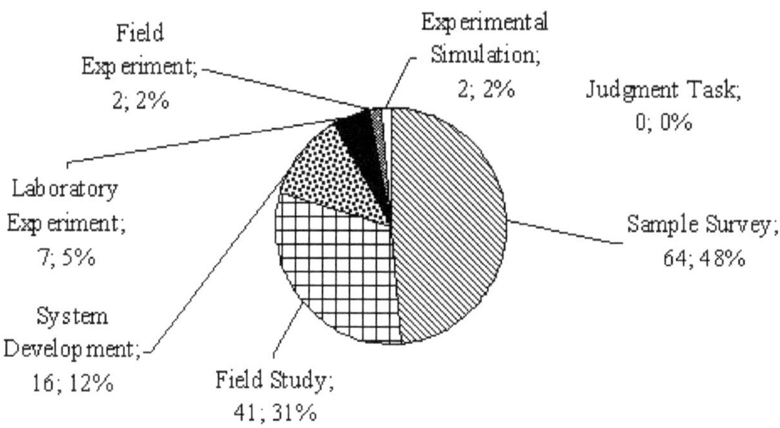

Figure 8 *Overall distribution of research methods in empirical research*

Fourthly, the findings show that *the constructs underlying the research are valid.* This permits a degree of confidence in the findings on constructs that were directly measured (Figure 2) as well as those that are integral to the logic, but were not directly measured.

This finding is the evidence required to demonstrate a certain pattern in the above findings, viz, that researchers align their choices of a research para-

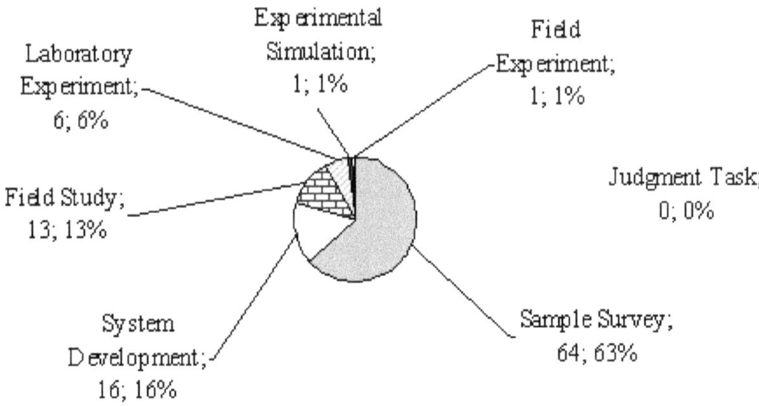

Figure 9 *Research methods for positivist research*

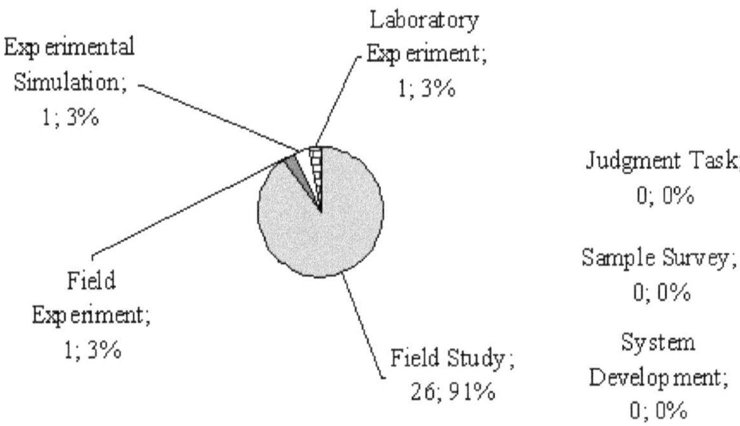

Figure 10 *Research methods for interpretive research*

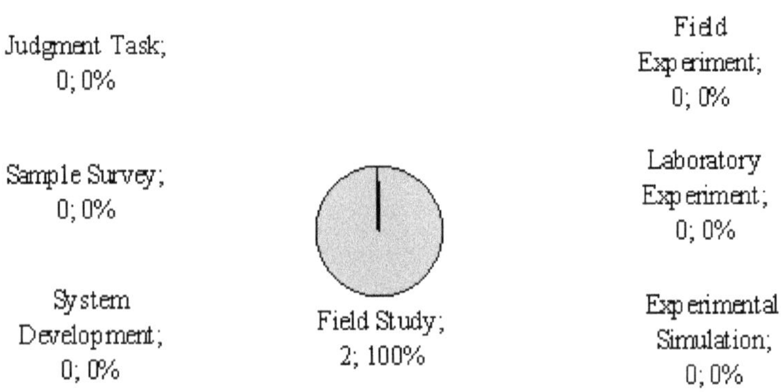

Figure 11 *Research methods for critical research*

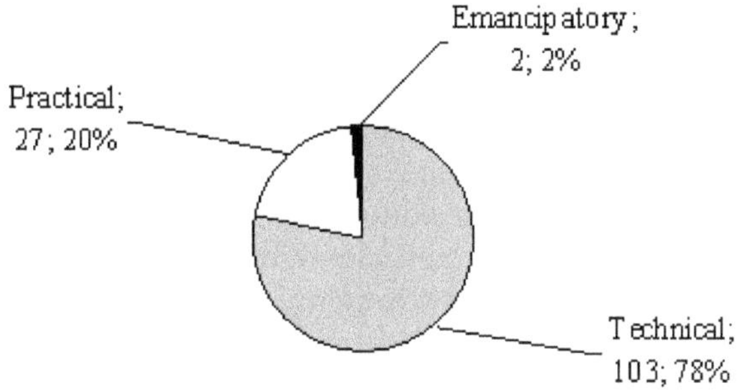

Figure 12 *Distribution of research interests*

digm, a research method, and a research interest. For example, the positivist paradigm, with its deductive reasoning, champions objectivity and generalizability (Table 2). It is not surprising that sample survey, which is adjacent to the point maximizing generality in McGrath's circumplex (Figure 1), is the most widely used method for positivist researchers (63%). The use of the positivist paradigm (76%) and research methods such as sample survey *can be linked with confidence to the reviewers' identification of a technical research interest* (78%). This logic also holds true for interpretive research and critical pluralist research.

6. Recommendations for knowledge management research

The long-standing problem in KM research of lack of cumulativeness is addressed by an intensive review of the KM literature that identifies gaps and recommends how these should be filled.

Major gaps are found. Current research on KM is unbalanced in its methods of inquiry. The majority of researchers embrace technical and/or practical knowledge interests, leaving emancipatory interest largely unattended.

Firstly, Habermas' *personal rationality* is barely recognized. KM research that overemphasizes technical rationality and social/organizational rationality leads to an unquestioned reinforcement of existing social systems. This constitutes a colonizing force of the life-world; it leaves little room for ethical consideration.

Secondly, the treatment of Churchman's role of *client* is cursory. KM research that focuses on measures of performance to serve the *decision maker* largely ignores the will of the individual, reducing him/her to a functional unit valued only as an organizational resource.

Thirdly, attention to personal aspects of *knowledge creation* is limited. Research that is only interested in organizational knowledge, as it is normalized, in relation to its perceived value at the technical level fails to fully comprehend the process in which personal knowledge is transformed into organizational knowledge. KM research associated with sense-making and organizational change will be curtailed.

Fourthly, the lack of *interpretive* and *critical pluralist* KM research needs to be addressed. Overuse of the positivist paradigm and its dominant research method (sample survey) prevents the exploitation of the highly relevant insights available via the interpretivist and critical pluralist paradigms. Approaches such as Grounded Theory and Participatory Action Research are recommended to correct this imbalance.

Finally, increased use of both *interpretive and critical field studies* is required to make KM research more relevant and value-sensitive. It is significant that the only two critical research articles found in this survey were published in a non-USA journal. None of the editors of the 10 prestigious US-based journals published any critical research articles in the period 2000-2004. However editors of MISQ and perhaps other prestigious US journals currently welcome a broad range of methodological approaches including those that surface personal experience via first-person accounts that directly addresses the subjective nature of deciding on the nature of the problem to be addressed *("Why I feel?")*.

In summary, the dominant knowledge interest is *technical*, focusing on objective measures of performance.

A more balanced recognition of all three knowledge interests is required. The triangulation of multiple paradigms and methods produces a more holistic appreciation of knowledge-related phenomena, and contributes more to a management science that "searches for truths about the ways in which men work and live together". (Churchman, 1955)

References

Note: References to *all* of the 174 knowledge management articles that form the database for this literature review are omitted.

Adams, L. A., & Courtney, J. F. (2004). *Achieving relevance in IS research via the DAGS framework.* Paper presented at the 37th Hawaii International Conference on System Sciences, Hawaii, USA.

Alavi, M., & Carlson, P. (1992). A review of MIS research and disciplinary development. *Journal of Management Information Systems,* 8(4), 45-62.

Arnott, D., & Pervan, G. (2005). A critical analysis of decision support systems research. *Journal of Information Technology, 20*(2), 67-87.

Burrell, G. (1994). Modernism, postmodernism and organizational analysis 4: The contribution of Jurgen Habermas. *Organization Studies, 15*(1), 1-19.

Burrell, G., & Morgan, G. (1979). *Sociological paradigms and organizational analysis: Elements of the sociology of corporate life.* Brookfield: Ashgate Publishing Company.

Cavana, R. Y., Delahaye, B. L., & Sekaran, U. (2001). *Applied business research : Qualitative and quantitative methods.* Milton, Qld, Australia: J. Wiley.

Chen, W. S., & Hirschheim, R. (2004). A paradigmatic and methodological examination of information systems research from 1991 to 2001. *Information Systems Journal, 14,* 197-235.

Churchman, C. W. (1955). Management science. *Management Science, 1*(2), 187-188.

Collis, J., & Hussey, R. (2003). *Business research: A practical guide for undergraduate and postgraduate students* (2nd ed.). New York, NY: Palgrave Macmillan.

Courtney, J. F. (2001). Decision making and knowledge management in inquiring organizations: Toward a new decision-making paradigm for DSS. *Decision Support Systems, 31*(1), 17.

Courtney, J. F., Croasdell, D. T., & Paradice, D. B. (1998). Inquiring Organizations. *Australian Journal of Information Systems,* 6(1), 3-14.

Croasdell, D. T., Jennex, M., Yu, Z., Christianson, T., Chakradeo, M., & Makdum, W. (2003). *A meta-analysis of methodologies for research in knowledge management, organizational

learning and organizational memory: Five years at HICSS. Paper presented at the 36th Hawaii International Conference on System Sciences, Hawaii, USA.
Crotty, M. (1998). *The foundations of social research: Meaning and perspective in the research process.* St Leonards, Australia: Allen & Unwin.
Deetz, S. (1996). Describing differences in approaches to organization science: Rethinking Burrell and Morgan and their legacy. *Organization Science, 7*(2), 191-207.
Guo, Z., & Sheffield, J. (2006). *Habermasian Inquiring System: Toward a General Framework for Knowledge Management Research.* Paper presented at the 39th Hawaii International Conference on System Sciences, Hawaii, USA.
Habermas, J. (1968). *Knowledge and human interests.* Boston: Beacon Press.
Hopp, W. J. (2004). Fifty years of *Management Science. Management Science, 50*(1), 1-7.
Marshall, N., & Brady, T. (2001). Knowledge management and the politics of knowledge: Illustrations from complex products and systems. *European Journal of Information Systems, 10,* 99-112.
McGrath, J. E. (1982). Dilemmatics: The study of research choices and dilemmas. In *Judgment Calls in Research* (pp. 69-102). Beverly Hills: Sage Publications.
Morgan, G., & Smircich, L. (1980). The case of qualitative research. *Academy of Management Review,* 5(4), 491-500.
Nunamaker, J. F., Chen, M., & Purdin, T. D. M. (1990-91). System development in information systems research. *Journal of Management Information Systems, 7*(3), 89-106.
Panagiotidis, P., & Edwards, J. S. (2001). Organisational learning - A critical systems thinking discipline. *European Journal of Information Systems, 10*(3), 135.
Peachey, H., & Hall, D. (2005). *Knowledge Management and the Leading IS Journals: An Analysis of Trends and Gaps in Published Research.* Paper presented at the 38th Hawaii International Conference on System Sciences, Hawaii, USA.
Pervan, G. P. (1998). A review of research in group support systems: Leaders, approaches and directions. *Decision Support Systems, 23*(2), 149-159.
Richardson, S. M., & Courtney, J. F. (2004). *A Churchmanian theory of knowledge management system design.* Paper presented at the 37th Hawaii International Conference on System Sciences, Hawaii, USA.
Richardson, S. M., Courtney, J. F., & Paradice, D. B. (2001). An assessment of the Singerian inquiring organizational model: Cases from academia and the utility industry. *Information Systems Frontiers,* 3(1), 49-62.
Rubenstein-Montano, B., Liebowitz, J., Buchwalter, J., McCaw, D., Newman, B., Rebeck, K., et al. (2001). A systems thinking framework for knowledge management. *Decision Support Systems, 31,* 5-16.
Schultze, U., & Leidner, D. E. (2002). Studying knowledge management in information systems research: Discourse and theoretical assumptions. *MIS Quarterly, 26*(3), 213-242.
Sheffield, J. (2004). The design of GSS-enabled interventions: A Habermasian perspective. *Group Decision and Negotiation, 13*(5), 415-436.
Sheffield, J. (2005). *The evaluation of GSS-enabled interventions: A Habermasian perspective.* Paper presented at the 38th Hawaii International Conference on System Sciences, Hawii, USA.
Ulrich, W. (2001). A philosophical staircase for information systems definition, design, and development: A discursive approach to reflective practice in ISD (Part 1). *Journal of Information Technology Theory and Application,* 3(3), 55-84.

Jim Sheffield graduated with a PhD from the University of Arizona in 1990. His doctoral work in group support systems led to the creation of a Decision Support Centre at the University of Auckland and participation in major policy development exercises. Action research initiatives include the development of economic strategy, science policy, and comprehensive regional planning. He has published in Journal of MIS, Group Decision and Negotiation and International Journal of Business Information Systems. Current interests include: Churchmanian and Habermasian inquiring systems, knowledge management, collaborative learning, and research methods.

Zining Guo holds a Mater of Commerce (First Class Honors) in Information Systems from the University of Auckland. He is currently serving as Associate Consultant at Xelocity Limited, Auckland, New Zealand. He has published in Hawaii International Conference on System Sciences. His primary areas of research interest include knowledge management, systems approaches, organization sciences, research methods, and philosophical foundations of inquiry.

Reflecting on the Dynamics of Our Intervention

Richard Monypenny
James Cook University, AUS

We are working with three senior staff of an association of local service providers to help then make concrete steps towards closer cooperation and increased value adding, as part of a medium term process to successfully develop a local industrial cluster. In providing this help to senior staff we use project facilitation skills in two complementary ways:

- **First, we use facilitation skills as the main building block for creating closer cooperation and increased value adding for local service providers.**
- **Second, we use facilitation skills as the framework with which the association integrates its activities in order to focus effort on creating closer cooperation and increased value adding for the association.**

In this paper we report on our reflections on the dynamics of our intervention in the association that are relevant to the Systems Conference. Reflections related to other aspects of our work are reported elsewhere.

Introduction

We are working with three senior staff of an association of local service providers to help then make concrete steps towards closer cooperation and increased value adding, as part of a medium term process to successfully develop a local industrial cluster. The benefits (and the costs) from this process will accrue both to local service providers and to service recipients. Participation in the process has two stages: Initial participation will include local service providers. Later, in a second stage, participation is expected to naturally extend along the supply chain to both suppliers and to service recipients. The paper has two parts:

- A very brief outline of the context for our work with the three association staff.
- An outline of a few aspects of our reflections that are most relevant to the participants at the Systems Conference:
    1. Maintaining staff support for our intervention, given that the dynamics from our intervention range from negative to very positive.
    2. Keeping a boundary between those staff who participate in "walk the talk" and those who do not.
    3. Maintaining a manageable demand on resources.
    4. Prioritizing projects to work on.

The context for our work with the three association staff

In this section we outline the context of our work with the three staff. The aim of this section is to give the background and the context for our reflections in the next section. The main take home message from this context section is that the context within which we were interacting with the three staff is, and always has been, a very good working environment with high staff morale, in which all staff are strongly encouraged to be good project managers. In practice this environment can be divided into two parts:

- That part that is relatively stable, like for example salaries.
- Second, that part that is very dynamic with significant opportunity to be proactive to unexpected opportunities, like for example to be offered extra funding from an unexpected source.

The main aspects of the context are:

1. The association of local service providers reported in this paper is the Townsville Division of General Practice. Its members are most of the General Practitioners who practice in Townsville.
2. The Division has been in operation successfully since 1994. Its core activity to date has been largely about teaching technical skills to General Practitioners and to staff that work with the General Practitioners.
3. We now want to place the Division in such a position so that it will have an increasing role, along with a few other stakeholders, in being the main drivers of the success of our future industrial cluster.
4. To date, most of the successful industrial clusters, both in Australia and overseas, have been in manufacturing industry.
5. One of the next challenges for industrial clustering proponents, practitioners and researchers is the development of industrial clusters of organizations that are mainly service providers, rather than mainly manufacturers.
6. We want our Townsville cooperation between the Townsville Division of General Practice, the State Development Centre Townsville (which is part of the Department of State Development Trade and Innovation, Queensland) and the School of Business at James Cook University, to be at the forefront of these developments.
7. When we stared this project in mid 2005 our challenge was how do we help the Division to awaken itself to the realization of the extraordinary opportunity that it had within its grasp?
8. And to do this without triggering the negative side effects that often come associated with such undertakings.
9. Some of the important aspects of why we felt that "the Division had an extraordinary opportunity within its grasp" are:
    a. A well-established reputation of being a reliable organization in the delivery of its core business.

   b. A well structured and experienced governance structure that includes a Board of Directors, a CEO and an Operations Manager.
   c. A recently established partnership arrangement with four other Regional Divisions of General Practice covering all of North Queensland to pursue projects of mutual interest.
   d. A well-established and experienced local Regional Economic Development agency, the State Development Centre Townsville.
   e. A well-established and experienced local Regional University, James Cook University.

10. From the start we acknowledged that our challenge to successfully develop a local industrial cluster was unlikely to succeed if we used a reductionist approach.
11. Rather we started from our evidence from our observed reality that the combination of health and regional economic development was likely to be about the most complex system that we could have picked to work on.
12. But, we were very lucky:
   a. To have access to well developed human capital in the form of the appropriate combination of staff skills and staff experience and
   b. To be operating in a work environment with the appropriate level of organizational flexibility, executive decision making and senior staff buy-in that got us to where we are now.
13. What really makes the work that we do in the Townsville Division of General Practice, different is that the Division's relationships are maintained irrespective of the 'Grant or Project' funding, and that the relationships help us move towards solutions as a team, providing mutual support as required, and picking up grant and project funding as it becomes available, to progress already identified and agreed objectives.
14. The literature related to the various aspects of our work in the Townsville Division of General Practice is relatively extensive. However, we are not aware of any literature that reports on projects similar to ours. One significant spillover of our work is the writing of a number of papers related to the current interests of the Division. As of July, 2006 10 of these papers are in various stages of being submitted to the SEGRA Conference for refereeing. The Conference link is http://www.managementsolutions.net.au/segra/index.html

For details on the papers please see the technical appendix of Monypenny (2006 - paper titled "Mentoring three local association staff" in this proceedings).

To close this context section, we would like to reinforce the idea of a very dynamic and invigorating context within which we were interacting with the three staff. The next section looks at a few aspects of our reflections.

A few aspects of our reflections

In this section we outline our main reflections. The main take home message from this reflections section is that any valid evaluation of our intervention needs to include the real dynamics of the system that underlie the normal activities of The Townsville Division of General Practice. We acknowledge that it is probably very difficult to create such a dynamic system from scratch but it is certainly very invigorating to have the opportunity to actively participate in such an intervention. The main aspects of our reflections are:

1. Maintaining staff support for our intervention, given that the dynamics from our intervention range from negative to very positive.
    a. Intervention in any organization can have expected and unexpected positive and negative outcomes and side effects.
    b. When we started working at the Division we looked at ways to help reduce the negatives from our intervention. After reflection we chose to work with a small group of staff (three) and to use the well developed "group skills" that two of the people have. We also used these group skills to decide not to change the people in the group for the duration of the intervention. In practice this was from about October 2005 until June 2006.
    c. We also used our group skills to reinforce discernment within the group. Initially this was only discernment directly related to the intervention. However, over time the discernment naturally extended to most aspects of our work activities.
    d. Now looking back, we realize that nearly all the negative aspects that arose were converted into positives or were neutralized, some more easily than others.
    e. Now, after reflection, we are ready to start a similar process with the rest of the staff to bring them up to speed on how they can take advantage of what we have learnt during the intervention, in their own project. We expect this process to run relatively smoothly.
2. Keeping a boundary between those staff who participate in "walk the talk" and those who do not.
    a. Intervention in an organization always creates a divide between those who participate and those who do not. This divide can create significant negative side effects from an intervention. The Division has two characteristics that we used to our advantage in this case.
        - First, the Division has an open work office design. This helps "walk the talk" conversations to be very inclusive.
        - Second, many of the Division's project staff only spends certain, and largely fixed days in the office. The rest of their time is spent in the field. Upon reflection initially we restricted intervention time to Thursdays before lunch. However, over time the intervention time naturally ex-

tended to times that were appropriate to the activities undertaken.

b. Now, after reflection, we are ready to start a similar process with the rest of the staff. Our current plans are to fold the two processes (the negatives and the walk the talk) into one regular and relatively widely inclusive process that, with time, we expect will run smoothly and just naturally evolve as required.

2. Maintaining a manageable demand on resources.

    a. Let me re-cap:

        i. The work that we are currently doing in the Townsville Division of General Practice is part of a medium term process to successfully develop a local industrial cluster.

        ii. The Division has a well established core activity that is largely about teaching technical skills to General Practitioners and to staff that work with the General Practitioners.

    b. Participating in the development of an industrial cluster and the on-going core activities both regularly need resources for their activities. The management of both demand expectations and of actual demand on resources require considerable time, staff energy and actual resources it's self.

    c. When we started working in the Division we made a special effort to regularly discuss the issue of realistic resource expectations. On reflection, we appear to have been successful.

4. Prioritizing projects to work on.

    a. Intervention in any organization usually creates the need for increased effort to be allocated to prioritizing of projects so that the organization can keep a manageable demand on resources. This is especially the case when a given intervention has already been allocated additional resources. The reality is that these additional resources, especially with the benefit of hindsight, often turn out to be insufficient.

    b. We spend considerable effort in reflection about the congruence, between the demand on resources and the wider Division goals and priorities. This item is an important item for reflection because historically the Division went through difficult times because the Division's efforts were significantly driven by a grant based existence. That is, you apply for a grant because you can. Once you have a grant you need to produce the deliverable. But, there was little concern about how or if the activities undertaken did or did not fit into an overall medium term plan or goal.

    c. Now, after reflection, we are very focused on keeping appropriate processes in place to maintain our congruence between our demand for resources and our wider Division goals. Our current process involves ongoing communication between relevant parties. This is usually staff, senior staff, the Chair of the Board and the regular Board meetings.

d. I have used the word "ongoing" rather that regular because of our observed reality that the Division is a complex, dynamic and invigorating system to work in. This usually requires relatively short response times.

e. One aspect of this issue upon which we need to keep a running brief is balancing our effort spent on current core activities and effort spent on current peripheral activities that could become core activities in the medium term.

f. Our current process is an informal but on-going monitoring of resource demands by senior staff in consultation with others as required.

g. Now, after reflection, our current plans are to continue with this same process because it works well and it is well understood by the main stakeholders in the process. We expect this aspect of the Division to run smoothly into the foreseeable future.

5. In summary, the take home messages from our reflections are:

   a. Dynamic and invigorating. Given that the Division is such a complex, dynamic system and it is very invigorating to work in, the main outcomes from our reflection are very different from the main outcome that we would expect from our reflections from a less dynamic and invigorating work place.

   b. Congruent management. Given that the Division is a complex, dynamic and invigorating work place, the management and leadership style and all intervention activities need to be congruent. This is especially important for internal staff in relation to external consultants that are brought into the Division to do a specific task.

   c. Learning from our experience. Given that the Division is such a complex, dynamic and invigorating work place, we are successful largely because we are part of a well functioning system. We did not have to create the well functioning system we just had to help it to be more successful.

   d. Our advice to others. Given that the Division is such a complex, dynamic and invigorating work place, upon reflection, we acknowledge that success comes not just from our own skills and efforts but from the summation of all aspects of the system. Our contribution is usually, but a small part. In the medium term it is unwise to let success go to our head.

Acknowledgements

This paper was developed as part of collaboration between the Townsville Division of General Practice, the State Development Centre Townsville in the Department of State Development Trade and Innovation (Queensland) and the School of Business at James Cook University.

Richard Monypenny, BIng. Agr., MS, PhD, is Associate Professor in the School of Business, James Cook University, on the Townsville Campus. His teaching responsibilities are in the two third year subjects EC3420:03 Regional Economic Development and EC3401:03 International Trade. His research supervision responsibilities are for Honours theses and for Ph D theses. He is the Postgraduate Liaison Officer in the School of Business. In this role he spends considerable time and energy in mentoring students and growing them into their Honours and Ph D programs. He is always interested in helping individuals who are looking for academic mentoring and supervision. Richard's areas of research interest include: Local Primary Healthcare in collaboration with the Townsville Division of General Practice; Regional Economic Development in collaboration with The State Development Centre, Townsville (Queensland State Government), and; International Trade and the Local Implications from International Trade.

History Repeats Itself: Current Traps in Complexity Practice from a Systems Perspective

Ray Ison & Sandro Luis Schlindwein
Open Systems Research Group, The Open University, UK
Department of Rural Engineering, Federal University of Santa Catarina, BRA

This paper discusses the history of systems scholarship and how this has been translated into particular forms of purposeful action, like complexity practice. Both systems and complexity approaches have something to offer when the situation is no longer amenable to analysis based on linear causality or reductionist approaches. In the hands of aware practitioners both offer epistemological devices for shifting our mental furniture and both are rich sources of metaphors, which have the capacity to trigger new and emergent understandings. In the last 70 or so years of systems scholarship those involved have diverged into a plethora of traditions or lineages, conserving, knowingly or not, one of two epistemological positions: the objectivist or positivist position and the constructivist or interpretivist position. These two epistemological positions constitute two language communities even though many who participate in them are unaware that they do. The trap in all of this is that so many people act without awareness of the positions that they hold or uphold and the historicity of their thinking and acting, resulting in conflict, rejection, lack of valuing of difference, bifurcation into smaller and smaller communities of practice, unethical practice, etc. Based on examples coming from academic practice, research management, modeling practice, policy praxis, among others, the implications of this lack of awareness are discussed.

Introduction

If we accept that "systems is a seminal word for complexity" (Morin, 1999), then systems thinking and systems practice are fundamental competencies in the learning process involved in managing situations of complexity of the 'real-world'. As academics we have been engaged with various traditions of systems thinking and practice, and this paper is based on our engagement with, and understandings of, the history of Systems scholarship, and how this has been translated into particular forms of purposeful action, like complexity practice. This has been for us also one of the main reasons to engage with the rapidly expanding 'complexity' literature.

As we engaged with both the systems and complexity material and activities of those who were enthusiastic about both we began to become aware of a number of things:

i. History seemed to be repeating itself, as for example through the repetition of concepts and methods coming originally from cybernetics;
ii. A lot of the scholarship was, frankly, appalling –most obvious of all was the lack of scholarly appreciation of the intellectual traditions on which particular claims rested;
iii. Yet another opportunity to deal seriously with the important things that Systems and Complexity perspectives had in common seemed in danger of being lost – in the sense of the fragmenting discourses, interest groups, lack of institutionalization – especially in Universities and in such pragmatic arenas as skills and competency frameworks for the professions. This seemed particularly tragic to us as these sorts of skills seem more needed now than at many other historical moment.

Similar concerns have been expressed also by others. For example, on Principia Cybernetica (http://pcp.lanl.gov/DEFAULT.html) the authors observe:

"Some recent fashionable approaches have their roots in ideas that were proposed by cyberneticians many decades ago: e.g., artificial intelligence, neural networks, complex systems, human-machine interfaces, self-organization theories, systems therapy, etc. Most of the fundamental concepts and questions of these approaches have already been formulated by cyberneticians such as Wiener, Ashby, von Bertalanffy, Boulding, von Foerster, von Neumann, McCulloch, and Pask in the 1940's through 1960's."

These fashionable approaches are usually labelled as:

"the emerging "sciences of complexity", also called "complex adaptive systems", studying self-organization and heterogeneous networks of interacting actors (e.g. the work of the Santa Fe Institute), and associated research in the natural sciences such as far-from-equilibrium thermodynamics, catastrophe theory, chaos and dynamical systems. A third strand are [sic] different high-level computing applications such as artificial intelligence, neural networks, man-machine interaction and computer modelling and simulation."

As we had done, they noted that:

"Unfortunately, few practitioners in these recent disciplines seem to be aware that many of their concepts and methods were proposed or used by cyberneticians [and systems practitioners] since many years. Subjects like complexity, self-organization, connectionism and adaptive systems have already been extensively studied in the 1940's and 1950's, by researchers like Wiener, Ashby, von Neumann and von Foerster, and in discussion forums like the famous Josiah Macy meetings on cybernetics. Some recent popularizing books on "the sciences of complexity" (e.g. Waldrop, 1992) seem to ignore this fact, creating the false impression that work on complex adaptive systems only started in earnest with the creation of the Santa Fe Institute in the 1980's."

As an example, for Capra (2002: 93) concepts like feedback and emergence have been developed by complexity theory. We think however the issues are more serious than even those on Principia acknowledge – they are issues which are common to both the systems and emerging complexity traditions. Therefore, in this paper we want to explore what is at issue and why we think they have arisen. We also want to give some practical examples of why it matters. To do this we want to start by exploring some aspects of what it is to be human so as to invite you to reflect on what it is that you, and others, do when you/they do what you/they do!

Explaining explanations

We start by inviting you to consider 'explanations' as a social phenomenon: When is an explanation an explanation? For example how many of you would accept Tony Blair's explanation of the reasons for going to war with Iraq? What about Einstein's equation $E=mc^2$ as an explanation of the quantum world? Or the account in Genesis of the origins of human beings?

We will argue that all explanations arise in social relations; following Vickers (1965) they give rise to our standards of fact and value. Moreover all explanations arise this way but some are conserved over longer time frames than others. Explanations also 'enter our bloodstream' as we live our lives – throughout our biological and social development. Maynard Keynes' oft quoted remark that in his experience those who claimed to be practical men (sic) were usually victims of some theory [or explanation] 30 years out of date exemplifies our point. Humberto Maturana (see Maturana & Poerkson, 2004) also draws attention to the role that explanations play in the lives of human beings – as young people we seek explanations, inquisitiveness is a feature of children raised in supportive contexts, and explanations are something we find satisfying, or not; they thus trigger emotional reactions. This is an important qualitative feature that rises from the explainer – listener – explanation relationship.

Living in language

Our explanation of explanations leads us to language. Everything said is said by someone. My world is different to your world and that is always so – what we have in common are our capacities for perceiving and languaging which both have an evolutionary basis. This leads to claims that we live in language, or languaging as Maturana (Maturana & Poerkson, 2004) would have it. This is a big issue and we only want to provide passing mention to it here. But we raise it because we would like to invite you to reflect on what happens in professional life when people belong to different language communities (e.g., the systems and complexity language communities).

Considering language also warrants a brief mention of metaphor. In the research carried out by one of us at the OU a comprehensive review of metaphor theory and practice concluded that (McClintock, *et al.*, 2004):

- metaphors are linked to understandings, and can inform research that aims to work with diverse understandings;
- metaphors play an important role in how we use language;
- metaphors work by 'restructuring' domains;
- 'Systems' and 'complexity are a fertile ground for considering metaphors; and,
- a metaphor can be seen as a description, and recognized by the use of the words 'is' and 'as'.

One way of summarizing this is that if we agree that we live in language then we also live in metaphor. But recent cognitive science research (Lakoff & Johnson, 1999) suggests it could be put another way: we do not use language and metaphor but they use us!

We do what we do living in language and with a desire to find satisfying explanations – and associated relationship making and breaking. Although we live in an ever unfolding present, that present is determined by our history and context. We thus come to any situation with our own models of understanding. These grow out of traditions; a tradition is a network of prejudices that provides possible answers and strategies for action or, following Gadamer (1975), as a network of pre-understandings (because the word prejudices may be literally understood as a pre-understanding). Traditions are not only ways to see and act but a way to conceal; they arise in the doing and are not labels for ways of acting and constitute the basis of praxis.

Sir Geoffrey Vickers (1972: 15) conceptualized this same phenomenon using the metaphor of the lobster pot as a man trap:

"Lobster pots are designed to catch lobsters. A man entering a lobster pot would become suspicious of the narrowing tunnel, he would shrink from the drop at the end: and if he fell in, he would recognise the entrance as a possible exit and climb out again – even if he were the shape of a lobster. A trap is a trap only for the creatures which cannot solve the problems it sets. Man traps are dangerous only in relation to the limitations of what men can see and value and do..."

Different traditions and distinct epistemologies

From our perspective one of the main characteristics of the last 70 or so years of systems scholarship is the extent to which those involved have diverged into such a plethora of traditions or lineages. Each lineage represents to us the conservation of a particular language community. Some lineages are more robust than others and some have stronger institutional capital than others (e.g., bigger societies, strong figure-heads etc). From our perspective those engaged in these diverse traditions conserve, knowingly or not, one of two epistemological positions. What is more these two positions are already apparent with the equally broad and bifurcating 'complexity' domain.

The epistemological positions are:

i. Objectivist or positivist – for these people 'systems' exist in the world and thus are describable, discoverable, able to be modelled etc. This is the legacy of general Systems Theory and can be typically spotted whenever the phrase 'systems science' is written. It is Peter Checkland's 'hard' systems tradition (Checkland, 1999). It has cultural dimensions – more common in North America and within aspiring or existing 'big science' communities. Within the field of complexity the term 'complexity science' is also a warning sign. This epistemology characterizes the Santa Fe group and its publications. It is what we describe as descriptive complexity (Schlindwein & Ison, 2004), i.e., complexity is in the world, describable, able to be modelled etc.
ii. Constructivist or interpretivist – for these people the role of the observer is crucial – systems are always brought forth by someone in a context. The act if formulating or seeing system is a way of knowing about the world, not an entity in it. A 'system' and the act of constructing a system thus becomes an epistemological device – a way of understanding and learning about situations in the world. It is no longer sciencia but praxis. In the field of complexity we have described this as perceived complexity – something which arises in the relationship between someone and a situation (Schlindwein & Ison, 2004). As such it is a choice we can make. It is the *as if* position – let us engage with this situation as if it were complex, or let us learn about this situation through a process of formulating systems of interest.

Now these two epistemological positions constitute two language communities even though many who participate in them are unaware that they do. They are language communities in the sense that we use the term because members of each community bring forth different traditions of understanding when they do what they do. They admit different claims and thus accept or reject particular explanations. What is more they often act without awareness; some act in ways that abrogate responsibility by claiming that their explanation is 'the truth' or 'objective'- the path of objectivity to use Maturana's term (Maturana, 1988). In social relations this is at the same time a claim to do as we say!

Why does this matter? For us the good news is:

i. Both systems and complexity approaches have something to offer when situations are no longer amenable to 'mainstream' practice of analysis based on linear causality or reductionist approaches;
ii. In the hands of aware practitioners systems and complexity approaches both offer epistemological devices for shifting our mental furniture – for 'managing' complexity;
iii. Both are rich sources of metaphors – and these metaphors have the capacity to trigger new and emergent understandings.

But what is the bad news? The trap in all of this is not that one position or the other is right or wrong, better or worse, but that so many people act without awareness of the position that they hold or uphold and without awareness of the historicity of their thinking and acting. The product of lack of awareness is conflict, rejection, lack of valuing of difference, bifurcation into smaller and smaller communities of practice, unethical practice and struggles to build institutional capital or brand around particular groupings or discourses.

Let us briefly give some examples from our recent experience:

Academic practice – peer review. Academic publication rests on peer review. Those who have been through the process know that it can be a mixed experience. At its worst one can become the victim of contradictory reviews which, we suggest, can often be seen as contrasting but undisclosed epistemological positions. It is rare in our experience for reviewers to state their own epistemological preference before reviewing a paper. Take the following points made by reviewers of a 300 word abstract for an international complexity conference:

- Drawing upon views of French philosopher, Gaston Bachelard, the paper develops a poetic, metaphorical framework to understand the complexity of human life;
- The authors need to clarify their use of the term 'complexity' and its cognates, in particular its relationship to non-linear, dynamical 'complexity theory';
- The idea is original but, in my opinion, too far-fetched and taking for granted in very different thinkers similarities in conceptual and semantic contents that seem more like only analogies in formulations.

The recommendations ranged from enthusiastic acceptance to rejection.

Research management. This example comes from recent European (Framework 5) experience of one of us as being the coordinator of a five country, 30 researcher project (see http://SLIM.open.ac.uk). The Framework programmes are significant not least in monetary terms: Framework 6 totalled €17.5 billion and Framework 7 is projected to be €40 billion. What is clear is that DG (Directorate General) Research is unable to orchestrate an epistemological conversation nor is it aware of the implications of not doing so. We have empirical evidence of this, having interviewed a range of staff from DG Research as well as Brussels-based NGO and industry representatives (SLIM, 2004a). Traditionally research projects are largely set up as blueprints with outcomes specified in advance (contract deliverables). This has some advantages in some types of research and as an accountability mechanism but does not guarantee the best research outcomes or that the research addresses the espoused issues of concern. At the heart of this conundrum are epistemological issues and projects can be at the mercy of the epistemology of their supervising project officer (similar roles operate in most projects). In practice the scope for negotiation and renegotiation of research deliverables depends on the relational capital (SLIM, 2004b) that is built up between the research project, especially the coordinator, and the scientific officer.

From our interviews we identified two prevailing epistemological predispositions: one associated with '*Transfer of Knowledge*' with features such as (i) policy blueprints; (ii) seeing policy implementation as distinct from policy development; (iii) role of 'standards'; (iv) need for new science (facts); (v) questioning the role of social research; (vi) need for better communication i.e. a first-order conception of communication; (vii) doing research on (people); (viii) complexity as knowable; (ix) consumer and thus market focus. The contrasting predisposition was '*Knowing in Action*' with features such as (i) policy as process; (ii) implementation integral to development (of policy); (iii) policies adaptive to local contexts; (iv) need for new stories or narratives; (v) recognized a role for social research; (vi) learning; co-construction; (vii) research with....; (viii) complexity not knowable – new ways?; (ix) citizen focus- ecoliteracy.

The fallacy of the toolbox. Over several encounters between SLIM researchers and research managers in DG Research they experienced the pervasive use of the term 'tools' and gained a strong perception that for many, particularly in DG Research, their desire was to have research produce new tools which could be used by the clients of research (e.g., other DGs, national actors etc). The way the metaphor was used had several entailments: the apparent desire by those who used the metaphor to contribute to making the EU work for which tools (levers, instruments etc.) were needed; 'tools' to fix or remedy particular problems; tools as something that anyone could use. What the metaphor concealed however was that a tool (e.g., a hammer) requires a user (a hammerer) and something to which the tool is applied (something that has to be hammered). Our experience was that those who use this metaphor focus on the hammer but rarely the hammerer or the hammered, thus failing to see that hammering, as a practice is context sensitive and is ultimately a performance usually conducted for some purpose. Within systems and complexity approaches all to frequently a 'system' or a CAS can be seen as the hammer – a tool not associated with devoid of understanding of its historical and contextual derivation and how in context it can become part of a performance.

Concluding remarks

As we have discussed by means of several examples, the implications of lack of epistemological awareness in systems and complexity practices are far reaching, and not restricted to particular domains of human activity. In our view the objectivist or positivist position is the dominant position in the complexity sciences and its practices, although there seems little awareness of this dominance amongst those who claim to be complexity practitioners. We assume that the huge effort that has been made to understand (by modelling, for example) situations of complexity in their details so as to better control and manipulate them, is an explicit manifestation of what we are claiming here. For us however, purposeful action in such situations can be carried out through systems practices designed to learn how to manage them, which is also a way to refute Rescher's claim (Rescher, 1998) that the complexity of real problem-situations is beyond our capacity to act. To avoid the epistemological trap in complexity practice requires awareness, which is also an act of responsibility. Being epistemologically aware also opens up more choices for action (Schlindwein & Ison, 2004).

References

Capra, F. (2002). *As conexões ocultas. Ciência para uma vida sustentável*, São Paulo: Cultrix, ISBN 8531607485.

Checkland, P. (1999). *Systems thinking, systems practice*, Chichester: John Wiley & Sons, ISBN 0471986062.

Gadamer, H. G. (1975). *Truth and method*, London: Sheed and Ward, ISBN 082647697X.

Lakoff, G. and Johnson, M. (1999). *Philosophy in the flesh: the embodied mind and its challenge to western thought*, New York: Basic Books, ISBN 0465056741.

Maturana, H. (1988). "Reality: the search for objectivity or the quest for a compelling argument", *Irish Journal of Psychology*, ISSN 0303-3910, 9: 25-82.

Maturana,H. and Poerkson,B. (2004). *From being to doing. The origins of the biology of cognition*, Heidelberg: Carl-Auer-Systeme, ISBN 3896704486.

McClintock, D., Ison, R.L. & Armson, R. (2004). "Conceptual metaphors: a review with implications for human understandings and systems practice", *Cybernetics and Human Knowing*, ISSN 0907-0877, 11(1): 25-47.

Morin,E. (1999). *Ciência com consciência*, Rio de Janeiro: Bertrand Brasil, ISBN 8528605795.

Rescher, N. (1998). *Complexity. A philosophical overview*, New Brunswick: Transactions publishers, ISBN 1560003774

Schlindwein, S.L. and Ison, R.L. (2004). "Human knowing and perceived complexity: implications for systems practice", *E:CO*, ISSN 1521-3250, 6(3): 27-32.

SLIM (2004a). WP 10 Report. Experiencing social learning: a process designed for Brussels-based policy makers. SLIM (see http://slim.open.ac.uk)

SLIM (2004b) The Role of Conducive Policies in Fostering Social Learning for Integrated Management of Water, SLIM Policy Briefing No. 5. SLIM (see http://slim.open.ac.uk)

Vickers, G. (1965). *The art of judgement*, London: Chapman and Hall (Republished by Harper and Row, 1983 and Sage, 1995), ISBN 0803973632.

Vickers,G. (1972). *Freedom in a rocking boat*, London: Penguin Books, ISBN 0140212051.

Waldrop, M. M. (1992). *Complexity: the emerging science at the edge of order and chaos*, Simon & Schuster, ISBN 0671872346.

Ray Ison joined the Open University as Professor of Systems in January 1994 where he led a process of organizational change resulting in the formation of the Centre for Complexity and Change. He was foundation Director of the Postgraduate Program in Environmental Decision Making and fostered the launch of the MSc in Information Systems in 2004. He is also foundation Director of the Open Systems Research Group comprising 20 researchers (see http://systems.open.ac.uk), one of the largest Systems research group in the world with research foci on systems thinking and practice, information systems and sustainable development. In the period 2001-4 he coordinated a major European project entitled Social Learning for the Integrated and Sustainable Use of Water at Catchment Scale (SLIM) involving researches from Sweden, France, Italy, the Netherlands and the UK. He is also co-Director of the EPSRC funded SPMC (Systems Practice for Managing Complexity) Network which after three years has become self funding and has made significant contributions to formulating new research directions for Systems and for raising awareness of the role Systems practice can play in areas such as Information Systems, organizational change, project management, etc (see http://www.spmc.open.ac.uk). Ray's research involved developing and evaluating systemic, participatory and process-based environmental decision making, natural resource management, organizational change and R&D methodologies. Examples discussed in his publications include second-order R&D; systemic inquiry, social learning, soft systems methodology, systemic action research; information systems; modeling; communities of practice and participatory institutional appraisal. He is the author or editor of three books, 16 book chapters, 40 refereed journal and 27 refereed conference papers and has been invited as keynote speaker at a range of conferences in Australia, Brazil, Ecuador, France, Germany, Ireland and Sweden. He is on the editorial board of three journals. At the present time he is on sabbatical leave at the Melbourne Water Research Centre, University of Melbourne.

Sandro Luis Schlindwein joined the Federal University of Santa Catarina, in Florianópolis, Brazil, as Adjunct Professor in 1993, after concluding his Ph.D in Soil Science at the University of Göttingen, Germany. He was foundation coordinator of the Postgraduate Programme in Agroecosystems, where currently he teaches systems thinking and practice, and supervises M.Sc. students. Sandro has been actively engaged in the development of indicators of sustainability, the main feature of which is to adopt an approach to overcome the dualism human-nature. Furthermore, his currently research interests are related to the adoption of systems thinking and systems practices in agricultural contexts. Sandro has been also visiting scientist at the Leibniz-Centre for Agricultural Landscape Research (ZALF), in Müncheberg, Germany. From June 2003 to June 2004 he was a Visiting Research Fellow at the Centre for Complexity and Change, Systems Department, The Open University, Milton Keynes (UK).

Proceedings of the 12th ANZSYS conference - Sustaining our social and natural capital
Katoomba, NSW Australia, 3rd-6th December, 2006

A Boundary Based Typology of Language Games: Making Sense of Systemic Interventions

Jorge Vélez-Castiblanco
Victoria Management School, Victoria University of Wellington, NZ

In general, Systems Thinking/Management Science Intervention Methodologies are designed to work in some predefined ways. However, authors such as Flood, Romm and Midgley make the case that methodologies can also be used in non standard ways. This paper reports initial findings of my research, looking at the role of intentions in these non standard uses. Towards that aim, it is proposed a typology of "language games" or ways in which the boundary of what is considered relevant in a situation can be affected. This typology draws on Wittgenstein's philosophy, language pragmatics and the notion of boundary in systems thinking.

Non standard uses of methodologies

Systems Thinking/Management Science methodologies are conceptual tools developed and used by consultants, academics and management practitioners. They aim to address a huge range of organizational aspects such as processes, design, culture and politics (Flood, 1995), and guide undertakings such as the appreciation, analysis, assessment and action (Mingers & Brocklesby, 1997).

There is a great variety of methodologies, so deciding what to use in a specific situation can be confusing. Consequently a lot of work has been done trying to find what is the right way to take advantage of methodology's intrinsic characteristics - such as philosophical underpinnings- and based on these, declaring *standard uses.*

This line of research have influential proposals such as Flood (1995), Flood & Jackson (1991) and Jackson (1991, 2000). Mingers & Gill (1997) contains a comprehensible collection on different views on the matter.

On the other hand, authors as Kay & Halpin (1999), Checkland (2000) and Taket (1994) have agreed that the practitioners and context are crucial factors that shape how methodology is used, making difficult to recognise an intervention as carried out from a particular methodological stand point.

Some proposals have been advanced to take advantage of these non standard uses. For example Watson, *et al.* (1995) from an hermeneutic perspective explain that methodologies behave as metaphors, for this reason we can interpret and use them in un-prescribed ways. Flood & Romm (1995) argue that is possible to take the principles of one methodology and apply them to a second methodology, generating in this way a non standard use that they call the "oblique use". Midgley (1997) commenting on Flood & Romm, suggests that there is more explanatory power if we consider that methodologies and principles are combined, achieving a synthesis in which a new method is created.

While Flood & Romm and Midgley propose non standard uses, adapting methodologies by replacement or synthesis of methodological components, Watson et al. obtain non standard uses based on interpretations not necessarily derived from methodological grounds.

In some way I want to combine these ideas. The freedom in diverse uses from different sources as in Watson et al. (1995), but with the possibility to have some guidance to intervene in the process as in Flood & Romm (1995) and Midgley (1997).

In order to conceptualise different uses I have decided to look how agents through their agency and specifically intentions affect methodologies. Inquiring about how intention can affect methodology use requires abandoning the idea that there is an intrinsic or "true" nature in methodologies that have to be expressed in unique ways. It requires a philosophical position where factors like the context and specially the actor, can have a place in understanding such use.

I propose that a path rooted in the philosophy of language, specifically in Wittgenstein's philosophy can offer us such possibilities.

Methodologies as language games

"Language is an instrument. Its concepts are instruments. Now perhaps one thinks that it can make no great difference which concepts we employ ... the difference is merely one of convenience" (Wittgenstein, 2001, §569)

Wittgenstein conceptualise language as a tool and rejects the idea that words carry intrinsic meanings: "the meaning of a word is its use in the language" (Wittgenstein, 2001, §43). Meaning arises from what is possible to do with any such word, giving us a window to conceptualise non standard uses of methodologies and how that use is affected by user's intention.

Wittgenstein approaches language using the notion of language games as "objects of comparison which are meant to throw light on the facts of our language" (Wittgenstein, 2001, §130). Language game is defined as "the whole, consisting of language and the actions into which it is woven" (Wittgenstein, 2001, §7). Additionally "the term 'language-game' is meant to bring into prominence the fact that the speaking of language is part of an activity, or of a life-form" (Wittgenstein, 2001, §23).

Now we can turn to the question of if it is possible to apply the concept of language games to methodology use. Some arguments to support this possibility include:

- If language games are "objects of comparison" it follows that we can use them to learn and compare against methodology use.
- The use of methodologies as language games involves a "whole, consisting of language and the actions into which it is woven".

- Inventing methodologies is like "invent[ing] a language" that "could mean to invent an instrument for a particular purpose" (Wittgenstein, 2001, §492)
- Examples of language games include giving and obeying orders, describing objects, reporting, speculating, forming and testing hypotheses, translating, asking, (Wittgenstein, 2001). What is more, Mauws & Phillips (1995) suggest that the concept is powerful enough to enable an understanding of managerial practice in terms of collections of diverse language games. The use of methodologies seems to sit between these two sets.
- Intervention methodologies seem to share with games the variety and the absence of common features among them. Compare games such as board games, card games, Olympic Games, ball games, ring-a-ring-a-roses, or bouncing a ball against the wall. Next look at how intervention methodologies can include elements as diverse as psychodramas, computer simulations, or methods to encourage debate. Wittgenstein suggests that there is not a central concept running throughout all language games but rather a "family resemblance".

In addition, there are some interesting gains from working with the notion of language games applied to methodologies. First, for Wittgenstein rules in language games cannot be private, so the understanding of a methodology use needs similarly to be seen as a social construction. This contrasts with the options to look at non standard uses (on the last section) that draws on explanations focused on individuals.

Secondly, apart from methodology use, the interactions, languages, activities and "forms of life" in the intervention context can also be considered in terms of language games. So when we are intervening what we are trying to do using methodologies (as a language game), is to affect the language games already in place, which is to say, we are using a tool to modify the tools that people in that context had developed in order to interact. What is more, because we are talking about tools that modify tools, it is also likely that the language games in place will modify the methodology in use.

Rules

"In philosophy we often compare the use of words with games and calculi that have fixed rules, but cannot say that someone who is using language must be playing such a game" (Wittgenstein, 2001, §81)

If we play with fixed rules, the likely outcome is that we will refer to standard uses of methodologies. However, Wittgenstein is suggesting that that is not to say that we must play in such a way. But what exactly can it mean not playing with fixed rules?

According to Wittgenstein, rules can be changed, created, or eliminated, "as we go along" (Wittgenstein, 2001, §83). Under those circumstances the rules on many occasions are not clear, incomplete, or even incoherent, but nevertheless people are able to accomplish tasks (Fogelin, 1996). This idea is also suggested by empirical research (Galantucci, 2005).

Wittgenstein's view also involves the notion that rules have multiple uses. So it is possible to use rules to communicate and to condition behaviors, also sometimes we are aware of their existence but we choose not to follow them, and even if they are not explicit, they can be inferred from the context. In this picture:

"A rule stands there like a sign-post ... is there only one way of interpreting them? – So I may say, the sign-post does after all leave no room for doubt. Or rather: it sometimes leaves room for doubt and sometimes not." (Wittgenstein, 2001, §85)

So basically rules guide but not rule. They help to make sense of what is happening and not necessarily are determining an outcome.

Language pragmatics, which is a field in which Wittgenstein is regarded as a pioneer (Bertucelli Papi, 1996), has proposals that work along this lines. One of them looks to move the understanding of communication from *rules to code-decode information* to *principles that help to infer it.*

In the first model, in order to understand a message, the people involved need to code and decode based on a common code (Littlejohn, 1999; Van Der Wiele, 1995). In contrast, in the inferential model proposed by Grice (1989), the speaker tries to show her/his intentions, and the hearer will try to make inferences based on his/her knowledge, the context and the evidence provided.

We can take advantage of inferential communication without talking. For instance, if work has to be done, I can leave the documents on somebody's desk. In fact it is also possible to take advantage of inferential communication by breaking rules when trying to convey a meaning. For example, people can be aware of X's knowledge about etiquette, so if X is not following the etiquette possibly X is using that knowledge to convey a special meaning, perhaps a joke, irony or a low opinion of the host. So etiquette here does not behave as a rule but as a principle that conveys Wittgenstein's idea that rules guide my behaviour.

Understanding communication based on inferences and principles, offers us the possibility of meanings conveyed in different ways and consequently with different uses. Furthermore, it show us a way to understand how intention can guide meaning. So I am approaching my research understanding the use of methodologies as language games, and the rules in terms of pragmatic principles.

The question now is how to find principles suitable to understand intentions in methodology use? To give an answer to this question, now I turn to the notion of boundary.

Boundary

"When one draws a boundary it may be for various kinds of reason. If I surround an area with a fence or a line or otherwise, the purpose may be to prevent someone from getting in or out; but it may be also part of a game and the players be supposed,

say, to jump over the boundary; or it may shew where the property of a man ends and that of another begins; and so on. So if I draw a boundary line that is not yet to say what I am drawing it for" (Wittgenstein, 2001, §499)

Wittgenstein is suggesting that a boundary can have many different uses. This is reflected in how the notion of boundary has been conceptualised. In the past the tendency was to consider boundaries as attached to the structure of reality (for example the skin is the boundary that distinguishes our bodies from the environment).

More recently and thanks to the work of Churchman (1968; 1979), authors such as Ulrich (1983), Midgley (2000) and Yolles (2001), work on the idea that boundaries are "social and personal constructs that define the limits of the knowledge that is taken as pertinent in an analysis" (Midgley, 2000: 35). Moreover these constructions are defined by "*contexts of meaning*" (Ulrich, 1983: 330).

Boundaries are relevant to non-standard uses because where exactly they "are constructed, and what the values are that guide the construction, will determine how issues are seen and what actions will be taken" (Midgley, 2000: 36). What is more Midgley also states that different boundaries will give rise to different methodological choices. Furthermore I would add that different boundaries imply also different uses in those methodologies.

It follows from the concept of language games that we can see what is inside the boundary as language games. In the same way the use of methodologies also can be seen as language games. It can be argued then that when used, these methodologies can have an impact on the boundaries of what is considered relevant. This is in some way in line to systemic intervention as "purposeful action by an agent to create change in relation to reflection on boundaries" (Midgley, 2000: 129).

What we need now is a way to consider the effects that methodologies can have on the boundary and inferential communication can assist us in it.

Sperber & Wilson (1995), building on Grice (1989), propose that an individual possesses a cognitive environment that encompasses all the assumptions that s/he uses to make inferences about communicative stimulus. Each time that a new stimulus arrives the cognitive environment change. New stimulus (that I will treat as information) can weaken or strengthen old assumptions according to their relevance.

The relevance of an input to an individual is understood then as:

a. Other things being equal, the greater the positive cognitive effects achieved by processing an input, the greater the relevance of the input to the individual at that time.

b. Other things being equal, the greater the processing effort expended, the lower the relevance of the input to the individual at that time.

(Wilson & Sperber, 2002: 252)

So basically something is relevant to somebody if it is possible to obtain many inferences from the stimulus and it is not difficult to reach such inferences. Therefore when we are trying to communicate something, we show others that they can connect our messages with their background and that they do not have to work much in order to do it.

Now my proposal is to treat the boundary of what is seen as relevant as the cognitive environment in Sperber and Wilson's terms. From there we can now propose different ways to affect the boundary and consequently different ways in which a methodology can be use.

Typology

"One learns the game by watching how others play. But we say that it is played according to such-and-such rules because an observer can read these rules off from the practice of the game" (Wittgenstein, 2001, §54)

I have derived this typology working with my theoretical framework and afterwards comparing and refining with my observations of a team of researchers planning interventions. Through my observations I have been seeing how others *play intervention games.*

These games can be very complex, so as a first step I am following Canfield (1993) who inspired by Wittgenstein, suggests that is possible to look for basic games that can later help us to understand more complex interactions. At this stage the important characteristic that I am looking for is that these games behave as interventions, that is as "purposeful action[s] by ... human agent[s] to create change (Midgley, 2000: 113).

So I am working on the premise that intervening can be carried out through complex methodologies or small interactions that some times require just a few words or actions. In any case the interventions are going to *play* in relation to the boundary of what is seen as relevant to tackle the problem.

The classification that I propose, looks to reflect "a rather abstract property of the speaker's informative intention: the direction in which the relevance of the utterance is to be sought" (Sperber & Wilson, 1995: 254). So let's look now that different directions, assuming that they are successful.

Setting the boundary

The first operation with relation to a boundary is to set one. Basically new information that is not possible to infer from old information is introduced. This information enable actors to achieve big contextual effects with not a big processing effort. If there is already a boundary in place, a new operation of setting boundaries also, will give strength to that previous information.

Following the boundary

This involves obtaining combinations or consequences of already present information.

The contextual effects are small and easy to obtain. Basically it is following the rules, like when people are acting accordingly to a plan or examining the consequences of it. Old information grows stronger due to new internal connections.

Probing the boundary

It enables the actors to realize "where" is the boundary. Actors propose new information to see if is derivable or agrees with old information. If is derivable we have now a following. If it agrees, new information is introduced, the contextual effects tend to be small and the processing effort is small.

If the information it is not derivable and does not agree, it will stay out of the boundary but it tends to strength the old information.

Challenging the boundary

New information is introduced. This information is not derivable and does not agree with old information. In fact this new information weakens old information. There are big contextual effects and the processing effort tend to be big due that it also have to replace old accepted assumptions. The intended outcome is to change the boundary.

Wandering

It is like a combination of challenging plus following. Challenging because new information is proposed. Following because inferences are obtained from that new information. The big difference is that this new information does not pretend to replace old one. It is used to emphasise what is not going to be done. In that sense what it does is to strength the already present boundary.

Conclusions

If the aforementioned games behave as pragmatic principles perhaps it will be possible to observe something similar to what has been proposed by Leech (1983: 8):

a. Principles/maxims apply variably to different contexts of language use.
b. Principles/maxims apply in variable degrees, rather than in an all-or-nothing way.
c. Principles/maxims can conflict with one another.
d. Principles/maxims can be contravened without abnegation of the kind of activity which they control.

The typology proposed is based on communicative intentions. The idea is that through the *games* proposed it is possible to appreciate a systemic intervention process in terms of the dynamics in relation to the boundary. This possibility do not depend of the actors following a specific methodology. Also, the aim is to be able to use it as a way to reflect on the intentions and the effects that we want to achieve on the boundary, using gestures, words, phrases or methodologies.

The usefulness of such ideas still have to be improved and tested particularly those regarding the use of methodologies, but I think there is a case to show that the use of language games concepts can be useful to make sense of interventions.

References

Bertucelli Papi, M. (1996). *Que es la pragmática*. Barcelona: Ediciones Paidos, ISBN: 84 - 493 - 0208 - 0.

Canfield, J. V. (1993). “Wittgenstein’s Intentions”. In J. V. Canfield & S. G. Shanker (Eds.), *Wittgenstein’s Intentions* (pp. 1-33). New York: Garland Publishing, ISBN: 0-8153-0067-0.

Checkland, P. (2000). “Soft Systems Methodology: A Thirty Year Retrospective”. *Systems Research and Behavioral Science, 17*, 11-58, ISSN: 1099-1743.

Churchman, C. W. (1968). *The Systems Approach*. New York: Dell Publishing, ISBN: 0440384079 (1984).

Churchman, C. W. (1979). *The Systems Approach and Its Enemies*. New York: Basic Books, ISBN: 0-465-08342-0.

Flood, R. L. (1995). *Solving Problem Solving: A Potent Force for Effective Management*. New York: Plenum Press, ISBN: 0-471-95590-6.

Flood, R. L., & Jackson, M. C. (1991). *Creative Problem Solving: Total Systems Intervention*. Chichester: Wiley, ISBN: 0-471-93052-0.

Flood, R. L., & Romm, N. (1995). “Enhancing the process of choice in TSI, and improving chances of tackling coercion”. *Systems Practice*, 8(4), 377-408, ISSN: 0894-9859.

Fogelin, R. J. (1996). “Wittgenstein’s Critique of Philosophy”. In H. Sluga & D. G. Stern (Eds.), *The Cambridge Companion to Wittgenstein* (pp. 1-33). Cambridge UK: Cambridge University Press, ISBN: 0-521-46591-5.

Galantucci, B. (2005). “An Experimental Study of the Emergence of Human Communication Systems”. *Cognitive Science, 29*, 737–767, ISSN: 1364-6613.

Grice, H. P. (1989). *Studies in the way of words*. Cambridge MA: Harvard University Press, ISBN: 0674852702.

Jackson, M. C. (1991). *Systems Methodology for the Management Sciences*. New York: Plenum, ISBN: 0306438771.

Jackson, M. C. (2000). *Systems Approaches to Management*. New York: Kluwer/Plenum, ISBN: 030646506X.

Kay, R., & Halpin, D. (1999). “Redefining the role of the practitioner in critical systems methodologies”. *Systems Research and Behavioral Science, 16*(3), 273, ISSN: 1099-1743.

Leech, G. N. (1983). *Principles of Pragmatics*. New York: Longman, ISBN: 0 582 55110 2.

Littlejohn, S. W. (1999). *Theories of Human Communication* (6th ed.). Belmont, CA: Wadsworth Publishing Company, ISBN: 0-534-54819-9.

Mauws, M. K., & Phillips, N. (1995). “Understanding Language Games”. *Organization Science, 6*(3), 322-334, ISSN: 1047-7039.

Midgley, G. (1997). “Developing the Methodology of TSI: From the Oblique Use of Methods to Creative Design”. *Systems Practice, 10*(3), 305-319, ISSN: 0894-9859.

Midgley, G. (2000). *Systemic Intervention: Philosophy, Methodology and Practice*. New York: Kluwer Academic/Plenum Publishers, ISBN: 0-306-46488-8.

Mingers, J., & Brocklesby, J. (1997). “Multimethodology: Towards a framework for mixing methodologies”. *Omega, 25*(5), 489-509, ISSN: 03050483.

Mingers, J., & Gill, A. (Eds.). (1997). *Multimethodology: The Theory and Practice of Combining Management Science Methodologies*. Chichester: Wiley, ISBN: 0-471-97490-0.

Sperber, D., & Wilson, D. (1995). *Relevance: Communication and Cognition* (2nd ed.).

Oxford: Blackwell Publishing, ISBN: 0-631-19878-4.
Taket, A. (1994). "Undercover agency? - Ethics, responsibility and the practice of OR". *The Journal of the Operational Research Society, 45*(2), 123, ISSN: 01605682.
Ulrich, W. (1983). *Critical Heuristics of Social Planning: A New Approach to Practical Philosophy*. Chichester: Wiley, ISBN: 0-471-95345-8.
Van Der Wiele, T. (1995). "Quality management in a teaching organization". *Total Quality Management, 6*(5&6), 497-508, ISSN: 0954-4127.
Watson, H., Wood-Harper, T., & Wood, B. (1995). "Interpreting Methodology Under Erasure: Between Theory and Practice". *Systems Practice, 8*(4), 441-469, ISSN: 0894-9859.
Wilson, D., & Sperber, D. (2002). "Relevance Theory". *UCL Working Papers in Linguistics, 14*, 249-287, ISSN: 0956-7194.
Wittgenstein, L. (2001). *Philosophical Investigations* (G. E. M. Ascombe, Trans. 3rd ed.). Oxford: Blackwell Publishing, ISBN: 0-631-23127-7.
Yolles, M. (2001). "Viable boundary critique". *The Journal of the Operational Research Society, 52*(1), 35, ISSN: 01605682.

Jorge Vélez-Castiblanco is a lecturer at Universidad EAFIT – Colombia - Management School, and a PhD Candidate at Victoria University of Wellington, New Zealand. He has a MA in Management Systems from the University of Hull, England. He also has Hons Semiotics of the Communicative Interaction, EAFIT University. Colombia and BSc Systems Engineering, EAFIT University. Colombia. Jorge's areas of research interest include: Systems thinking; language pragmatics; organizational theory; Action Research; Autopoiesis; Complexity theory.

Ethical Inquiry in Knowledge Management

James Sheffield[1] & Zining Guo[2]
[1] University of Auckland, NZ
[2] Xelocity Limited, NZ

Modern organizations are characterized by knowledge-intensive activities and complex relationships among individuals and groups, both within and outside of the organization. A study of such a system requires both an appreciation of what might constitute knowledge, and some collective notion of the limits of practical knowledge. Consequently the defining feature of systems thinking is reflection on the boundaries of inclusion and exclusion. This paper reviews concepts of knowledge (and ignorance) provided by Kant, and three twentieth century authors - Churchman, Ulrich, and Habermas. We propose a normative theoretical framework comprised of four architectural elements - knowledge perspectives, sources of boundary judgment, system roles and knowledge dynamics. The purpose of the proposed framework is to identify the knowledge concepts that must be addressed to enable those in a position of power (because they are involved with the knowledge management system) to impose order on the behavior of others (who are affected but not involved) by means of ethical inquiry.

1. Introduction

"Once the rockets are up who cares where they come down"
"That's not my department" said Wernher von Braun.
From a 1965 song by Tom Lehrer.

Modern organizations are characterized by knowledge-intensive activities and complex relationships among individuals and groups, both within and outside of the organization. Knowledge management is therefore a complex endeavor that may be understood from the perspective of organizational systems, and systems thinking. From a systems perspective, business organizations are seen as interlocking networks of complex, adaptive human activities. The properties of the system only emerge when discrete components of the system, such as humans, activities, structures, intentions and outcomes, interact with one another and with the environment. A study of systems is impossible without an appreciation of what might constitute knowledge, and some collective notion of the limits of practical knowledge. A critique of knowledge (and ignorance) requires the concept of a system (and boundary) so as to guide systemic learning and justify the resulting design.

Many authors have commented on the conceptual difficulties associated with managing knowledge across boundaries. While practical knowledge management approaches are necessarily selective, it is impossible for those concerned to "see" the universe (or multi-verse) of issues that they are selected from. It is therefore difficult to settle rival claims about holistic issues (i.e., those related to the whole of the system) without a normative scheme to support systemic inquiry.

A parallel difficulty confronts a knowledge management researcher in the face of a confusing variety of approaches to perspectives, paradigms and frameworks from which to identify constructs, build theories, and test them against empirical data. This difficulty is succinctly expressed by Kay and Goldspink "Knowledge management, as a field of study, has suffered from issues of definition since before the term caught the attention of practicing managers and IS professionals. Put simply, it is unclear what it is that everyone is trying so hard to manage." (Kay & Goldspink, 2005)

Judgments about the appropriate conceptualization of the boundary of the system become central to the design of knowledge management systems. In an age where major business organizations (Enron, Worldcom) are renowned for their cupidity it is relevant to ask what theories of knowledge management and inquiry are used in practise. Do those involved draw system boundaries narrowly, as in Lehrer's song about Wernher von Braun, the Nazi rocket scientist-turned-NASA star recruit? In designing knowledge management systems those involved are in the position of imposing order on the behavior of others. It would be inappropriate if those involved assumed that they were master, and those affected (but not involved) were slaves to their design. Natural justice suggests that any formulation of the problem must be open to critique by those affected. Problem formulation and system design in such a context therefore has at its core an ethical dimension.

This paper approaches knowledge management from the viewpoint of ethical inquiry based on an appreciation of the concepts of knowledge (and ignorance) provided by Kant, and three twentieth century authors - Churchman, Ulrich, and Habermas. The research seeks to illuminate those issues which constitute a necessary part of the critique of judgment about the proper boundary of a knowledge management system. The purpose of the proposed framework is to identify the knowledge concepts that must be addressed to enable those in a position of power (because they are involved with the knowledge management system) to impose order on the behavior of others (who are affected but not involved) by means of ethical inquiry.

2. Literature review and research objectives

The purpose of the literature review is to provide a brief indication the nature of the debt that current conceptions of ethical inquiry in knowledge management research owe to Kant, Churchman, Ulrich, and Habermas. Because the subject matter is rich and diverse the author's project overlaps the work of many others, the review is selective, and reflects the authors' training in information systems and group decision making and their experiences in facilitating strategic interventions in the New Zealand context. The section concludes with a statement of research objectives.

2.1 Kant

"Without sensibility no object would be given to us, without understanding no object would be thought. Thoughts without content are empty, intuition without concepts are blind. It is, therefore, just as necessary to make our concepts sensible, that is, to add the object to them in intuition, as to make our intuitions intelligible, that is, to bring them under concepts. These two powers or capacities cannot exchange their functions. The understanding can intuit nothing, the senses can think nothing. Only through their union can knowledge arise" (Kant, 1787:93)

An overview of knowledge management and systems thinking grounded in Kant's knowledge concepts (or inquiry into the nature of knowledge) is provided by Lehaney, Clarke, Coakes and Jack (2004). Kant's *Transcendental Idealism* identifies the relative priority of subjective knowledge (e.g., of personal intentions and outcomes) and objective knowledge (e.g., of physical inputs and outputs); and the relative priority of concepts we are born with (what he calls *a priori* concepts) and those empirical observations that we access through our senses. The objectively real and our subjective understanding are interlaced, each providing the conditions for the other. (ibid: 95) Inquiry by Kant underpins theories of knowledge and organizational learning in which subjective and objective elements are tightly coupled – neither could be represented without the other. This suggests that knowledge management research based on a single perspective – values *or* facts – can provide only a partial truth. This has a leveling effect in that the practical consequences of belief – of scientist and citizen alike – are open to critique.

2.2 Churchman

C. West Churchman is part of the tradition of American Pragmatism, a school of thought associated with the names of C. S. Pierce, William James, John Dewey, and A. E. Singer Jr. This distinctively American school is marked by the doctrines that the meaning of conceptions is to be sought in their practical bearings, that the function of thought is to guide action, and that truth is pre-eminently to be tested by the practical consequences of belief. According to Matthews (2006:183) "pragmatism accepted the interplay of observation (empirical datum) and concepts (*a priori* thought), however, it gave neither an autonomous, fixed, or foundational character". That is, the school follows Kant in asserting that no *a priori* worldview can be possibly sufficient (and thus fully justifiable) and the legitimacy of any social system designed from *any* pre-given or *a priori* worldview is subject to critique from other worldviews. Based on this premise, Churchman believes that it is impossible to apprehend the whole system. Instead, Churchman advocates a "sweep-in" approach by which the system designer collects as much information as possible about (conflicting) *a priori* worldviews.

Churchman (1971) draws upon Leibniz, Locke, Kant, Hegel, and Singer to show how complex human systems may be designed to support critique. These insights have major practical consequences for the design of knowledge management systems. For example, a purely objective or technical approach cannot by itself determine an appropriate system design because system goals must address intertwined *technical, organizational* and *personal* issues. (Mitroff & Linstone, 1993) Strategies must be adopted to avoid solving the wrong problem. (Chae, *et al.*, 2005) While complexity demands respect for role-specific perspectives, "progress can be measured in terms of the degree to which the *client, decision maker* and *designer* are the same." (Richardson & Courtney, 2004)

It is clear that Churchman's "sweep-in" approach is fundamentally distinct from the positivist empirical approach. The positivist approach believes in the objectivity of a world "out there" and deploys scientific methods to represent this objective reality. While the positivists believe in absoluteness, Churchman acknowledges contingency and contextuality. System design must be approached from a holistic perspective that "sweeps in" the perspectives of those inside the system (and involved in system design) and those outside (and affected but not involved). That is, pragmatism accepts the need for a pluralism based to a significant degree on an ethical process for building – and critiquing - consensus about both facts and theories, and about facts and values. (Putnam 1995:14)

2.3 Ulrich

Ulrich draws upon Churchman (1971) and Kant to identify the normative issues underlying a critique of judgment about the proper boundary of complex human systems, including systems of knowledge management. A critique of knowledge (and ignorance) that focuses on the concept of a system (and boundary) tests the practical consequences of the existence or non-existence of the system (Pierce, 1877, 1878). Ulrich's *critical system heuristics* (Ulrich, 1983, 1987, 1993, 2000, 2002a, 2002b, 2003, 2005) constitute a system for boundary critique. Boundary judgments about 'facts' (observations) and 'values' (evaluations) are analyzed, both from the perspective of what '*is*' and what '*ought to be*'. Three basic boundary issues are identified for those *involved* in the system: Source of *motivation*, source of *power*, and source of *technical knowledge or expertise.* The basic boundary issue for those affected but not involved is source of *legitimacy.* Ulrich created a new system role – that of *witness* - to pursue legitimacy. Three questions are posed from the perspective of each of four system roles (*witness, client, decision maker, and designer or planner*), to produce 12 questions. Discourse on these questions is intended to render explicit the normative premises that inevitably flow into social system design. (Matthews, 2006)

There is alignment between three knowledge perspectives or bases for rationality (*technical, organizational or interpersonal,* and *personal)* employed by some of Churchman's disciples (e.g., Chae, *et al.*, 2005; Mitroff & Linstone, 1993) and the sources of boundary judgments (*technical knowledge or expertise, power, and motivation*) for those *involved* in the system. Ulrich sees these three knowledge perspectives as a staircase where the organizational or interpersonal level builds on the technical level, and the personal level builds on the organizational or interpersonal level. (Ulrich, 2001a, 2001b)

2.4 Habermas

Habermas has contributed many theories related to *technical, organizational or interpersonal* and *personal* perspectives on knowledge. In writings such as *Knowledge and Human Interests* (Habermas, 1968) these perspectives are the cognitive interests associated with technical, practical and emancipatory knowledge. Habermas's treatment of these perspectives could be interpreted as arguing in part for a separate focus on each perspective as though it was independent of the others. This led some systems theorists in the context of the 'paradigm wars' of the 1990's to argue that research paradigm and methodologies based on each knowledge perspective produced concepts that were incommensurate across paradigms. If this was the case, no research paradigm exists to support Churchman's 'sweeping-in' process. Ethical concerns would be removed from the information systems discipline and removed to its own separate (bounded) discipline. If this were the case, no paradigm would exist to support the notion of making common sense via *consensual validation*, that is 'the things people agree upon because of their common sensual apparatus and deeply common interpersonal experiences make them seem objectively so.' (Weick, 1979)

In *The Theory of Communicative Action* (Habermas, 1984, 1987) Habermas argued that all knowledge is interdependent in that it is mediated by language. The human instinct for language provides the site of, and the precondition for, knowledge. Loosely speaking, Habermas locates Kant's *Transcendental Idealism* at the boundary of language (thought) and action (experienced through the senses) and grants it form via a theory of *Universal Pragmatics*. Critical discourse analysis based on a Habermas (1984, 1987) excavates 'meaning' via examining claims that technical knowledge is based on *objective truth*, examining claims that interpersonal knowledge is based on normative *rightness*, and examining claims that personal truth is based on *sincerity*. Knowledge management researchers currently apply Habermas's theory of communicative action to promote diversity under the banner of 'critical systems meta-methodology' (Jackson, 2000), methodological pluralism (Midgley, 2000; Mingers, 2001, 2005) and pragmatic pluralism. (Taket & White, 1996)

2.5 Research objectives

Midgley (2000: 171-216) interprets twentieth century debates on systemic interventions as occurring in three waves, linked respectively to Habermas's technical, interpersonal and personal worlds. Ulrich's critical system heuristics and Habermas's theory of communicative action are third wave theories that deal with power relations via dialogue. Yet in practice coercive situations are generally characterized by the closure of debate (Midgley, 2000: 208). A fourth wave theory may be characterized as one that is explicitly ethical in nature. It focuses on value clarification by both those involved (and included) in dialogue about the system and those that are affected (and excluded). (Bawden, 2005)

The aim of the research is to develop a fourth wave theoretical framework for ethical inquiry in Knowledge Management that builds upon two existing theories – a *Habermasian inquiring system* developed by Guo & Sheffield (2006a), and Ulrich's *critical systems heuristics* (Ulrich, 1983, 1987, 1993, 2000, 2003, 2005).

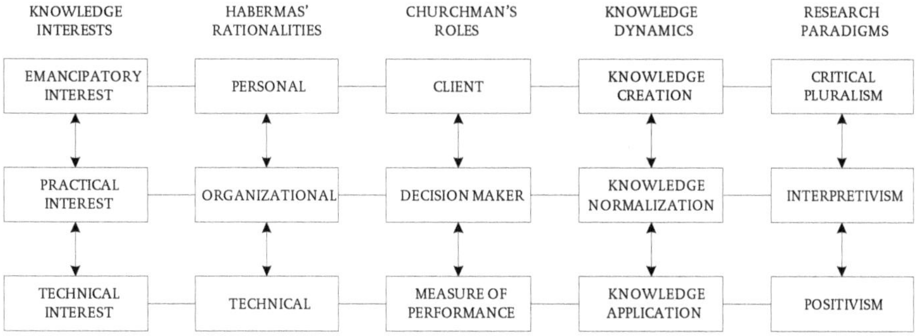

Figure 1 *Habermasian inquiring system (Guo & Sheffield, 2006a)*

In Guo & Sheffield's (2006a) Habermasian inquiring system, the key architectural element is Habermas's knowledge constitutive interests. (Habermas, 1968) Habermas's three knowledge interests (technical, practical, and emancipatory) form a three-wave or three-level integrating structure. The development of the Habermasian inquiring system consists of describing how four other design elements (Habermas' three rationalities, Churchman's roles, knowledge dynamics, and research paradigms) are positioned within this integrating structure. (Figure 1)

The Habermasian inquiring system was applied by the authors to critique the paradigms and methodologies used in the knowledge management literature. (Guo & Sheffield, 2006b) The key finding is that inquiry in Knowledge Management is starkly unbalanced. Overuse of the positivist paradigm and its dominant research method (sample survey) prevents the exploitation of the highly relevant insights available via the interpretive and critical pluralist paradigms and the field study method. Because, as is argued above, neither the subjective nor the objective aspects of knowledge can be represented without the other, the knowledge management literature is seen as reframing issues of rightness and sincerity as issues of objective truth. This focus on facts rather than values obscures the need for critique, and makes those affected by the 'system' vulnerable to those who are involved in its design and maintenance.

In the section devoted to Ulrich's critical systems heuristics it was noted that there is alignment between three knowledge perspectives (*technical, organizational or interpersonal,* and *personal)* employed by some of Churchman's disciples (e.g., Chae, *et al.*, 2005; Mitroff & Linstone, 1993) and the sources of boundary judgments (*technical knowledge or expertise, power, and motivation*) for those *involved* in the system. The basic boundary issue for those affected but not involved (a system role that Ulrich called *witness*) is source of *legitimacy*. This suggests that the power of Habermasian discourse by those involved in the system to 'sweep in' the interests of those affected but not involved may be strengthened by directly addressing the system role of witness.

In summary, the proposed framework is a fourth wave theory that identifies the knowledge concepts that must be addressed to enable those in a position of power (because they are involved with the knowledge management system) to impose order on the behavior of others (who are affected but not involved) by means of ethical inquiry. Elements of the proposed framework are selected from the Habermasian model illustrated in Figure 1 and Ulrich's critical systems heuristics to meet two objectives. Firstly the framework must identify the elements involved in the sweeping in of those affected (but excluded) from the system. Secondly the framework must identify the architectural elements that link those inside the system (involved) and those outside the system (affected).

3. The proposed framework

The proposed framework is illustrated in Figure 2. (The development process has been omitted due to space restrictions.) Normative theory to support boundary analysis and the surfacing of the interests of those affected (but otherwise excluded) is provided via the knowledge creation aspect of knowledge dynamics, and three concepts from Ulrich's critical systems heuristics: an *ethical* perspective on knowledge interests and rationalities; source of *legitimation* of boundary judgment; and the system role of *witness*. (Research objective 1) Four architectural elements link those inside the system (involved) and those outside the system (affected): Knowledge perspectives, sources of boundary judgment, system roles and knowledge dynamics constitute. (Research objective 2) The proposed framework is seen as a fourth wave normative theory that identifies the knowledge concepts that must be addressed to enable those in a position of power (because they are involved with the knowledge management system) to impose order on the behavior of others (who are affected but not involved) by means of ethical inquiry.

4. Discussion and conclusion

The development of a framework for ethical inquiry in knowledge management was motivated by the conceptual difficulties associated with managing knowledge across boundaries. The framework illuminates issues which constitute a necessary part of the critique of judgment about the

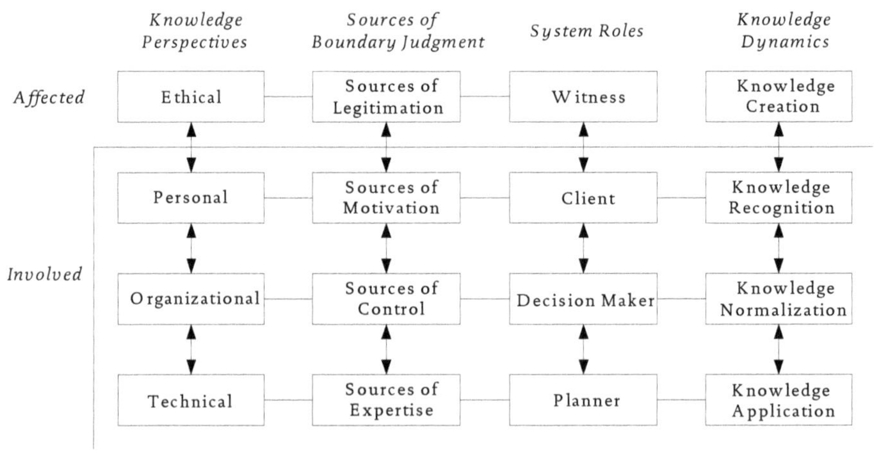

Figure 2 *Framework proposed for ethical inquiry in knowledge management*

proper boundary of a knowledge management system. To achieve this purpose the framework drew on concepts of knowledge (and ignorance) provided by Kant, and three twentieth century authors - Churchman, Ulrich, and Habermas. Key concepts and linkages are as follows:

Kant's *Transcendental Idealism* identified that the objectively real and our subjective understanding are interlaced, each providing the conditions for the other. Knowledge management research based on a single perspective – values *or* facts – will therefore provide only a partial truth. Churchman identified that the whole system cannot be apprehended and that system design must therefore be approached from a holistic perspective that "sweeps in" the perspectives of those inside the system (and involved in system design) and those outside (and affected but not involved). Ulrich identified a systematic set of issues that support the sweeping in process. Habermas identified the cognitive interests associated with technical, practical and emancipatory knowledge, and the nature of the communicative process that gives rise to each.

The framework was designed to 'sweep-in' to the system the interests of those *affected* (but not involved. It was developed from two existing theories - a Habermasian inquiring system (Guo & Sheffield, 2006a, 2006b) and Ulrich's critical systems heuristics. The former is seen as identifying the knowledge perspectives, sources of boundary judgments, system roles, and knowledge dynamics of those *involved* in the system. The latter is seen as identifying the knowledge perspectives, sources of boundary judgments, system roles, and knowledge dynamics of those *affected* (but not involved).

The value of the framework, and its Churchmanian and Habermasian roots, are reflected in the categories of knowledge dynamics – *knowledge creation* is seen as that ethical awareness that secures legitimation by virtue of the communicative action of a witness who "sweeps in" the voice to those affected (but not involved); *knowledge recognition, knowledge normalization and knowledge application* are the communicative actions that constitute systemic learning and design justification by those already involved.

Movement from the first to the fourth levels of the proposed framework represents "stages" or "waves" of intellectual and moral development. At the first level knowledge is applied when planners set system boundaries (narrowly) to source the expertise that solves technical problems. At the second level knowledge is normalized when the collective decision maker sets system boundaries (less narrowly) to obtain control of organizational activities. At the third level knowledge that is recognized by those inside the system informs the wishes of the client and system boundaries are set (broadly) motivated by an exploration of some of the more personal issues in the life-world of the client. At the fourth level the creation of knowledge about those affected (but not involved) by the act of witnessing sets systems boundary (even more broadly) so as to seek the sources of legitimation required to address ethical issues.

The framework thus provides a fourth wave theory describing the knowledge concepts that must be addressed to enable those in a position of power (because they are involved with the knowledge management system) to impose order on the behavior of others (who are affected but not involved) *by means of ethical inquiry.*

References

Bawden, R. (2005). Systemic Development at Hawkesbury: Some Personal Lessons from experience. *Systems Research and Behavioral Science* 22, 151-164.

Chae, B., Paradice, D., Courtney, J. F., & Cagle, C. J. (2005). Incorporating an Ethical Perspective into Problem Formulation: Implications for Decision Support Systems Design. *Decision Support Systems, 40*, 197-212.

Churchman, C. W. (1971). *The design of inquiring systems: Basic concepts of systems and organizations.* New York, NY: Basic Books, Inc.

Guo, Z., & Sheffield, J. (2006a). *Habermasian Inquiring System: Toward a General Framework for Knowledge Management Research.* Paper presented at the 39th Hawaii International Conference on System Sciences, Hawaii, USA.

Guo, Z., & Sheffield, J. (2006b). *A Paradigmatic and Methodological Examination of KM Research: 2000 to 2004.* Paper presented at the 39th Hawaii International Conference on System Sciences, Hawaii, USA.

Habermas, J. (1968). *Knowledge and human interests.* Boston: Beacon Press.

Habermas, J. (1984). *The theory of communicative action Volume 1: Reason and the rationalization of society* (T. McCarthy, Trans.). Boston, MA: Beacon Press.

Habermas, J. (1987). *The theory of communicative action Volume 2: Lifeword and system* (T. McCarthy, Trans.). Boston, MA: Beacon Press.

Jackson, M. C. (2000). *Systems Approaches to Management.* New York, NY: Kluwer Academic / Plenum Publishers.

Kant, I. (1787). *Critique of Pure Reason*: Palgrave Macmillan.

Kay, R., & Goldspink, C. (2005). *Organizational knowledge and autopoiesis: Implications for knowledge management.* Paper presented at the 11th ANZSYS / Managing the Complex V Conference, Christchurch, New Zealand.

Lehaney, B., Clark, S., Coakes, E., & Jack, G. (2004). *Beyond Knowledge Management.* London: Idea Group Publishing.

Matthews, D. (2006). Pragmatism Meets Systems Thinking: The Legacy of C. West Churchman. In J. McIntyre-Mills (Ed.), *Rescuing the Enlightenment from Itself: Critical and Systemic Implications for Democracy.* New York, NY: Springer.

Midgley, Gerald. (2000). *Systemic Intervention: Philosophy, Methodology, and Practice.* London: Kluwer Academic.

Mingers, J. (2001). Combining IS research methods: Towards a pluralist methodology. *Information Systems Research, 12*(3), 240-259.

Mingers, J. (2005). More dangerous than an unanswered question is an unquestioned answer: A contribution to the Ulrich debate. *Journal of the Operational Research Society, 56*, 468-474.

Mitroff, I. I., & Linstone, H. A. (1993). *The Unbounded Mind: Breaking the Chains of Traditional Business Thinking.* New York: Oxford University Press.

Pierce, C. S. (1877). The Fixation of Belief. *Popular Science Monthly, 12*, 1-15.

Pierce, C. S. (1878). How to Make Our Ideas Clear. *Popular Science Monthly, 12*, 286-302.

Putnam, A, E. Jr. (1995). *Pragmatism: An Open Question.* Oxford UK and Cambridge USA: Blackwell

Richardson, S. M., & Courtney, J. F. (2004). *A Churchmanian theory of knowledge management system design.* Paper presented at the 37th Hawaii International Conference on System Sciences, Hawaii, USA.

Taket, A., & White, L. (1996). Pragmatic Pluralism: An Explication. *Systems Practice,* 9(6), 571-586.

Ulrich, W. (1983). *Critical Heuristics of Social Planning: A New Approach.* Berne: Paupt.

Ulrich, W. (1987). Critical heuristics of social systems design. *European Journal of Operational Research, 31*, 276-283.
Ulrich, W. (1993). Some difficulties of ecological thinking, considered from a critical systems perspective: A plea for critical holism. *Systems Practice, 6*(6), 583-611.
Ulrich, W. (2000). Reflective practice in the civil society: The contribution of critically systemic thinking. *Reflective Practice, 1*(2), 247-268.
Ulrich, W. (2001a). Critically systemic discourse: A disursive approach to reflective practice in ISD (Part 2). *Journal of Information Technology Theory and Application, 3*(3), 85-106.
Ulrich, W. (2001b). A philosophical staircase for information systems definition, design, and development: A discursive approach to reflective practice in ISD (Part 1). *Journal of Information Technology Theory and Application, 3*(3), 55-84.
Ulrich, W. (2002a). Boundary Critique. In H. G. Daellenbach & R. L. Flood (Eds.), *The Informed Student Guide to Management Science*. London: Thomson.
Ulrich, W. (2002b). Critical Systems Heuristics. In H. G. Daellenbach & R. L. Flood (Eds.), *The Informed Student Guide to Management Science*. London: Thomson.
Ulrich, W. (2003). Beyond Methodology Choice: Critical Systems Thinking As Critically Systemic Discourse. *Journal of the Operational Research Society, 54*(4), 325-342.
Ulrich, W. (2005). *A Brief Introduction to Critical Systems Heuristics (CSH)*. Retrieved 22 June, 2006, from http://www.geocities.com/csh_home/csh.html
Weick, K. (1979). *The Social Psychology of Organizing*. New York: Random House.

Jim Sheffield graduated with a PhD from the University of Arizona in 1990. His doctoral work in group support systems led to the creation of a Decision Support Centre at the University of Auckland and participation in major policy development exercises. Action research initiatives include the development of economic strategy, science policy, and comprehensive regional planning. He has published in Journal of MIS, Group Decision and Negotiation and International Journal of Business Information Systems. Current interests include: Churchmanian and Habermasian inquiring systems, knowledge management, collaborative learning, and research methods.

Zining Guo holds a Master of Commerce (First Class Honors) in Information Systems from the University of Auckland. He is currently serving as Associate Consultant at Xelocity Limited, Auckland, New Zealand. He has published in Hawaii International Conference on System Sciences. His primary areas of research interest include knowledge management, systems approaches, organization sciences, research methods, and philosophical foundations of inquiry.

Identifying the Philosophical Assumptions of Knowledge Management Perspectives

Jim Sheffield & Edward Lau
University of Auckland, NZ

The importance of Knowledge Management research and practice in the knowledge economy is well-recognized. However, successful development of KM as a recognizable discipline requires progress on conceptual, methodological and theoretical issues. The authors have chosen to approach the goal of conceptual integration by investigating the philosophical assumptions of KM research. The purpose is to promote conceptual integration in the KM discipline via awareness of one's epistemic commitments and those of others. In this initial paper we identify plausible philosophical assumptions of KM perspectives. The KM literature is explored and different perspectives on knowledge identified. Initial support is found for integration via a simple three-dimensional framework based on the work of Jürgen Habermas. The application of a Habermasian framework to the surfacing of philosophical assumptions of popular knowledge management frameworks will be reported in a subsequent paper.

1. Introduction

Knowledge Management (KM) is a discipline that has emerged in tandem with the establishment of the "knowledge economy" – the emergent economic era in which intellectual, rather than physical, capital is the principle source of wealth and power. The belief that putting intellectual capital effectively to work in organizations will create unique competitive advantage motivates KM researchers and practitioners to unlock the potential of knowledge which is supposedly lying dormant within the organization. 'If only we knew what we know,' is an often cited phrase (Davenport & Prusak, 1998: xii) in KM that captures practitioners concerns of their limited ability to identify their most important intellectual resources and furthermore how to utilize those resources in ways which positively impacts the performance of their firm.

While these broad principles of KM are straightforward, the concept of 'knowledge' has proved to be far more elusive. The term has been used rather loosely in the KM literature. It is for this reason that KM has been criticized as management fad. Some have argued that in much of the literature 'knowledge' is synonymous with 'information' and that 'knowledge management' is merely marketing rhetoric which replaced 'information management' when 'information' fell out of fashion with consultants (Wilson, 2002).

Certainly there is evidence that KM is not just another passing management fad. Ten years after KM first became popular there continues to be significant literature output (Peachy & Hall, 2005). Furthermore, a preliminary time-series analysis of KM article counts suggest that KM is not following the same "boom-bust" cycle as Quality Circles, Total Quality Management, and Business Process Reengineering (Ponzi & Koening, 2002). Therefore, this paper is motivated by the hope that KM is a useful endeavour and work is needed to achieve a satisfactory level of conceptual coherence about research output in order to promote mutual understanding, integration, consistency, clarity, and focus within the discipline.

There still remains the problem that the elusive nature of knowledge makes it is difficult to make sense of the KM literature as a whole. As a discipline, KM is replete with frameworks, theories, and definitions that *altogether lack cohesiveness and conceptual integration* (Guo & Sheffield, 2006). One consequence of this is that debates among KM researchers appear to focus almost exclusively on empirical studies and the reporting of facts. With a few exceptions (primarily from outside of the USA), there is no discussion about values and no attempt reach consensus on the shape of the discipline. In particular, there is little attempt to surface the assumptions underlying KM research. The authors are not aware of any papers in the KM discipline that focus on ethics. In this environment debate on conceptual linkages among facts, organizational norms and values quickly becomes brittle and acrimonious.

This research is part of a larger study that *promotes conceptual integration in the KM discipline via awareness of one's epistemic commitments and those of others*. The objective of the study is to analyze the philosophical assumptions of knowledge management perspectives and frameworks, and to draw implications from this analysis. In other words, the study aims to interpret the way in which those perspectives and frameworks treat knowledge and then examine how that affects their use in guiding KM research and practice. Mapping KM perspectives and KM frameworks onto the different knowledge interests will allow a consistent comparison useful for identifying common concepts and gaps. The results should be useful in building a conceptual foundation for KM.

This paper represents the first step of the journey. In this initial paper we identify plausible philosophical assumptions of KM perspectives. The KM literature is explored and different perspectives on knowledge identified. The purpose is to demonstrate initial support for integration via a simple Habermasian three-dimensional framework. The application of this framework to the surfacing of philosophical assumptions of popular knowledge management frameworks will be reported in a subsequent paper.

The remainder of this paper is organized as follows. Sections 2 and 3 give a brief overview of different perspectives and types of knowledge found in the KM literature. Section 4 introduces Habermas' knowledge interests and knowledge worlds. *Section 5 demonstrates the relevance and potential usefulness of the Habermasian categories by mapping into them the different knowledge perspectives and types.* Section 6 concludes this first step of the journey and introduces the second step.

2. Data, information, knowledge

Early writings in KM defined knowledge by distinguishing it from data and information and placing them all onto a three-level hierarchy as shown in Figure 1. Essentially the hierarchy is an extension of the relationship between data and information established in Information Systems: data are raw facts and figures; information is processed data made meaningful by placing in a context relevant to the recipient of that information. Knowledge is then information somehow transformed to make it more valuable than the original information, for example by way of an individual personalizing the information by evaluating it against, and then incorporating it into, his or her previous knowledge (Alavi & Leidner, 2001: 109).

Underlying this hierarchy appears to be the perspective of knowledge as an object – an object which is the result of processing two more fundamental objects lower in the hierarchy. This assumes that knowledge can be universal. For example, two individuals both possessing the same mental framework (e.g., relevant professional training) can obtain the same knowledge if they both have access to the same information, which was previously derived from the same data. The implication of this for organizations is the imperative to generate knowledge by building systems which can move up the hierarchy. Of course, such an imperative presupposes that knowledge creates the capability for action, i.e., individuals, and collectively, organizations will know what to do with the knowledge once it is obtained.

Alavi & Leidner (2001) identify some additional knowledge perspectives which are summarized in Table 1. Knowledge as an object, access to information and a capability has been alluded to above. In contrast to the assumption of knowledge as an object, Stenmark (2002) notes that the perspectives of knowledge as a state of mind, process and capability alternatively regard knowledge as action-oriented, socially-situated and context-dependent. This idea will be developed further in the discussion of knowledge taxonomies below.

3. Knowledge taxonomies

A second useful approach to understanding knowledge is to classify different types of knowledge (Alavi & Leidner, 2001: 110). Perhaps the most widely-adopted taxonomy of knowledge in KM is the distinction between 'tacit' and 'explicit' knowledge as popularised by Nonaka (1994) and

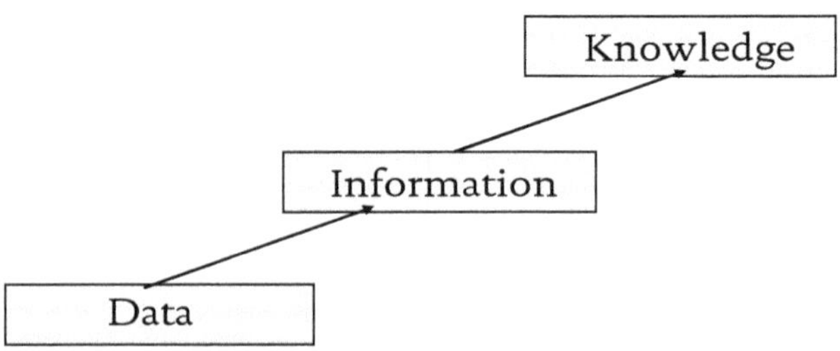

Figure 1 *The data, information, knowledge hierarchy (source: Stenmark, 2002, p. 3)*

Perspectives	Definitions	Implications for Knowledge Management (KM)
Knowledge vis-à-vis data and information	Data is facts, raw numbers. Information is processed/ interpreted data. Knowledge is personalized information.	KM focuses on exposing individuals to potentially useful information and facilitating assimilation of information
State of mind	Knowledge is the state of knowing and understanding.	KM involves enhancing individual's learning and understanding through provision of information
Object	Knowledge is an object to be stored and manipulated.	Key KM issue is building and managing knowledge stocks
Process	Knowledge is a process of applying expertise.	KM focus is on knowledge flows and the process of creation, sharing, and distributing knowledge
Access to information	Knowledge is a condition of access to information.	KM focus is organized access to and retrieval of content
Capability	Knowledge is the potential to influence action.	KM is about building core competencies and understanding strategic know-how

Table 1 *Knowledge perspectives and their implications (source: Alavi & Leidner, 2001: 111)*

Nonaka & Takeuchi (1995). Tacit knowledge can be thought of as that which we inherently know but find difficult to articulate and explain. Alavi & Leidner (2001), quoting Nonaka (1994), describe tacit knowledge as "rooted in action, experience, and involvement in a specific context" comprised of cognitive elements, which include "mental models, beliefs, paradigms, and view-points", and technical elements, which include "concrete know-how, crafts, and skills which apply to a specific context" (p. 110). On the other hand, explicit knowledge has been "articulated, codified, and communicated" in some form (ibid: 110).

Table 2 summarizes further approaches to classifying knowledge as identified by Alavi & Leidner (2001). These include: mode of knowledge creation/existence (mind of the individual, norms of the social collective), knowledge-orientation (know-about, know-how, know-why, know-when, know-with) and pragmatic classifications.

While the above taxonomies are useful, their simplicity does not fully capture the aspects of knowledge phenomena that are "multi-faceted and complex" (Blackler, 1995: 1032). A more sophisticated classification scheme

Knowledge Types	Definitions	Examples
Tacit Cognitive tacit: Technical tacit:	Knowledge is rooted in actions, experience, and involvement in specific context Mental models Know-how applicable to specific work	Best means of dealing with specific customer Individual's belief on cause effect relationships Surgery skills
Explicit	Articulated, generalized knowledge	Knowledge of major customers in a region
Individual	Created by and inherent in the individual	Insights gained from completed project
Social	Created by and inherent in collective actions of a group	Norms for inter-group communication
Declarative	Know-about	What drug is appropriate for an illness
Procedural	Know-how	How to administer a particular drug
Causal	Know-why	Understanding why the drug works
Conditional	Know-when	Understanding when to prescribe the drug
Relational	Know-with	Understanding how the drug interacts with other drugs
Pragmatic	Useful knowledge for an organization	Best practices, business frameworks, project experiences, engineering drawings, market reports

Table 2 *Knowledge taxonomies and examples (source: Alavi & Leidner, 2001: 113)*

is offered by Blackler (1995) who identified five knowledge types (embodied, embedded, embrained, encultured, encoded) in his review of literature on organizational learning. An overview of those knowledge types is given in Table 3.

Analyzing these various classifications shows that they also have commonalities and underlying perspectives just as the different knowledge perspectives had in Section 2.1. To begin, the examples of knowledge under the knowledge-orientation (know-about, know-how, know-why, know-when, know-with) and pragmatic classifications appear mostly oriented towards describing knowledge either contained within or associated with rules. These could actually be subsumed under the embrained knowledge type – knowledge associated abstract thinking. As such, these classifications share in common an underlying perspective of knowledge that is generalized and universal – which corresponds to knowledge as an object. This perspective also underlies the explicit/encoded knowledge type (the two terms are synonymous).

In contrast, the underlying perspective of both the social and encultured

Knowledge Type	Description
Embodied	"Knowledge that is action oriented and is likely to be only partly explicit" (p. 1024); practical thinking.
Embedded	"Knowledge which resides in systemic routines" (p. 1024); emerges through relationships and material resources.
Embrained	"Knowledge that is dependent on conceptual skills and cognitive abilities" (p. 1023); abstract thinking.
Encultured	"The process of achieving shared understanding" (p. 1024); emerges specifically through inter-personal interaction within groups.
Encoded	"Information conveyed by signs and symbols" (p. 1025); explicit knowledge.

Table 3 *Descriptions of Blackler's (1995) knowledge types*

knowledge types (which are almost synonymous, differing only in that the former is static and the latter is dynamic) is that knowledge is action-oriented, socially-situated, and context-dependent. Key assumptions of this perspective are that knowledge is emergent and that it is only meaningful when interpreted in a specific context. There are several implications of this perspective. Firstly, organizations cannot hope to 'know what they know' until an event occurs that calls for putting organizational knowledge into action. Secondly, an intervention that attempts to manage knowledge as a 'thing' to be extracted, transferred, stored and applied is misguided. Instead the focus should be an appreciation of the interpersonal relationships and social conditions ('the way we do things around here') which facilitate shared understanding and meaning between individuals. Thirdly, an appreciation of how existing organizational knowledge both enables and constrains the creation and normalization of new organizational knowledge could provide a useful perspective on how knowledge evolves within particular organizations.

Another underlying perspective on knowledge that has not been discussed yet is that knowledge is inherently value-laden. This perspective can be interpreted as consistent with the popular definition of knowledge as a "justified true belief" (Nonaka & Takeuchi, 1995: 21) because beliefs are founded on feelings which in turn are associated with values. While this perspective underlies both the social and individual knowledge types, it is arguably more relevant to individual knowledge considering that knowledge by way of mutual understanding within a collective does not guarantee equivalent positions regarding the validity of that knowledge among individuals (Marshall & Brady, 2001). In other words, an individual does not have to personally accept what is known in the organization. An implication of this perspective is that it allows for ethical considerations in KM if individuals are free to examine and challenge the values and interests motivating knowledge. Another implication is that an organization will become more open to the possibility of radical change if individuals at all levels of an organization feel they can challenge the established 'way we do things around here' without repercussion.

The preceding discussion on knowledge taxonomies has identified three underlying perspectives on knowledge: objective, social and personal. So far, the authors have presented each knowledge type as linked to just one of these underlying perspectives. However, it should be noted that these perspectives are neither mutually exclusive nor incompatible with each other. This is illustrated, for example, by the tacit knowledge type, which can be interpreted as incorporating all of perspectives. Firstly, we can perceive tacit knowledge as an object because it has the potential to be articulated as if it were a heuristic. A heuristic can be considered a generalization because it can be applied in different contexts. However, a heuristic will never be a rule because its application depends upon the judgement of the actor who is applying it. In this way the social perspective also underlies tacit knowledge because tacit knowledge is grounded in experience and emerges through action sensitive to context. Secondly, because tacit knowledge exists within the mind it must also incorporate values and therefore be guided by the interests of the knower. Finally, while the social knowledge perspective strongly underlies both the embodied and embedded knowledge types they can also be seen to partially incorporate the knowledge as an object perspective as well.

4. Habermasian knowledge interests and knowledge worlds

In Sections 2.1 and 2.2 a framework which identifies three fundamental perspectives of knowledge has begun to develop. This section will continue to develop this framework by adopting the three 'knowledge interests' examined by Habermas in *Knowledge and Human Interests* (1987) and located in the 'knowledge worlds' described his *Theory of Communicative Action* (1984, 1987). The reason for drawing on Habermas is because he provides a more precise articulation of the three underlying perspectives of knowledge already identified.

Jürgen Habermas is a prominent German philosopher and sociologist. His Theory of Communicative Action can be broadly described as concerned with "discourse ethics" (Endres, 1996). Within his theory, Habermas defines the 'system-world' and the 'life-world'. The system-world is the totality of social systems whereas the life-world is the totality of an individual's lived-experiences. Each is defined via its relationships to three perspectives on knowledge or knowledge worlds: "the objective world, which represents facts independent of human thought and serves as a common reference point for determining truth; the [interpersonal or] social world, comprised of inter-subjective relationships; and the subjective world of private experiences" (ibid.). Within each knowledge world Habermas' recognises a different knowledge interest, viz. technical, practical and emancipatory, respectively. The technical interest is concerned with instrumental interventions for controlling humans' interaction with the physical world. The practical knowledge interest is concerned with achieving shared interpretation and meaning between individuals. The emancipatory knowledge interest is concerned with self-reflexivity and recognizing "systematic communicative distortions" (Guo & Sheffield, 2006: 2) that are the result of the "colonization" of the life-world by the system-world.

Knowledge Interest	Knowledge World	Knowledge Perspective	Research Paradigm
Technical	Technical	Objective	Positivist
Practical	Interpersonal	Social	Interpretivist
Emancipatory	Personal	Personal	Critical

Table 4 *Habermasian categories and corresponding KM perspectives and research paradigms*

The three Habermasian knowledge interests (technical, practical and emancipatory) and knowledge worlds (technical, interpersonal and personal) correspond to the objective, social, and personal perspectives of knowledge, as identified in Sections 2.1 and 2.2, respectively. This is illustrated in Table 4 (also listed are corresponding research paradigms which may help in understanding knowledge interests and perspectives for those readers who are already familiar with these different paradigms). *Aligning each knowledge perspective with the appropriate knowledge interest/world surfaces the researcher's purpose and underlying theoretical perspective and epistemological commitment.* For example, the declarative, procedural, causal, conditional and relational knowledge types presented in Table 2 all have a common technical interest in explaining, controlling, and predicting phenomena concerned with human 'work' as defined by Habermas.

5. Knowledge perspectives and types mapped onto the Habermasian categories

This insight, together with the concepts discussed in Sections 2.1 and 2.2, lead to the mapping of knowledge perspectives and types defined by Alavi and Leidner (2001) and Blackler (1995) onto the knowledge worlds described by Habermas. A key assumption is that these knowledge interests and knowledge worlds are broad enough to provide a comprehensive categorization of knowledge. Because knowledge phenomena are inherently complex and multidimensional these categories are not mutually exclusive. A knowledge perspective/type may be associated with multiple knowledge interests or knowledge worlds. At the risk of oversimplifying the richness of the relationships the result is presented graphically in Figure 2. The depth of shading represents the strength of association between a particular knowledge perspective/type.

6. Conclusion

The aim of this paper was to address two key problems in KM research, namely a certain lack of cohesiveness and conceptual integration, and difficulty in discussing the relationships among facts, social consensus and values. The solution to both problems is approached by investigating the philosophical assumptions of KM research. The purpose is to promote conceptual integration in the KM discipline via awareness of one's epistemic commitments and those of others. In this initial paper we identified plausible philosophical assumptions of KM perspectives. The KM literature was explored briefly to iden-

	Habermasian Category		
Knowledge Interest	T	P	E
Knowledge World	T	I	P
Knowledge Perspective			
Knowledge, data, information	■		
State of mind		■	
Object	■		
Process		■	
Access to information	■		
Capability		■	
Knowledge Types (Alavi & Leidner, 2001)			
Tacit	░	░	░
Explicit	■		
Individual			■
Social		■	
Declarative	■		
Procedural	■		
Causal	■		
Conditional	■		
Relational	■		
Pragmatic	■		
Knowledge Types (Blackler, 1995)			
Embodied	░	■	
Embedded	░	■	
Embrained	■		
Encultured		■	
Encoded	■		

Figure 2 *Knowledge perspectives and types mapped onto Habermas's knowledge interests and knowledge worlds*

tify different perspectives on knowledge. Initial support is found for integration via a simple three-dimensional framework based on the three 'knowledge interests' (technical, practical and emancipatory) that Habermas examined in *Knowledge and Human Interests* (1987) and located in each of the three 'knowledge worlds' (technical, interpersonal and personal) described in his *Theory of Communicative Action* (1984, 1987). The application of a Habermasian framework to the surfacing of philosophical assumptions of popular knowledge management frameworks will be reported in a subsequent paper.

References

Alavi, M., & Leidner, D. (2001). Knowledge management and knowledge management systems: Conceptual foundations and research issues. *MIS Quarterly, 25*(1), 107-136.

Blackler, F. (1995). Knowledge, knowledge work and organizations: an overview and interpretation. *Organization Studies, 16*(6), 1021-1046.

Davenport, T. H., & Prusak, L. (1998). *Working knowledge: How organizations manage what they know.* Boston, Massachusetts: Harvard Business School Press.

Endres, B. (1996). *Habermas and Critical Thinking.* Retrieved 14/06/2006, from http://www.ed.uiuc.edu/EPS/PES-Yearbook/96_docs/endres.html

Guo, Z., & Sheffield, J. (2006). *Habermasian Inquiring Systems: Toward a general framework for knowledge management research.* Paper presented at the Thirty-Ninth Annual Hawaii International Conference on System Sciences.

Habermas, J. (1984, 1987). *The theory of communicative action* (T. McCarthy, Trans.). Boston: Beacon Press.

Habermas, J. (1987). *Knowledge and human interests* (J. J. Shapiro, Trans.). Cambridge: Polity.

Marshall, N., & Brady, T. (2001). Knowledge management and the politics of knowledge: Illustrations from complex products and systems. *European Journal of Information Systems, 10*(2), 99-112.

Nonaka, I. (1994). A dynamic theory of organizational knowledge creation. *Organization Science, 5*(1), 14-37.

Nonaka, I., & Takeuchi, H. (1995). *The knowledge-creating company: How Japanese companies create the dynamics of innovation.* New York, NY: Oxford University Press.

Peachy, T., & Hall, D. (2005, January 3-6). *Knowledge management and the leading IS journals: An analysis of trends and gaps in published research.* Paper presented at the 38th Hawaii International Conference on System Sciences, Hawaii, USA.

Ponzi, L. J., & Koening, M. (2002). Knowledge management: another management fad? *Information Research, 8*(1), http://informationr.net/ir/8-1/paper145.html.

Stenmark, D. (2002, January 7-10). *Information vs. knowledge: The role of intranets in knowledge management.* Paper presented at the 35th Hawaii International Conference on System Sciences, Hawaii, USA.

Jim Sheffield graduated with a PhD from the University of Arizona in 1990. His doctoral work in group support systems led to the creation of a Decision Support Centre at the University of Auckland and participation in major policy development exercises. Action research initiatives include the development of economic strategy, science policy, and comprehensive regional planning. He has published in Journal of MIS, Group Decision and Negotiation and International Journal of Business Information Systems. Current interests include: Churchmanian and Habermasian inquiring systems, knowledge management, collaborative learning, and research methods.

Edward Lau holds a Bachelor of Commerce (First Class Honors) in Information Systems from the University of Auckland. He is currently researching knowledge management in a hospital context. He has published in Hawaii International Conference on System Sciences. His primary areas of research interest include knowledge management, systems approaches, organization sciences, research methods, and philosophical foundations of inquiry.

Social Capital and Health: Preliminary Work on a System Dynamics 'Concept Model'

John Cody[1], Robert Y. Cavana[2] & David Pearson[3]
[1] Social Worker, NZ
[2] Victoria Management School, Victoria University of Wellington, NZ
[3] Humanities and Social Sciences, School of Social and Cultural Studies, Victoria University of Wellington, NZ

Health disparity is identified as a public issue. A short review of literature in public health and sociology is used to create a conceptual framework for discussing disparity. Work in progress on a small System Dynamics model based on the framework is outlined. The model may be useful for discussing the extent to which civic action can address health disparities.

Introduction

This discussion presents work in progress identifying concepts from sociology and system dynamics that might help structure public discussions about health disparities. The aim of the project is to create a generic framework that both captures the essential features of each case and allows comparisons among populations.

The concept of social capital is current in discussions of population health status. High social capital has been associated with good population health and strategies to reduce health disparities. Others criticise references to social capital because the concept is poorly defined, it shifts attention from public health policies to community and household activities, and it implicitly assumes capitalist economics and rational action theory. The discussion has been summarized by Blaxter (2004: 117-121).

Both arguments seem to have merit. Citizens can assess the relative merits if we can discuss and share our understandings of the features of social structure that affect health. This type of discussion is problematic for many reasons including differences of language, experience, commitments and power. This paper does not review the wider aspects; it has the limited objective of exploring the use of system dynamics (SD) in deliberative discussions.

The sections that follow: briefly describe a context for the issue; propose a general cause of health disparity (Wilkinson, 2000; Wilkinson, 2005); summarize some social theory that seems relevant (Runciman, 1989; Coleman, 1990; Runciman, 1999); reviews relevant work by SD modelers; and speculates about the application of this material.

Local setting

By New Zealand (NZ) standards the Porirua Basin is geographically divided between high and low deprivation areas (Crampton, *et al.*, 2000) (see Appendix). The population of the area shown in the Appendix is about 62,000; approximately 50,000 people live in Porirua City. The main point is the relatively high contrast between the neighborhoods shown in green (low deprivation) and those in red (high deprivation). The administrative boundary of the City is at the south end of the areas shown in red. The position of the boundary is indicative of the political dynamics in the Basin. The implications of these indicators for relative health and health services have been documented over the last three decades (Salmond, 1975; Reinken, *et al.*, 1980; Gould, 1992; Central Regional Health Authority, 1994; National Research Bureau Ltd, 1994; Porirua City Council, 1999; Porirua Kapiti Healthlinks Project, 2000).

Recently 'health disparity' has been defined as a national public issue. Conceptually discussion of 'health disparities' in NZ follows a similar course to that in the United Kingdom, the United States, the World Health Organisation, and the European Union. The distinctive difference is that Maori (the indigenous people of NZ) and Pacific people (migrants from Pacific nations or their descendants) are the main ethnic populations identified as adversely affected by disparity. Recent legislation, policy frameworks and substantial research programmes have shaped the responses. For example the NZ Public Health and Disability Act 2000 s.22(f) directs District Health Boards to 'to reduce, with a view to eliminating, health outcome disparities between various population groups within New Zealand by developing and implementing, in consultation with the groups concerned, services and programmes designed to raise their health outcomes to those of other New Zealanders.' This illustrates two features of current policy: ambitious objectives and localized responses – localized both geographically and within the health sector (Director-General of Health, 1997b; 1997a; Ministry of Health, 2002)].

This approach has produced a series of initiatives to improve access to community health services in high deprivation areas, particularly reorganization of primary health services and development of an integrated continuum of care. Local participation in health issues has been reviewed. The City Council included improved health status among its strategic goals; previously health was regarded as the sole responsibility of central government. There was local mobilisation to retain and improve local health services, particular re-thinking the role of the community hospital as a hub for integrated care. There have been instances of 'clustering' to create local capacity across the public health spectrum. This was intended to introduce 'common cause' into the activities of health professionals and interested members of communities by creating social capital (Williams, 1997b) and offset the limiting effects of competitive tendering with government agencies and tightly defined programme delivery.

This has evolved into a proposal to create a Centre of Excellence (Blakeley, 2006) with a major goal 'to eliminate disparities in health status'. This is a substantial attempt to create the local infrastructure required to address this issue and sharpens questions about feasible strategies and system design. Communities adversely affected by disparity have not engaged with the process. The most recent report on the topic concludes current processes are 'too hard'. Energy is reduced by 'mistrust and tension', 'views that have gone unchallenged', 'communities passive in seeking advice', and the 'relentless' action of organizations (Smith, 2006). Emphasizing 'disparities' is probably not going to address those points. The comparisons and imposed indicators implicit in the term are often rejected, similar to responses to naming the issue as 'inequality' (Blaxter, 1997) or 'racism'. There is also a question of scope; for example to what extent with housing, income and employment be included in the agenda of the Centre? Are those factors to be regarded as indicators or causes?

Health disparities

Wilkinson and Marmot provide a considered explanation of health disparities that reaches similar conclusions to lay 'common sense' about causes of poor health (Cody, 1999: 57; Milstein, 2005: 17 quoting JM Mann 1999). They have concluded that the social gradient in health (at least in 'developed economies') is due to the distribution of stress created by hierarchies of dominance, social isolation and disorder. The processes that translate social relations into physical effects are probably associated with fight-flight responses. This capacity evolved when the response required was intense, short-lived, and essential for survival. Social conditions have changed over the last 15,000 years and this type of stress has become chronic, particularly for those lower in hierarchies. The effect is, in a sense, physical 'wear and tear' and is reflected in different rates of mortality and chronic disease among stratified populations. Peer relations can offset the impact of dominance to some extent. The framework adopted in the NZ study 'Decades of Disparity' (Williams, 1997a; Fawcett, *et al.*, 2006: 3) seems to adopt a similar line of argument (see Figure 1).

Both Wilkinson and Marmot reflect on the implications of their position. Their views are repeated here, partly because they are informed commentators and partly because they reflect common responses by commentators seeking to reduce inequality. In an early overview Wilkinson asserted that:

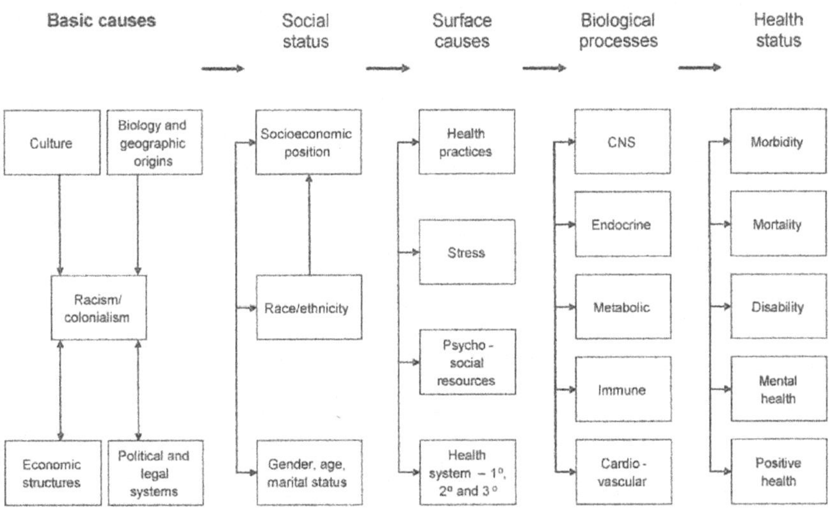

Source: Adapted from Williams 1997

Figure 1 *Health inequalities framework in 'Decades of Disparity'*

"If it is possible for some people to have death rates as low as those in upper social classes, then it should be possible to achieve equally low death rates in all groups" (Wilkinson, 1996:59).

Later Marmot takes a different position:

"All societies will have social rankings; ergo all societies will have health gradients... I ask myself if I can envisage a society where all are equal. My answer is not in real life. Hence, health gradients are inevitable." (Marmot, 2004: 25)

Further on Marmot seems to shift his position. Firstly he qualifies his comments by introducing a delay:

"There is no reason why the health of today's lowest social group should not, tomorrow, be as good as the health of today's highest group." (Marmot, 2004: 255)

But then he concludes:

"There will always be inequalities in society but the magnitude of their effects on health is within our control." (Marmot, 2004: 266)

Can sociology help structure a response to questions about the inevitability or necessity of inequalities in health?

Social theory

Dominance, isolation and disorder were identified as the main factors creating inequalities in health status among sub-populations of a society. For these purposes physical stress in taken to be a direct consequence of relatively high effort and risk.

Coleman makes some precise comments about how social capital can be introduced into this discussion. In his use of the concept:

"The power of an actor at equilibrium ... is a direct measure of the social capital available to the actor within that system" (Coleman, 1990: 315)

And when discussing public policy decisions based on the criterion of efficiency he says:

"The calculation of economic efficiency can be carried out only after a particular distribution of power or resources is taken as given ... all persons' benefits and costs are not counted equally." (Coleman, 1990: 799)

So social capital is essentially a measure of power and when 'efficiency' is a primary criterion it is likely that those with less power will be required to commit proportionally more effort or resources than those with greater power.

Lin outlines some applications of the concept. His contribution provides the basis for distinguishing between 'lateral' relations based on trust or a freely given mandate to exercise authority, and 'vertical' relations when there are structural differences in power (Lin, 2001: Chap 10).

Coleman described the value of the concept of social capital in these terms:

"Whether social capital will come to be as useful a quantitative concept in social science as are the concepts of financial capital, physical capital, and human capital remains to be seen; its current value lies primarily in its usefulness for qualitative analyses of social systems and for those quantitative analyses that employ qualitative indicators." (Coleman, 1990: 305-6)

That describes what is being attempted here. Coleman concluded that he had exhausted the potential of linear modelling of social interaction. He anticipated evolutionary modelling would address some of the limitations (Coleman, 1990: 931). Turner subsequently indicated how the central element of Coleman's theory (the development of norms) related to his general theories of societal evolution (Turner, 2003: 15). Lenski (1966, 2005) provides a link to the other major contemporary contributor to evolutionary sociology, Runciman.

Runciman seems to be the sociologist who has framed the modelling issues most succinctly. He proposes a scheme based on two modes of interaction between institutionalized roles – dominance and cooperation. Unique social configurations emerge from combinations of selection pressures and random variations. The elementary structure of the system is depicted in Figure 2.

This figure represents the three dimensions of power – economic, ideological and coercive – defined in terms of institutionalized roles (Runciman, 1999: 71). The depth of the diagram indicates the extent of inequality in the system; roles at the same level can interact laterally. This scheme seems to be generally applicable. For example local discussions are drawn to work through

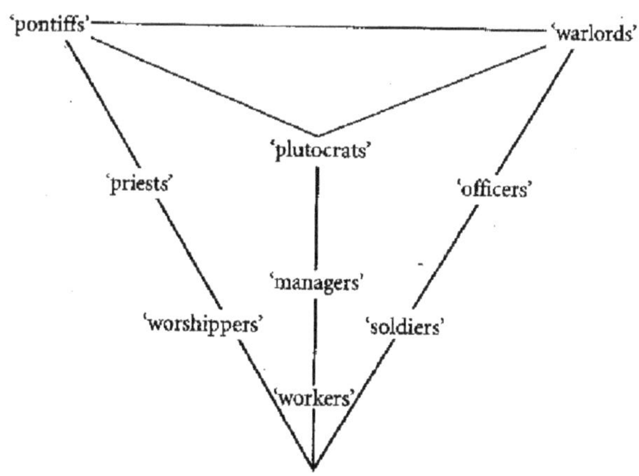

Figure 2 *Runciman's 'device' for visualizing the social space created by power (Runciman, 1999: 72)*

the interactions involved in investing, owning, employing and purchasing.

Sympathetic commentators have noted two particular problems with Runciman's approach. Anderson (1992:218-224) is not convinced that Runciman has adequately defined and applied the concept of 'selection pressure'. That problem is also a feature of the approach outlined here. (See Beinhocker, 2006, for an attempt to more specific about this process.)

Carling (2004) points to problems with conflating the orientation of role holders and the nature of the relation. Taking Carling's point Runciman's scheme has been amended here so that his 'interaction' is interpreted as the 'orientation' of a role. That implies that there are four forms of social capital: cooperation (two cooperators); conflict (two dominators); dominance (a dominator and a cooperator); and an absence (when no combination exists). This is similar to many other schemes, for example the 'boundary relations' identified by Tilly (1998) as the basis of durable inequality, and the cycles of asibya, conflict and imperial dominance that Turchin (2006) identified in the historical dynamics of empires. This framework might provide a basis for sketching a concept model.

System dynamics modelling

Sociologists seem to have made relatively little use of system dynamics models and the weight of opinion favours agent-based modelling (Halpin, 1999; Macy & Willer, 2002; Sawyer, 2003; Cederman, 2005; Gilbert & Abbott, 2005; Gilbert & Troitzsch, 2005). The main examples that informed this project are an early model by Phillips (1972) with Senge, a substantial body of work by Hanneman and colleagues (Hanneman & Collins, 1987; Hanneman, 1988; Jacobsen & Hanneman, 1992; Hanneman, 1995; Hanneman, *et al.*, 1995; Collins & Hanneman, 1998), and other work by Jacobsen (1987, 1990; 1995, 1999a, 1999b). The SD literature contains one extended argument suggesting that SD modelers made greater use of social theory, Giddens in particular (Lane, 2001a; 2001b). However there is a great deal of implicit social theory in SD models developed to address issues of public policy, economics and environmental impacts. While this makes models practical and useful to specific users it may constrain the options that can be considered and, in a situation such as this, limit the possibilities that can be considered.

Two sets of SD models are directly relevant to this project. Firstly, cumulative discussions of health service organization and health outcomes (Koelling & Schwandt, 2005) that has recently address the political dimension of system change (e.g., Hirsch, 2006; McDonnell, *et al.*, 2006). Those models are designed for 'decision makers' who are familiar with the sector. For these purposes the models are important because they introduce political factors and have a generic core that can be used with the module outlined below. The core is an 'ageing chain' depicting relative health in a population as flows through levels of those in good health, at risk, afflicted, and afflicted with complications (Homer, *et al.*, 2004; Rees, 2005; Homer, *et al.*, 2006; McLeroy, *et al.*, 2006). The single published SD model that addresses the dynamics of social capital, defined as trust (Dudley, 2004) supplements this set.

Secondly, there are the 'small' and 'concept models' provided by Richardson (Richardson, 2006). Rahn (2005) has drawn on other aspects of Richardson's work (Weaver & Richardson, 2006) to propose an SD representation of a political archetype. The archetype (see Figure 3) consists of two balancing loops oscillating around a threshold. The loops represent constituencies that seek to either lower or raise the value of the threshold.

This project is seeking to establish whether this dynamic can be applied to social structure more generally. If it can it would emulate a form of social theorizing advocated by sociologists such as Fararo (1989: 72), Mackenzie (2005:57) and Faia (1986; 2002). It also offers the possibility of reducing the complexity of numerous causal loops and intervening variables that can stifle users' responses (Woog, *et al.*, 2006).

An unanswered question at this stage is whether making greater use of the tacit knowledge of participants can offset the loss of detail. The working assumption is that people who are active in their communities develop common sense knowledge that is well grounded. Ford and Sterman (1998) provide a way to introduce this type of knowledge into SD models through structured development of non-linear functions. The wider issue of how tacit knowledge can be elicited from a wide range of participants and organized (Maani & Cavana, 2000; Woog, *et al.*, 2006) is not included in this discussion.

Application

Can this selection of concepts be drawn together to structure a systematic response to the local impacts of the issues raised by Wilkinson and Marmot? The following CLD (in Figure 4) takes the Rahn archetype and extends it to include three sources of pressure.

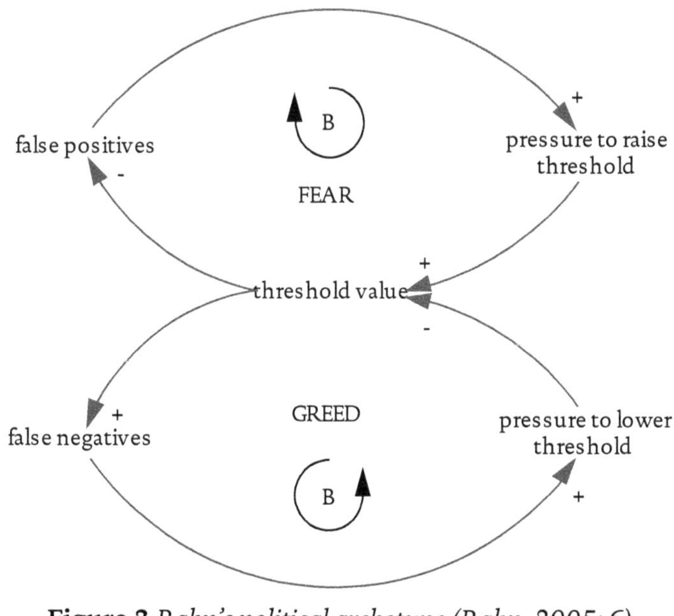

Figure 3 *Rahn's political archetype (Rahn, 2005: 6)*

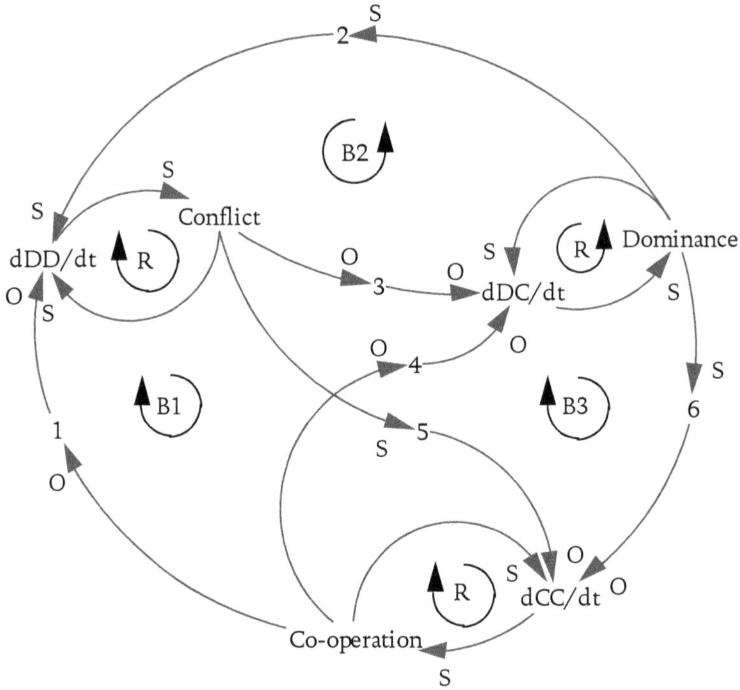

Figure 4 *The basic Causal Loop Diagram for the proposed model*

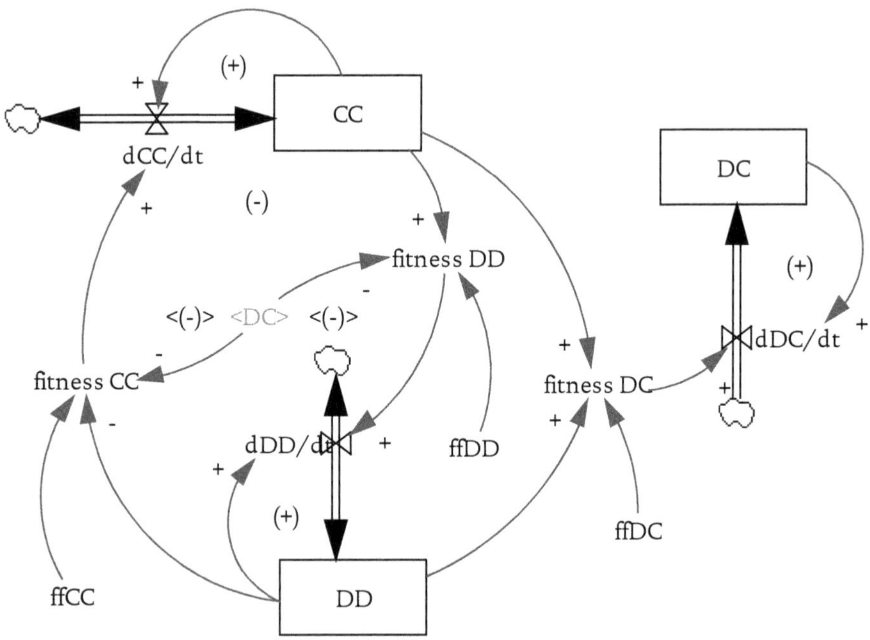

Figure 5 *The Stock-Flow Diagram for the proposed model*

If this has some merit the model could be progressively developed by refining endogenous and exogenous 'selection pressures', bearing in mind Wolstenholme's (2003) distinction between problem and solution archetypes. The stock-flow representation of that CLD model is provided in Figure 5.

The values of the variables are derived by:

- 'Mapping' main institutionalized roles (see Figure 2);
- Estimating current levels (or ratios) of interaction (initial values);
- Estimating 'fitness functions' for each type of interaction.

The illustrations here are exaggerated for effect and are produced using linear functions. The reference mode for the model is provided in Figure 6.

The interpretation is that the maximum number of interactions in the system is about 10,000 per iteration. Each iteration is the same time period, between a week and a month. The system oscillates. Current levels are a quarter of the maximum amount with equal proportions of cooperation and dominance. The system is sensitive to conflict, which is actively managed. Cooperation has a tendency to increase when conflict is low. Increasing interaction increases conflict. Dominance rises to contain and conflict and depresses the level of cooperation as well so total interaction falls.

A contrasting scenario is a 'cluster' option that maintains higher levels of interaction with less dominance and conflict is contained.

Conclusion

The discussion outlined the exploratory phase of a project that aims to use system dynamics modelling to represent the composition of social capital. Many issues have not been addressed such as concepts of causality, the varied and changing meaning of variables, and logical implications of a 'genetic' model.

The initial motivation for the study was to develop a structured approach to statements made in public discussion about population-level health disparities. Typically statements have been recorded on paper and written up for reference after the meeting. This restricts consideration of the implications and seems to lead to a premature sense of consensus or, alternatively, argument on poorly defined issues and options. System Dynamics modelling might contribute to more deliberative public discussion.

References

Anderson, P. (1992). *English Questions* London, Verso. ISBN 0 86091 591 3.

Beinhocker, E. D. (2006). *The Origin of Wealth: Evolution, Complexity, and the Radical Remaking of Economics* Boston, Harvard Business School Press. ISBN 1-57851-777-X.

Blakeley, R. (2006) *Centre of Excellence of Health for Porirua Basin: Strategic Framework* Porirua

Blaxter, M. (1997). "Whose Fault Is It? People's own conceptions of the reasons for health inequality." *Social Science and Medicine* 44(6): 747-56.

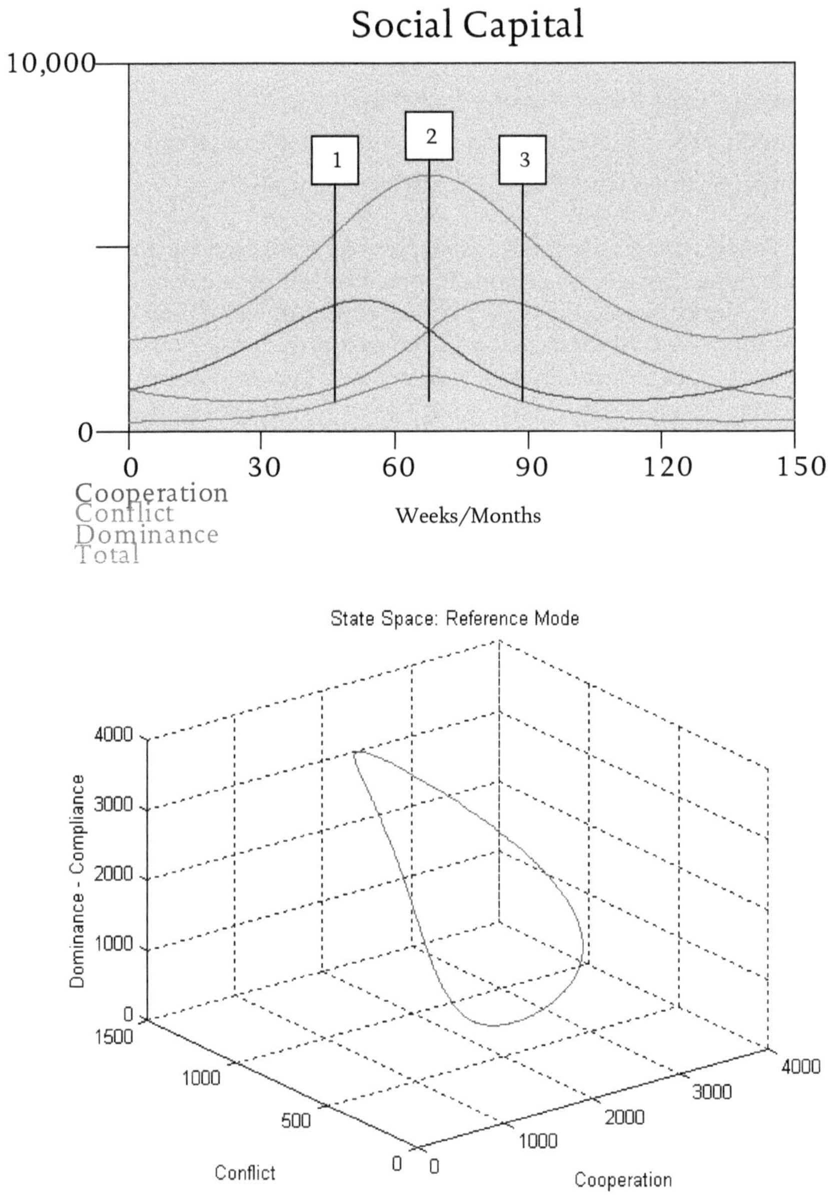

Figure 6 *Reference mode - behavior over time and state-space graphs*

Blaxter, M. (2004). *Health* Cambridge, Polity Press. ISBN 0-7456-3082-0.
Carling, A. (2004). "The Darwinian Weberian: W.G. Runciman and the Micofoundations of Historial Materialism." *Historical Materialism* 12(1): 71-95.
Cederman, L.-E. (2005). "Computational Models of Social Forms: Advancing Generative Process Theory." *American Journal of Sociology* 110(4): 864-93.
Central Regional Health Authority (1994) *Strong Links: Building better services to meet the health and disability support needs of people in Porirua* Wellington
Cody, J. R. (1999). Social Capital and Health: Community Health Activity in Porirua. Social Capital in Action. D. J. Robinson, Ed. Wellington, Institute of Policy Studies, Victoria University of Wellington.
Coleman, J. S. (1990). *Foundations of Social Theory* Cambridge, Mass., Belknap Press of Harvard University Press. ISBN 0-674-31226-0.
Collins, R. and R. Hanneman (1998). Modelling the Interaction Ritual Theory of Solidarity. The Problem of Solidarity: Theories and Models. P. Doreian and T. Fararo, Eds. Amsterdam, Gordon and Breach.
Crampton, P., C. Salmond, R. Kirkpatrick, R. Scarborough and C. Skelly (2000). *Degrees of Deprivation in New Zealand: An atlas of socioeconomic difference* Auckland, N.Z., David Bateman. ISBN 1-86953-460-3.
Director-General of Health (1997a). *Strengthening Public Health Action: The background paper* Wellington, Ministry of Health. ISBN 0-478-20818-9.
Director-General of Health (1997b). *Strengthening Public Health Action: The strategic direction to improve, promote and protect public health* Wellington, Ministry of Health. ISBN 0-478-20817-0.http//www.moh.govt.nz
Dudley, R. G. (2004). *The Dynamic Structure of Social Capital: How Interpersonal Connections Create Communitywide Benefits.* 22nd International Conference of the System Dynamics Society, Oxford, England. http://www.systemdynamics.org/conf2004/indexpapers.htm
Faia, M. A. (1986). *Dynamic functionalism: strategy and tactics* Cambridge, Cambridge University Press. ISBN 0 521 32657 5.
Faia, M. A. (2002). "Differential Equation Modeling as a Source of Theoretical Insight: Four Disparate Examples." *Quality and Quantity* 36: 169-95.
Fararo, T. J. (1989). *The meaning of general theoretical sociology: Tradition and formalization* Cambridge, Cambridge University Press. ISBN 0-521-27258-5.
Fawcett, J., T. Blakely, B. Robson, M. Tobias, R. Harris and N. PakiPaki (2006) *Decades of Disparity III: Ethnic and socioeconomic inequalities in mortality, New Zealand 1981–1999* Wellington http://www.moh.govt.nz
Ford, D. N. and J. D. Sterman (1998). "Expert knowledge elicitation to improve formal and mental models." *System Dynamics Review* 14(4): 309-40.
Gilbert, N. and A. Abbott (2005). "Introduction." *American Journal of Sociology* 110(4): 859-63.
Gilbert, N. and K. G. Troitzsch (2005). *Simulation for the Social Scientist* Maidenhead, Open University Press. ISBN 10 0 335 21600 5.
Gould, J. (1992) *Porirua City Council Health Report* Porirua
Halpin, B. (1999). "Simulation in Sociology." *The American Behavioural Scientist* 42(10): 1488-508.
Hanneman, R. and R. Collins (1987). A Dynamic Simulation of Marx's Model of Capitalism. The Marx-Weber Debate. N. Wiley, Ed. Newbury Park California, Sage.
Hanneman, R. A. (1988). *Computer-assisted Theory Building: modeling dynamic social systems* Newbury Park, Cal, Sage. ISBN
Hanneman, R. A. (1995). "Simulation Modeling and Theoretical Analysis in Sociology." *Sociological Perspectives* 38(4): 457-62.
Hanneman, R. A., R. Collins and G. Mordt (1995). "Discovering Theory Dynamics by

Computer Simulation: Experiments on State Legitimacy and Imperialist Capitalism." *Sociological Methodology* 25: 1-46.
Hirsch, G. B. (2006). *Expanding Health Coverage and Access for the Uninsured: A Model of Common Factors in the Experience of Several States.* 24th International System Dynamics Conference, Nijmegen, The Netherlands, System Dynamics Society. http://www.albany.edu/cpr/sds/conf2006/proceed/index.htm
Homer, J., G. B. Hirsch, M. Minniti and M. Pierson (2004). "Models for collaboration: how system dynamics helped a community organize cost-effective care for chronic illness." *System Dynamics Review* 20(3): 199-222.
Homer, J., B. Milstein, W. Dietz, D. Buchner and E. Majestic (2006). *Obesity Population Dynamics: Exploring Historical Growth and Plausible Futures in the U.S.* 24th International System Dynamics Conference, Nijmegen, The Netherlands, System Dynamics Society. http://www.albany.edu/cpr/sds/conf2006/proceed/index.htm
Jacobsen, C. (1999a). "The Process of Crescive Legitimation: Theory, Simulation Model, and Three Empirical Tests." *Adaptive Behaviour* 7(3/4): 255-68.
Jacobsen, C. (1999b). Social Dynamics: Theory and Research. Technion
Jacobsen, C. and R. Bronson (1987). "Defining sociological concepts as variables for system dynamics modeling." *System Dynamics Review* 3(1): 1-7.
Jacobsen, C. and R. Bronson (1995). "Computer Simulations and Empirical Testing of Sociological Theory." *Sociological Methods and Research* 23(4): 479-506.
Jacobsen, C., R. Bronson and D. Vekstein (1990). "A Strategy for Testing the Empirical Adequacy of Macro-Sociological Theories." *Journal of Mathematical Sociology* 15(2): 137-48.
Jacobsen, C. and R. Hanneman (1992). "Illegal Drugs: Past, Present and Possible Futures." *The Journal of Drug Issues* 22(1): 105-20.
Koelling, P. and M. J. Schwandt (2005). *Health Systems: A Dynamic System-Benefits from System Dynamics.* 2005 Winter Simulation Conference.
Lane, D. C. (2001a). "Rerum cognoscere causas: Part I— How do the ideas of system dynamics relate to traditional social theories and the voluntarism/determinism debate?" *System Dynamics Review* 17(2): 97-118.
Lane, D. C. (2001b). "Rerum cognoscere causas: Part II— Opportunities generated by the agency/structure debate and suggestions for clarifying the social theoretic position of system dynamics." *System Dynamics Review* 17(4): 293-309.
Lenski, G. (2005). *Ecological-Evolutionary Theory: principles and applications* Boulder, Paradigm. ISBN 1-59451-100-4.
Lenski, G. E. (1966). *Power and Privilege: A Theory of Social Stratification* New York, McGraw-Hill Book Company. ISBN
Lin, N. (2001). *Social Capital: A Theory of Social Structure and Action.19.* Cambridge, England, Cambridge University Press. ISBN 0-521-47431-0.
Maani, K. E. and R. Y. Cavana (2000). *Systems Thinking and Modelling: Understanding Change and Complexity* Auckland, Prentice Hall. ISBN 1-877258-00-8.
Mackenzie, A. (2005). "The Problem of the Attractor: A Singular Generality between Sciences and Social Theory." *Theory, Culture and Society* 22(5): 45-65.
Macy, M. W. and R. Willer (2002). "From Factors to Actors: Computational Sociology and Agent-Based Modeling." *Annual Review of Sociology* 28: 143-66.
Marmot, M. (2004). *Status Syndrome: How Your Social Standing Directly Affects Your Health and Life Expectancy* London, Bloomsbury Publishing. ISBN 0 7475 7049 3.
McDonnell, G., J. Dewdney and S. Wilson (2006). *Exploring the Political and Economic Dynamics of Health Policy.* 24th International System Dynamics Conference, Nijmegen, The Netherlands, System Dynamics Society.
McLeroy, K., S. J. Leischow and B. Milstein, Eds. (2006). American Journal of Public Health.

Special Issue on Systems Models and Systems Thinking. ISBN
Milstein, B. (2005). Selected Concepts in Health System Dynamics
Ministry of Health (2002) *Reducing Inequalities in Health* Wellington
National Research Bureau Ltd (1994) *Assessment of Health Needs in the Porirua Area* Porirua
Phillips, B. S. (1972). *Worlds of the Future: Exercises in the Sociological Imagination* Columbus, Charles E. Merrill Publishing Company. ISBN 0-675-09085-7.
Porirua City Council (1999) *A new hospital for the Wellington region at Kenepuru* Porirua
Porirua Kapiti Healthlinks Project (2000). *Porirua City Health and Disability Report and Plan* Wellington, Ministry of Health. ISBN 0-478-23981-5.
Rahn, R. J. (2005). *FEAR & GREED - A Political Archetype.* 23rd International System Dynamics Conference, Boston, System Dynamics Society. http://www.albany.edu/cpr/sds/conf2005/proceed/index.htm
Rees, D. (2005). *System Dynamics Modeling as a Tool in Healthcare Planning.* 11th Australia New Zealand Systems Society Conference, Christchurch, New Zealand, ANZSYS. http://isce.edu/ISCE_Group_Site/web-content/ISCE%20Events/Christchurch_2005/Papers/Rees.pdf
Reinken, J., A. de Lacey and C. Salmond (1980) *Community Attitudes to Sickness and Health: Stimulus and Response* Wellington
Richardson, G. (2006). *Concept Models.* 24th International System Dynamics Conference, Nijmegen, The Netherlands, System Dynamics Society. http://www.albany.edu/cpr/sds/conf2006/proceed/index.htm
Runciman, W. G. (1989). *A treatise on social theory.Volume II: Substantive social theory.* Cambridge, Cambridge University Press. ISBN 0 521 36983 5.
Runciman, W. G. (1999). *The Social Animal* London, Fontana. ISBN 0 00 686382 5.
Salmond, G. C. (1975) *Maternal and Infant Care in Wellington: A health care consumer study* Wellington
Sawyer, R. K. (2003). "Artificial Societies: Multiagent Systems and the Micro-Macro Link in Sociological Theory." *Sociological Methods and Research* 31(3): 325-63.
Smith, L. (2006) *Porirua Healthlinks Communityy Voices Annual Report: 1 July 2005 to 30 June 2006* Porirua http://www.phics.org.nz/
Tilly, C. (1998). *Durable Inequality* Berkeley, University of California Press. ISBN 0-520-21171-5.
Turchin, P. (2006). *War and Peace and War: The Life Cycles of Imperial Nations* New York, Pi Press. ISBN 0-13-149996-3.
Turner, J. H. (2003). *Human Institutions: A Theory of Societal Evolution* Lanham, Rowman & Littlefield. ISBN 0-7425-2559-7.
Weaver, E. A. and G. Richardson (2006). "Threshold setting and the cycling of a decision threshold." *System Dynamics Review* 22(1): 1-26.
Wilkinson, R. G. (1996). *Unhealthy Societies: The Afflictions of Inequality*, Routledge. ISBN
Wilkinson, R. G. (2000). *Mind the Gap: Hierarchies, Health and Human Evolution* London, Weidenfeld & Nicolson. ISBN 0 297 64648 6.
Wilkinson, R. G. (2005). *The Impact of Inequality: How to Make Sick Societies Healthier* New York, The New Press. ISBN 1-56584-925-6.
Williams, D. R. (1997a). "Race and Health: basic questions, emerging directions." *Annals of Epidemiology* 7: 322-33.
Williams, I. F. (1997b). Choral Societies and Export Growth. Social Capital and Policy Development. D. J. Robinson, Ed. Wellington, Institute of Policy Studies, Victoria University of Wellington.
Wolstenholme, E. F. (2003). "Towards the definition and use of a core set of archetypal

structures in system dynamics." *System Dynamics Review* 19(1): 7 - 26.
Woog, R., R. Y. Cavana, R. Roberts and R. Packham (2006). "Working at the Interface between Systems and Complexity Thinking: Insights from a Market Access Design Project for Poor Livestock Producers." *Systems Research and Behavioural Science.*\

John Cody is a social worker. He has been involved with child care and protection services, community health programmes, community organization and research on income adequacy. His research interests relate to social stratification and processes of public deliberation. He has a B.A. (Sociology) and Dip. Soc. Wk from Victoria University.

Bob Cavana is a Reader in Systems & Logistics with the Victoria Management School at the Victoria University of Wellington, New Zealand. He holds a PhD in management science (system dynamics) from the University of Bradford, England and is a Chartered Member of The Institute for Logistics & Transport in NZ. He has coauthored two books: *Systems Thinking and Modelling: Understanding Change and Complexity* (Pearson Education NZ, 2000) and *Applied Business Research: Qualitative & Quantitative Methods* (Wiley, Brisbane, 2001; Chinese edition 2004). He is currently the Vice President at Large of the System Dynamics Society and a Managing Editor of *System Dynamics Review.*

Appendix 1: The Porirua Basin showing relative deprivation for small areas

THEME FIVE
Social Sustainability and Community

Wendy Gregory, Environmental Science & Research
Jack W. Meek, University of La Verne

While the theme 'Social sustainability and community' embraces a broad area of interest, we were seeking papers for this conference from two areas that would add value to a discussion in the field of social sustainability and community. First, there are new sustainability domains emerging around environmental and social sustainability, fuelled to some extent by the development of concepts like the Triple Bottom Line (or 'Four Well-beings', as we have in New Zealand, incorporating cultural sustainability along with environmental, social and economic sustainabilities). Some authors argue that these are 'mutually exclusive' domains, but we see them as interrelated: we argue that social sustainability is an ecosystems problem, and ecosystems sustainability is a social problem. Second, we were interested in the broader area of research and discussion as to how communities seek sustainable relations within governance structures – in urban, suburban and rural contexts – in the effort to enhance participation and community.

The four papers presented here offer different approaches to and concerns about 'community and social sustainability'. Two offer significant empirical evidence on trends that effect community and social sustainability (one on contextual learning and one on coastal wealth patterns). Two offer significant advances in framing philosophical approaches to community and social sustainability.

In each of the four papers, sustainability is treated with care and concern, but each teases out the challenges toward achieving sustainability and community. The question of whether communities that are seeking to sustain themselves and their settings, either economically, culturally, or socially, should take account of the possible environmental impacts of their activities is one that all of the authors in this stream have tackled in different ways. We will start with a paper reflecting on some locally based community projects, followed by a paper that uses financial analysis to investigate where costs associated with environmental amenity should be located, then move on to a paper which looks at a more global conflict and asks whether there is scope for developing new ways of framing issues to overcome the conflict between 'the West and the rest'. Finally, the last paper is concerned with investigating whether some of the emerging concepts in social sustainability (for example, those of social capital) really help us to challenge norms or whether they contribute to the perpetuation of the status quo.

The paper by Bronwen Maxwell and Cheryl Lin-Yo, "The Value of Community Participation," summarizes a number of projects based in the Greater Blue Mountains Heritage area. These projects have served to bring together people from diverse communities in order to address environmental issues through process of cultural appreciation and development. Using various art forms, ranging from story telling and oral histories through to films, music and visual arts, community members were empowered to consider the role of culture and cultural differences in informing environmental decision-making. As a side product of all the projects that Cheryle Lin-Yo describes in her section of the paper, the development of civic or social capital and the development of greater environmental awareness were achieved. The section of the paper that Bronwin covers addresses a schools based project (Earth Journeys) that is designed to link schools in the local heritage area with other schools 'on the doorstep'. This is intended to ensure that learning about local environmental issues is able to "ripple out" to the wider community. As a whole the paper provides evidence that community and sustainability are powerful concepts that can be learned. The paper draws lessons from two programs – the Community Cultural Development program (CCD) and 'Earth Journey' – that are illustrative examples of human learning potential. Based on the logics of interdisciplinarity and contextuality, the protocols outlined in both CCD and Earth Journeys offer evidence that these learning protocols are transformative. Indeed, Bronwen Maxwell and Cheryl Lin-Yo have answer the question regarding educational strategies that are based on emergent patterns that recognize and honor complexity: they work!

The second paper by Jeff Pompe, "Residential Property Values in Gated and Nongated Barrier Island Communities," focuses on the free market, and observes that there has been a developing phenomena of gated communities, many of which are island or coastally based. The paper highlights how those who benefit from the access to aesthetic value derived from a pristine environment should also pay the costs of securing the coastal surrounds that they enjoy. Looking at the role of private neighborhood associations in sustaining the natural ecosystems, the author shows how premiums for gated communities could assist with this. Although one can disagree with the rightness of such communities being able to externalize negative impacts normally felt in other communities (e.g., free-riders), there is a value in the analytical work undertaken by the author to highlight and inform the fairer distribution of costs for costal protection by allocating these to those who benefit from the enhanced natural environment. After a valuable review of the implications of PNAs, including inequities in housing and land access, the paper drills down on home prices with a model that reveals inequities in wealth accumulation. The implications of these findings for shared contributions to sustaining common shorelines and for sustaining and integrating communities are apparent but emerge in a more hidden and evolutionary form.

Arnis Vilks, in "Sustaining the World Against Religious Maniacs and a Selfish Superpower," offers a warning regarding the global logics lurking behind western leadership armed with logical positivism and a mistaken and misattributed love affair with economic liberalism. His central concern is on sustaining the world and he sees such hope lost on this generation of Western leaders hopelessly enamored with ideas that have lost both their meaning and insight for human affairs. Focusing on a global conflict that all of us are aware of - that between the Muslim world and the West as led by Bush - Vilks looks for the roots of this conflict in philosophical and economic tenets, suggesting that efforts to enhance and sustain dialogue between the two communities is an imperative for those concerned with avoiding future global disasters. Pointing out many of the inequalities and inequities between the 'West and the rest', Arnis highlights the decline in 'alms' from its post-war peak in the US at nearly 3% to the current figure of less than 0.15% of GDP. Further, Arnis points out that neither religion, nor arts, nor ethics could be deduced from observation plus mathematics (the empirical and analytical sciences). Hence a new language is required that will supercede both Logical Positivism and the Economic Liberalism that lead to a focus on 'national interests' rather than concern for others.

Elaine Hartley, in her paper, "A Complex Systems Perspective on Incommensurable Evaluations Within Triple-Bottom-Line Reporting," picks up this theme of looking at incommensurable positions, pointing out that these are common in Triple-Bottom Line (TBL) Reporting. Hartley similarly believes that the media has played a role in shaping people's attitudes about issues of concern in which they have no direct experiential understanding. Like Vilks, Hartley points to the need for concern for others, highlighted in Kofi Annan's 2005 address to the UN. The paper discusses the difficulties associated with TBL reporting - for example, that the indicators used for social, economic and environmental assessments can not easily be aggregated, leading to people formulating preferences in a relative vacuum despite the abundance of information available. Hartley questions whether mechanisms like TBL reporting can genuinely regulate socio-environmental relations, or whether they provide the illusion of progress without disturbing the status quo. Focusing on how economic liberalism shapes our framing of the sustainability issue, leads Hartley to conclude that the notions of social and natural capital contribute to sustaining rather than reformulating the current situation. Hartley raises concern about the almost blind adoptions of 'abstract' logics like TBL that are derived from mental states that are far separated from the more meaningful contextually based (post formal-cognitive) learning possibilities addressed in the paper by Maxwell and Yin-Lo. As a result, these abstract logics are misguided and are likely to fail.

All four papers thus have something to say on the question of whether and how social sustainability is commensurable with environmental sustainability. A number of complex issues were raised in these papers in relation to this theme:

- How can post-formal cognitive approaches be employed for the many? (CCD and Earth Journey focus on limited numbers and CCD on the youth);
- How do we know these lessons will be sustained (over time)? And not be subjected to market and other influences;

- How are we to reduce the inequalities that arise from the privatization of space?

Residential Property Values in Gated and Nongated Barrier Island Communities

J. Pompe
Francis Marion University, USA

We examine how location in a gated community affects the value of single-family homes and whether beach valuation varies between communities. Using data on 2,358 sales on barrier islands near Charleston, South Carolina, the empirical results indicate that in a gated community, homeowners pay a premium of 37 percent for a house. In gated communities, a beach creates greater value for houses not on the oceanfront, relative to houses in nongated communities.

Introduction

In the past three decades, the private neighborhood association (PNA) has become the predominant legal form of housing developments in many of the United State's fastest growing areas[1]. Members of PNAs use explicit contracts that allow the community to make many of the decisions that local governments generally make. Among other activities, PNAs regulate property use, tax members to provide for local services, and collectively own shared areas, such as beaches. PNAs are popular in many countries throughout North and South America, Europe, Africa, and Asia (Webster, *et al.*, 2002). In the United States, PNAs govern four-fifths of new housing developments and approximately 18 percent of the nation's population live in one of the 274,000 housing developments (Community Association Institute, 2005). Approximately 46,000 PNAs are gated communities (GCs), which are communities that restrict access to residents and their guests by building walls around the development[2].

A real estate developer would design a GC, which would be more costly to construct, if doing so would enhance revenues sufficiently. Some suggest that developers can earn more revenue in a gated community because housing is denser (McKenzie, 1998). In addition, if property owners are willing to pay a premium for a house in a gated community, revenues will be higher. Residents may be willing to pay a premium for property in gated communities for several reasons. Homebuyers may prefer voluntary, private contracts to zoning, which use government policing power rather than contract law to enforce land controls. In addition, covenants may offer better protection for a homeowner's investment by controlling negative externalities more effectively. Creating a gated community may lead to greater homogeneity among members which would lower the costs of controlling negative externalities. This self-selection

1 We follow Robert Nelson's (2005) use of the term "private neighborhood association," although other terms such as "common interest developments" are often used.

2 In Australia, as many as 40,000 people live in gated communities, and the number is growing exponentially (Burke, 2001).

may lower the costs of free-riders, which would lead to outcomes that are more efficient. Thus, property values in gated communities may be higher because public goods such as open space may be better protected than in nongated communities (Pompe & Rinehart, 1997).

Some empirical analysis of a gated community's effect on housing value is available. Blakely and Snyder (1997) suggest that gates have no influence on property values, although they provide little empirical analysis. They suggest that any benefits created by gates may be reduced because GC members suffer from intrusive covenants and gates may signal an area with crime problems. Bible and Hsieh (2001) find that gates increase property value by approximately six percent in Shreveport, Louisiana neighborhoods. They believe that reduced traffic, increased prestige, and perceived greater safety create higher values in GCs. LaCour-Little and Malpezzi (2001) find that in a St. Louis neighborhood, houses in gated communities command a 26 percent premium over houses on unrestricted streets; they estimate that 17 percent of the premium is for the homeowner's association (HOA) and 9 percent for the gate.

In this study, we examine the effect that gates have on property values and factors that may influence a property owner's valuation of the beach.

Study area

The population in South Carolina's coastal counties, which has increased dramatically in recent decades, is now 1 million people. Housing prices have risen considerably in coastal areas, where four of the five fastest growing counties in the state are located. The median sale price of a coastal home increased by 25 percent from 2004 to 2005, compared to a state and national average of 10 and 6 percent, respectively (Young, 2005). We examine the housing markets for Dewees Island, Isle of Palms, Sullivan's Island, Folly Beach, Kiawah Island, and Seabrook Island, which are six barrier islands in the Charleston area. Barrier islands are elongated landforms that parallel, but are separate, from the mainland.

We examine four gated – Dewees Island, Wild Dunes, Kiawah Island, Seabrook Island – and three nongated - Isle of Palms, Sullivan's Island, Folly Beach - coastal communities. Development on the four gated communities began in 1991, 1975, 1976, and 1970, respectively. For three of the gated communities, (Dewees, Kiawah, and Seabrook), the community is the complete island. Wild Dunes is the only gated community on the Isle of Palms. The communities are distinct from each other geographically, because the islands separate each community, except for Wild Dunes and Isle of Palms, which are separated by the security gate. A bridge or causeway connects each island in this study to the mainland, except for Dewees Island, which has a single, community boat dock as the only entrance to the island. Each community is within 30 miles of Charleston, although the three nongated communities, which began development in the 19th century, are closer to the city than the gated communities are. Dewees Island is 12 miles northeast of Charleston, but because one must take a boat to the Isle of Palms, the length of travel time to Charleston is approximately 1 hour.

Each community has characteristics that make it unique. For example, Dewees Island, in order to minimize damage to the natural environment, enforces extensive restrictions on development such as allowing construction on only 35 percent of the island and not allowing automobiles. One difference between the gated and nongated communities is the age distribution in the community. Approximately 80 percent of the residents in Seabrook and Kiawah are over the age of sixty. In the Isle of Palms, Sullivan's Island, and Folly Beach, approximately 27 percent of the residents are over sixty. However, many demographic characteristics such as income, education, and property values are similar for the communities. Residents in each community, where all houses are within two miles of the shore, enjoy a "beach town" lifestyle. Residents' "lifestyle" describes an important feature of each community in our study and creates a homogeneous housing stock[3].

Coastal homeowners, who may be exposed to shoreline erosion, value wider beaches for their protection and recreational benefits (Pompe and Rinehart, 1999). In order to measure shoreline changes, South Carolina monitors 400 coastal survey markers, including 103 for the islands in this sample. Some sections of the six islands' shorelines have suffered severe erosion in the past five years[4]. For example, on northern end of the Isle of Palms, 150 feet of shoreline has been lost in recent years (State of the Beach, 2005). Although some areas of the barrier islands' shorelines experience severe erosion, the majority of the shoreline in this study is stable and some is accreting.

We obtained the sale price of a house and numerous other characteristics from the multiple listing service that is available to realtors. Our sample consists of 2,378 single-family homes that sold on the six barrier islands between January 2002 and December 2005. Although the average price of a gated house is 18 percent higher than a nongated house, the average price of a Sullivan's Island house is 83 percent higher than the average Seabrook house. Indeed, houses on Sullivan's Island, which are among the most expensive second-home real estate in the country, are on par with houses in Aspen, Nantucket, and Palm Beach (Higgins 2005). Although such comparisons do not adjust for numerous characteristics such as distance to the beach, our sample does not support the common perception that houses in gated communities are more expensive than nongated.

---

3 Blakely and Snyder (1997) define three categories of gated communities: "lifestyle," where leisure activities such as golf are important; "elite," where prestige is a prime motivator; and "security," where fear of crime is the prime motivator. It is unlikely that residents value the gated communities in our sample for increased safety because crime rates are low on the six barrier islands.

4 For our sample period, survey markers near erosive areas include: Wild Dunes 3178, 3180; Isle of Palms 3167, 3175, 3180; Sullivan's Island 3085, 3090, 3092; Folly Beach 2860, 2863, 2865, 2867 2873, 2880, 2883, 2885, 2890. Shorelines are very dynamic and current erosion patterns can vary from long-run patterns. Five observations are erosive oceanfront and gated.

Hedonic model and empirical results

In order to determine the marginal contribution that a gate and other characteristics have on property values, we use hedonic regression analysis. Hedonic theory explains that because a house is a bundle of characteristics such as location and size, implicit prices of the individual characteristics can be estimated by multiple regression (Rosen, 1974). The hedonic model for the study is:

$$Pi = f(Hi, Ni, Ci)$$

where *Pi* is the residential sale price of the *ith* property, H*i* is a vector of structural characteristics associated with the house (e.g., square footage), N*i* is a vector of neighborhood/location variables (e.g., waterfront location), C*i* is a vector of community variables (e.g., gated development).

We define the variables in Table 1 and summary descriptive statistics are in Table 2. In the semi-log functional form of the housing price function, which is reported in Table 3, most variables are significant and of expected sign. Of interest for our study is the variable indicating location in a GC (GATE), which estimates the value that homeowners place on locating in a gated community, holding other variables constant. The coefficient on GATE, which is significant, indicates that homeowners pay a 36.7 percent premium for a house in a GC[5]. Although this result does not explain why gates increase the property values, interaction variables combining location and beach quality provide some clues.

Beaches in the GCs are less crowded than public beaches because property owners and people renting housing are the only ones who have access to the beach in a GC. We use several interaction variables to determine if the increased privacy on GC beaches affects property values. GATEDBCH, which is an interaction variable between the distance that a house is from the beach (DBCH) and GATE, indicates that house farther from the beach are lower priced in gated communities, as would be expected. However, the distance from the beach decreases property value in GCs by a lesser amount than in nongated (NONGTDBCH) communities. We hypothesize that a GC property owner places greater value on locations near the beach because the GC restricts access, which decreases beach congestion and creates a more pleasant beach experience.

GTOCNFR and NONGTOCNFR, which indicate the value of an oceanfront location in a gated and nongated community, respectively, are positive and significant. For property in a GC, location on the oceanfront adds 113 percent to house value; for nongated communities oceanfront location adds 132 percent to sale price. A prime location, such as a GC oceanfront, may be lower value, relative to nongated, because the beach is more private and less congested in a GC[6]. Therefore, prime locations in GC are not as valuable as locations in

5 When the dependent variable is in log form, a dummy variable must be adjusted by $100^*(e^{b1}-1)$ percent, where b1 is the coefficient of the dummy variable (Kennedy, 1981).

6 In a separate equation, the interaction variables entered were GTOCNFR, GTDBCH, and GTSQFT. In this equation, GTOCNFR was negative and GTDBCH and GTSQFT were positive. This supports the above results that in a GC, oceanfront property and non-oceanfront property are lower and higher valued, respectively, than similar

Variable	Definition
PRICE	Sale price of house
DI	Dummy variable: 1 if Dewees Island, 0 otherwise
WD	Dummy variable: 1 if Wild Dunes, 0 otherwise
IOP	Dummy variable: 1 if Isle of Palms, 0 otherwise
SI	Dummy variable: 1 if Sullivan's Island, 0 otherwise
FB	Dummy variable: 1 if Folly Beach, 0 otherwise
KI	Dummy variable: 1 if Kiawah Island, 0 otherwise
SEA	Dummy variable: 1 if Seabrook Island, 0 otherwise
GATE	Dummy variable: 1 if gated, 0 otherwise
MRSH	Marsh location
WTVW	Waterview
WTRFRNT	Waterfront location
SQFT	Square feet of house
AGE	Age of house (years)
AGESQUARE	Age of house squared
YR(TWO)	House sold in (2002), 2003, 2004, or 2005
EROS	Located near severe erosion
DBCH	Distance from beach (miles)
EROSOCNF	Severe erosion near oceanfront house
EROSDBCH	Distance from severe erosion
GTOCNFRNT	Oceanfront location in GC
NONGTOCNFRNT	Oceanfront location in nongated community
GTDBCH	Distance from beach for gated property
NONGTDBCH	Distance from beach for nongated property

Table 1 *Variable definitions*

nongated locations.

We include interaction variables for proximity to a severely erosive section of beach for oceanfront location (EROSOCNFRNT) and for properties that are near the erosive section but not on the oceanfront (EROSDBCH). EROSDBCH is significant and negative, but EROSDBCH is not significant. This suggests that although severe erosion is of concern to oceanfront property-owners, those located farther away do not discount house value because of a nearby erosive shoreline. This result suggests that because oceanfront owners would value shoreline protection more than those farther back from the ocean, oceanfront residents should pay more for shoreline protection.

We include a dummy variable for each community, except for Folly Beach and Dewees Island, which are omitted and are reflected in the constant term. The community variables show that residents prefer some communities to others. Houses on SI enjoy a premium over FB houses and houses on SEA are discounted relative to DI, for example. Most other variables are significant and perform as would be expected. The location variables WTFRNT, WTVW, and MARSH, are each positive and significant. WTVW adds 37 percent to housing values, for example. SQFT, the amount of square footage, is significant and pos-

properties in nongated communities.

Variable	Mean	Std.Dev.	Minimum	Maximum
PRICE	899394.626	701910.220	103492.000	7500000.00
DI	.424088210E-02	.649976051E-01	.000000000	1.00000000
WD	.191263783	.393379683	.000000000	1.00000000
IOP	.190839695	.393046334	.000000000	1.00000000
SI	.720949958E-01	.258700000	.000000000	1.00000000
FB	.144189992	.351356788	.000000000	1.00000000
KI	.249787956	.432982030	.000000000	1.00000000
SEA	.147582697	.354761072	.000000000	1.00000000
GATE	.592875318	.491402669	.000000000	1.00000000
NONGATE	.407124682	.491402669	.000000000	1.00000000
MRSH	.729431722E-01	.260098358	.000000000	1.00000000
WTVW	.780322307E-01	.268279192	.000000000	1.00000000
WTFRNT	.377438507E-01	.190616530	.000000000	1.00000000
OCNFR	.699745547E-01	.255158239	.000000000	1.00000000
YRTWO	.198897371	.399255317	.000000000	1.00000000
YRTHRE	.234944869	.424054286	.000000000	1.00000000
YRFOUR	.292196777	.454868736	.000000000	1.00000000
YRFIVE	.273960984	.446083796	.000000000	1.00000000
SQFT	2373.40797	1013.31669	200.000000	7189.00000
AGE	22.8235793	19.8938913	.000000000	179.000000
EROS	.958439355E-01	.294439538	.000000000	1.00000000
DBCH	.340567396	.329318665	.000000000	2.00000000
AGESQ	916.514843	1944.43676	.000000000	32041.0000
EROSOCNF	.106022053E-01	.102441442	.000000000	1.00000000
EROSDBCH	.168023749E-01	.753408037E-01	.000000000	.500000000
GATEDBCH	.262925785	.330842085	.000000000	1.71000000
NONGTDBC	.835703647E-01	.209963868	.000000000	2.00000000
GATEOCNF	.220525869E-01	.146885739	.000000000	1.00000000
NONGTOCN	.479219678E-01	.213646461	.000000000	1.00000000

Table 2 *Descriptive statistics*

itive. We include AGE and AGESQ because the age of a house may affect house prices nonlinearly. We include a dummy variable for year of sale (YEARTHREE, etc.) to control for price inflation and changes in the housing market. The base year is 2002.

Conclusion

In this study, we examined factors that affect property values in three nongated and four gated communities on barrier islands near Charleston, South Carolina. Of particular interest was the impact GCs and beaches have on property value. We find that after controlling for other factors, a house in a GC will have a 37 percent higher value, than a nongated community house. For our sample, if the average nongated house, which sold for $810,779, were located in a GC, the house would sell for $1,110,767. The premium for a GC house is significantly higher in this study than the two studies noted earlier, which showed premiums of 6 and 26 percent. We expect that a primary reason that GC homeowners are willing to pay more for a house is that voluntary agreements in GCs reduce the free-rider problem and homeowner investment risk. In addition, we find that residents in GCs place higher value on proximity to the

Variable	Coefficient	t-statistic
Constant	12.47713890	301.885
WD	-.1796550614	-1.523
IOP	.2004762588	7.221
SI	.3223665945	8.294
KI	-.1923151052E-01	-.163
SEA	-.4460133802	-3.698
GATE	.3893558172	3.252
MRSH	.1792233789	5.795
WTVW	.3139982345	10.338
WTFRNT	.4311908627	10.403
YRTHRE	.9567326836E-01	4.143
YRFOUR	.2231158622	10.080
YRFIVE	.4755082477	21.069
SQFT	.3103641310E-03	33.013
AGE	-.9911530802E-02	-9.181
AGESQ	.7656542885E-04	7.379
EROSOCNF	-.1605600942	-1.986
EROSDBCH	.3143732757E-01	.283
GATEDBCH	-.2232049540	-5.696
NONGTDBC	-.3054456534	-6.869
GATEOCNF	.7570403512	13.416
NONGTOCN	.8401226867	20.252

Observations=2358, Adjusted R-squared=.68

Table 3 *House price function estimates*

beach than residents in nongated communities. The results suggest that less congested beaches add to a GC's value.

Attention on PNAs and GCs has been increasing recent years. For example, the entire March 2005 issue of *Housing Studies* was devoted to gated communities, and international conferences have been devoted to the issue. Our study provides information about GCs and coastal communities that may be beneficial to policy makers as they deal with the problems associated with growing populations along shorelines. For example, the National Oceanic and Atmospheric Administration website that discusses "Alternatives for Coastal Development," suggests that in order to achieve "smart growth," community development should be "fair and cost-effective." Beach renourishment can be a continual maintenance cost for coastal communities. Our results suggest that an equitable distribution of shoreline protection costs would place higher payments on oceanfront property owners, both for gated and nongated communities. Using the model discussed above could create a fairer system to pay for beach nourishment by allocating the costs of nourishment to those who receive the benefit.

References

Bible and Hsieh, (2001) "Gated Communities and Residential Property Values," *Appraisal Journal*, (April): 140-146.

Blakely, Edward and Mary Gail Snyder. (1997) *Fortress America.* Washington, DC: Brookings Institution.

Burke, Matthew. (2001) "The Pedestrian Behavior of Residents in Gated Communities," Australia: Walking the 21st Century, Perth, Western Australia, February 20-22.

Community Association Institute. (2005) Data on U.S. Community Associations. http://www.caionline.org/about/facts.cfm.

Higgins, Michelle. (2005) "Prices on Sullivan's Island are among the least laid-back in America," The State, Columbia, South Carolina, September 18, p. G8.

Kennedy, Peter. (1981) Estimation with Correctly Interpreted Dummy Variables in Semilogarithmic Equations" *American Economic Review* v. 71, 802.

LaCour-Little, Michael and Stephen Malpezzi. (2001) Gated communities and property values. Wisconsin-Madison CULER working paper.

Nelson, Robert H.. (2005) *Private Neighborhoods and the Transformation of Local Government.* Washington, DC: Urban Institute Press.

Pompe, Jeffrey and James Rinehart. (1997) Entrepreneurship and Coastal Resource Management, *The Independent Review*, v. 1, no. 4, Spring, 543-559.

Pompe, Jeffrey and James Rinehart. (1999) *Establishing Fees for Beach Protection: Paying for a Public Good,* Coastal Mgmt. Vol. 27 at 59.

Rosen, Sherwin. (1974) Hedonic prices and implicit markets: Product differentiation in pure competition. *Journal of Political Economy,* 82 (1) 34-55.

South Carolina Ocean and Coastal Resource Management. Annual State of the Beach, Charleston, SC, March 2006.

Webster, Chris' Georg, Glasze; Klaus Franz. (2002) The Global Spread of Gated Communities. *Environment and Planning B*, v. 29, 315-20.

Young, Tanya Fogg. (2005) "Aiken, Myrtle Beach boost entire state," *The State*, April 19, p. B9.

Sustaining the World: Against Religious Maniacs and a Selfish Superpower?

Arnis Vilks
Department of Microeconomics and Information Systems,
Leipzig Graduate School of Management, GER

I argue that the current escalation of violence between the US-led West and the Muslim world bears the grave risk of extreme disasters - including nuclear conflicts. To counteract these risks, scholars in all parts of the world ought to use their knowledge, institutions, and networks to deepen the dialogue among the 'clashing' civilizations. As a contribution to this dialogue, I argue that the world-views of Western elites very often rest on two outdated philosophical and economic tenets – the legacies of Logical Positivism, and Economic Liberalism, respectively. Positivism has generated a widespread contempt for ethics and religion, while Liberalism accounts for the view that 'free markets' ensure an optimal allocation of the world's riches. I argue that unless ethics gets rehabilitated among scholars in the West, its political leaders are likely to continue with international policies characterized by selfishness and disrespect for others.

1. Our final century?

The Physicist Martin Rees (2003) has argued forcefully that humankind might easily eradicate itself during this century. He estimates the odds that this might happen to be no better than 50:50. One of the main risks is that the unhampered technological progress may soon produce and make easily available the germs, the nuclear, or more "innovative" weapons which – by sheer error or by malign intent – can destroy and make uninhabitable large parts of the world.

In fact, the likelihood of "megaterrorism" by extremists from the Muslim world seems to be rather multiplied by the policies that the US-led West has pursued since 9/11. While the invasion of Afghanistan to topple the Taliban regime was seen as inappropriate only by few in the West, it gave rise to massive protests in many parts of the Muslim world. Much more so did the invasion of Iraq – which was not only opposed by major Western allies of the US, but has caused protests, anti-Bush, and anti-Blair campaigns throughout the world. Even before the Iraq war, it had to be expected that at least as many innocent Iraqis would be dying from Western bombs and bullets, as Americans died on 9/11. Moreover, one could predict almost with certainty that more Muslims would be drawn into fanaticism, and the already fanatical ones would turn into terrorists in greater numbers than before.

The terrorist attacks in Bali, Madrid, and London that followed the invasion of Iraq provided sad confirmation for that expectation. That the primary targets were Australians, Spaniards, and Brits - rather then Germans or Frenchmen - indicates that these terrorist attacks were indeed triggered by the Iraq war, and might not have occurred without it.

Now, with Iran denying Israel the right to exist, the Bush administration accusing Iran to strive for the atomic bomb, and the view being discussed that this may require the pre-emptive use of tactical nuclear weapons (Hersh, 2006), the risk of even further escalation seems to increase. A so-called 'dirty' nuclear bomb - detonated in, say, New York or Washington again – cannot be regarded as unlikely, and could easily trigger vengeful reaction by the Bush administration. Pakistan, which already has nuclear weapons, and an illiteracy rate of more than 50%, can hardly be viewed as a politically stable ally of the West.

Sustaining the world seems to require every attempt to try and counteract the confrontation among Islam and the West.

2. What can we do?

It is obvious that most of us have only indirect means of diminishing the probability of further violence between 'the West and the rest', to use Huntington's (1997) phrase. But indirect means we do have.

We live in a 'global village'. We easily travel form Europe to Australia, from America to Asia, and back again. What is said and written in Washington, London, or Sydney - is heard in Marrakech, Tehran, and Cairo. Not only an invasion of Iraq by a huge US army, but even caricatures in a Danish newspaper can trigger violent demonstrations in places as far away as Jakarta or Islamabad.

Less spectacular, but, no less important is what John Maynard Keynes(1973: 383) called the 'gradual encroachment of ideas':

"... the ideas of economists and political philosophers, both when they are right and when they are wrong, are more powerful than is commonly understood. Indeed the world is ruled by little else. Practical men, who believe themselves to be quite exempt from any intellectual influences, are usually the slaves of some defunct economist. Madmen in authority, who hear voices in the air, are distilling their frenzy from some academic scribbler of a few years back ..."

In fact, one can take it for granted that what is discussed and said in conferences like this one, what is written in books and journals, what is taught today in universities and business schools – is not only influencing the world-views of tomorrow's leaders of the Western world, but it is also noticed by Muslim scholars, and it will thus shape the West's image among future Muslim generations.

Thus, if there is any hope of gradually reducing the confrontation between Islam and the West, we need to discuss the issues that cause confrontation and hatred. We need to discuss them with Muslim colleagues and scholars, but also among ourselves.

3. What went wrong?

This is the question asked by the American Islam scholar Berhard Lewis. He explains in careful historical detail that "for many centuries the world of Islam was at the forefront of human civilization and achievement"

(Lewis, 2002:3), but has clearly lost that status to the West – in military terms as well as in science, technology, commerce, and in terms of material prosperity. Today, the Western world produces and consumes about 50% (20 trillion USD in 2004) of the world's annual gross product, the Muslim world (the members of the Organization of Islamic Countries) with about twice the population produces and consumes just 5% (2 trillion USD in 2004). Extreme poverty, hunger and illiteracy are rather rare in the West, while they govern large parts of the Muslim world.

But it is not just this inequality that breeds hatred of the West. In the age of the Internet, it is known around the globe that the repeatedly made promise by the rich Western countries to increase official development aid (ODA) for the poor to 0.7 % of GDP has been ignored for decades by most of the Western governments. In the case of the US, the official alms share to the poor has more or less continuously fallen from its peak post-WWII level of almost 3% to a mere 0.15% of GDP in 2004. With 20.000 people perishing every day from extreme poverty, this is plainly not consistent with the compassion demanded by Islam, Christianity or any other somewhat compassionate religion or system of ethics. One can hardly disagree with Jeffrey Sachs (2005:288) that this behavior of the West is tantamount "to announce brazenly to a large part of the world, 'you count for nothing'". In the age of the Internet, this announcement is heard and understood around the globe.

The connection between poverty and illiteracy on the one hand, and terrorism on the other hand, has often been doubted by pointing out that the terrorists themselves are often neither poor nor illiterate. But again, one must agree with Sachs (2005: 330): "Whether terrorists are rich or poor or middle class, their staging areas ... are unstable societies beset by poverty, unemployment, rapid population growth, hunger, and lack of hope."

It seems very likely that the world would be a much better and safer place, if the powerful West had taken the Millennium Development Goals seriously, and if Western leaders would have explained to their constituencies that it is a matter of enlightened self-interest to eradicate poverty and illiteracy.

Instead, the US-led West is continuing to act, as if the pursuit of happiness would be an unalienable right not for all men and women, but only for our kin.

4. The world-view of today's Western leaders

Some see the 'clash' as the confrontation of two religions – Christianity and Islam. However, if one reads bin Laden's 'letter to the American people' of November 2002, or Ahmadi-Nejad's more recent letter to the US president, it is quite clear that neither of them are attacking Christianity as such – on the contrary, the name of Jesus is always followed by the respectful phrase "pbuh". And although the US president seems to listen to people who think and say that Islam is "an evil and wicked religion", Bush himself has often described mainstream Islam as "peaceful".

Clearly, the main religion of the Western civilization is Christianity, but the role of Religion is quite different here from what it is in most parts of the

Muslim world. Particularly among the Western elites, many consider religion *as such* to be old-fashioned, superstitious nonsense. To be sure, in the Bush administration there are many who consider themselves devout Christians, but for many, religion remains a matter for sunday prayer, and has little to inform professional life.

Professional life is rather governed by an attitude that is characterized by such epithets as "realistic", "no-nonsense", "sober-minded", "down-to-earth", "objective", "facts-based", or just "professional".

Of course, there is the other side of the dichotomy. What is "unprofessional", "nebulous", "unrealistic", or "subjective", is not banned outright – one is free to have opinions, and to spend ones spare time with all sorts of diversions, but what is unprofessional is by definition not suitable for the workplace.

Although things begin to shift slowly, and business ethics, corporate social responsibility (CSR), and sustainability enter the curricula of more and more business schools, these are topics which are typically categorized as "subjective" or "nebulous" by the currently ruling generation of "sober-minded, down-to-earth professionals".

This attitude is not too difficult to explain. Western academia of the 1950s and 60s, where the older ones of today's political and business leaders formed their world views, was deeply shaped by the philosophy of Logical Positivism that originated in the Vienna Circle in the 1920s. This philosophy was an attempt to establish a 'scientific world-view', and to ban everything from this *weltanschauung* that was only pseudo-scientific, or metaphysical. The influence of this renewed Positivism can hardly be exaggerated. It was regarded as the "received view" of scientific theories until the 1970s (Suppe, 1974). In a jargon that has become quite standard among Western academics, the basic tenet of Logical Positivism can be expressed as "all knowledge is either empirical or analytical". The empirical is identified with the observable, and the analytical with linguistic or mathematical convention. The important thing is that every kind of discourse that is neither empirical nor mathematical, was regarded as, strictly speaking, nonsense.

Arguably, neither religion, nor art, nor ethics can be deduced from observation plus mathematics. For the CEO or political leader, the implication is that ethics, CSR, and religion may be a matter for the PR or communications department, but that they have no place in a strategy meeting.

In a somewhat similar way, today's Western elites have come to believe in "free markets". They were educated at a time when economists only gradually mastered the mathematics required to analyze multi-market interaction. Not many economists at that time understood Debreu's (1959) mathematical reformulation of Adam Smith's claim that 'the invisible hand' of the market turns private vice into public virtue. Today's leaders of the West were brought up to believe that a system of perfectly competitive markets would ensure a socially optimal allocation of goods. Today's economic neo-liberalism can only be attributed to a somewhat naïve reading of some theorems in mathematical economics which were proved in the 1950s – but not well understood even by an average professor of Economics at that time.

As it happens, the 'invisible hand' thesis combines very well with Logical Positivism in creating and supporting an attitude of selfishness. Laissez-faire liberalism allows one to ignore how one's decisions will affect others, and reduces ethics to the maxim: Do what maximizes your material wealth – competition and the markets will ensure that the global allocation resulting from selfish behavior will nevertheless be socially optimal (cf. Kelly, 2002).

It is small wonder that the egocentric and selfish world-view generated by Positivism and Economic Liberalism translates into world politics as frequent disrespect for the concerns of other countries' inhabitants. It is the 'national interest' that counts, not concern for others.

5. What Western leaders need to learn

The simple fact is that within the respective scientific communities of philosophers, and economists, respectively, neither Logical Empiricism nor the invisible hand thesis have survived criticism. Today, after Thomas Kuhn (1962) and Paul Feyerabend (1975), Ethics, Metaphysics, and Religion should be just as respectable fields of enquiry and scholarship as quantum mechanics, cosmology or string theory in physics.

And well-trained economists today understand clearly that there are many problems that free markets cannot solve even in a highly stylized hypothetical model of 'perfect competition'. The notion of 'social optimum' was eventually replaced by 'efficiency' (Arrow & Hahn, 1971), and there is a huge literature on market failure. Moreover, the perfect competition paradigm is being gradually replaced by Game Theory, which treats cooperation as no less important than competition (c.f., Nalebuff & Brandenburger, 1996).

Maybe we cannot update the world-views of current Western leaders' - but at least we can hope that the next generation will take over soon, and be better prepared to sustain the world.

References

Arrow, K., and Hahn, F. (1971). *General Competitive Analysis*, Amsterdam: North-Holland, ISBN 0444854975.

Debreu, G. (1959). *Theory of Value. An Axiomatic Analysis*, New Haven and London: Yale University Press, ISBN 0300015593.

Feyerabend, P. (1975), *Against Method*. London: New Left Books, ISBN: 0860916464.

Huntington, S. (1997). *The Clash of Civilizations and the Remaking of world Order.* London: Schuster &Schuster, ISBN 074323149X.

Kelly, M. (2002). "Reflections on Adam Smith's 'The Wealth of Nations' and its Effect on our Contemporary World," *International Journal of Management Literature.* ISSN 0972-4079, 2,1&2: 81-92.

Keynes, J.M. (1973). *The General Theory of Employment, Interest and Money,* London and Basingstoke: Macmillan Press, SBN 333009428.

Kuhn, T. S. (1970). *The Structure of Scientific Revolutions,* 2nd ed., Chicago: University of Chicago Press, ISBN 0226458040.

Lewis, B. (2002). *What Went Wrong?Western Impact and Middle Eastern Response.* London: Phoenix, ISBN 075381675X.

Lewis, B. (2004). *The Crisis of Islam: Holy War and Unholy Terror.* London: Phoenix, ISBN 0753817527.

Nalebuff, B., and Brandenburger, A. (1996). *Co-opetition.* London: HarperCollinsBusiness, ISBN 0006387241.

Rees, M. (2003). *Our Final Century: A Scientist's Warning: How Terror, Error, and Environmental Disaster Threaten Humankind's Future in This Century*. Arrow, ISBN: 0099436868.

Suppe, F. (1974). "The Search for Philosophic Understanding of Scientific Theories," in F. Suppe (ed.), *The Structure of Scientific Theories.* Urbana: University of Illinois Press, ISBN 0252006348, pp. 3-232.

Arnis Vilks, born in 1956, studied Economics, Philosophy, Mathematics, and Political Science at the University of Hamburg, where he earned a doctoral degree in Economics in 1983. He has held various teaching and research positions in Hamburg, Cologne, Cambridge (UK), Riga, Stanford, and Hamilton (NZ). Since 1995 he is a Professor of Economics at Germany's oldest Business School, HHL (Leipzig Graduate School of Management), where he served as Dean from 2000 to 2005. His research focuses on philosophical and logical issues in economics and management, in particular formal logic approaches to game theory, the notions of rationality, utility, and happiness, and the *weltanschauung* of the Western world, its roots and effects in the global economy.

A Complex Systems Perspective on Incommensurable Evaluations within Triple-Bottom-Line Reporting

Elaine Hartley
Faculty of Life and Social Sciences, Swinburne University, AUS

The adoption by corporations and government agencies of Triple-Bottom-Line Reporting practices suggests that the complex choices which arise in "world risk society" bring social and environmental assessments into our calculations of economic advantage. Applying the TBL concept can be deeply problematic, however, in that these different components of assessment are situated in incommensurable gestalts. From a complex systems perspective, social, environmental and economic factors are complicit in generating social order as a whole with emergent properties, not discernible through measuring each component of the TBL. This paper explores whether the emergence of post-formal cognitive styles, grappling with constructs as embedded in, and emergent from, contextual factors, will foster a more reflexive understanding of the conventions we construct for measuring progress towards sustainability.

Introduction: Global risk society – socialized assessments, mediated experience

Post-modern global cultures are beset with uncertainties, fluid identities and simultaneous transformations in social structure and context. Citizens are constantly informed through the media about issues for concern regarding which they have no direct experiential understanding. According to Giddens (1991: 119), public discourse is saturated with expert profiling of risk, giving rise to a proliferation of 'abstract systems' for managing public response i.e. the common concerns of Network Information Society are filtered through numerous social mechanisms for constructing the significance of, problematizing and measuring progress in addressing these social issues.

Beck (2002: 3) suggests, however, that many of these expert systems are quite seriously divorced from an actual understanding of the uncertainties which exponential technological and social change is generating. This situation will be discussed by using an example of one such "abstract system" – "Triple Bottom Line Reporting"[1] – which has received an unexpected level of support from corporations, non-profit groups and government agencies (Elkington, 1999).

After the Rio Declaration on the Environment and Development in 1992, business embraced the idea of combining eco-efficiency in resource use with economic growth, but the third pillar of Sustainable Development – Social Progress – has remained largely neglected until Kofi Annan's 2005 address to the UN highlighted the growing gap between rich and poor, despite international commitments to reducing poverty and inequality (United Nations,

2005: *"In Larger Freedom")*.

The speed with which concepts of Corporate Social Responsibility[2], the advantages of promoting Social Capital and TBL reporting practices entered corporate thinking has been attributed by many to a backlash after the 1999 Seattle protests against exploitative free-market practices.. Multi-national corporations realised a need to address just how negative their public image had become. The Business Council of Australia views greater transparency and better public relations as good for profits and has thus strongly supported Triple Bottom Line Reporting as:

"a solid foundation for measuring our progress ... in relation to financial, environmental and social performance... and broader aim of improving the negative public perception of business in Australia" (Group 100, 2003: foreword).

Giving consumers comparable data on corporations will guide their ethical choices and thus corporate responsibility is allied with green consumerism (Chapman, 2006). However, Norman and MacDonald (2004) consider these expectations fallacious - as ratings on the basis of social indicators have no comparative validity across companies or over time (being discursively constructed and qualitatively and contextually evaluated). Neither is it admissible to aggregate ratings of social, economic and environmental performance in order to establish a clear ranking of companies – so that consumers can develop a preference of one company over another in the way we assess other products.

Consequently, they conclude that what is sound about the concept is not novel (companies have always taken employee welfare, customer perceptions of the company ethics, and shareholders' beliefs about the company's soundness into account) and what is novel about the concept (packaging this information and making it publicly available through measurable, comparable reporting) is not sound, and might in fact provide a smokescreen behind which companies can avoid genuine accountability.

This paper asks whether promoting accountability mechanisms such as TBL Reporting can genuinely regulate the defective socio-environmental relationships of global capitalism or whether they provide the illusion of "progress" without disturbing the status quo? In the words of Max Weber, are "formal" rationalities (i.e., abstract mechanisms) taking over from "substantive" rationalities (Stryber, 2001)?

The self-organization of evaluative mechanisms

Through the lens of economic liberalism[3], "governance at a distance" requires that the governed (as free rational individuals) become complicit in the processes whereby they are governed (Higgins, 2005). For pragmatic reasons, however, it is also imperative that public response be channelled in appropriate directions. One way of encouraging this is to set up auditing processes, standardised decision-making instruments and self-regulating programs which mould the sensibilities of independent agents.

Under the influence of these mechanisms, promulgated through trade networks, international conventions or social mimicry, well-defined conceptual highways start to shape strategies for posing and addressing problems - compounded through social dynamics, not any intrinsic fitness of the strategies themselves. For example, Sustainability Reporting Guidelines developed by the Global Reporting Initiative of the North-American Coalition for Environmentally Responsible Economies (CERES) became a de facto international standard after the United Nations Environment Programme (UNEP) partnered CERES in 1999 (Global Reporting Initiative, 2002). Then the Australian Bureau of Statistics disseminated these standards through numerous reports on Social Capital[4] and how to measure it (ABS, 2001, 2004; Stone and Australian Institute of Family Studies, 2001; Stone, *et al.*, 2002). The more prevalent the use of these standards, the less visible the ideological framing of ideas such as Social Capital becomes; it appears self-evident.

Institutional shaping of evaluative frameworks

Consequently, it matters **whose** construction of concepts is dominant in how we make sense of the world. The western habit of abstracting ideas and generalizing them to new domains conceals how implicit priorities and metaphoric framing from an economic domain may be transferred unwittingly through such mechanisms and start to influence our ideas about human relationships and social contracts. Framing social assessments through economic language - such as developing Social or Human Capital to build the "intangible assets" of corporations – unconsciously assimilates economic conventions such as methodological individualism (which views society as the sum of atomistic actions). The frames we adopt shape what kinds of conversations can be held and the type of conclusions which emerge (Lakoff, 2004).

Triple Bottom Line Reporting might be considered a 'category error' i.e. a statement which is figurative and rhetorical has been taken literally to fit corporate practices. Using the language of accountancy in incommensurate domains acquires a semblance of meaningfulness through blurring the semantic distinction between the concept of a financial bottom line and the notion of primary significance: i.e., making a profit is what ultimately matters (economic growth); having resources is an absolute precondition for production (environmental capital) and (in the words of Kofi Annan) if society collapses then so do markets (social capital).

Thus "social capital" viewed as a subset of "economic capital" sets up an expectation that "social and natural capital" should be preserved – just as we accumulate financial capital - without really exploring the meaning of that statement. Conventions, widely shared, provide a comforting sense of "orderliness" – thereby disguising just how mutable, contingent and contextually-embedded human modes of sense-making are in actuality (Varela, *et al.*, 1991). It becomes difficult to see just how much regulatory mechanisms shape our sense-making habits.

The failure of instrumentalism

Economic liberalism, resting on the premise that human rationality is limited and not to be trusted in directing the course of social institutions, could be seen as an acknowledgement that the epistemological habits of western cultures are failing to render comprehensible or to regulate the complex, self-organizing dynamics of globalism. Both public and private modes of problem-solving in capitalist economies are deeply dependent on reductionist, instrumental modes of thought, i.e., drawing clear boundaries and setting overriding objectives permits consensus, determines how tasks are prioritized and simplifies assessments.

Public policy developers and environmental activists alike have realized, however, that isolated interventions – conceived through the lens of selective institutional goals – cannot accommodate the highly indeterminate, deeply complicit issues of post-modern cultures. Freyfogle (2006) in fact claims that "special interest" environmental campaigns are actually counter-productive, breeding public misconceptions about the aims of environmentalism and obfuscating how these are relevant to the health and wellbeing of society.

Calls for integrated action and "holistic" approaches fall foul of the selective instrumental frameworks through which different groups construct their understanding of situations; multidisciplinary overviews claiming a privileged over-perspective turn out to be partial and unsupported interpretations. It has become painfully apparent that these selective filters in our construction of meaning are endemic to human understanding; this throws into doubt and confusion the artifacts we create as orienting points around which to form collective assessments.

Capitalist cultures, transformed by the cultural diversity of Network Society, are caught in these contradictions between instrumental and contextualized, emergent modes of knowing (Capra, 2002). Kegan believes that this contradiction is never resolved., but if we are able to move into "a bigger space where we can look AT the contradiction", we learn to hold the tension between these incongruent and ambiguous constructions of issues (Kegan & Lahey, 2001). Evolve post-formal styles of thought (i.e., a capacity to articulate insights into how meaning is being framed) helps to emancipate groups from the perceptual boundaries created by their unconscious sense-making traditions. Reflexive awareness of these thought patterns open up alternative responses. Kegan describes this developmental process as a subject/object transformation: developing the capacity to observe more of our thoughts and emotions , shifts the ratio between thoughts which just "have us" - and those we can reflect upon - towards the "objective" end of the continuum.

Capitalist economies appear to be in a transition period between managing dissent through setting up foundational truths and awareness of how these abstract mechanisms rule us. We can step back from the "agent-design perspective" for some of the time, but then succumb to a desire to have intractable issues resolved for us through self-organized outcomes.

Self-organizing information cultures in a sustainable global system

In society is viewed as a complex self-organizing system (with emergent properties not predictable as a sum of the parts), the pursuit of a sustainable social order does not stop at making each part energy efficient, non-polluting etc.; it requires that we grapple with the non-linear, self-organizing dynamics which entrain our identity, aspirations, modes of assessment, habitual strategies, to the dominant political economy. The social dynamics of the system *as a whole* has an environmental impact: levels of conflict, capacity to resolve incompatible, short-sighted objectives, maintaining complex capabilities with smaller populations – these types of dynamics are implicated in being able to promote positive cycles of social and environmental health and wellbeing.

An exponential leap in ability to construct meaningful patterns and respond appropriately to situations through harnessing the distributed "intelligence" of citizens demands, however, that complex self-organizing processes are appreciated as a double-edged sword – beneficial for some purposes and fostering ideological rigidity at other times. Evolutions in the nature of the social contract have changed the course of history (Schama, 1987). Greater prevalence of post-formal styles of thinking (internalizing a sense of self as structurally coupled to others and redefining our understanding of meaning-making) might permit forms of cooperation to emerge which are not amenable to groups operating through atomistic individualism.

Conclusion: The triumph of "substantive rationalities" over "formal rationalities"?

In conclusion, the question this paper highlighted is whether grappling with incommensurable Triple Bottom Line assessments will expose our social constructions as embedded, contextual or whether these accounting practices will become an opaque social convention, reproduced without reflection.

We might also ask why (given the conceptual and operational difficulties of Triple Bottom Line Reporting) corporations, government agencies and civil organizations have been so ready to adopt the TBL concept?

Wallerstein (2000) notes that the unexpected degree of resistance from anti-systemic movements, protesting the gap which global capitalism is driving between haves and have-nots, threatens the ability of government to guarantee stable conditions for investment. Destabilization of a neo-liberal market-based order might be adverted through legitimizing the social accountability of corporations – no matter what the actual outcomes. Stryber (2001) - referring to Max Weber's concept of legitimacy - suggests that "people's cognitive and normative orientations to some legitimacy source—whether tradition, charisma, or legality—can counteract the tendency of interest-based conflicts to undermine social order. When legitimacy results from collective orientation to, and belief in, a particular set of 'rules of the game', people will support and comply with decisions that run counter to their socioeconomic, political, or legal interests."

Can we escape this determination by our social technologies to foster a political economy more responsive to socio-environmental challenges than a monopolistic dependence on the market? This paper suggests that understanding the evolutionary processes shaping our sense-making of issues brings a capacity to transcend the "abstract mechanisms" through which our evaluations are channelled. However, this may depend on the degree to which post-formal thinking styles become prevalent in the population of global cultures.

Notes

[1] John Elkington claims to be the innovator of Triple Bottom Line Reporting in his 1999 book.
[2] Corporate Social Responsibility: legitimating the accountability of corporations to more than just their shareholders.
[3] Based on the assumption that social order self-organizes most efficiently through the rational self-interest of individuals, expressed through market choices. See Hayek, 1976, for more about the assumptions underlying economic liberalism.
[4] Social Capital: *"Networks, together with shared norms, values and understandings which facilitate cooperation within or among groups"* (Australian Bureau of Statistics, 2004).

References

Australian Bureau of Statistics (2001). *Measuring Wellbeing: Frameworks for Australian Social Statistics,* Report No. 4160.0, Canberra: ABS.

Australian Bureau of Statistics (2004). *Measuring Social Capital; an Australian Framework and Indicators,* Canberra: ABS.

Beck, U. (2002). "The Silence of Words and Political Dynamics in the World Risk Society." *Logos*, ISSN 1543-0820, (Fall): 1-14.

Capra, F. (2002). *The Hidden Connections*, London: HarperCollins, ISBN 0002570475.

Chapman, G. and Australia. Parliament. Joint Committee on Corporations and Financial Services. (2006). *Corporate Responsibility: Managing Risk and Creating Value,* Canberra, A.C.T: Commonwealth of Australia.

Elkington, J. (1999). *Cannibals with Forks : the Triple Bottom Line of 21st Century Business,* Oxford: Capstone, ISBN 1841120847.

Freyfogle, E. T. (2006). *Why Conservation is Failing and How it can Regain Ground,* London: Yale University Press, ISBN 0300110405.

Giddens, A. (1991). *Modernity and Self-identity : Self and Society in the Late Modern Age,* Cambridge: Polity, ISBN 0745609325 .

Group 100 (2003). *Sustainability: a Guide to Triple Bottom Line Reporting,* Melbourne: Group of 100, ISBN 097502050.

Hayek, F. A. (1976). *Law, legislation and Liberty; a New Statement of the Liberal Pinciples of Justice and Political Economy,* London: Routledge and Kegan Paul.

Higgins, W. (2005). "Standardisation, Globalization and the Rationalities of Government", Paper delivered at the conference on *Organizing the World: rules and rule-setting among organizations*, Stockholm: Stockholm Centre for Organization Research.

Global Reporting Initiative (2002). *Sustainability: Reporting Guidelines*, Coalition for Environmentally Responsible Economies, www.globalreporting.org, Accessed 25 July 2006.

Kegan, R. and Lahey, L.L. (2001). *How the Way We Talk Can Change the Way We Work: Seven Languages for Transformation*, San Francisco: Jossey-Bass, ISBN 0787955353.

Lakoff, G. (2004). *Don't think of an elephant! : progressive values and the framing*

wars : the progressive guide to action. White River Junction: Chelsea Green, ISBN 1931498717.
Norman, W. and MacDonald, C. (2004). "Getting to the bottom of the 'Triple Bottom Line'", *Business Ethics Quarterly.* ISSN 1052-150X, 14, (2): 243-263.
Schama, S. (1987). *The embarrassment of riches : an interpretation of Dutch culture in the golden age.* London: Collins, ISBN 000217801X.
Stone, W. and Australian Institute of Family Studies. (2001). *Measuring Social Capital: Towards a Theoretically Informed Measurement Framework for Researching Social Capital in Family and Community Life,* Melbourne: Australian Institute of Family Studies, ISBN 0642394865.
Stone, W. et al. (2002). *Social Capital : Empirical Meaning and Measurement Validity,* Melbourne: Australian Institute of Family Studies, ISBN 0642394946.
Stryber, R. (2001). "Legitimacy", *International Encyclopedia of the Social & Behavioral Sciences,* Minnesota, USA: University of Minnesota, pp. 8700-8704.
United Nations (2005). "In larger freedom: towards development, security and human rights for all", *Report of the Secretary-General, Kofi Annan,* United Nations.
Varela, F. J. et al. (1991). *The Embodied Mind: Cognitive Science and Human Experience,* Cambridge, Mass.: MIT Press, ISBN 0262720213.
Wallerstein, I. (2000). "Globalization or the Age of Transition?", *International Sociology,* ISSN 0268-5809, **15**(2): 251-267.
World Business Council on Sustainable Development (2006). *From Challenge to Opportunity; the Role of Business in Tomorrow's Society,* ISBN 294024087640, http://www.wbcsd.org/DocRoot/C77uYrHIIJcti1FWAn1s/tomorrows-leaders.pdf, accessed 8 August 2006.

Elaine Hartley has B.A. Hons (Psychology), M.A. Prelim (Neuropsychology., multivariate stats), Master of Environmental Science, and is currently completing a Ph D under the supervision of Dr Arran Gare, environmental and process philosopher, Faculty of Life and Social Sciences, Swinburne University and Joseph Needham Centre for Complex Processes. Her research investigates changes in self-concept, aspirations and collaboration as post-formal cognitive styles co-evolving with distributed forms of online organization. She suggests that "sustainability" be reframed from an objective which we pursued to a complex, self-organizing transformative social process which is generated, not designed. Elaine's interests include: Social ecology, urban sustainability, economic reform, process philosophy, cultural transformation through globalizing networks, online democracy. Environmental science and social research has been a mid-life career change after Elaine spent many years providing community-based situational counseling and support in the psychiatric disability field. She works as an independent research consultant from a home office in the beautiful Dandenong Ranges on the outskirts of Melbourne.

The Value of Community Participation

Bronwen Maxwell[1] & Cheryle Yin-Lo[2]
[1]Earth Journeys, AUS
[2]Community Cultural Development, Blue Mountains World Heritage Institute, AUS

Introduction

This paper will be presented in two parts. It will explore issues of community cultural development (CCD) practice and Earth Journeys, an enviro-cultural schools program and working collaboratively with communities in artistic and cultural expression to explore environmental issues. Different models and interdisciplinary approaches will be illustrated through work currently undertaken through these two programs and our work at the Blue Mountains World Heritage Institute, a non profit and independent organization promoting the conservation of the natural and cultural heritage of the Greater Blue Mountains World Heritage area.

Our presentation will focus on combining the key pillars of social, economic, educational, environmental and cultural outcomes in building social and natural capital and the role of community capacity building and cultural development in a regional area.

Implementation and evaluation of pilot projects and CCD processes within the local community in an interdisciplinary approach combine shared knowledge and skills with artists, scientists, Indigenous, academic and communities to explore issues impacting on the of the Greater Blue Mountains World Heritage area.

Through this approach multiple perspectives are presented which highlight the capacity for communities to understand better issues of difference and commonalities and to create dialogue to potentially discover the synergy or common ground to solve complex issues and development of partnerships.

Issues such as raising community's awareness through CCD projects and public programs and engaging diverse communities of different ages in a holistic and cross cultural approach provide a rich insight to explore people's values, attitudes and behavior to the environment and their relationship to it. Providing more creative and interactive opportunities for individuals in participatory processes potentially enhances community ownership and responsibility to encourage a more active role in local, regional and global issues especially in issues such as urban development and sustainable environmental issues.

The introduction and CCD program will be presented by Cheryle Yin-Lo, Community Cultural Development worker at the Blue Mountains World Heritage Institute and

Bronwen Maxwell, Blue Mountains World Heritage Institute Education Program Coordinator and Project Manager of Earth Journeys will present about the program's model of working with schools.

Community Cultural Development Program (CCD)

The role of the community cultural development worker based in a research institute such as the Blue Mountains World Heritage area was an experiment and pilot program initially funded by the Community Partnerships Program of the Australia Council for the Arts for 18 months in 2005 and now funded by Arts NSW.

The position was to develop a program of strategic and interdisciplinary projects that address environmental issues in promoting the conservation of the natural and cultural heritage of the Greater Blue Mountains World Heritage area and to explore how CCD processes can be used a research tool in an integrated part of the BMWHI organizational approach and its projects and partners to assist in informing policy and management issues of both the natural and urban development of the region .

Why is the process of CCD so important to communities and what are the main principles of CCD and how can it be applied and what are the benefits and outcomes?

CCD is a broad term covering an enormous range of activities through artforms such as music, visual arts, performance, puppetry, photography, film & video, sculpture, storytelling and oral histories that give communities the opportunity to tell their stories, build their creative skills and be active participants in shaping the development of culture. It also encourages communities to participate in collaborations between communities and artists as equal partners using CCD projects to explore more complex social and environmental issues and using creative problem-solving and seeing things from a different perspective. More importantly, the process of CCD encourages cultural development and practice to empower communities to initiate, direct and manage their creative ideas and projects.

It is possible to achieve a wide range of social, artistic, cultural, economic and environmental outcomes and how it can raise new perspectives and change perceptions and attitudes in understanding environmental issues. I will illustrate a few examples of projects which I have been involved in nad the different approaches.

Case study 1: *Nature through fresh eyes*

This project was initiated by the BM Multicultural Residents Association and was in partnership with the Mountains Community Resource Centre and BMWHI with funding from BMCC Cultural Partnerships Program and the Australia Council for the Arts.

- It was a cross cultural environmental arts project to increase the awareness of multicultural residents and refugees of the World Heritage area and the Australian natural environment. It aimed to break the social isolation of residents from emerging and established communities who were 1st and 2nd generation.
- The project ran for over 6 months and achieved the following outcomes:

- Over 50 residents and their families participated from over twenty countries as part of the Blue Mountains community. The meeting of these cultures and of Aboriginal culture was a very enriching in acknowledging the cultural diversity of the Blue Mountains.
- A working committee with professionals and community people met regularly in planning and delivering the project
- Five bushwalks for residents and their families with Indigenous interpretive guides & arts tutors in the BM World Heritage area. Participants were supplied art materials with visual diaries and disposable cameras
- Structured practical art workshops to record experiences and responses with art tutors small individual canvasses and to produce a collaborative mural
- CCD worker was involved in mentoring component to build up arts project management skills with people from culturally diverse background living in the Blue Mountains.
- CCD worker involved in research & evaluation conducted pre, post and during the project with participants and partners. Quotes from participants about their association and values about the World Heritage included comparisons to Australia and their home country, their experiences living in the 'bush 'and the World Heritage area and their experiences of bushfires and bushfire management. Comments ranged from exhilaration to overcoming the fear and mystery of 'the bush'. Evaluation of project indicated increased understanding of their immediate environment, increased social skills and confidence and making new friends, created interest to want to participate in future projects, reconnecting with their arts practice, created local employment and increased confidence to work cross culturally.
- An exhibition was organized at the Katoomba Civic Centre with adult and children's artwork.
- A DVD was produced which documented the process and participants experiences and responses to the project.
- Interest by other environmental entities in the region have expressed interest to display artworks at other community centre venues and environmental events throughout the year.
- SBS TV came to film a segment of project and broadcast on *Here Comes the Neighbourhood series.*

Case study 2:Giant dragonfly project

This project was initiated by a group of concerned individuals who wanted to raise the awareness about the threat to endangered habitats and species such as hanging swamps and Giant Dragonfly and the Leura skink and the human impact of the relationship of ground water issues .The group comprises of people from various background and experiences- from artists, retirees, teachers, nurses, bush care, scientists, Indigenous and conservationists.

The group work with the CCD worker as facilitator to meet regularly as a group and explore visually how they could communicate their message to the wider community.

Through the learning of lino cut printing and working with an artist, the group made individual prints informed through various environmental and scientific visuals and talk by a scientist and drawing from their own direct experiences. These prints were then mounted in a group work and shown in Paddy Pallins shop window as part of Artstreet and the Winter Magic Festival in Katoomba. Through this event many community people were interested in the issue and interest of commissions of prints.

The group continue to still meet to explore ideas to work on a larger scale and to find and refine effective ways to communicate important environmental messages to the wider community. Projects such as this provide the opportunity for sharing interdisciplinary knowledge and skills (existing and new), research and build up resources and encourage leadership and community building skills in working collaboratively and as a team.

Case study: Working with local Aboriginal communities

Working collaboratively with local Indigenous artists and communities and creating opportunities to work more closely with policy and management bodies and with environmentalists and scientists has created a stronger knowledge and skills base. Through developing exhibitions for 2006 NAIDOC Week and Rock Art Lives On! It has resulted in short and long term outcomes-building trust with the community and BMWHI, producing new and shared cultural knowledge within and with other language groups and increased respect for the diverse Indigenous language groups in the region. Increased profile of Aboriginal culture and artists and contemporary work generating increased skills development, employment, tourism and cultural opportunities.

Enriching our social capital through caring for our natural capital

A schools program model conducted in the Blue Mountains World Heritage Area and Penrith regions

Aims and approach

Living in a World Heritage Area comes with the responsibility of maintaining certain requirements laid out in the WH Convention. The responsibilities are to value our natural and cultural landscapes by conserving and protecting our heritage and to pass on knowledge to the next generation.

Earth Journeys is a leading, award winning World Heritage "envirocultural" education schools program. It is an integrated, high quality cross-cultural and cross-curricular education program that engages students in experiential learning and creative application. Working at a "grassroots" level, it is a program about teaching and learning and which involves a real and relevant community interface. *Earth Journeys* aims to engender within school students a sense of place, a positive connection with the natural environment and cultural landscapes and the knowledge, skills and understanding that will foster responsible stewardship of their local environment and, in general, the earth.

The model structure

Endorsed by UNESCO, *Earth Journeys* is an initiative of Earth Partners Inc, a not-for-profit association of Blue Mountains professionals with diverse specialisations as artists, storytellers, environmental scientists, educators and filmmakers. Earth Partners Inc. is a recipient of a Keep Australia Beautiful "Sustainable Cities" Award (2005), is endorsed by the Blue Mountains City Council Cultural Partnerships Program and has the support of the Penrith City Council, Sydney Water and the Dept. of Environment and Heritage (the Asia-Pacific Focal Point for World Heritage of Australia).

The success of *Earth Journeys* is interdependent on the strength of its cooperative partnerships. To this end the program is currently supported by and works in partnership with the Department of Education (DET), the NSW National Parks and Wildlife Service (NSW NPWS) and the Blue Mountains World Heritage Institute (BMWHI). Corporate sponsorship also plays an essential role in helping the program to achieve its overall aims.

Our target audience is school students and includes teachers and, more broadly, families and local communities. In the initial development stages of the program the target is primary schools, with the objective of developing the program to extend into the middle stages of secondary schools. The *Earth Journeys* "model" has been piloted and is being established initially in primary schools within the Blue Mountains and Penrith Districts. There is a symbiotic relationship between the two areas, one being a World Heritage Area and the other, an area directly on its doorstep. The bigger picture message about caring for our place extends beyond both areas, as the responsibility to protect the gum trees, the ecosystems and most importantly our water and catchment systems, is a responsibility we all have to share.

With the support of the Department of Education, the program has run annually in the Blue Mountains and Penrith schools districts since 2004. *Earth Journeys* has been conducted as a highly successful and visible program in a number of primary public schools throughout the Blue Mountains and Penrith LGA, and has to date involved over 2, 850 students. Students' artwork of sculptural gum tree creations have been showcased at a number of public exhibitions, including the venues of Songlines Festival, Mt. Tomah Botanic Gardens, the NSW NPWS Blackheath Visitors' Centre and the National Eco-Tourism Conference held at the Fairmont Resort, Leura.

The method

The *Earth Journeys* program is designed to run for the duration of a school term or more, thus providing an opportunity for a concentrated focus on environmental and cultural matters. This creates a platform interface through which there can be more integrated community involvement with other cultural and environmental groups and services such as local Councils, National Park agencies, bush walking groups, historical societies and, especially, local Aboriginal communities.

The integrated package presents four components:

1. The *Earth Journeys* Work-Units are cross-curricular work-units for primary Stages 1-3 (at this point). Based on three books about people and place by Australian authors they are written in partnership with the NSW Department of Education (Penrith Schools District) applying quality teaching through the integration of key learning areas of literacy, HSIE, science and creative arts.
2. The "Gum Tree Stories" creative art units and workshops focus on the eucalypt. The tree becomes a metaphor for the students themselves, on their journey of development through school and then 'branching out' into the world. The gum tree is the central link of the workshops and reflects the interconnection of all the processes involved in the workshops. In the 'Meet Your Tree' workshops students engage with ecology stories, from an Aboriginal and non-Aboriginal perspective, about their particular gum tree native to their area. Their tree is "painted up" with ochre; a small ceremonial fire is lit; the wood makes charcoal. This charcoal is then used in the enviro-art sculptural workshops as part of the observation drawing exercise. This process of exploring form, color and light aids and informs the creative crafting process of sculptural creatures and gum trees. Woven into these activities is an introduction to art history and vocabulary. A further integration experience occurs through connecting the body to the dance and music of the Aboriginal creation stories of the land and the creatures.
3. The "Boundaries" Field Trip is an experience of an "immersion in the bush" journey. This component is run in partnership with the NSW NPWS Discovery Rangers. Based on methodology of the Earth Education Institute the program is designed to stimulate a confidence in and a curiosity and respect for our natural landscape.
4. Documentation and research is an additional component and service offered by Earth Partners. A DVD is made of each year's program process, which serves as both as a tool for promotional purposes and a visual documentation record. A current PhD research work is investigating and measuring the outcomes of the process to establish a benchmark. This research work will be supported through a video documentary of interviews and responses from selected students.

The focus on a specific "iconic" gum tree is designed to act as a catalyst for behavioral change, encouraging students to develop an appreciation not only of the characteristics of their gum tree but also the water systems and the surrounding biodiversity and habitat. Through developing an attachment to their particular gum tree, students may be better placed to engage with environmental issues such as impacts of urban encroachment, pollution of water systems, threatened species and the importance of sustainability and maintaining and protecting a natural biodiversity.

Whether one is living in the middle of a World Heritage listed Area or on its doorstep, the relevance and objective of *Earth Journeys* is to provide children in the 21st century with the tools necessary for their future in being responsible citizens caring for place.

Conclusion

In tackling problems affecting `community wellbeing' governments have begun to find that an integrated and whole-of government approach is necessary, one which links policy, planning and delivery of mechanisms between government agencies and between different spheres of government and partnerships with government and non-government agencies.

"This re-examination , the balanced and integrated approach is gaining recognition and a vital step towards increased community wellbeing and ecological sustainability. This new balancing act often requires the reformation and/or extension of ideas and information, a shared understanding for a 'suspended moment' for working towards a shared vision. This results in the establishment of shared understandings of the meaning of sustainability and the building of shared commitment to solutions."[1]

CCD processes can help break down some of the barriers in centralizing power and the over-reliance on scientific and technical language and knowledge only which can act to prevent communities from expressing their values, realising their self-determination and acknowledging their sense of identity.

Community engagement is now gradually realized and acknowledges the potential for 'active citizenship' or 'participatory democracy' as a building block for sustainability and wellbeing to increase ownership in shaping our community, culture and society.

1 Art & Wellbeing: p.7 securing the connections Paul Brown and Deborah Mills. Published by the Australia Council for the Arts, 2004.

THEME SIX
Livelihood, Innovation and Opportunity

Ian Knowd, University of Western Sydney, Australia

The four papers under this theme explore sustaining social and/or natural capital in specific contexts and through mechanisms of livelihood. Each represents a case example of situations where a systems perspective not only helps to build understanding and inform decision-making, but also highlights a range of challenging questions as to the sustainability of the human/environmental processes with which they deal.

We have two papers exploring issues in farming communities via a systems perspective. Lucia Boxelaar and her colleagues explore resilience in a dairy farming context and the challenges that investigating this presents to practitioners and researchers alike. My own paper on farming in the Sydney basin presents the case of Hawkesbury Harvest, a community-based initiative that emerged in response to pressures of urbanisation and changes in global food systems affecting farmer's viability. Both cases advocate for a better understanding and present insights into the systems dynamics of rural communities in the Australian context.

Jae Eon Yu's paper takes us into an industrial context where he explores issues of interdependence and the use of Soft Systems Methodology to develop a case showing rhizomatic systems within a well-defined enterprise structure and system of management in a Korean context. The international perspectives build with Mulej and colleague's paper which explores the shift to an ethic of interdependence within enterprises in a European context. Tracking the evolution of this ethic through recent history, the paper reveals a Eurocentric set of perspectives on social responsibility and sustainable societies.

The papers in this theme present a challenge for system thinkers as many of the issues canvassed have relevance in the other theme sessions presented at ANZSYS 2006, which the authors have shown great discipline in resisting the temptation to address. What is interesting in these papers are the nuances that context brings to human activity systems and the challenges for practitioners of system thinking in tackling the problems of the constituencies they seek to assist.

Sustaining Our Social and Natural Capital: Enhancing the Resilience of Dairy Farmers

Lucia Boxelaar, Michelle Sharma & Mark Paine
Faculty of Land and Food Resources, University of Melbourne, AUS

In recent years, a number of industry and government initiatives have been implemented to develop the knowledge and skills of dairy farmers so they are better able to adapt their farming practices to align with society's expectations for improved natural resource management. The literature suggests that what causes people to positively adapt to a challenging situation is not just education to improve performance, but a degree of resilience, i.e., the capacity for renewal in the face of significant change. This paper discusses a research project that has been instigated to develop a better understanding of the resilience of dairy farmers in the face of critical issues, including natural resource management challenges. It outlines a conceptual framework that includes the attributes of resilience and questions these give rise to for our research into the resilience of dairy farm businesses.

Introduction

Increasingly society expects of dairy farmers that they manage natural resources in a sustainable way. Natural resource management poses a significant challenge to the dairy industry as it often requires considerable changes in dairy practices and systems. A number of industry and government initiatives have been implemented to develop the knowledge and skills of dairy farmers to adapt their farming practices in order to align with society's expectations for improved natural resource management. The literature suggests that what causes people to positively adapt to a challenging situation is not just education to improve performance, but a degree of resilience, i.e., the capacity for renewal in the face of significant change, or as Coutu (2003: 13) defines it, the 'skill and the capacity to be robust under conditions of enormous stress and change'.

This paper reports on a research project that has recently been established with the aim of developing a better understanding of the factors that contribute to the resilience of dairy farmers. The research is funded by Dairy Australia and is embedded within their 'People in Dairy' and 'Natural Resource Management' programs. Working within these programs allows us to inform the development of a range of extension and capacity building initiatives. Below we discuss the conceptual framework that we have developed and the research questions this gives rise to.

Conceptual framework on resilience

Dairy farmers are operating in an increasingly uncertain environment where often the only available strategy for management is 'to accept the inevitability of surprise, and concentrate on building the capacity of the system – society, economy and environment – to cope with whatever happens' (Beratan, *et al.,* 2004: 182). In the face of uncertainty and complexity, the conventional command and control approach that aims to predict what lies ahead in order to plan for it, is increasingly inadequate. The uncertainty and complexity of the current rural change context demands that farmers are able to cope with whatever emerges. Beratan *et al.* (2004: 182) refer to this capacity to cope as resilience. In Adger's words, resilience is the 'ability of a system to evolve in order to accommodate perturbation or to expand the range of variability within which it can cope' (in Armitage 2005: 706).

In what follows we will outline the attributes that have been associated with resilience in the literature. A distinction is made between individual attributes, on the one hand, and attributes of the social setting in which people operate on the other (cf. Sonn and Fisher 1998). However, while we recognise that certain attributes are manifest within individuals, and others within social settings, we do not consider the attributes of the individual as inherent, essentialist characteristics of a person, but as the product of the social setting in which the person operates. Consequently, we expect their manifestation within people to be highly variable and inconsistent. Our aim is therefore not to develop a typology of farmers that would allow us to 'diagnose' a farmer's inherent and stable resilience attributes, but to develop a better understanding of the contextual and social factors that contribute to a farmer's resilience in a particular situation.

Individual attributes

Willingness to face 'reality' of uncertainty and ambiguity

According to Coutu (2003:6), resilient people are characterized by a staunch acceptance of reality. Organization theorists have argued that the capacity of organizations to adapt to changing circumstances is dependent on their ability to become 'learning organizations' (Leeuwis, 2003: 15), which means that they accept that uncertainty and ambiguity are inescapable. Organizational learning is based on reflective practice, which ensures that uncertainties and ambiguities are not suppressed or circumvented, but are worked out. An organizational learning approach brings these to the surface in order to deal with them (Schön, 1991: 254). As Leeuwis (2003:15) argues, this sounds very simplistic, yet in many organizations ambiguities and difficult situations are seen as a sign of weakness and therefore they are obscured and ignored (Schön 1991:69). This leads to a failure to learn and adapt.

Ability to make meaning of events in a way that builds a bridge to the future

Related to the above, is the suggestion that resilient people are able to make meaning out of emerging events in a way that builds a bridge to the future (cf. Coutu, 2003: 10). The shift from production focused agriculture to an emphasis on sustainable development has challenged the existing agricultural narratives or meaning schemes that construct the farmer as an autonomous agent in the rural landscape, whose main concern is agricultural production. Within a complex rural landscape farmers are now one stakeholder group among many;

they operate in an arena they share with global economic interests, broader national interests, conservationists and indigenous interests (Halpin & Martin, 1999). As Lane *et al.* (2004: 110) argue, '"rurality" in Australia is now a space inhabited by diverse communities pursuing diverse practices; the rural landscape is a mosaic not a monoculture.'

The post-modern landscape forces upon farmers a learning process that they cannot avoid if they are to build their bridge to the future. They have to learn to negotiate their way within a growing number of communities of meaning (cf. Dessewffy, 1998), and this involves what has been referred to as 'double loop learning' (Ison, *et al.*, 2000: 39). Double loop learning is a process whereby people become aware of and learn to deal with the contingency of their own representations of the world. It involves '...the deliberate breaking down and restructuring of mental models that support worldviews' (*ibid.*: 39). What this means is that for farmers to build a bridge with the future, their sensemaking needs to extend well beyond existing meaning schemes.

In terms of assessing the resilience of dairy farm businesses it is important to explore what meaning schemes or cultural narratives are constructed within a dairy community or dairy farm business and how these shape the way farmers make sense of and respond to the challenges they face.

A concept of self that is compatible with the current structural changes in agriculture

We mentioned above that the learning process that is required involves farmers learning new roles in a way that challenges their identity and cultural foundations (Dessewffy, 1998). In fact, as farmers are required to take on multiple roles, they may need to develop the capacity to live with a more fluid sense of self and community (cf. Shaw, 2002). Burton and Wilson (2006) argue that farmers' self-concepts and attitudes are not compatible with the current structural changes in agriculture. Their research in the United Kingdom demonstrates that farmers' self-concepts are still dominated by production – oriented identities. While the Australian situation is likely to be very different, this nevertheless raises questions about the identity narratives dairy farm businesses may have access to or construct within a specific social setting; the extent to which these are compatible with current structural changes in agriculture; and how these contribute to their response to dealing with significant issues.

Sense of self-efficacy

The literature suggests that a strong sense of self efficacy plays a significant role in ensuring the well-being of people during significant change events. Self efficacy is associated with peoples' beliefs about their ability to influence positive outcomes. A strong sense of self efficacy positively affects the ways in which people behave, feel, think and motivate themselves, and often results in difficult tasks or problems being approached as challenges (Bandura, 1997). A low sense of self efficacy usually results in problem avoidance, low aspirations, possible depression and a weak commitment to any goals that may want to be pursued (*ibid.*).

Inventiveness

A further aspect of resilience is the ability 'to make do with whatever is at hand' (Coutu, 2003: 14). It requires inventiveness to improvise a solution without necessarily having all the right tools and materials; it is an ability to muddle through.

Social and institutional connectedness

The literature suggests that the degree to which people are connected to their neighbors, their community and the institutional infrastructure is another attribute of resilience (Beratan, *et al.*, 2004). Coutu (2003: 6) suggests resilient people are able to form attachments to others and are able to get others to help them out. Social and institutional connectedness is particularly important in dealing with natural resource management issues as these require a multi-stakeholder response.

Environmental efficacy

The literature suggests that farmers will only adapt and change their practices if they have confidence in the functioning of the broader change process and the networks, organizations and institutions that manage and support this process. Leeuwis (2003: 157) refers to this as environmental efficacy. Farmers also need to have confidence that their peers will engage in the change process. Leeuwis (2003: 157) suggests that in the context of multi-stakeholder issues such as natural resource management, if farmers believe their peers will not change their practices, then they are unlikely to do so themselves.

Anecdotal information suggests that farmers' confidence in the broader social system may be low. For example, farmers often express a lack of confidence in government institutions and processes. It is possible that the lack of confidence in the broader social system around natural resource management is quite widespread amongst dairy farmers, and this is likely to have significant implications for the way in which farmers adapt to natural resource management issues.

Attributes of the social system

Effective natural resource management is dependent on the coordinated activities of a number of diverse stakeholders. The change management process associated with natural resource management must therefore be understood in terms of a '... complex, social process – one occurring among a variety of stakeholders' (Engel, 1997: 125).

Below we discuss four attributes that contribute to the resilience of the broader social system in terms of dealing with natural resource management issues: effective networks and relationships that connect people and organizations; effective institutional arrangements that support the social system; a recognition of mutual interdependence between people in the social system; and finally, diversity within the system.

Networks

Social capital theorists have argued that networks and social ties provide people with a resource that can be used to achieve specific goals (Field 2003). In the agricultural context, Engel's analysis of innovation revealed that it involves a social process that '... emerges from diffuse social interactions among many different actors' (1997: 11). He argues that farmers innovate and change their practices through networking with other actors. Through the interplay between various actors in a network, the influence of available information is mediated, i.e. the network mediates the meaning making process described as critical to resilience above. Engel emphasises the value of *networking* as a process, rather than networks as a resource (*ibid.*:110) and in doing so highlights the emergent nature of networks. He challenges the notion of networks that is implicit in much of the social capital literature that considers networks to exist prior to the social interaction between stakeholders. Instead he emphasizes how these networks are the product of this interaction.

Institutional arrangements

The effectiveness of the networks and social system is influenced by the institutional arrangements within which they are embedded (Macadam , *et al.*, 2004). Institutional arrangements refer to 'the complex of laws, customs, markets, norms and associated organizations that channel our energy toward social goals and the way we relate to others' (*ibid.*: 23). Nettle (2003: 77) suggests that current institutional relationships between research, extension, farming and the community are 'a significant hindrance to eco-agricultural progress'. In Boxelaar (2004:172) it is argued that the natural resource management context is characterized by fragmented institutional arrangements, where relations of mutual engagement and accountability are yet to be established between the diverse yet interdependent stakeholders and agencies involved.

Recognition of mutuality and interdependence

Because improved natural resource management is dependent on the activities of a number of different, yet interdependent stakeholders, it is important that each of these stakeholders recognise that they have a role to play in addressing the issue. In addition, each of the stakeholder groups needs to recognise the legitimacy of other parties to participate in the change process. What is required then is recognition of mutuality (Aarts & van Woerkum, 2002: 421), or mutually experienced interdependence (Leeuwis, 2003: 157).

Sampson refers to this mutually experienced interdependence as collective efficacy, which includes a number of qualities, including affinity, shared identity, reciprocity, trust, informal social control and willingness to act for the good of the group (in Vinson, 2004: 32).

Diversity

A final attribute of the social system that we discuss here is diversity. The literature suggests that social resilience is enhanced with the availability of a diversity of programs or pathways to adapt (Beratan, *et al.*, 2004). For example, Adger

	Attribute	Questions to inform our research
Individual	Willingness to face 'reality' of uncertainty and ambiguity	• To what extent do farmers and extension officers acknowledge the fact that uncertainty and change are inescapable? What stops them from acknowledging this? • To what extent do extension programs acknowledge that they need to build the capacity of farm businesses to become learning organizations; rather than provide farmers with a once off set of skills and knowledge to deal with a particular issue?
	Ability to make meaning of events in a way that builds a bridge to the future	• What meaning schemes or cultural narratives are available within a specific dairy community and how do these shape the way farmers make sense of and respond to the challenges they face? • How can extension practitioners ensure that a setting is created that provides engagement with positive and constructive cultural narratives?
	A sense of self that is compatible with current structural changes in agriculture	• What identity narratives are available to farmers within the case study regions and how do these shape the way farmers make sense of and respond to the challenges they face? • What change is needed in terms of how farmers perform their identities, if they are to be more resilient in dealing with critical issues? • How can extension practitioners ensure that a setting is created that facilitates engagement with identity narratives that are compatible with the current structural changes in agriculture?
	Self-efficacy	• How confident are farmers in their ability to affect change and to influence or manage issues around natural resource management? • What can extension practitioners do to increase farmers' confidence, so that they are more likely to engage with these issues and change their practices?
	Inventiveness	• To what extent are farmers able to make do with whatever is at hand? • To what extent do extension strategies build the capacity for improvisation and inventiveness?

	Social and institutional connectedness	• How well connected are farmers within their region and how does this shape the way in which they respond to critical issues? • How can extension programs enhance the social and institutional connectedness of dairy farmers?
	Environmental efficacy	• Do farmers have confidence in the functioning of the broader social system, including their peers, the organizations and other stakeholders involved – to achieve natural resource management outcomes? • What can be done to improve levels of environmental efficacy amongst dairy farmers?
Social system	Networks	• How are farmers socially organized around natural resource management issues? • How can the networking between farmers and other stakeholders be improved to enhance the capacity of the broader social system and of farmers themselves to address natural resource management issues?
	Institutional arrangements	• How do current institutional arrangements support the farmers and the broader social system to achieve natural resource management outcomes? • What can be done to improve the institutional arrangements?
	Sense of mutuality/ collective efficacy	• Do stakeholders and farmers recognise that they have a role to play in natural resource management? • Does each of the stakeholder groups recognise the legitimacy of other parties to participate in decision making around natural resource management? • What needs to be done to ensure a sense of mutual interdependence?
	Diversity of pathways to adapt to critical issue	• What can be done to improve the diversity of available pathways to respond to a critical issue?

Table 1 *Attributes of resilience*

(2000:352) suggests that communities that rely on a single underground mineral resource have severely limited capacity to adapt to shocks or challenges.

Overall, a number of attributes have been linked with resilience. Table 1 summarizes these attributes and outlines the questions they raise for our research.

Research design

The framework above provides the basis for empirical work which will be conducted over the next two years. Our research will utilise a case study approach, and comprises an in-depth exploration of four dairy regions in Australia, where farmers are dealing with a variety of critical issues, including succession planning, water reform, and natural resource management challenges. It is anticipated that initial scoping of the issues within each case study will assist in prioritizing specific research questions from the table above. We will be observing meetings of extension projects and conducting interviews and focus groups with farmers and service providers in each of the four regions. Transcripts and notes of these events will be analyzed using NVivo software to identify key themes in the data. In addition, a narrative strategy (cf. Bojé, 2001; Boxelaar, 2004) will be used to examine the stories that people invoke or tell explicitly about themselves and the way in which they make sense of the critical issues they are dealing with. This will allow for an exploration of the systems of meaning that operate in dairy communities.

In order to ensure that our research findings are implemented and utilised by program managers, extension officers and service providers, we will be implementing a participatory action research approach (Foote-Whyte, 1991; Reason & Bradbury, 2001). This approach was developed in response to traditional academic research, which separates the development of new knowledge from its implementation in practice. Action researchers ensure that the link with practice is built in as a structural component of the research process. The link with practice is achieved by actively involving research 'clients' (in this case, program managers, service providers and extension practitioners) in the research process. Within each case study region a participatory action research team will be established that comprises relevant service providers, extension staff, consultants, milk company field staff, rural counsellors, financial service providers and others. The team will assess the extent to which the current mix of services enhances the resilience of diverse dairy farm businesses, and it will develop and trial strategies to address any gaps.

Conclusion

In this paper we have discussed a framework that outlines the attributes of resilience, which will provide the basis for field work to be conducted in four case study dairy regions in Australia. It is anticipated that this integration of social research within existing dairy extension programs is likely to make a significant contribution to unlocking some of the constraints to change.

References

Aarts, N. and van Woerkum, C. (2002). "Dealing with uncertainty in solving complex problems," in C. Leeuwis and R. Pyburn (eds.) *Wheelbarrows full of frogs: Social learning in rural resource management*, Assen: Koninklijke Van Gorcum, ISBN 90-232-3850-8, pp. 421-435.

Adger, W. (2000). "Social and ecological resilience: Are they related?" *Progress in Human Geography,* ISSN 0309-1325, 24(3): 347-364.

Armitage, D. (2005). "Adaptive capacity and community-based natural resource management," *Environmental Management,* ISSN 0364-152X, 35(6): 703-715.

Bandura, A. (1997). *Self Efficacy: The exercise of Control,* ISBN 0716726262, New York: Freeman.

Beratan, K., Kabala, S., Loveless, S., Martin, P. and Spyke, N. (2004). "Sustainability indicators as a communicative tool: Building bridges in Pennsylvania," *Environmental Monitoring and Assessment,* ISSN 1573-2959, 94: 179-191.

Bojé, D. (2001) *Narrative methods for organizational and communication research,* ISBN 0761965874, London: Sage Publications.

Boxelaar, L. (2004). Diversity and convergence in platforms for change: Building social capability for land management. PhD Thesis, Institute of Land and Food Resources. University of Melbourne.

Burton, R. F. and G. A. Wilson (2006). "Injecting social psychology theory into conceptualisations of agricultural agency: Towards a post-productivist farmer self-identity?" *Journal of Rural Studies,* ISSN 0743-0167, 22: 95-115.

Cary, J., Webb, T., and Barr, N. (2002). *Understanding landholders' capacity to change to sustainable practices: Insights about adoption and social capacity for change*, Canberra, Bureau of Rural Sciences.

Coutu, D. (2003). How resilience works. *Harvard business review on building personal and organizational resilience,* ISBN 1591392721, Boston, Harvard Business School Press: 19-38.

Dessewffy, T. (1998). Strangerhood without boundaries: An essay in the sociology of knowledge, in S. Suleiman (ed.), *Exile and creativity: Signposts, travelers, outsiders, backward glance,* ISBN 0-8223-2187-4, London: Duke University Press: 353-370.

Engel, P. (1997). *The social organization of innovation: A focus on stakeholder interaction,* Amsterdam: Royal Tropical Institute.

Field, J. (2003). *Social capital,* ISBN 0415257549, London: Routledge.

Foote-Whyte, W. F., Ed. (1991). *Participatory action research,* ISBN 0803937431, London: Sage Publications.

Halpin, D. and P. Martin (1999). "Farmer representation in Australia: Avenues for changing the political environment," *Australian Journal of Public Administration,* ISSN 03136647, 58(2): 33-46.

Ison, R., High, C., Blackmore, C.P. and Cerf, M. (2000). "Theoretical frameworks for learning based approaches to change in industrialised country agricultures," in Cerf, M., Gibbon, B., Hubert, B., Ison, R., Jiggins, J., Paine, M., Proost, J. and Röling, N. (eds.) *Cow up a tree: Knowing and learning for change in agriculture. Case studies from industrialised countries,* ISBN 2738009298, Paris: Institut National de la Recherche Agronomique: 31-53.

Lane, M. B., McDonald, G.T. and Morrison, T.H. (2004). "Decentralisation and environmental management in Australia: A comment on the prescriptions of the Wentworth Group," *Australian Geographical Studies,* ISSN 00049190, 42(1): 103-115.

Leeuwis, C. (2003). *Communication for rural innovation,* Third Edition, ISBN

063205249-x, Oxford: Blackwell Publishing.
Macadam, R., Drinan, J.P., Inall, N, and B. McKenzie (2004). *Growing the capital of rural Australia - the task of capacity building,* ISBN 0642587442, Barton, ACT: Cooperative Venture for Capacity Building for Innovation in Rural Industries.
Nettle, R. A. (2003). The development of a national dairy extension strategy: A literature review. Parkville, Institute of Land and Food Resources, University of Melbourne.
Reason, P. and Bradbury, H. (2001). *Handbook of action research: Participative inquiry and practice,* ISBN 0761966455, London: Sage Publications.
Schön, D. (1991). *The reflective practitioner: How professionals think in action,* ISBN 1857423194, London: Arena.
Shaw, P. (2002). *Changing conversations in organizations: A complexity approach to change,* ISBN 0415249155, London: Routledge.
Sonn, C. C. and Fisher, A.T. (1998). "Sense of community: Community resilient responses to oppression and change," *Journal of Community Psychology,* ISSN 00904392, 26(5): 457-472.
Vinson, T. (2004). *Community adversity and resilience: The distribution of social disadvantae in Victoria and New South Wales and the mediating role of social cohesion,* ISBN 0975191500, Richmond: The Ignatius Centre for social policy and research.

Acknowledgements

The authors would like to thank Dairy Australia for the funding received for this research. We would also like to thank Pauline Brightling, coordinator of the People in Dairy sub-program and Cathy Phelps, coordinator of the Natural Resource Management Program for their support and assistance in developing the research.

Lucia Boxelaar is a Research Fellow in the School of Agriculture and Food Systems of the Faculty of Land and Food Resources at the University of Melbourne. Her research interests include community capacity building methodologies and the organizational alignment required to implement these successfully. She teaches social research methods and community management of land and natural resources. Lucia previously worked as a coordinator of a community development program with the Country Fire Authority. Lucia holds a Bachelor of Arts (Honours) degree (1994) from Monash University and a PhD from the University of Melbourne (2004).

Michelle Sharma is a Research Fellow in the School of Agriculture and Food systems of the Faculty of Land and Food Resources at the University of Melbourne. Her research interests include understanding peoples' capacity for change in rural and remote communities. Michelle previously worked as a Project Officer for the Centre for Rural and Remote Mental Health, University of Newcastle. Michelle holds a Bachelor of Landscape Management and Conservation (Honours) degree (2000) from the University of Western Sydney and a Masters Degree from Charles Sturt University (2005).

Mark Paine is a Principal Research Fellow in the School of Agriculture and Food Systems of the Faculty of Land and Food Resources at the University of Melbourne. His research focuses on change and innovation management in the agricultural sector. Mark moved to Melbourne in January 2000 after managing the social systems research group based at Ruakura, Hamilton New Zealand. Mark has previously worked in private sector consultancy, public sector extension and university teaching since graduating from Lincoln College in 1981. He holds a Masters degree in Agricultural Systems from Massey University and a PhD from Wageningen Agricultural University (The Netherlands).

Ethic of Interdependence as a Basis for Ethic of Social Responsibility and Sustainable Society

Matjaz Mulej, Jozica Knez-Riedl & Vojko Potocan
Faculty of Economics and Business, University of Maribor, SLO

Sustainable society is based on sustainable enterprises. Their social responsibility is high, based on ethic of interdependence rather than individualism; it implies systemic thinking by holistic perception, thinking, decision-making, and action.

The selected problem and viewpoint of our dealing with it

A 20%-part of humankind established entrepreneurship and innovation by freeing the individual initiative and competition instead of previous monopolies of guilds, clergy, and nobles. It became the world's advanced part. The other 80% of humankind did not do so or did not do so then. (Rosenberg & Birdzell, 1986). In World Bank data the span of national per capita incomes grew from 3:1 before freeing the individual initiative, to 100:1 in one century before 1970, and to +500:1 now. This process changed the prevailing knowledge and ethic. Individualism prevailed. Specialization grew along the growing amount of humankind's knowledge. Besides their good consequences, individualism and specialization make people think narrowly. After 1950, this dangerous habit caused establishment of systems theory; its practice is still too limited for humankind to avoid current problems. Now, knowledge and ethic of sustainable development enhance it by ethic of interdependence, social responsibility, and sustainable society.

One-sidedness, general systems theory, sustainable development paradigms

One-sidedness of knowledge and ethic of humankind reached a very dangerous triple pick in 1914-1945: two world wars and the economic crisis. People kept dying in millions in wars etc. later on, too, due to one-sided decisions of the influential ones. Scientists, who lived in these terrible times, offered humankind systems theory to prevent re-appearing of such crises.

The General Systems Theory was first, now there are many systems theories and cybernetics (François, 2004). Many disciplines that have emerged from them tend to forget their roots (Umpleby, 2005). Not all systems theorists/cyberneticians accept that Bertalanffy (1979, VII) has created systems theory against over-specialization, not as one of many disciplines, but as culture of holism and methodology supporting it. Other pioneers also created 'the science of synthesis' (Hammond, 2003). Elohim (1999) quoted the reason: 'to survive, people must behave as citizens of the entire world and consider the entire biosphere; the world is full of interdependences'. United Nations, the humankind's highest political body, supported this attitude: sustainable development links care for socio-economic development and nature (UN, 1992; WBCSD, 2004, 2005; WCED, 1987/1998).

Systemic thinking and sustainable development are poorly accepted for three crucial reasons:

1. People are unavoidably narrow specialists to know something about a selected profession and topic. They mostly apply systems theory to describe their selected topics from their own single selected viewpoint; few use systems theory as the culture and methodology of holism by interdisciplinary creative cooperation (Mulej, 2006).
2. Market and democracy replaced the solidarity of many millennia and feudal monopolies. But the 20th century crises showed expectations are too optimistic. Market experiences many more interventions, than the first authors expected. Democracy practices changing of one-sided parties in power, less a society-wide holism and creativity enhancer.
3. After the 2nd World War the speed of development and innovation, supporting survival in competition exceeds human capability to adapt to changing conditions.

Figures 1-3 illustrate points 1-3.

After WWII, competition keeps causing lower cost, including a poor care for nature, if short-term and one-sided views prevail in ethic and action. Costly eco-remediation, health care, organizational, managerial, business and technological innovation concerning e.g. emissions in air and water and their prevention under ISO 14000, become necessary. Therefore, the sustainable enterprise is the best solution in economic terms (Ecimovic, *et al.*, 2002; Knez-Riedl, *et al.*, 2001; Potocan & Mulej, 2006). It enables development toward the sustainable society where requisite holism will include ethic of interdependence of specialists, needing each other.

Thus, ethic of interdependence is replacing ethic of individualism and social irresponsibility of individuals, companies, and countries. Earlier, interdependence and solidarity lost ground, because the most individualistic people created prevailing values, culture, ethics, and norms. Ethic, supported with 'external economics', which makes people forget about the broader social consequences of their actions helped them in a short term only. For selfish reasons one may not be too selfish.

Global economy demands enterprises to innovate worldview toward corporate citizenship (McIntosh, *et al.*, 1998: 4, 61) with its rather holistic combination of managers' competences and sensibility for problems that used to be found unimportant for business.

Humankind has, for the first time in modern history, a strong and broadly accepted feeling that things will not turn better and we have no exit (Chomsky, 1997: 294). Humankind must face its own individual mortality and possibility of its collective death (Keane, 2000: 134).

There is no need to wish, that a citizen would be either too good or too subjective, nor too senseless or too realistic: "He or she must act by considering

←→		
Fictitious holism/realism (inside a single viewpoint)	Requisite holism/realism (a dialectical system of all essential viewpoints)	Total = real holism/realism (a system of all viewpoints)

Figure 1 *The selected level of holism and realism of consideration of the selected topic between the fictitious, requisite, and total holism and realism*

Type of Market	Basic Relation/s Between Production and Consumption	Humans' Impact on Sustainability
RANDOM MARKET	Producers' own consumption and occasional exchange of random surpluses	Minimal impact, growing as humankind grows in number and requirements
PRODUCERS' MARKET	Growing production for poorly considered, customers, lacking impact over suppliers	Specialization and narrow thinking grow; so does the humans' detrimental impact over nature, especially by industrialised production
BUYERS'MARKET	Growing impact of customers requiring total quality of products, services, and conditions of life	Specialisation and its bad one-sided impact over nature keep growing, so does biased application of science, causing need for interdisciplinary cooperation
STATE SUPPORTED BUYERS' MARKET	Increasingly organised / legalised impact of customers demanding total quality of products, services and conditions of life	Growing awareness about the terrible impact of humankind's one-sided impact over nature and its dramatic consequences for humans' survival

Figure 2 *Development of market relations and environmental care quality*

oneself, subgroups and society. This moral is eternal; it is common to most developed societies, societies of near future, and less developed societies, as much we can imagine them. Those are bases." (Mauss, 1996: 141). We find them close to what we call ethic of interdependence. Globalization keeps requiring it. Managers, who consider the new reality, can (McIntosh , *et al.*, 1998: 39-40):

- Think in terms of global citizens,
- Be interested in various ideas and opinions,
- Work with people with various experiences and different perspectives about the world,

Decade	**Market & Social Requirements**	**Enterprises' Ways To Meet Requirements**	Type of Enterprise
1950	Covering of post-war conditions (lack of goods, rebuilding)	Supply anything; supply lower than demand	Supplying
1960	Suitable price (customers judgement)	Internal efficiency, i.e., cost management	Efficient
1970	Suitable price X[1] quality (customers judgement)	Efficiency X technical & commercial quality management	Quality
1980	Suitable price X quality X range (customers judgement)	Efficiency X technical & commercial quality X flexibility management	Flexible
1990	Suitable price X quality X range X uniqueness (customers judgement)	Efficiency X technical & commercial quality X flexibility X innovativeness management	Innovative
2000	Suitable price X quality X range X uniqueness X care for nature (customers judgement)	Efficiency X technical & commercial quality X flexibility X innovativeness X care for nature	Sustainable

Figure 3 *From a supplying to a sustainable enterprise*

- Create relations and construction of new societal and organizational structures,
- Predict other societal realities,
- Lead in complicated and confused surroundings,
- Lead over geographical limits,
- Understand individuals' values and the fact that we are all involved in business with values.

This is why Figure 3 exposes sustainable enterprise; it is a step to sustainable society.

This is where the notion of social responsibility enters the scene. But beware: promotion of interdependence, including social responsibility, faces serious difficulties (Midgley, 2004; Mulej, *et al.*, 2004) despite of old roots of the notion of interdependence (Smith, 1759, 1776). Social responsibility must not be fictitious or old-fashioned, e.g., writing nice codices of conduct per professions without following them, forgetting about the tacit rules of behavior in any profession or situation, etc.

Social responsibility in practice

We tried to define social responsibility in practice at two conferences (Hrast, *et al.*, 2006; Rozman & Kovac, 2006). Selected outcomes:

- It requires decisive persons to 'think globally, while acting locally' and 'to do good in order to do well'. One should do so to prevent consequences of one-sidedness – oversights that can hit the one-sided person back with bad consequences, which can be prevented with more systemic thinking, covered under social responsibility (Knez-Riedl, *et al.*, 2006).

- It requires balancing of several alternatives such as:
    * Individuals, who do not buy avoidable cloths, cars etc., do good for nature, but not for employment.
    * Individuals, who do good by subsidizing fire-brigades, sport and culture associations, do good for the quality of leisure-time of people, but endanger their own investment, innovation and competition capability. Then, they cannot support the said activities.
    * It is bad, if youngsters must work in e.g., factories in less advanced countries for a rather small pay. But it is even worth, if they have nothing to live on.
- Several professions developed their codes of ethics.
- Smaller enterprises meet social responsibility by donating time and money to local actions.
- Practice of 'fair trade' helps marginalized producers in developing countries etc.
- More education and communication about social responsibility practice is needed.
- 'Social Accountability Standard' exists since 1997, based on conventions of the International Labour Organization, the Universal Declaration of Human Rights and the UN convention of the Rights of the Child.
- The motive spans from economic, legal, ethical, to philanthropic backgrounds.
- Individuals can fight socially irresponsible behavior: (1) as consumers they make their choices, (2) as citizens they can conduct and/or join protests, boycotts, etc. Such actions can make the influential persons more requisitely, causing more social responsibility. (Knez-Riedl, 2003a, 2003b, 2003c, 2004).
- Unpleasant press, detecting and publicizing bad behavior, abuse and misuse of influence and power can help humankind develop social responsibility. There are many misuses around (Rant, 2006).
- Perhaps corporate social responsibility provides excuse for one-sided behavior of the influential ones (Tavcar, 2006), rather than a way to better life of all (Potocan & Mulej, 2006).

Conclusions

Sustainable society grows from sustainable enterprises and individuals. If they lack ethic of interdependence, they tend to lack social responsibility and feel/think in too narrow horizons for humankind to survive.

References

Bertalanffy, L. (1968/1979): *The General Systems Theory.* Braziller, New York.

Bolwijn, T., Kumpe, T. (1990): Manufacturing in 1990's. *Long Range Planning*, 23(4), 12-26.

Chomsky, N. (1997): Growing dark of democracy. In Slovenian. Ljubljana: Studia humanitatis.

Ecimovic, T., Mulej, M., Mayur, M. (2002): *System Thinking and Climate Change System.* Korte. SEM Institute for Climate Change. www.institut-climatechange.si

Elohim, J. L. (1999): *Letter from Prof. Elohim.* 10th WOSC Congress. Uxbridge. WOSC

Embley, L. (1993): Doing well while doing good. New Jersey, NJ: Prentice Hall, Englewood Cliffs.

François, Ch., ed. (2004): *International Encyclopedia of Systems and Cybernetics. 2nd Ed.* Saur, Munich

Hammond, D. (2003): *The Science of Synthesis. Exploring the Social Implications of General Systems Theory.* University Press of Colorado, Boulder, CO

Hrast, A., Mulej, M., Knez-Riedl, J., eds. (2006): *Social Responsibility and Challenges of the Time 2006. In Slovenian.* Maribor, IRDO.

Keane, John (2000): Civil society: old pictures, new visions. In Slovenian. Ljubljana: Znanstveno in publicisticno sredisce.

Knez-Riedl, J., Mulej, M. and Zenko, Z. (2001): Approaching sustainable enterprise. In Lasker, G. E., Hiwaki, K. (Eds.): *Sustainable development and global community.* International Institute for Advanced Studies in Systems Research and Cybernetics

Knez-Riedl, J. (2003a): Corporate social responsibility and communication with external community. *Informatologia (Zagreb)*, 36, 166-172

Knez-Riedl, J. (2003b): Social responsibility of a family business. *MER,* 5(3), 90-99

Knez-Riedl, J. (2003c): Corporate social responsibility and holistic analysis. in: Chroust, G., Hofer, Ch. (eds.). *IDIMT-2003: proceedings,* (Schriftenreihe Informatik, Bd 9). Linz: Trauner, 187-198

Knez-Riedl, J. (2004): Slovenian SMEs: from the environmental responsibility to corporate social responsibility. In: Sharma, S. K. (ed.). *[An enterprise odyssey: building competitive advantage. 2004,* (Zagreb International Review of Economics & Business), 127-139

Knez-Riedl, J., Mulej, M., Dyck, R. (2006): Corporate social responsibility from the viewpoint of systems thinking, Kybernetes, 35, 441 - 460

Mauss, M. (1996): Essay on talent, and other essays. Introduction to work of Marcel Mauss/ Claude Levi-Strauss. In Slovenian. Ljubljana: SKUC: Znanstveni institut Filozofske fakultete – Studia humanitatis.

McIntosh, M., D. Leipziger, K. Jones, G. Coleman (1998): Corporate citizenship. London: Financial Times Management.

Midgley, G. (2004): Systems Thinking for the 21st Century. *International Journal of Knowledge and Systems Sciences,* 1, 63-69

Mulej, M. (1979): *Creative Work and Dialectical Systems Theory (in Slovenian).* Razvojni center, Celje.

Mulej, M. (2006): Systems theory – a worldview and/or a methodology. In: Trappl, R. (ed), *EMCSR '06.* Vienna. Austrian Society for Cybernetics Research, XXV-XXVII

Mulej, M., Potocan, V., Zenko, Z., Kajzer, S., Ursic, D., Knez-Riedl, J. (2004): How to restore Bertalanffian systems thinking. *Kybernetes,* 33, 48-61.

Potocan, V., Mulej, M., Kajzer, S. (2005): Business Cybernetics. *Kybernetes*, 34, 1496-1516.

Potocan, V., Mulej, M. (2006): Social responsibility of a sustainable enterprise In: Rozman, Kovac, quoted here; 41-44. In Slovenian

Rant, M. (2006): Negative impacts of governance structure on business decision and

socially responsible action; current trends in Europe. In: Rozman, Kovac, quoted here; 45-50. In Slovenian
Rosenberg, N., Birdzell, L. E. (1986): *The Past: How the West Grew Rich.* Basic Books, New York.
Rozman, R., Kovac, J., eds. (2006): *The 7th conference on organization: Social responsibility and ethic in organizations.* In Slovenian. Zveza organizatorjev Slovenije etc., Kranj,
Smith, A. (1759/2000): *The Theory of Moral Sentiments.* Prometheus Books, New York.
Smith, A. (1776/1998): *An Inquiry into the Nature and Causes of the Wealth of Nations.* Modern Library, New York.
Tavcar, M. (2006): Social responsibility or corporate ethic. In: Rozman, Kovac, quoted here; 65-72
Umpleby, S. (2005): Some Influences on the Cybernetics Movement in the United States. In: Mulej, M. et al, eds (2005): *Proceedings of The WOSC 13th International Congress of Cybernetics and Systems.* WOSC, with SDSR, and University of Maribor, FEB, Institute for Entrepreneurship and Small Business Management. Maribor
UN (1992): *Rio Declaration.* UN, Rio de Janeiro.
WBCSD (2004): *A Business Guide to Development Actors.* WBCSD, Geneva.
WBCSD (2005): *Business for Development.* WBCSD, Geneva.
WCED (1987/1998): *Our common future on environment and development.* Oxford University Press, Oxford.

Notes

[1] X denotes interdependence. No attribute is avoidable. The original (Bolwijn & Kumpe, 1990) did not contain X, but +. Experience shows summation is an oversimplification. Decades of 1950 and 2000 were not contained.

Exploring the Real-Time Enterprise From Systemic Perspective: The Case of Korean Enterprise

Jae Eon Yu
Department of Information Management, Seoul University of Venture & Information, KOR

There is a growing awareness that the social and cultural aspects of the Real-Time Enterprise (RTE) are the critical factor of good business practice. This fact shows that sustainable development and long-term profitability of business enterprise will not continue through only having with capital and technology. A new form of business strategy is necessary for complementing the availability of 'social capital' and social network analysis for maintaining the viability of the business enterprise in the era of a knowledge-based economy. We explore the nature of the RTE from rhizomatic systems perspectives. Rhizomatic systems perspectives enable us to appreciate the holistic understanding of social, cultural and political dimensions of the RTE in the form of social networks or communities. In this paper, 'rhizomatic systems' thinking is demonstrated through the application of Checkland's Soft Systems Methodology (SSM) to appreciate the process of problematization, and generate systemic knowledge about the transformational processes of Korean enterprise.

Introduction

For real-time organizations attempting to be successful in the 'knowledge society', the concept of social capital is useful. While there are many success factors for creating values and maintaining relationships, building effective networks through constitutive partnerships, knowledge sharing and learning has attracted significant research attention (Parkhe, *et al.*, 2006). Briefly, the field of social network analysis focuses on the social process that makes sure people have time and encouragement to participate from the network of relationships posses by an individual or social unit (Nahapiet and Ghoshal, 2000: 132-140). Social network analysis focuses on value created by fostering connections between individuals (Lesser, 2000: i). Acknowledging the fact that social network will increasingly be influenced by human networks and communities that make cooperative action possible, the aim of research is to help practitioners and managers to understand the value of soft/ rhizomatic systems thinking and concept of social networks for the successful execution of the Real-Time Enterprise (RTE). The case study presented and analyzed in details throughout the application of Soft Systems Methodology (SSM) within Korean company.

Exploring the real-time enterprise from systemic perspective

How the real-time organization functions, that is to say how they control their actions, how individuals learn from errors and mistakes that make in their workplace? Put differently, how do we understand the cultural aspect of the Real-Time Enterprise (RTE)? Systems theory provides the conceptual basis for discussing the dynamics and inner workings of any real-time enterprise. It enables us to understand and reinterpret how effective real-time organizations work. Systems thinking, which is based on holism, feedback, self-regulation, self-reflection, creating values and maintaining relationships, is useful to explore the complex nature of the RTE within changing environments (Hugos, 2005). For instance, as we try to make a sense of a new organizational form that is suited for maintaining organizational identity within constantly changing environments than the traditional hierarchical forms, the Viable System Model (VSM) and Soft Systems Methodology (SSM) offer useful insights, ideas and techniques (Hugos, 2005: 41-58). Roots of the RTE can be trace back to the 1930s at organization as Bell Laboratories and other research labs and universities (Hugos, 2005: 15). Historically, the principles and practices that define the RTE have been evolved over the last 80 years or so. Gartner, a prestigious research firm defines the idea of the RTE as a 'enterprise nervous system' or 'living nervous system' which is concerned with collaboration between the RTE and suppliers, and linking with trading partners to produce better design and demand forecasts for real-time applications (McCoy, 2002). In this sense, the RTE can be viewed as an open, adaptive, living system or social network systems. To be open systems, according to Ashby's Law of Requisite Variety, the RTE must be at least as great as the variety of environmental turbulence. In this vein, a living nervous system must have sufficiently minimum and necessary varieties of actions in order to ensure to maintain organizational viability from the environmental changes. To be an adaptive whole system, the RTE must have capacity to learn from experience and perform anticipatory learning, in order to achieve adaptation at all level: from system functioning to systems constitution. In this way, an understanding the RTE in terms of systems theory, would provide a theoretical basis for supporting the learning processes of self-transformation or self-organization towards higher states of systems stability and the complex emergence of the whole living systems. The living nervous systems will have real-time principles to their internal processes, achieve the continuously improving key business processes, and adopting them to changing environments (Flint, 2002). From systemic perspective, there are some fundamental principles that characterize why and how time-based organizational transformation will happen within the RTE. These principles are as follows (Raskino, 2002). Firstly, the simplicity and openness of time-based improvements will promote broader understanding within the RTE across, its value network and within its stakeholders' base. Secondly, in the RTE, the management of elapsed time is given much higher priority, visibility and management information support. The setting of headline goals for the end-to-end delay reduction will cause questioning of all new, time-consuming tasks and a rational debate about their value. Eventually, wasting elapsed time will become as culturally unacceptable as wasting money. There is a need for the structure of the RTE to make control of elapsed time. Thirdly, the RTE is event-driven. As soon as a requirement arises, the RTE can judge its performance on how quickly it can satisfy the requirement. Requirements come from customers, supply chain partners, shareholders and other organizations, in the case of automated processes and activities, from other systems. Each of these entities has a window, from the time the original request is notified, to the time the response comes. Beyond a certain time, the wait causes irritation and frustration, or the response becomes less valuable, and end eventually, irrelevant. With the ever-quickening response on events that happen in environments, these 'tolerance windows' that is, how long the recipient will wait for the response, will get shorter. The managers must understand the current tolerance window of the recipients and the rate at which these windows are shortening. Fourthly, although we refer to the RTE, it is not sufficient to consider a single enterprise in isolation. Elapsed-time saving should focus on processes that span organizations horizontally and vertically. It will have little impact if initiatives are only local and uncoordinated. Therefore, most end-to-end process cycles that are worth tackling will cross value networks. In many cases, the RTE collaboration will involve suppliers and supply chain partners in shrinking process times. Finally, business ethics and integrity are being questioned in ways not seen for a decade or more. In order to make transparency amongst supply chain partners, customers, and shareholders, ethical governance is needed for better decisions, moral behaviors and ethical conducts. Creating ethical governance through improvements of transparency and speed will increase trust amongst people. This will contribute to increase a synergy of various working groups and organizations, and lower internal costs by improving inefficient management processes. Dealing with complex situations, knowledge management is necessary to activate and synchronize various working units and viable organizations within the RTE in order to achieve a synergy of various working groups and organizations.

Understanding social networks from rhizomatic systems thinking

Understanding the nature of complexity from systemic perspectives, complexity means not only nonlinearity but a huge number of elements with many degrees of freedom that leads to unpredictable, complex, irregular, and chaotic states in a given situation (Mainzer, 1994: 3). Dealing with complexity, as Yu (2001, 2006) has shown, the concept of 'rhizomatic systems thinking' is useful. It deals not only with the knowing about problem situations, but also with thinking about (and doing something about) problem contexts. Thinking about problem contexts, the formation of a community of inquiry and practice is useful when it related with knowledge management and organizational learning, which are the interdisciplinary frameworks of systemic interventions. We now live in a society where the mechanisms of power and knowledge are greatly transformed. Building community of inquiry and practice focused on organizational process and in the creation of new knowledge.

In turn, studying the processes of social participation in which learning takes place both individually and collectively within these communities. While there are many kinds of communities, we find what we call 'rhizomatic networks' relevant and consistently useful to our particular interests, the way of understanding micropolitics and micropower. For Deleuze and Guattari (1988) micropolitics is the essence of what rhizomatic networks are based on the idea of learning and acting as social participation. Groups of people share a concern, a set of problems, or a passion about *desire*, and pursue common pursuit of solution through the process of problematization and subjectivization. Rhizomatic networks are, in effect, constituted by flows of desire, belief, micropolitics and micro-power, in which rhizomic decision-making can occur in the process of learning and acting as social participation. Key feature of rhizomatic networks include: shared visions and interests, shared desire, belief and knowledge, voluntary participation, autonomy in setting goals and facilitating learning and outcomes, awareness in learning that can be central or peripheral to the process of developing connection which, as a totality, because everything is affected, effects everything (Jackson and Carter, 2000). We have said that rhizomatic networks embeds social networks, which relates to complex relations with multiple 'real interests' of the various actors involved within such networks (Coleman, 2000). As Foucault argued, the social forms of power and knowledge can define relations of force in terms of the everyday life of men (Foucault, 1977: 205). In this sense, we argue that social networks are not only concerned with knowledge creation and sharing, but concerned with power relations are emerged as relations of force which are the very basis of a power analysis. Having said that, we argue that local form of power/knowledge exists in 'regional' and 'local' forms in any social networks, which embed social capital that is the network of relationships that individuals have throughout the organization (Lesser, 2000).

Case Study: The project for the EA manufacturer

Project outline

As part of the requirement of the work in the project for the EA manufacturer, participatory action research (PAR) was carried out from October 2005 to September 2006. EA is a Korean semiconductor manufacturer which in 2006 have around 2,100 employees and had a sales turnover $ 360 millions for the financial year 2005. EA is a pseudonym for the company in order to protect the confidentiality of the informants. The 'direct' manufacturing division mainly performed the corporate functions of EA, which is responsible for production and production development, and the sales department that is directly concerned with the selling the semiconductors. The 'indirect' manufacturing division mainly functioned to facilitate the overall activities of 'direct' manufacturing within EA. The 'Managerial Innovation Centre' (MIC) is mainly concerned with the strategic planning and its operational activities which are mainly concerned with organizational change and learning within EA. The author, who was the systems analyst of the research project, acted as a facilitator for carrying out participatory action research within EA.

Application of the 'problem-solving' method

Peter Checkland (1981) proposed Soft System Methodology (SSM) in the 1980s and 1990s. SSM regards as a 'problem-solving' approach that is suitable for dealing with 'human activity systems' in organizational and social contexts (Checkland, 1981; Checkland & Scholes, 1990). SSM is to make a systemic inquiry into the complexity of problematic situations in which Morgan (1986) sees organizations as culture. Dealing with cultural and social contexts of RTE within EA, soft systems methodology (SSM) was used to explore problem situations within EA. The basic process of the systemic intervention using SSM is summarized by the following stages.

Stage 1: Finding out "key issues"

The methods of finding out a 'rich picture' of EA were conducted by the study of the written documents, official records, social network analysis, observation and informal and formal interviews. It was identified that a centralized decision-making process was formed through a top-down hierarchy of EA. There were a tendency for poor communication, lack of trust and understanding between managers and workers throughout the divisions in EA.

Stage 2: Express the problematic situation

In addressing any messy problem situations concerning the successful operation of the RTE within EA, the various perceptions were identified. These perceptions were expressed as follows.

Person A

"The leaders of the working groups (including MIC) lack a vision for the successful operation of the RTE that functions in the form of social network to react to environmental changes."

Person B

"There is a need for exchange a good quality of information between managers, working staff and customers at a regular basis in order to make fast decision-making for improving the efficiency of work."

Person C

"Taking into consideration the semiconductor industry, what can we do about corporate culture that gives us pride in our work?"

Person D

"There are no effective evaluation systems and procedures for the measurement of operational performance of individual workers and working groups within the 'indirect' manufacturing division."

Person E

"Communication flow is top-down and managers exercise a directive leadership. Our corporate policy focuses mainly on the operational performance in terms of the financial criteria. Workers do not wish to interact with others unless they have to do. Having to deal with a hierarchical structure within EA, it is difficult to share the information and knowledge amongst all the levels of the employees in the various divisions and working teams in EA... I would like to see a good relationship and more interactions between managers, workers, suppliers, large and small customers within EA in order to make the successful implementation of the RTE within EA."

There were other 'areas of concern' that identified by the social network analysis. These are summarized as follows.

- Leaders of working groups lack of leadership with a new way of thinking;
- Managers and leaders fail in gaining people's sympathy;
- It needs to change employees' mind and attitude;
- There are no good understanding and communication between managers and working staff;
- -There is a tendency to evade responsibilities, and difficulty in clarifying where responsibilities lie.

There were consensus about the 'knowledge management system' and 'the effective managerial activities system' that had to be established to facilitate the implementation of the RTE within EA.

Stage 3: Preparing 'root definitions'

Stage 3 was concerned with the preparation of 'root definitions', which were precise definitions of a notional system within a given situation. Having clarified the root causes of the problem contexts, root definitions were formulated, which seemed 'relevant' to the problem situations within EA. These were the 'knowledge management system' and the 'effective managerial activities system', which are described as follows.

Root definition 1: The knowledge management system

A private owned system that produces and distributes a good quality of information and knowledge amongst employees within EA in order to make up three levels of management, which are concerned with operational efficiency at production, the corporate effectiveness at the strategic level, and ethical management at the normative level, through cooperative problem-solving, participatory decision-making, and having with leadership, technology and learning culture within EA.

Root definition 2: The effective managerial activities system

It is a system to facilitate the formulation of corporate visions, strategies, effective managerial activities and capacity to achieve these activities with the aim of implementing corporate philosophy of the RTE by establishing a new form of social network which is necessary for complementing the availability of social capital and networks within EA.

Stage 4: Building conceptual models of the perceived reality

There are two conceptual models which seem to relevant to problematic situation in EA. These are the conceptual models of 'the knowledge management system', and of 'the effective managerial activities system'. These models are shown in Figures 1 and 2. The development of the conceptual models had proved very useful to identify new aspects of the situation and possible areas for change. The comparison of these models with the situation revealed, as has to be done in SSM, which was actually not happened in a given situation in which consensus did not occur during the process of PAR.

The process of probelmatization

When a new thought begins with knowing about the problem situations within EA, the systems analyst and participants focus on "what happens" rather than "what is (the case)". A new thought begins not with merely dependent on the imposition of systems methods or an 'intellectual framework' that makes sense of both the situation and the researcher's purposeful activities in it, but by considering micropolitics, in preference to micro-power, where everyone in an organization was making decisions. In this sense, "the decision-making which goes on in organizations, ubiquitous and universal, is itself rhizomic and produces activity and behaviour which is rhizomic - unspecifiable, unpredictable, uncapturable" (Jackson and Cater, 2000, p. 253). This leads to new insights concerning 'what happens here and now'. We can speak of 'what happens' in terms of the Deleuzian sense of an event (Deleuze, 1990). The process of problematization can be divided into three distinctive stages in which participatory learning had happened within the problem-solving practice. These stages can be summarised as follows.

Stage1: Collecting 'information'

In an actual study the archive of 'information' collected can include written documents and records, questionnaire-based survey and notes taken from formal and informal interviews with the member of staff (i.e., Chief Information Officer, senior managers) within EA. Only the facts which are considered to be of relevance for creating and sharing an appreciation in a given situation in stage 2 are given as follows.

The company

The EA enterprise, though at core business of manufacturing semiconductors, is engaged in many businesses which include chemicals, construction, and engineering. It is essential that EA's technology is continually being assessed and updated. For this reason, EA has been cooperated with American and Japanese companies for researching and developing 'non-memory' semiconductors since early 2000.

The tasks of the MIC

- Monitoring the process of manufacturing activities in production units;
- Monitoring the performance of production units;

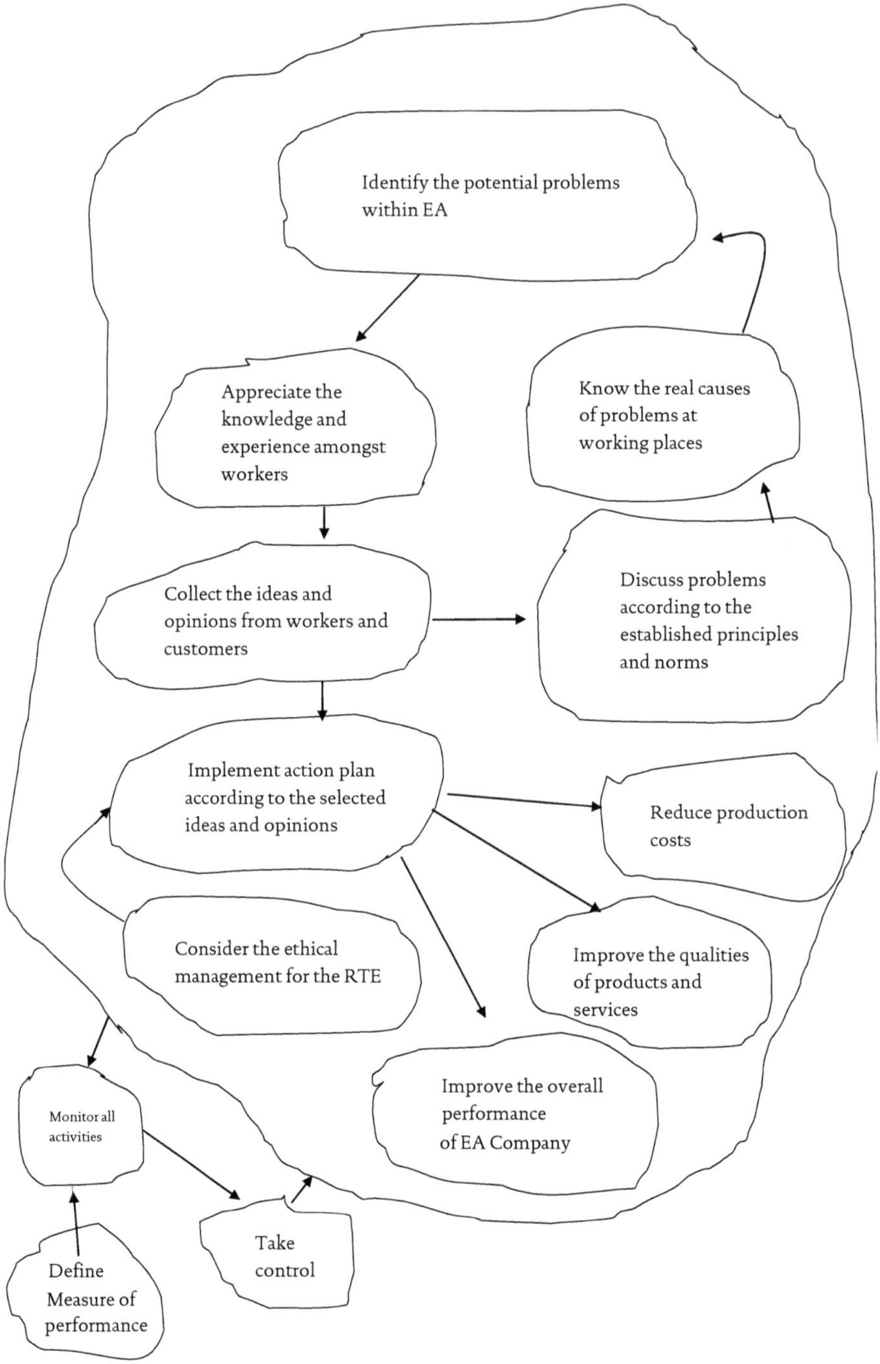

Figure 1 *The knowledge management system in EA*

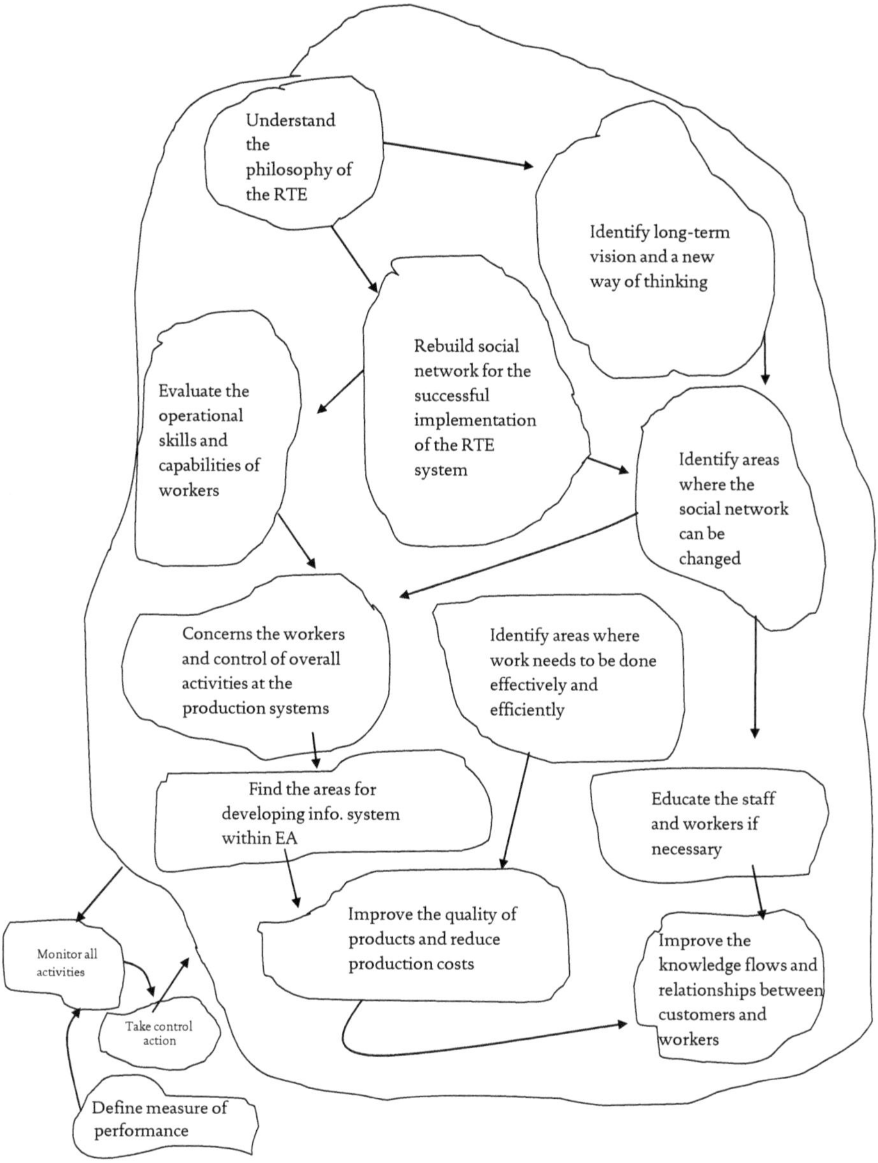

Figure 2 *The effective managerial activities systems*

- Establishing and implementing real-time enterprise (RTE) system within EA;
- Making the unique corporate culture for the EA enterprise

Ideas and questions expressed by participants

- The high degree of autonomy and responsibility will be necessary for improving the overall performance of EA;
- How to create and develop the unique corporate culture for the EA enterprise?

Stage 2: Creating and sharing an appreciation amongst participants

In stage 2, it is concerning about creating and sharing an appreciation amongst participants in order to building the community of inquiry and practice that aimed to create 'open space' which generate the semantic communication to discuss and reappreciate the problem situation within EA. It is about 'response', which is partly for the sake of its connection with responsibility as Vickers (1970: 128) reminds us, "Life consists in experiencing of relations, rather than in seeking goals." This included the process of an appreciation in discriminating form from contingent contexts, in appreciating the manifold relations between the problem situation (as the entities observed) and with participants (who are observers). In this way, an appreciation of the collected 'information' took place to understand problem situations being dealt with micro-power as indeed, no problem situations or decision-making process can be understood without considering the complex dynamics of cultural, social and 'political' dimensions within an action research context (Yolles, 1999).

Stage 3: Making sense of 'problematizing fields'

As Tsouvalis (1995: 220) has shown, problematization creates the space, the conditions, within which various responses can be given. It is through making sense of 'problematizing fields' which make up the question of the exercise of micropolitics amongst participants. This question will be examined in details as participatory action research will be carried out further in due course.

Conclusions and discussion

Exploring the Real-Time Enterprise from rhizomatic systems thinking that we have demonstrated in this paper, what are main contributions of SSM that has been applied into the problematic situations within EA? Our contention is that rhizomatic systems thinking can help us to explore the social, cultural and political dimensions of the RTE which can grasped by the metaphor of a 'living nervous system' that was perceived in the continuous process of transformation within regional and local contexts. Using SSM, the semantic communication are vital to carry out participatory action research which facilitates debate by which actors or participants were committed to the learning process of participatory action research within EA, and to create interventions to create or explore the shared norms and values amongst participants. During the process of problematization, participants were appreciated open-ended process of learning in order to deal with 'problem-solving practice' as cultural change went on within EA. This process was subjected to the conscious thought and actions of people in dealing with complexity in contingent contexts. Whereas the notion of 'real-time' is much considered in RTE, our research focuses on the process of problematizationm which the research process focuses on "what happens" rather than on "what is " or "what is the case." In the process of appreciation of "what happens", it is possible to appreciate complex situations which are concerned with power and conflicts amongst participants. Further research will then be required to dealing with issues of power, conflicts, politics and ethics concerning the sustainability of RTE within shared environments. To do so, we aware that rhizome kind of open systems exist upon the intense space of 'outside' thoughts as Deleuze and Guattari (1988) have proposed for dealing with unpredictable and chaotic situations.

References

Adams, G. B., and Catron, B. L., (1994). Communitarianism, Vickers, and Revisioning American Public Administration, in Blunden, M., and Dando, M., (Eds.) *Rethinking Public Policy-Making: Questioning Assumptions, Challenging Beliefs*, London: Sage, 44-63.

Checkland, P. B. (1981). *Systems Thinking, Systems Practice*, Chichester: Wiley, ISBN 0-471-27911-0.

Checkland, P. B. (2005). Webs of Significance: The Work of Geoffrey Vickers, *Systems Research and Behavioral Science*, ISSN 1092-7026, **22**(4), 285-290.

Checkland, P. and Casar, A. (1986). Vickers' concept of an appreciative system: a systemic account, *Journal of Applied Systems Analysis*, **13**, pp. 3-17.

Checkland, P. B., and Scholes, J., (1990). *Soft Systems Methodology in Action*, New York: Wiley, ISBN 0-471-92768-6,

Coleman, J. S. (2000). Social Capital in the Creation of Human Capital, in Lesser, E. L. (Ed.), (2000). *Knowledge and Social Capital: Foundations and Application*, Oxford: Butterworth Heinemann, ISBN 0-7506-7222-6, pp. 17-41.

Deleuze, G. (1990). *The Logic of Sense*, (trans, M. Lester with C. Stivale), London: Althlone, ISBN 0-485-30063-X.

Deleuze, G. and Guattari, F. (1988). *A Thousand Plateaus: Capitalism and Schizophrenia*, (trans. Massumi, B.), London: The Athlone Press, ISBN 0-485-12058-5.

Flint, D. (2002). Real-Time Operations Demand Real-Time applications, *Research Note*, Gartner, Inc., 1 October, 2002

Foucault, M. (1977). *Discipline and Punish: The Birth of the Prison*, London: Allen Lane, Penguin.

Harris, K. (2002). Commentary: Knowledge Management: RTE Processes and Technology, *Research Note*, Number COM-18-1382, Gartner, Inc., 26 September, 2002.

Hugos, M., (2005). *Building Real-Time Enterprise: An Executive Briefing*, Hoboken: Wiley, ISBN 0-471-67829-5.

Jackson N., and Carter, P., (2000). *Rethinking Organizational Behaviour*, Harlow: Financial Times/ Prentice Hall, ISBN, 10: 0-273-63007-5.

Lesser, E. L. (Ed.), (2000). *Knowledge and Social Capital: Foundations and Application*, Oxford: Butterworth Heinemann, ISBN 0-7506-7222-6.

Mainzer, K. (1994). *Thinking in Complexity: The Complex Dynamics of Matter, Mind, and Mankind*, Berlin: Springer, ISBN 3-540-60637-8.

McCoy, D. (2002). Software Infrastructure: Present Status, Future Directions, *Letter from the Editor*, Gartner, Inc., 22 August 2002.

Morgan, G., (1986). *Images of Organization*, Beverly Hills, CA: Sage, ISBN 0-8039-2830-0.

Nahapiet, J. and Ghoshal. S. (2000). Social capital, intellectual capital, and the organizational advantage, In Lesser, E. L. (Ed.), (2000). *Knowledge and Social Capital: Foundations and Application* (pp.119-157), Oxford: Butterworth Heinemann, ISBN 0-7506-7222-6.

Parkhe, A. Wasserman, S. and Ralston, D. A. (2006). New Frontiers in Network Theory Development, *Academy of Management Review*, ISSN, 0363-7425, **31**(3), 560-568.

Raskino, M. (2002). Commentary: Nine Key Principles of the Real time Enterprise, *Research Note*, Number COM-18-3299, Gartner, Inc., 3 October, 2002

Tsouvalis, C. (1995). *Agonistic Thinking in Problem-Solving: The Case of Soft Systems Methodology*, PhD thesis, University of Lancaster.

Vickers, G. (1965), *The Art of Judgement: A Study of Policy Making,* London: Sage, ISBN 0-8039-7362-4.

Vickers, G. (1970), *Freedom in a Rocking Boat: Changing Values in an Unstable Society*, Harmondsworth: Penguin.

Williams, G. (2005). Geoffrey Vickers: Philosopher of Responsibility, *Systems Research and Behavioral Science*, ISSN 1092-7026, **22**(4), 291-298.

Yolles, M. I. (1999). *Management Systems: a Viable Approach*, London: Financial Times/ Pitman, ISBN 0-273-62018-5.

Yu, J. E. (2001). *Towards rhizomatic systems thinking in management science*, DPhil Dissertation, University of Lincoln, U.K.

Yu, J. E. (2006). Creating 'Rhizomatic Systems' for Understanding Complexity in Organizations, *Systemic Practice and Action Research*, forthcoming, ISSN 1094-429X.

Jae Eon Yu graduated from the Lincoln School of Management, University of Lincoln, UK. He has a DPhil degree in Systems/Organization Sciences and is currently a full-time lecturer in the Department of Information Management, Seoul University of Venture and Information, Seoul, Korea. Previous publications include "reconsidering participatory action research for organizational transformation and social change" in Journal of Organizational Transformation and Social Change (2004), "Incorporating knowledge management as HRD strategy: the case of a Korean electronics company" in Journal of Organizational Transformation and Social Change (2005), "Creating 'Rhizomatic Systems' for Understanding Complexity in Organizations" in Systemic Practice and Action Research (2006). His research focused on the applications of Viable System Model (VSM), Soft Systems Methodology (SSM) and 'rhizomatic systems thinking' into the areas of understanding complexity, social capital, social network analysis, organizational development, human resource development, and Real-Time Enterprise.

Farming and Urbanizing Environments: Hawkesbury Harvest and the Cultural Landscape of Western Sydney

Ian Knowd
University of Western Sydney, AUS

Hawkesbury Harvest formed in 2000 in response to the pressures of urbanization and globalization on the farming and health constituencies of the Hawkesbury. Its initial response to these pressures was a Farm Gate Trail to support small-holding farmers, and a growers directory to enhance local access to fresh food and "reorient the food system" to increase food quality, safety and security. In the six years since its creation, Hawkesbury Harvest has evolved into an organization with wider community and land use interests. It is working to build a future for agriculture in the Sydney basin based on a resilience founded in economic diversity and interconnectedness with other industries in the basin, as a provider of ecological and cultural services for its host communities in the peri-urban fringe of Sydney, and as a catalyst for rethinking the role of agriculture in land use governance in an urbanizing environment.

Introduction

This paper draws on previous works to describe the history of Hawkesbury Harvest (Harvest). It then goes on to explore some of the systemic characteristics of the phenomenon, particularly the interaction between cultural landscape and urban development in the Sydney basin. While Harvest emerged out of the local context, it has evolved into a catalyst for innovation in small-holding farming in the basin and a contributor to potential features of resilience in agriculture in peri-urban Sydney.

Harvest's host community is the rural and farming community on the urban periphery of Sydney, Australia. The area is defined by the catchment of the Hawkesbury River and three Local Government Areas (LGAs) (Hawkesbury City, Hornsby and Baulkham Hills Shires) and is about 54% of the landmass of Greater Western Sydney. It is home to approximately 17% of the population of Western Sydney and is colloquially known as the 'food bowl' of Sydney. (WS-ROC, 2000)

The conceptual framework for this paper is that landscape is a human construct, that we have land and that land-scapes are what we see and understand as humans about land (Robertson & Richards 2003). Thus we can conceive of the idea of wilderness because it is land devoid of human interference. Heritage landscapes are attached values that make them heritage. City landscapes are attached values. Rural landscapes are attached values. These values and the aesthetics that go with them are founded on cultural constructs. Landscapes are a cultural phenomenon created when humans attach their values to what they see and experience in land. The concept of landscape is an ideological thing. The dominant ideologies of the time 'write' our landscapes.

Using this framework, we see that landscapes are meaningful because they are about land and people. Indigenous Australians have been articulating this concept in order to demonstrate their connection with country and the same applies to the efforts of our post-colonial farming families on their holdings in Greater Western Sydney. Out of this connection comes social capital, histories, heritage, economy, culture, recreation and other social and environmental goods. The assault on agri-culture needs to be seen as a systematic erasing of the cultural landscape that has defined Sydney in the past and will define Sydney in the future. What kind of place will we leave for our children's children?

The richness of the asset includes what has been in the past and what is in the present. It's about stewardship and a renewable and culturally relevant use of land for food and other primary products, in close proximity to Sydney that provides other benefits for the communities of the Basin. It's about the physical attributes AND the associated values we attach to them as human beings in the place we call home. It's about farming and heritage, the landscape as a cultural asset and people and their food.

An integrated, systemic view of land-use planning in the Sydney basin that reflects these values does not exist in any other form than rhetorical references in the long line of planning documents for Sydney since settlement. This absence produced creative tension and is a good example of the paradox that rural communities face in the global context – the pressures of urbanization create both threat and opportunity for farming. Within the region a range of forces created the conditions and interest in a tourism solution for three local constituencies (agriculture, health, and tourism) each with its own issues (Knowd, 2005). Figure 1 summarizes these forces or pressures and the states they generated. The Farm Gate Trail (FGT) was a solution driven by the farming and community health agendas, and one that dovetailed into the needs of the tourism sector in addressing fundamental, long-term problems with destination identity. Despite the prominence of the Farm Gate Trail in tourism promotion for the area, it remains the domain of the farming and community health constituency of the region.

The farming and health communities were interested in exploring solutions for the impacts of neo-liberal policies on the global scale and urbanization on the local scale. Their response created something that the tourism industry in the region had also been seeking for some time. A detailed exposition can be found elsewhere (Knowd, 2003, 2005, 2006a). Farming and community health agendas began to converge after the Rio Conference of 1992. Agenda 21 and the Healthy Cities program that grew out of sustainable development created a structural framework within which the issues of local food and food production could be addressed. The convergence occurred at a time (1998) when NSW Agriculture (now the Dept. of Primary Industry) was actively implementing a Strategic Plan for Sustainable Agriculture in the Sydney Basin. A public meeting was held in May 2000 and Hawkesbury Harvest (later Inc.) was formed. With financial support from Hawkesbury City Council and then Regional Partnerships Program funding from the Department of Transport and

Context ...a constituency	Pressure ...facing a problem	State ...feeling the strain	Response ...a tourism initiative
Farming	Urban Development Market Structures	Financial Distress Generational Crisis	Farm Gate Trail
Community Health	Food Health	Food Supply, Security, Safety, Quality, Access Sustainability	
Tourism	Destination Identity Destination/Transit Route Building Community Support for Tourism	Lack of Major Attractions/ Icons Planning Constraints Attitudes to Rural Living	

Figure 1 *Community pressures set the scene for a tourism solution*

Regional Services (DOTARS), a Farm Gate Trail (FGT) was established as one of two principal goals, the other being a growers directory. Hawkesbury Harvest's mission is:

"Hawkesbury Harvest is a community-based Association committed to improving the economic viability and sustainability of local agriculture." (Hawkesbury Harvest, 2002)

An evaluation of the project after its first year concluded that it was a success in terms of raising the region's profile as a tourism destination (Groth as cited in Orton, 2002), increasing the viability of participating farms, and generating employment. (Orton, 2002) By mid-2002 local newspapers were making naïve claims about role of tourism in the 'salvation' of farms in the region (Gibson, 2002).

In 2003, after gaining further funding from DOTARS and the Hawkesbury City Council, and additional funds from NSW State and Regional Development through Greater Western Sydney Economic Development Board, a Business Development Manager was contracted for Hawkesbury Harvest as part of the Paddock-to-Plate project. This Stage II initiative identified regional branding, growers markets, local retailing, developing specialty (small acreage, boutique) agriculture, provedoring and open farm events as possible agri-industry development and potential avenues for Hawkesbury Harvest's financial independence. Paddock-to-Plate saw the implementation of brand building, the growers market and open farm events. A key learning outcome for Harvest with this project was the impossibility of achieving all the industry development identified within a single-year time frame. This became one of the justifications used for a third round of funding applications in late 2004.

The third and most recent project commenced in March 2006 when Harvest received funding for a Food and Wine Coordinator project. This project focuses on the opportunities that Harvest did not have the time or resources to implement under the Paddock-to-Plate project with additional network-building and market development in the region. Again the Department of Transport and Regional services is a partner, with the Greater Western Sydney Economic Development Board, and a new corporate partner, Lend Lease/GPT at the Rouse Hill Regional Centre.

As the focus shifted from creating the Farm Gate Trail to developing a wider range of business opportunities, the role of Hawkesbury Harvest also evolved into a broader regional development role. A move into broader land-use advocacy, particularly for the role of agriculture in retaining and maintaining landscape (the commons) and the interdependence of tourism and agriculture in diversifying an economy within such a retained landscape, forced Harvest to engage in political processes and to take policy positions it had not previously had to articulate.

Harvest projects generally have fallen into the category of short-term fixes driven by the neo-liberal processes that they used to undertake them. The economic determinism reflected in contemporary government approaches

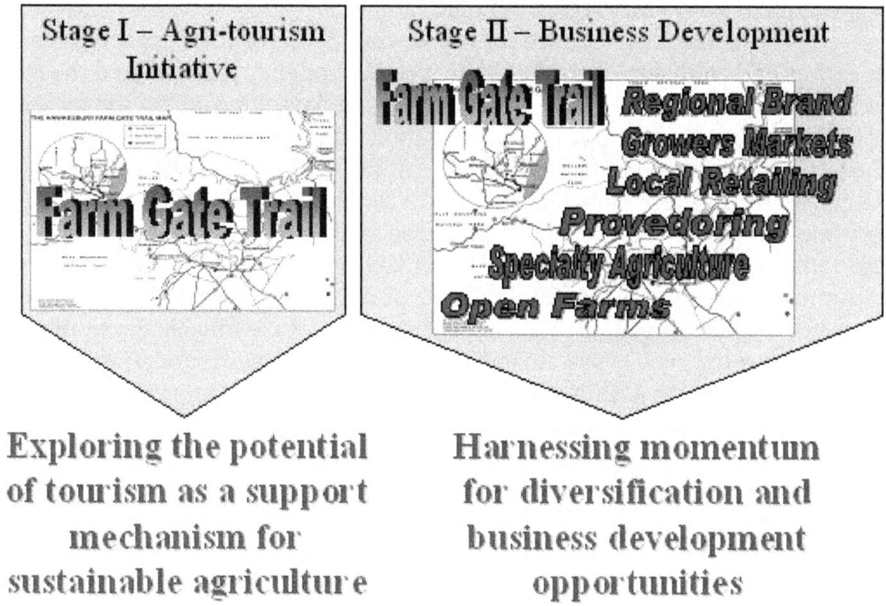

Figure 2 *Changing focii in the development of the Farm Gate Trail Initiative*

to community development means that each project must stand alone and be executed in a one year time frame. Outcomes are couched in economic terms (incomes, jobs) and this encourages groups like Harvest to look for panaceaic solutions. (Knowd, 2006b)

The pitfalls of panaceas arise from a failure to appreciate that panaceaic approaches 'fix on a destination and calculate what a group must do to get to get there, with no concern for member's preferences' (Mintzberg cited in de Wit & Meyer, 1998: 217) In Harvest's case, member's preferences, particularly the farmers, in terms of capacity and willingness to adapt to tourism, to adopt new skills, knowledge and attitudes, to deliver on the tourism promise, created serious challenges and an analysis of these can be found elsewhere (Knowd, 2003). Despite this, Harvest initiatives are innovative when considered in the context of mainstream farming models in the Sydney basin. They do add to system resilience by diversifying the economic system through which farmers survive.

Harvest is an example of the 'virtuous circles' of sustainable development wherein economic activity reinforces social and cultural practices, which reinforce environmental systems integrity and capacities for regeneration over time (Selman & Knight 2006). Hawkesbury Harvest and the Farm Gate Trail have successfully addressed many of the issues for those farmers who invested sufficient resources, and due to the linkages with tourism and allied industries, the phenomenon has revealed some important potentials for sustainable development and sustainable tourism in the region. These potentials were formally accepted by representative bodies for agriculture through Hawkesbury Harvest Inc. and tourism through Tourism Hawkesbury Inc. in May 2004 and Figure 3 articulates the mutual interest that tourism and agriculture have in the landscape of the Hawkesbury region (Knowd, 2004)

Tourism and agriculture constituencies identified their explicit interest in each other's industry, articulated the interdependency and accepted the role of both industries in the future development of the region and the importance of the natural and agri-cultural asset base upon which both industries depend in the face of the limited opportunities that mining and urban development might deliver for future generations of residents. It clearly describes the nexus between agriculture, tourism and development in the region. This nexus defines a set of relationships between the industries that have important cultural landscape, community identity and system stability potentials for both tourism and agriculture. Figure 3 shows a theoretical orientation of sustainable development based on a nested view of sustainability, with ecology as the foundation, society as the human system within which economy is a sub-system. Suitable and sustainable economies are a function of the social system, its values and priorities, and the underlying capacity of the ecology to support it. A land economy based on agriculture and tourism supports the regional society by conserving and enhancing regional amenity and health assets, while conserving and encouraging better stewardship of the natural and agriculture ecosystem of the Hawkesbury River catchment.

Despite Harvest initiatives establishing a range of innovative strategies for supporting agriculture, the situation of farmers in the basin exhibits features of 'over-connection' to their system variables – farmer age (older farmers resistant to change and attitudes, skills and knowledge), land-use the suburbs-in-waiting syndrome, the central food marketing system, (Alison & Hobbs, 2004

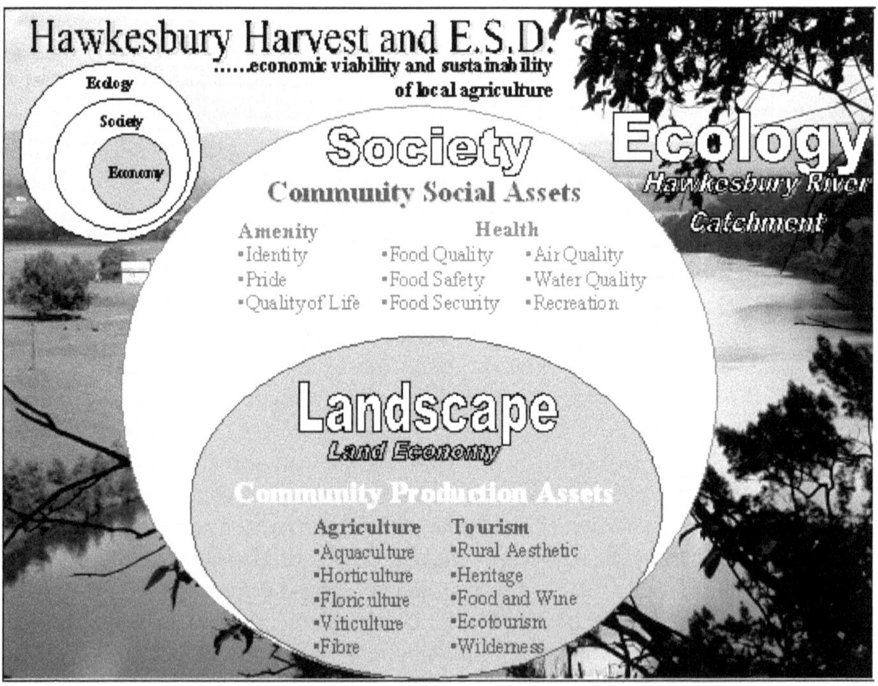

Figure 3 *Tourism and agriculture's common interest in landscape*

citing Holling, *et al.*, 2002). It also exhibits characteristics of 'lock-in' (Alison & Hobbs, 2004: 15) whereby farmers continue to liquidate their land asset in preference to changing their response to the external environment.

In this sense the existing situation in the basin implies that current modes of behavior exhibit considerable resilience (the ability to resist external disturbances), particularly the entreaties of organizations like Harvest and the Hawkesbury Sustainable Farmers Network to consider alternative 'post-productivist' (Halfacree, 1997; Hadjimichalis, 2003) market orientations and micro-climate production technologies. The sub-optimal performance of these organizations relates to the lack of willingness to engage with the issues in the farming community, to build the social capital that this social-environmental system requires. This resilience is a function of the extent to which the environment is degraded and is characterized by industry shifting from farming to land-banking, where the bankable asset can only be realized if the current trends in urbanization and development in the Sydney basin continue un-addressed. The debate in the Hawkesbury region regarding land use, particularly the continued zoning of land for farming, is raging with great pressure from developers and sub-dividers to convince the outer-urban councils to allow suburban lot sizes and restrict farming activity. (See Letters to the Editor in the Hawkesbury Gazette, Hawkesbury Independent and other regional newspapers 2005 and 2006). The situation in the Hawkesbury warrants the kind of resilience analysis that have been performed elsewhere (Alison & Hobbs, 2004, Walker et al. 2002) if for no other reason than to settle the debate, as the system may be beyond the threshold for sustainable agriculture in the region already.

Similarly, the issues of the capacity of the landscape to absorb shock events such as flooding is irreparably modified because of urban development in the catchment and the loss of land surface to housing and tarmac. The hydrological system was permanently changed with the building of the Warragamba Dam, and the Hawkesbury Nepean is now a constant flow watercourse with little additional capacity to absorb urban runoff, let alone major flooding of the reservoir that might result in flood surges with overflow events. It is likely that the Sydney basin has exceeded its ecological buffering capacity on this and many other fronts and we are thus left with the conclusion that Alison and Hobbs have drawn about the WA agricultural zone, that is, only 'regional economic, demographic, or social' (2004: 16) mechanisms are left in our arsenal for combating the potential collapse of the system in the basin. As in the WA case, the likelihood that the production system will change shape and potentially collapse in the face of urbanization and market forces is very high, and the only mechanisms currently in place to inject 'novelty' into the system are the initiatives developed through Hawkesbury Harvest. Existing land use planning mechanisms, specifically zoning controls applied at Local Government Area level, have not been effective in delivering resilience to the system, and could well be one of the primary mechanisms to have created the land-banking phenomenon. The system even exhibits features of 'policy resistance' (Sterman, 2001 as cited in Alison & Hobbs, 2004: 17) as the issues facing agriculture have been known since the County of Cumberland Plan attempted to address the 'promiscuous urbanization' (1948: 129) threatening to degrade Sydney's quality of life.

The lack of recognition of agricultural assets, and specifically the significance of food supply for the greater Sydney region, is symptomatic of a 'boiling frog syndrome'. Sydney still sources the majority of its perishable fruit and vegetables from its own backyard (Knowd, *et al.*, 2005) and there is an apparent complacency about food supply and food security in Sydney. It is noteworthy that jurisdictions in other regions of Australia and throughout the developed world have codified their responses to managing urbanization and agricultural protection and reform with formal planning and other instruments (Mitchell, 2005; Keating & Stevenson, 2006). However, in Sydney, these issues and the threat to rural lands are largely left to local governments and their planning controls, particularly zoning. (Sinclair, 2003; Mitchell, 2005). This lack of policy and planning for agriculture in the Sydney basin has created the conditions for community activism and at the same time deprived interest groups of a forum and policy framework within which to conduct the debates.

These problems with system governance are not unique and the debates about how to plan for the rural in urbanizing environments is one that continues wherever cities are expanding and farmlands are being seen as the next 'greenfield' for urban development (Hadjimichalis, 2003; Healey, 2002, 2004; Scott, 2006).

There are a range of imperatives that the recent versions of the Sydney Metropolitan Strategy are recognizing, recognition that has also come out of the land use conflict and Hawkesbury Harvest phenomenon. Imperatives exist for the future of agriculture in sustainable forms that will deliver benefits to future generations. Imperatives exist in choosing what character of place the Sydney basin will offer residents and visitors. Imperatives exist for designing-in economic, social and environmental resilience in the Basin. This requires integrated local management and the multiple knowledges that that process demands.

Using a sustainability framework we can overlay some essential planning outcomes that are required if agriculture and urban development are to be successfully integrated. In a social sense, planning frameworks and instruments need to deliver community engagement, communication and education. Economically, development, incentive schemes and infrastructure are required, and for the natural environment, the policy and regulatory settings need to be right.

In a practical sense this means that integrating social and economic planning outcomes means that economic development is a matter of choice, one that is determined by the environmental and social contexts, that economic incentives and infrastructure supports are used to engage the community with agriculture, to communicate the values of agriculture and educate consumers about the food and other primary production that makes it possible for them to have quality-of-life. Examples are found in the Hawkesbury Harvest model, where farmers choose to use alternative channels (Farm gate sales and Farmers Markets) to market their produce and consumers choose to seek out the benefits that this makes available to them by sourcing their food directly from the

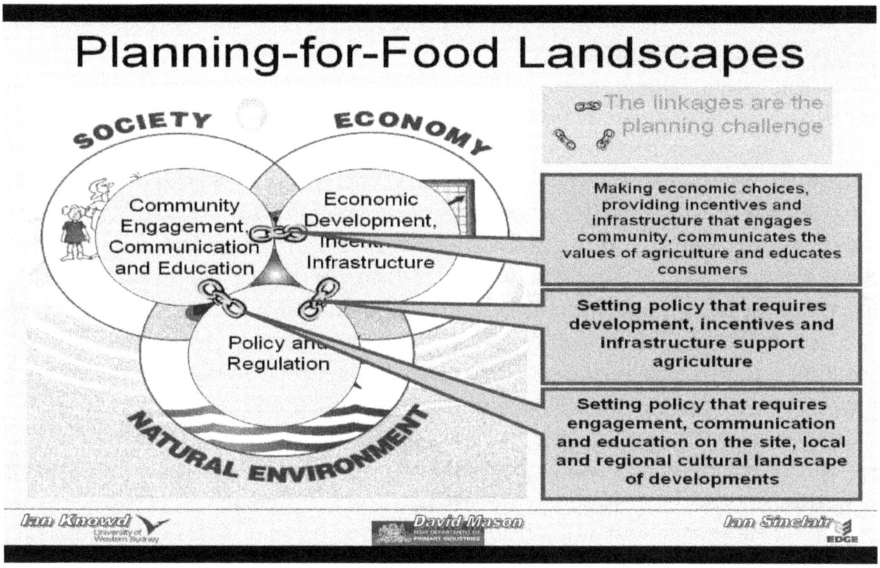

Figure 4 *Planning for system resilience (Source: Knowd, et al., 2006)*

farmer. It's also implemented by broadening out the planning terms of reference for developments by extending EIS and SIA processes to include landscape assets and engagement with the wider regional community, rather than the site specific approach that currently exists. There are many other ways that activate these linkages that could be explored.

The linkages between economic and environmental spheres can be practically activated by explicitly setting policy that requires developers to consider and propose the economic development strategies that are implied in place-making that includes relationships with agriculture and particularly food supply. This is starting to be done but planning frameworks do not require developers make their master plans regionally engaged and economically integrated. Examples of this can be found in the Hawkesbury which has amended LEPs and developed DCPs for farm gate sales, but problems still exist in terms of internal inconsistencies between aims of these instruments, and inconsistencies with regional and state level planning instruments. This leads to ambiguity and potential conflict.

The linkages between social and environmental spheres is activated through policy that requires engagement, communication and education as part of the proposal and development process. The practical implementation is again a matter of re-engineering the requirements of the planning process to make developers deliver these kinds of outcomes, rather than leaving it to individual developers to interpret such things as they see fit. An example of an area where this idea can be seen to have failed the community is in the Hawkesbury where, despite having an appropriate community consultation process, the recently adopted cultural plan fails to formally recognize agriculture as a cultural asset and process – it only recognizes the artifacts of agri-culture, the dead and usually heritage objects in a living landscape. If the one industry that has defined the character of the place and continues to deliver the principal quality-of-life benefits that residents enjoy is not recognized for its role in the cultural life of the Hawkesbury, how can policy and regulation that would adequately protect agriculture be identified and implemented?

This paper has sought to explore the implications of trying to integrate agriculture with urban development. The situation in the Sydney basin is somewhat unique because of the geo-physical characteristics and the consequent pressures placed on agriculture by the "rolling wave" of development. What has emerged is a potential way of re-thinking the planning priorities and processes for future urban expansion in the Basin that would integrate agriculture as a cultural asset in to the fabric of Sydney's urban landscape. This re-thinking requires that agriculture be seen as an essential element in the cultural imperative to create sustainable, liveable, and culturally rich places to live in Sydney. It's an imperative because agriculture isn't being pushed back, it's being squeezed out and we risk losing it forever. This is an issue for us because we owe it to our children's children to pass on something that reflects our values and the values of agricultural heritage in the food bowl of Sydney. It means we need to see that we write the landscape with our planning laws and development processes in ways that leave enduring legacies of resilience and innovation. We have a choice to make about the legacy we want to leave, but we're running out of the time and space to do it in.

References

Allison, H. E. and R. J. Hobbs. 2004. 'Resilience, adaptive capacity, and the "Lock-in Trap" of the Western Australian agricultural region'. *Ecology and Society* 9(1): 3. [online] URL: http://www.ecologyandsociety.org/vol9/iss1/art3

County of Cumberland, (1948) *County of Cumberland Planning Scheme Report*, presented to the Hon. J.J. Cahill, M.L.A., Minister for Local Government, on 27 July.

de Wit, B. and R. Meyer 1998. *Strategy: Process, Content, Context* 2nd Edition, International Thomson Business Press, London.

Gibson, C. 2002. 'Tourism is the key to survival for agriculture.' Hawkesbury *Independent* (August). Jemshaw P/L., South Windsor

Hadjimichalis, C. 2003. 'Imagining rurality in the new Europe and dilemmas for spatial policy', *European Planning Studies*, 11, pp. 103–113.

Halfacree, K. 1997. 'Contrasting roles for the post-productivist countryside, a postmodern perspective on counterurbanisation', in: P. Cloke & J. Little (Eds) *Contested Countryside Cultures, Otherness, Marginalization and Rurality* ,London: Routledge.

Hawkesbury Harvest 2002. *Hawkesbury Harvest - Regional Community-Based Action for Sustainable Agriculture and Rural Community Development: The Story So far*, PowerPoint presentation, Hawkesbury Harvest Steering Committee, Windsor, NSW, Australia.

Healey, P. 2002. 'Urban–rural relationships, spatial strategies and territorial development', *Built Environment*, 28, pp. 331–339.

Healey, P. 2004. 'The treatment of space and place in the new Strategic Spatial Planning in Europe', *International Journal of Urban and Regional Research*, 28, pp. 45–67.

Holling, C. S., L. H. Gunderson, and D. Ludwig. 2002. 'In quest of a theory of adaptive change'. In C. S. Holling, editor. *Panarchy*. Island Press, Washington, D. C., USA.

Keating, M. and L. Stevenson 2006. 'Rural Policy in Scotland after Devolution', *Regional Studies*, Vol. 40.3, pp. 397–407, Routledge.

Knowd, I. 2003. 'Grasping the Nettle: farmers and Tourism' in *Riding the Wave of Tourism and Hospitality Research*, Braithwaite, R.W. and R.L. Braithwaite eds Proceedings of the Council of Australian University Tourism and Hospitality Education Conference, Coffs Harbour, Southern Cross University, Lismore

Knowd, I. 2004. *Meeting with THI re ATDP project*, email correspondence 14/5/2004.

Knowd, I. 2005. 'Tourism As Panacea: Exploring The Role Of Tourism In Non-Tourism Development Agendas', in *Sharing Tourism Knowledge*, Proceedings of the Council for Australian University Tourism and Hospitality Education Conference, Alice Springs, Tremblay, P. and A. Boyle eds, Charles Darwin University, Northern Territory.

Knowd, I. 2006a. Tourism as Mechanism for Farm Survival, *Journal of Sustainable Tourism*, 14, (1), 24-42, Channel View Publications.

Knowd, I. 2006b. *Old Wine in New Bottles: the case of Hawkesbury Harvest*, refereed paper presented to the Governments and Communities in Partnership conference, Centre for Public Policy, University of Melbourne, 25- 27 September.

Knowd, I., D. Mason and A. Docking 2005. *Urban Agriculture: The New Frontier*, paper presented to the State of Australian Cities Conference, Griffith University, Brisbane, 30 Nov. – 2 Dec.

Knowd, I., D. Mason and I. Sinclair 2006. *Growing Food and Growing Houses: It's a Landscape Thing*, paper presented at the World Planners Congress, Vancouver, 17-20 June, Canada.

Kramer, R. M. 2006. Social Identity and Social Capital: The Collective Self at Work *International Public Management Journal*, 9(1), pages 25–45

Mitchell, H. 2005. *The Protection of Production on Rural Lands: A review of tools and techniques for rural planning*, The Centre for Rural and Regional Innovation – Queensland

Orton, V. 2002. *Regional Assistance Programme: Evaluation Report*. Department of Transport and Regional Services. Canberra

Robertson, I. and P. Richards (Eds) 2003. *Studying Cultural Landscapes*, Arnold, London

Selman P. and M. Knight, 2006. 'On the Nature of Virtuous Change' in *Cultural Landscapes: Exploring Sustainability through Qualitative Models* Vol. 31, No. 3, 295 – 307

Sinclair, I. 2003. *Growth Management and Rural Land*, University of NSW Faculty of the Built Environment Planning, Law and Practice Short Course.

Sterman, J. D. 2001. 'System dynamics modeling: tools for learning in a complex world.' *California Management Review* 43(4):8–25.

Walker, B., S. Carpenter, J. Anderies, N. Abel, G. Cumming, M. Janssen, L. Lebel, J. Norberg, G. D. Peterson, and R. Pritchard. 2002. 'Resilience management in social-ecological systems: a working hypothesis for a participatory approach.' *Conservation Ecology* 6(1):14. (Online.) URL: http://www.consecol.org/vol6/iss1/art14.

WSROC [Western Sydney Regional Organisation of Councils] 2000. *Western Sydney Regional State of the Environment Report*, produced for WSROC by the Regional Integrated Monitoring Centre, University of Western Sydney, Hawkesbury, Australia

Ian Knowd is a lecturer in tourism studies at the University of Western Sydney. His research interest is in the way rural communities engage with tourism, the role of agri-culture in forming landscapes, and tourism as a mechanism for farm survival and cultural engagement. He is past Chair and current treasurer for Hawkesbury Harvest Inc., a community-based, not-for-profit organization seeking to build the economic sustainability of agriculture in the Sydney basin.

THEME SEVEN
Language, Learning and Transdisciplinary Practice

Susan Goff, University of Western Sydney, Australia

I experience systemic thinking like a current spreading throughout an ocean, a trail of smoke coiling into the airways: it is fluidity in spirit challenging us, welcoming us and beckoning us on like a dragon. Like a "rainbow serpent". Systemic thinking is a way of seeing before it can be anything else. For systemic insight to come into being certain practices need to be developed that are not common to everyday western or non-systemic (fragmented) ways of being. The theme of "language, learning and transdisciplinarity" holds open a place for such practices to be named, developed and deployed. Such practices help us shift how we think and as a consequence more knowingly, more responsibly, be in relation with whole living systems.

This theme exists at the threshold of non-systemic thinking, at that moment when we see connections firing up across social systems, schools of thought and any other fragmented entity attracting our attention. At this moment, we practitioners are drawn into the extraordinary dynamics of new connections unfurling across layers of disconnection and what sometimes feels like eons of neglected relationships requiring attention, exploration and often, healing. If we are limited to the language of fragmentary thinking we cannot express and communicate what we are experiencing in the unveiling of such hidden relationships; if we are alienated from the vulnerability and depths of learning that new language can bring to us, then we cannot make the transformative shifts of insight, critique and healing that new connections might require of us; if we are locked into discipline-based assumptions regarding the nature of knowledge we are forever running up against that brick-wall that says "here connectivity ends, look no further, ask no more, accept this orthodoxy". Language, learning and transdisciplinary practices give us the means for our intuition and integrity to approach such arbitrary conveniences and act without them in braving whatever glorious mess of inspiration is waiting for us in the re-gathering of orders of thought.

In facilitating this theme, we were excited to know what the papers would look like when they arrived. Who would respond to such a theme? From what ground would they speak? With what language, what qualities of learning and what takes on the notion of 'transdisciplinarity'? How would the various works come together to form a whole entity if only for a moment? And who would we be as people together in that extraordinary meeting place of Gundungurra and Dharug countries in the Blue Mountains?

In respect of such spirit we presented seven papers. Eleven abstracts were received but not all their authors were unable to participate in the conference. Being transdisciplinary, each paper is grounded in its own construction of what knowledge is, or can be for the author(s). When looked at through the lens of disciplinary origins, each reflects a range of arts, sciences, epistemologies and fields of concern in the socio-political realm. Such a lens however, is only useful for stabilizing the reader's giddiness should they be feeling it, because in truth each author is stretching a very long tether from their positivist origins. In many cases they have broken free entirely, losing interest in the colorations of 'environmental science', 'family therapy' or 'primary health' for example, to look much more intently at what happens when new connections are not only made but given the means to have a life of their own: what practices do we create and work by when such a thing happens? How can we masterfully work alongside each other and the ever-evolving "substance" that comes into being?

"Manduway Yunupingu notes that the white fella tends to dislike the brackish waters of the mangrove. After all it is neither one thing nor the other - and malodourous to boot. The black fella on the other hand lives the metaphor - the mangrove is both a rich source of food as well as a potent model for an integral epistemology. So efficient!" (Peter Whitecross pers.comm.)

It feels unjust to dissect the papers in this introduction and risk creating a division between the aspects of our theme and the author's much more holistic invocation to consider their concerns. Aware of this liability, let me briefly consider each paper to provide a touchstone for your engagement, and then move towards an overview of the systemic field that they give together - as we the authors agreed and presented at the conference.

Andrew Gaines, in his paper "Communicating with people who are not accustomed to big picture thinking" describes a transdisciplinary practice of bringing conversation and graphic illustration together to show the connections between human behavior and environmental degradation at a systemic level. His paper is extraordinary for its insistence on being present in the actions of drawing the images on paper in concert with the unfolding understanding of the person being willing to learn of this relationship. Language for Andrew exceeds linguistics, to include image and relational knowing: knowledge is not on the page but exists in the ever-developing construction that comes into being between the two people as learning takes form.

Suzanne Regan and David Thorpe take an altogether different stance in their paper "How the common language of 'Child Abuse' makes enemies of parents". They show how language has systemic powers and reverberations. It reaches into the capacity of parents to have confidence in themselves, into the capacity of service providers to believe that there is such a thing as good parenting, and into whole systems of welfare practice in their construction of the field of "child protection". In this paper language is shown to interrupt and incise what could be a natural connection of individual and community healing. The authors make a powerful political claim on restoring community capacity for self-generation, beginning with a shift in language.

Jade Herriman, Juliet Willetts and Emma Partridge invite the reader into "Learning together for sustainability - the value of group based peer learning". Locating their work in Australian local government's commitment to support residents to live more sustainably, they discuss several different approaches to peer group learning. In so doing they highlight its advantages while also noting its lack of fit (connection) with such presumed, non-systemic management practices as short term funding, output focused evaluation and targeted, linear planning. The authors know that peer group learning is more than able to meet the need to transform our communities to reduce their environmental footprint, but this extraordinary capability is continually hampered by literally 'outmoded' managerial habits.

Jane Gregg and I coauthor a paper, "Principles of participatory practice", which shifts the focus of transdisciplinary practice away from any specific field of application such as environmental sustainability, child welfare or pedagogy, to look into the values that enable participation to come about at the systemic edges that transdisciplinary approaches make development priorities. When, for example, programme managers, funding body representatives, service providers and consumers alike shift their working assumptions from being evaluated by an outsider to self evaluate in collaboration with others, a junction appears that requires all participants to stop, see what has not been seen before, relate differently and progress in new ways. Participatory practitioners bring to such junctions sparse but potent resources that have the power to enable participation to form without predetermining what it will be. This paper describes such resources in practice.

"Theory, identity, language and practice: the value of participatory practices to sustainability" takes the scholarly obligations of reporting on systems thinking and practice into another form altogether. While we pay some attention to existing theories of practice the authors of this paper hold to individual narratives which encompass praxis, social capital, mythology and poetics in the braided and not so braided honesty of collaborative learning in the interests of whole system sustainability. In our shared refusal to be held to any orthodoxy for too long, Jane Fisher gives the writings of John Lavarack, Peter Whitecross and Lesley Porter a critical, self-reflexive turn to give meaning-making, and even the paper's structure, back to the participatory reader. This paper gives to 'living transdisciplinarity' a language and form, acknowledging that there are significant gaps to be crossed (or not) between the elements that make up a whole system.

Finally, we return to Juliet Willetts, this time writing with Cynthia Mitchell, in their paper "Learning to be a transdisciplinary sustainability researcher: a community of practice approach". The authors introduce the reader to some key concepts about transdisciplinarity and learning within communities of shared practices. They go on to describe their experiences of supporting post graduate students in the journey from discipline-based assumptions about the nature of knowledge, to other constructions as deployed in research practices particularly in the interests of sustainability. It is honest work, discussing the shifts that the educators make with their students in the difficult and rewarding joint venture that is "co-learning". The paper is inspiring, a beacon into the very near future and providing a strong scholarly basis for each of the other papers that precede it.

Together, all the papers illuminate an issue that becomes the binding meta-theme for our area of interest. Each paper describes the practical power of transdisciplinary work, while also commenting on the frustrating limitation it meets when confined to working at cellular levels within fragmented large-scale infrastructures. Each paper shows how a lack of understanding or regard for the inter-connectedness of things rubs up against the extraordinary potential of systemic practices. This is not unusual for any approach still being perceived as new: but the difference in our case is that systemic practitioners see the systems at work even here - even in the edges that make up the disconnections. As self interest or sheer energy allows, we may be forgiven for drawing our own boundaries around our systemic practice sites, or for living within those boundaries that are given to us. But as this theme illustrates we do not have to do so because we have the skills (language, learning and transdisciplinary capabilities) to venture into these edgy places and be in transformation with them. Moreover, we should persist in showing and critiquing such margin making, when we see, as Andrew Gaines, Suzanne Regan and David Thorpe so clearly show, their dreadful socio-ecological cost.

As each paper suggests, if ever there was a group of people whose time has come, whether that time be to embed sustainability or develop innovations from such a way of being, we are they. In this theme, join with us as individual authors and an emerging community of practice in seeing and understanding transdisciplinary vistas; in enriching our expressions of it; and in enabling systemic realities in the interests of social and environmental sustainability come into being so they can have their powerfully healing and energizing ways with us:

"No primacy is given to either the individual or the community: rather, both are considered valid boundaries for understanding agency and moral decision making, and the emphasis is shifted to how processes of boundary critique and systemic action can be enabled at all relevant levels… to provide a basis for moving towards more ecologically sustainable forms of development (Midgley, 2000, *Systemic Intervention: Philosophy, Methodology and Practice*, Kluwer Academic, New York, p. 400).

Proceedings of the 12th ANZSYS conference - Sustaining our social and natural capital
Katoomba, NSW Australia, 3rd-6th December, 2006

Principles of Participatory Practice

S. Goff[1] & J. Gregg[2]
[1]Centre for Cultural Research, University of Western Sydney, AUS
[2]Faculty of Science, Health and Education, University of the Sunshine Coast, AUS

In 2001 an Australian research consultancy conducted a national survey to inquire into the meaning of participation for individuals who had experience in participatory approaches to research, evaluation and learning. Thirty-three telephone interviews were held with participants from the three tiers of government as well as community, academic and small business sectors. The core business activities of participants included primary industries, social services, environmental management and moral campaigns. The survey results told a fascinating story about the emergence, valuation and application of participatory approaches in Australia. Three generic principles were identified: equity, transparency and inclusion. They were proposed as core to a generally held idea as to how to recognise "participatory practices" at work and as being necessary for authentic participation to manifest in any change strategy. The consultants compiled the information into a report, which was returned to survey participants. Having the benefit now of five years of application in a variety of contracts, these principles are discussed both as a model and as a case study. The paper concludes with a brief reflection on the practical implications of participatory practices that use realize these principles.

1. Background

1.1. Participatory practice

The term "participatory practice" is unfamiliar to many, even though an idea of participation as "consultation", "decision-making", "engagement" or "community development", for example, is approaching the commonplace. The actual practices necessary for participation to take place are somewhat muddled into specific disciplines as their proponents question traditional approaches to expertise, authority and the reification of forms of knowing from everyday life. For example; environmental education which was once preoccupied in educating others from an environmental science-based stance, is increasingly practiced in appreciating and augmenting what people already know about their local ecosystems; and crime prevention is as interested in working with a community's history, culture and social networks as imposing policing, target hardening and surveillance technologies. If we stand back and consider "participation" across the various disciplines, a "trans-disciplinary" pattern of participation emerges, recognizable to all and modifiable by each. It is this pattern in the form of "principles" and their inter-relationships in the hands of those who create the means by which participation takes place, that we understand "participatory practice" to mean, and which we are enthusiastically promoting for the benefits that participation offers.

1.2. Practical concerns

In our experience, the lack of a recognized participatory practice as a paradoxically generic kind of specialization means that workers, organizations and schools of thought are re-inventing their cognitive wheels unnecessarily, slowing down urgently needed reforms, collaborations and innovations. We do not argue against the value of learning from the ground up as a way of building local capacity for participation and enriching what we understand about it. We maintain a *principled* approach to participation allows learning and capacity building to take place with greater efficiency and communicative resonance, strengthening practice quality and the democratic qualities of our daily lives in the process of realizing a participatory outcome.

Participation is still establishing itself somewhere between pilot project, managerial system, philosophy of knowing and assumed societal right and responsibility. As practitioners navigating these delicate variations in specific sites we can waste precious resources in walking backwards from a strategic starting point, unlearning what is assumed to be true before moving forward with a shared understanding of participation. In participatory projects, resources are generally slim for the depth of work that needs to be done, so this recasting of knowledge, action, practice and development within a participatory frame can be an expensive double task for participants and practitioners.

For example, a participatory approach to issues related to family violence, not only needs to address the specific issues of abuse participatively, but to also recast a host of related issues and approaches such as forms of therapeutic intervention, organizational management, inter-agency collaboration, legal, criminological and even feminist and community development principles.

The scale of such a task is impossible in the capacity of any "pilot project" or "independent research" contract. When faced with such systemic challenges, determinations need to be made as to which and how many frame shifts take place and importantly, who is involved in such determinations.

How do we make such choices? Our proposition is that the three principles of participation can guide us through these turning points while also providing a kind of weather-check for the quality of participation that we are enabling through our practice.

Let us first revisit the survey so that the origin of the principles can be appreciated.

2. The survey

An eighteen-question interview protocol, of mostly open-ended questions was designed and trialled prior to contacting potential contributors. Respondents were sourced through our own business networks or ethical investment publications, and initially contacted via phone to seek their interest in being involved. Thirty-three telephone interviews were held with participants from the three tiers of government, community, academic and small business sectors. The core business activities of the participants included primary industries, social services, environmental management and moral campaigns. Table 1 below presents the range of contributors to the survey.

Common-wealth	State Gov	Local Gov	Industry	Academic	Consul-tancies	Commu-nity
1 rep each of 2 depart-ments	1 rep each of: 6 de-partments in Q, 1 in SA, 1 in WA	1 rep each of 2 coun-cils in Q, 1 in NSW, 1 in WA	1 rep each of 3 corporate interests: mining, petro-leum and commu-nications technolo-gies	5 reps of 4 unis in Sydney; 1 regional NSW TAFE; 1 uni in Victoria	3 Sydney consul-tancies 2 Brisbane consul-tancies	1 Victorian charity, 1 environ-mental or-ganization in NSW, 1 ACT char-ity, 1 moral campaigner in NSW, 1 moral cam-paigner Q

During interviews we refrained from prompting, only intervening for clarification's sake and recording key ideas in notes as we listened. When the interviewing process was complete, raw data was reviewed and organized into themes and categories of our own making, that we recognized as being "participatory".

Table 1 *The contributors to the survey*

2.1 A summary of findings

Participatory practices were described as being associated with decision-making by a broad spectrum of people with primary stake holding in the matter being addressed. Decision-making processes were associated with democracy, learning and human rights. Information about the meaning of participatory practices suggested that certain principles could be used knowingly to set the conditions, and characterise the actual environment for participatory practices to be introduced and nurtured.

All sectors reported mixed levels of familiarity with the term 'participatory practices'. For a few the term was entirely new, but not necessarily the practices themselves. Beyond that there was general familiarity and comfort with the term, and a view that, even if it was not used, there was some level of "buy in" to the concept somewhere in each respondent's organization.

Respondents saw participatory practices being applied in three domains: within an organization (internal), in the organization's stake holding groups/networks (external), and in between the two (bridging between and organization and its "constituency"). Key constraints to the application of participation included issues around resources for participation and personal preferences of particular approaches to strategy and practice.

Contributors' descriptions of the emergence of participatory practices over the past 10 years reflected changes in the following four areas: thinking, including the ways we have been thinking, and thinking about thinking; valuing the community as a source of expertise rather than trouble makers; the developing culture of business practice and diversification of how core business was being accomplished; and an increasing range of resources available for the development of participatory practices.

Benefits of the application of participatory practices were described in terms of credibility, social worth, strategy and resource sustainability, and kudos. Risks were described in terms of personal and organizational risks of working in this way when the majority of a work place culture was unfamiliar or uncomfortable with participatory principles.

Five main directions for the future of participatory practices were identified. In developmental sequence these were: building awareness of participatory practice; building or developing systems and structures for participation; finding ways of augmenting the benefits of participation; increasing the resources for participation; and strategizing the agenda for participatory practices. These future steps appeared to us as a good blueprint for any systemic development towards participatory approaches that an organization or even a particular strategy might want to take.

3. The participatory principles model (PPM)

3.1. The participatory principles

When we asked questions about what principles were at work in participatory practices, three generic values kept occurring throughout the data. The idea of "Inclusivity" was common to most respondents' views. By this we mean that participatory practices, by definition, enable people to be included in particular activities. Almost without exception, the concept of Inclusivity implies the inclusion of people who in the past have been overlooked or excluded from the named activities. Thus Inclusivity centres on the ideas of "insiders and outsiders", decision-making and relevance.

Two other strongly convergent values evident in the data and associated with participatory practices were "Transparency" and "Equity". From this convergence we proposed that when engaging people in processes of decision-making about matters that they see as directly affecting them and that they have not previously engaged with formally, then "transparency" and "equity" are critically important to the efficiency and effectiveness of the process of broadening boundaries and inviting people into new terrains of responsibility. These qualities open up the governing world to its internal and external stakeholders, be that world an organization, institutional power, defined community or network.

By such disclosure the principles of Transparency and Equity operate as means of information sharing, building trust and self-confidence, as well as introducing stakeholders to the subtle realities that underpin the issues being addressed. In short, Transparency and Equity enable learning for participation to take place once an individual or group is Included in the participatory opportunity.

Spontaneously, many contributors also described what they felt participatory practice to be and what it was not; they began to create boundaries around a recognizable "field" of practice. These ideas are assembled In Table 2.

3.2. A model

The three principles spell a three-bodied schema (see Figure 1 below) where each principle interacts with the other, focussing the practitioner's attention on specific tasks to establish a participatory environment and sustain its integrity

Participatory practice is not	Participatory practice is
Lip service, rhetoric	Listening, Valuing, Developmental ways of working, Participatory validation of claims
Isolated, formal consultation events	Key stakeholder and task sensitive approaches. Carrying out participatory action, research and learning
Becoming embroiled in reactivity	Co-mentoring, Reflection, Critical reflection, Using outcomes from reflection in participatory action
Self interest and individualism	Team work, Partnership work using clear criteria for participatory practices, Collaborative learning and personal development
Closed, directorial action	Making policy and recommendations for participation, Reflecting on participation, Participatory facilitation and facilitation of participation

Subsequent to conducting this research we developed a model of participatory practice deploying the three generic principles of Equity, Transparency and Inclusion. We have used the model in a variety of research, learning and evaluation strategies or environments; presented it at professional and network events, seen others using it and received a variety of local and international responses to both the report and model. It is with this developmental history behind us that we present this model for more general access.

Table 2 *Some recognizable boundaries of participation (opportunities for intervention)*

throughout an initiative's life cycle. The primary responsibility of a practitioner is to attend to the emergence, balance, qualities and consequences of these principles at work. Such attention draws us into boundary-making, strategic design, mediation and negotiation of rights and responsibilities to participate. It involves the active development of relationships between individuals and entities - including language, practices, identities and theories.

This practitioner focus frees participants to co-create their own means and benefits of participation so that their means of transition more closely reflect and develop their local context, while still growing from the three interacting and generic principles. All principles will do this if they are used in practice, but the specific nature of principles, the way they interact and are manifested in practice, the ways they are recognized by participants and authority and tracked for their presence in the multiple dimensions of interactive work, all these variants makes a difference to what comes about. We learned that good practice is not about doing the principles "to" others in an ideological manner. It is about realizing them within ourselves as qualities of attitude and behavior as much as principles for designing strategy or evaluating outcomes. We learned to create

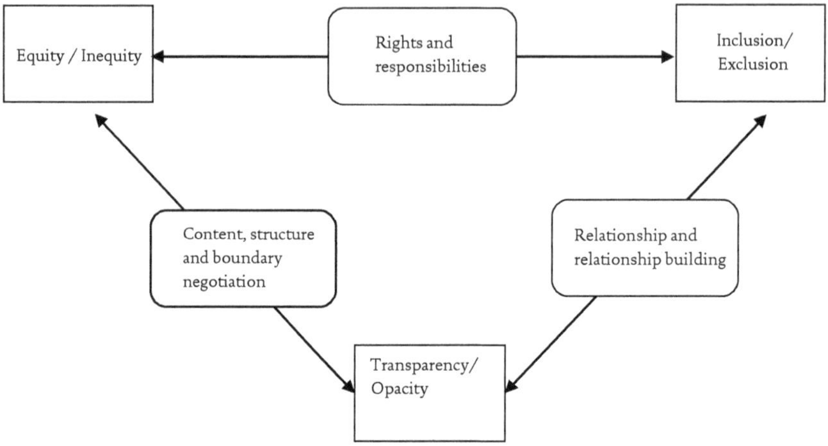

Figure 1 *The three participatory principles and core practice skills that they generate*

a principled environment with people; to bring them as a bare minimum that was both evocative and also left plenty of room for others to contribute their resource to the participatory entity; and to pause and pay attention to what is going on (or coming about) when the principles are having effect. Imbalances, invisible exclusions and hidden truths demand careful work.

As the schema shows, to assist with its appreciation, we present a counter-state of a non-participatory context in each principle's domain. When we see the antithesis of participation at work in a local context we have an indication of the possible limit of an initiative's influence, or the boundaries at which participatory intervention may be desirable. We came to understand that participation is never a fixed, ideal or definable state so much as the valuing of something that may otherwise be invisible; a movement between relative participatory potentials that any given system can foster or inhibit at different moments of participatory engagement. In short, participatory practice is an ongoing inquiry into the participatory potential of any aspect of practice, relationship, system or knowledge structure. These three principles are a guide towards understanding what this means and how to accomplish such work.

We move on to illustrate the principles at work in an evaluation strategy, where the practitioner created a participatory environment with the principles in relationship with the co-participants. In turn, the co-created the opportunity and means with which the participants designed their own evaluation framework, implemented it, determined the learning outcomes and evaluated the evaluation approach. In so doing, the evaluation created an outcome that reflected the participants' experiences and relationships, while also creating optimum conditions for individual learning and strategic development.

4. Application

4.1. Case study

The evaluand for this case study was a Commonwealth funded capacity building strategy working across select services in regional and remote Queensland to provide Primary Health Care services to their local communities. Examples of the emergence of the participatory principles - Inclusivity, Transparency and Equity - are considered at three points across the evaluation strategy, then discussed more generally with respect to the facilitators' developing understandings about their participatory practice.

4.2. Emergence of the Participatory Principles Model (PPM)

4.2.1. Key Event 1: Stakeholder Meeting 1

An initial face-to-face stakeholder meeting was held between the evaluation funding and auspicing organizations' representatives and the contracted evaluators[1]. The purpose of the initial meeting was to negotiate broad terms and conditions for the evaluation strategy's approach and scope, including the three principles. This meeting resulted in agreement on a participatory form of evaluation; however, the voices of participating services' were not included in this initial decision-making. The exclusion of these critical voices raised a participatory practice tension, in that those most affected by an initiative needed to have Equity in making decisions about the strategy that would affect them (e.g., Carr & Kemmis, 1986; Wadsworth, 1989; Hart & Bond, 1995). The evaluation facilitator made individual calls to service representatives to set up a collaborative workshop with all stakeholder groups to codesign the evaluation framework. Simultaneously, the Transparency principle transpired by means of the participants creating a structured communication system between stakeholders engaged with the initiative at this early point in time.

4.2.2. Key Event 2: Stakeholder Workshop 1

A one-day face-to-face collaborative workshop with all stakeholder groups codesigned the evaluation framework. The Inclusivity principle guided the facilitator's negotiations about what to include in the framework's broad structure. For example, participants' previous evaluation experiences, their expressed values for collaboration, their evaluation questions and criteria, data collection and analysis methods, and reporting processes were all important elements with which to construct the framework so that it could be used by them to produce evaluative outcomes that were relevant to them. The Transparency principle was alive in these deliberations and enabled participants to discuss and learn about their own and others' interests and needs with respect to the evaluation. This event was also effective in realizing the Equity principle in the inter-relationships between the workshop participants; but as the workshop found, the broader service stakeholder voices at the community level were still missing from the evaluation.

1 Contracted evaluators we refer to as "evaluation facilitators" because in a participatory environment, it is the participants who do the evaluation; the participatory facilitator provides the means for this to be done in ways that are recognised as trustworthy and reliable evaluation.

4.2.3. Key Event 3: Regional Service Onsite Workshops

Onsite workshops were held in each of the participating services' communities to implement the evaluation framework. This involved the evaluation facilitator using the tools that had been generated and endorsed by participants in previous stages to collect and interpret evaluation data. This event engaged the broader range of stakeholders at the service level, which was a tension that the Inclusion principle illuminated earlier. It also provided the opportunity for the facilitator to explore with this broader service level (rather than management level) stakeholder group their understandings and experiences of the strategy being evaluated - an action that is directly related to the Transparency principle. The application of the evaluation tools at workshops to collect evaluation data illuminated the Equity principle, in that the data represented the stake holding group most affected by the evaluand.

4.3. General discussion points in reference to participation and personal practice

The explicit application of the three participatory principles within and across these key events, and others in the evaluation strategy, worked to provide both very broad and specific guidance on the facilitator's participatory practice. At the broad level they provided a "back brain" participatory model to reflect iteratively on:

- The voices and interests that were or were not present - Inclusivity;
- The levels and reaches of input from stakeholders into decisions about the evaluation's critical components - Equity;
- The systems for ensuring opportunities for stakeholders' access to the evaluation's processes, outputs and outcomes - Transparency.

This broader guidance simultaneously enabled the evaluation facilitator to identify participatory practice tensions at a micro level (specific to a particular moment as compared to overall structure) and subsequently to devise the means for navigating their resolve. In this way the principles worked to enable transition within and across the evaluation's key events, and between ongoing cycles of unknowing (complexity) and knowing (pattern). The flexibility and interdependency of the three principles in action was evident in the way each principle emerged and interacted with others differently at different points in time. For example, at the micro level of practice and pending the nature of the tension, a principle would emerge as the leading principle and at the same time be dependent on the co-emergence of the other two principles to work through the tension. Consistently, the Equity principle emerged as an outcome whilst the principles of Inclusivity and Transparency emerged as both processes and outcomes upon which the experience of Equity was dependent.

5. Conclusion

In our view, the opportunities within the evaluation strategy that enabled participation to be realized included:

- Stakeholder participation in the design and implementation of an evaluation and the interpretation of evaluation data, including generation of recommendations;
- Face-to-face interactions with stakeholders to work through the evaluation process and direct its course iteratively;
- Faithful use of participant values in the design and implementation of the strategy and tools as made evident through the foundational use of the three participatory principles;
- Faithful document management to ensure balanced stake holding of information and outcomes.

Unless there is some cynical interest at work, participatory approaches to any form of work are drawn from a commitment to finding new ways of functioning that both support and enable a more prosperous future. When we refer to "prosperity" we do so in full recognition of the all-too-familiar systemic problems such as the rich are getting richer and poor are getting poorer, our being habitually and structurally wedded to old baggage (individual and global), and our being trapped in and even unwilling perpetuators of social and political systems that are not meeting the needs of those who are a part of them or who feel their consequences. It is our experience, cautionary notes stuck to our foreheads, that participatory approaches value the human condition and the environment in which we live as they enable recognition and valuing of local human and non human resources. That is what we mean by "prosperity".

The three principles are straightforward, easily recognizable in a participatory endeavour that builds the boat as it sails, and structurally conducive to co-evolve with a local community's way of doing, knowing and being in any mutual responsibility. As our case study shows, they allow for systemic dynamics of complexity, self-organization and regeneration to come about, without need for any one particular form of authority to exert itself and save the day by replicating old ways of doing. They are conducive to transformation.

The participatory principles certainly perturb old systems, but this is necessary when old systems fail to meet their responsibilities or adapt to new conditions. These particular principles have the following effects within an initiative:

- Redistribute traditional power structures according to current or perceived needs;
- Accordingly require and create different decision making processes and generate different consequences and responsibilities in any participatory setting;

- Involve transgression (at personal, inter-personal and systemic levels) that is the source of learning;
- Identify divergences by continually monitoring equity/inequity;
- Use divergence as a medium for moving on by building in inclusion and transparency when divergence comes about;
- Create and manage risk by enabling sensitivity to shifts and deploying them into open-ended and softly-bounded development scenarios;
- Work with chaos, emergent pattern, rhythm and transformation;
- Reveal hidden assumptions for exploration and incorporation, modification or abandonment;
- Provide for public accountability in reasoned and comprehensible ways;
- Create the basic building blocks for rigour in any mode of knowledge generation (research, evaluation, learning, design etc.);
- Value people's real life experiences and how they know them as the basis for social development;
- Purposefully generate new knowledge for change at individual, collective and systemic scales, and;
- Provide a "safe house" for change to seed, be learned of, establish and mature.

For us perhaps their most exciting attribute is that they are simple and easy to make our own in practice. In your hands, let them generate unique forms of participation that still resonate with qualities that the field recognises as authentic, without capitulating to either external orthodoxies or internal conservativism, both of which while useful tensions can if not recast by such a model, get in the way of participatory systems taking form.

References

Carr, W. and Kemmis, S. (1986). *Becoming critical, education, knowledge and action research*, London, Falmer Press.

Hart, E. and Bond, M. (1995). *Action research for health and social care: a guide to practice.* Open university Press, Buckingham, Philadelphia.

Wadsworth, Y. (1989). *Everyday evaluation on the run.* ARIA Publications, Melbourne.

Notes

[1] Since 2001 many other modes and discipline-based forms of participation have been developed

The Core Briefing: Helping People Understand the Big Picture Via Diagrams

Andrew Gaines
Futures Foundation, AUS

This paper describes the "Core Briefing" – a way of drawing a diagram on one page that shows the connection between ecology, our current mode of industrial production, economics and personal psychology – with reference to the two core values, partnership/respect relating and domination/control relating that will make or break Australia as a viable society. The Core Briefing was developed as a way of communicating with people who do not have the time, inclination, or possibly the intellectual capacity to read and absorb books on the different dimensions of sustainability. It helps people break out of their mental silos and begin to develop a much more comprehensive understanding of how everything is together. It also provides a way to highlight constructive points of change that can make us viable in the long run.

Helping people understand the big picture via diagrams

Arguably we must change our entire system if we are to actually become ecologically sustainable and socially healthy. But people tend to think in silos, and only a few people can read and integrate the books necessary to get a big picture understanding of the multiple levels of change we need to make to become sustainable.

So we have a major challenge: how to communicate a systems perspective to people who are not accustomed to big picture thinking?

The approach I am presenting here is to draw diagrams. I am going to show a process I have developed for drawing a diagram that shows how the environment, industry, economics and our personal psychology work as an integrated system. We get the main elements on one page!

I call this the Core Briefing. It illuminates the dynamics the unsustainable society we have today. Many of the elements of the diagram are known to people. What is new is seeing how they all fit together, and identifying leverage points for positive change.

Drawing the diagram step by step, rather than presenting the whole thing in one go, gives people the opportunity to think, argue, probe deeper and otherwise engage with the ideas. The Core Briefing has been used with Leadership Forums for the Project to Make Wellbeing a National Priority, WEA discussion groups and with individuals. Some people have been quite moved by it.

In this paper the key ideas have been presented as succinct assertions which may seem dogmatic. A fuller account is given in my book *Evolving a World That Works*, available as a free download on the Futures Foundation web site. Look under the Project to Make Wellbeing a National Priority.

My hope is that many people will use these diagrams in their own way to foster discussions that will accelerate our shift to an environmentally sustainable and socially healthy society.

Ecology and our industrial system

We start with basic industrial ecology. The process of obtaining raw materials through mining, industrial agriculture and cutting down forests produces environmental degradation.

Figure 1

The raw materials are processed in factories, which produce their own environmental damage through chemical toxins, acid rain and the greenhouse gasses that accelerate climate change.

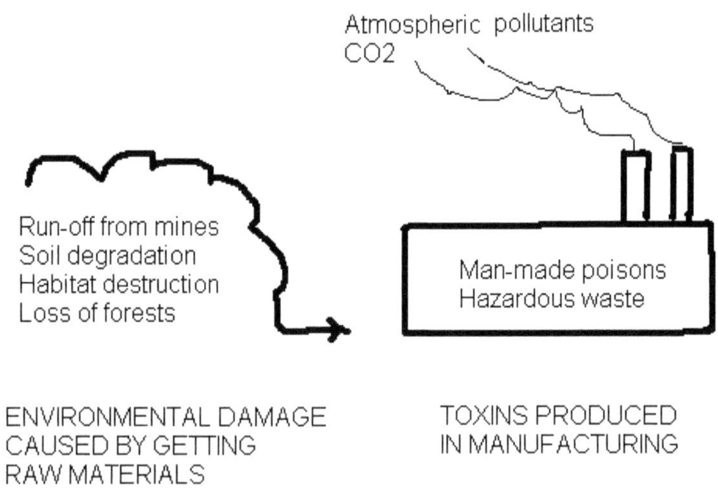

Figure 2

All of these are involved in the production of the ordinary things that we use.

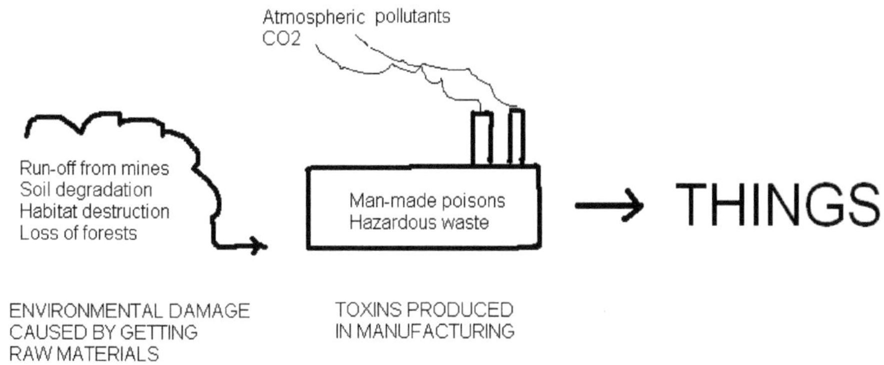

Figure 3

The more things we produce and consume, the more environmental damage is produced – vastly more than most of us imagine. In the following diagram, as the red arrow on the right representing increasing consumption goes up, the red arrow on the left representing environmental deterioration goes up even more.

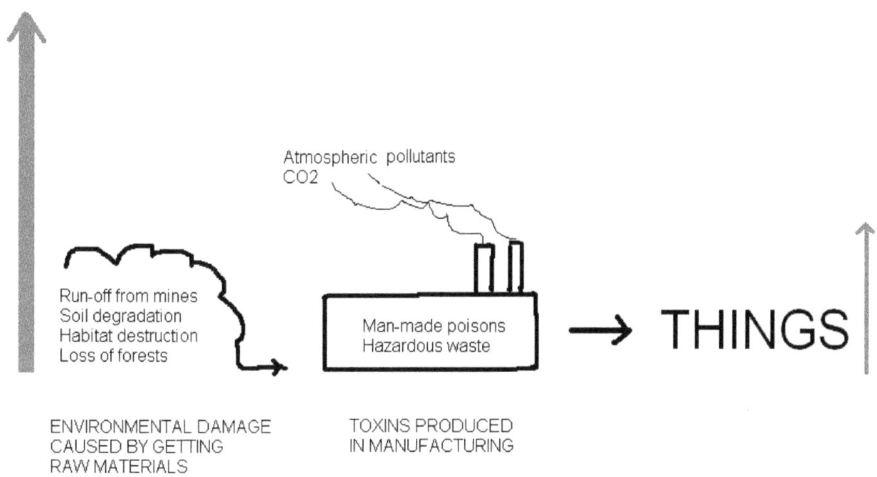

Figure 4

It is clear that if we are to become environmentally sustainable, we must reduce the cumulative environmental impact of the process of making and consuming things. By how much, we may ask.

If we cut trees from a forest plot, and we cut them faster than they regrow, gradually the forest will get smaller and smaller until it is gone. This is an example of *cumulative* damage. It adds up over time. If we intend to be ecologically sustainable, our goal must be to *reduce our cumulative rate of environ-*

mental degradation in each key area (water, biological diversity, fish stocks...) *to...* **zero.** Zero! This rigorous demand comes from the nature of reality itself. If the overall trend is of increasing deterioration, we end up producing a desert.

The technical hope is to reduce environmental degradation through improved design. A great deal can be done in this direction. Lovins and Hawken's *Natural Capitalism* (1999) shows that in every area from agriculture to architecture to manufacturing we can reduce energy use and material throughput 90 percent or more. This is exciting stuff, and more of us should know about it. It is a crucial key to our future sustainability.

But improved design may not be sufficient by itself. Sometimes improved design means that things are produced more cheaply, making it easier for people to buy more of them. And as affluence increases, many people tend to buy more things. Currently this is incompatible with becoming environmentally sustainable.

So let us ask a question: what factors tend to increase consumption?

The first answer most people give is *advertising*. And they are right.

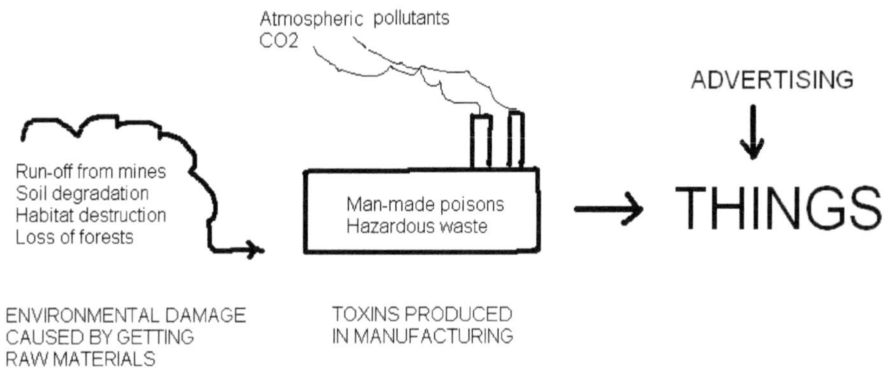

Figure 5

But advertising *per se* does not compel us to buy things. There are psychological drivers that affect our desire to purchase things. Here I will give one account of these dynamics. It is necessarily incomplete. My purpose is to bring a discussion of psychological dynamics into our considerations of sustainability.

Many people lack a feeling of inner wellbeing, and instead have a zone of pain inside. It is sometimes experience as an empty hole inside. This comes from the experience in childhood that their parents for whatever reasons (often reasons beyond their control) did not give adequate nurturing (Grille 2005, Slater 1980).

If this zone of pain were to be directly experienced it would be extremely painful. People avoid it by compulsively distracting themselves, taking drugs, or by stuffing themselves with things. Thus some stuff themselves with chocolates and others indulge in 'retail therapy'.

Desire for status also drives excess consumption. There is healthy status and dysfunctional status. Healthy status is earned; it arises because of one's contribution to the community based on competence and caring. Pathological status is based on a profound need to feel good about oneself by appearing to be superior, or at least not inferior. This finds expression as conspicuous consumption and keeping up with the Jones.

So *retail therapy* and *striving for economic status* are two ways of avoiding the zone of pain inside. They drive excess consumption beyond our real needs or genuine enjoyment. We will add them to our map.

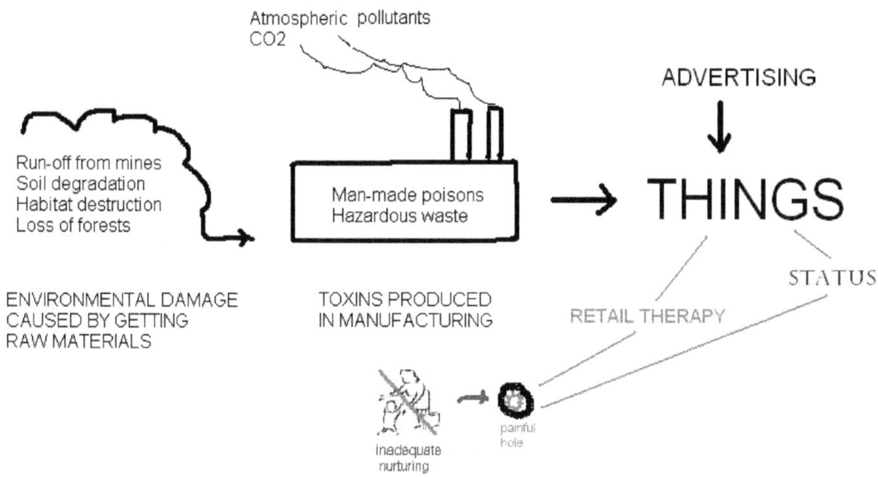

Figure 6

As you know, our economic system is oriented around *increasing shareholder value.* The vast majority of shares are not owned by ordinary people, but by a relatively small number of extremely wealthy people (Wolff, 2000). So we may say that our economic system is set up to help the wealthy get wealthier. They are assisted in this through government policies that they themselves have influenced – policies that emphasize increasing the Gross National Product. And while most businesses advertise, the most extravagant advertising is done by transnational corporations. So we may add a nexus of shareholders, TNOs and the WTO to our diagram.

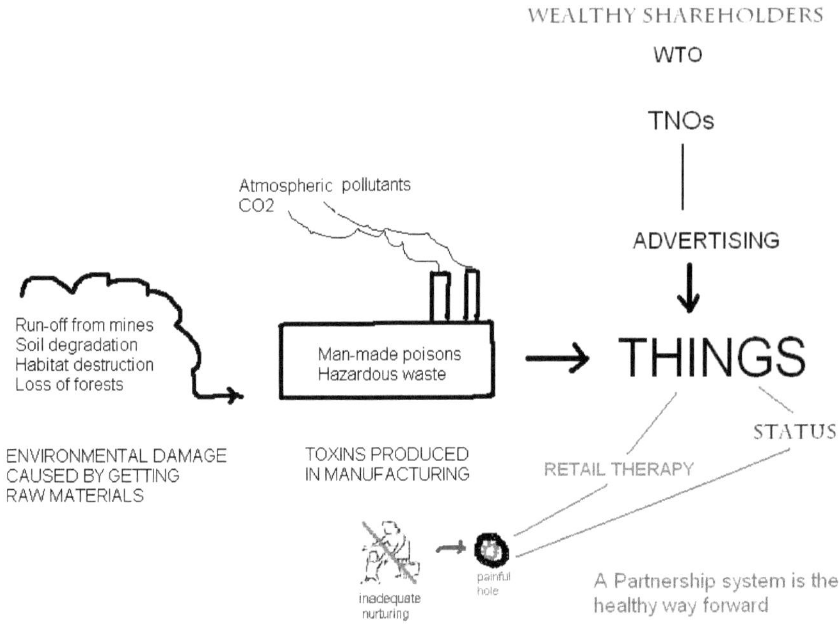

Figure 7

The behavior of people directing the large transnationals is at times extremely aggressive, to the extent of even being closely aligned with wars such as the invasion of Iraq. The WTO and the pattern of increasing transnational corporate power can be seen as the current expression of the 5000 year old pattern of empires, where wealth is forcibly made to flow from the periphery to the centre.

There may be more to this aggression than just the desire for wealth. People who have been physically abused in childhood tend to repeat or 'act out' that abuse when they become adults (Grille, 2005; deMause, 2002). This acting out finds expression at many levels, including abusing one's own children or spouse, workplace bullying, and adopting policies that hurt groups of people or entire countries. Yes, our personal psychology plays out on the world stage! From this point of view, it appears that some aggressive corporate behaviour is driven by an (unconscious) desire to hurt others. So when we include these dark psychological factors our diagram looks like this:

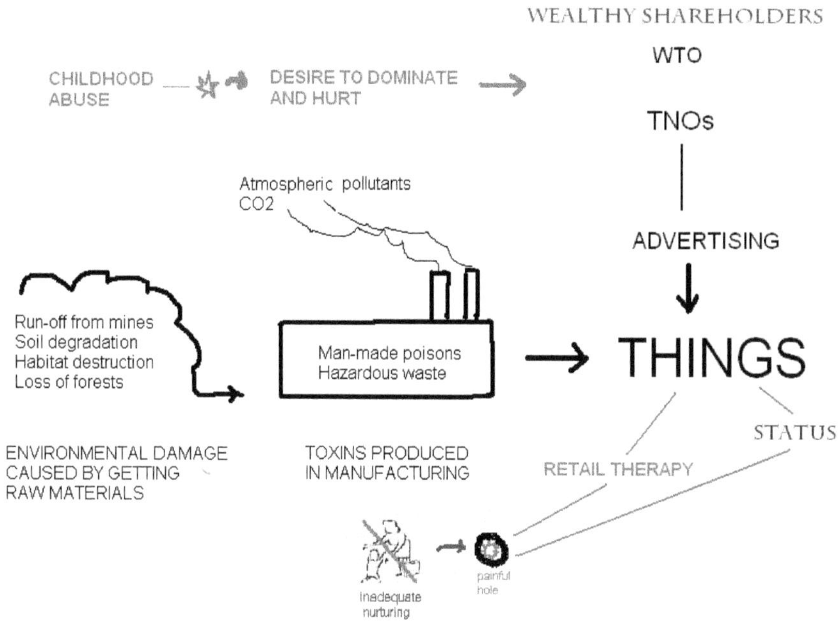

Figure 8

We have talked about inner pain and lack of felt well-being, as well as about responses to childhood abuse finding expression as large-scale corporate aggression. These psychological aspects, although they are rarely discussed, are actually key drivers of environmental deterioration in developed countries. They lead people to over-consume, on the one hand and to an attitude of *attacking* nature and society on the other. Thus to a significant extent our environmental deterioration is driven by unhappy people.

It follows that a key point of change for creating a positive future is that *people should become happier in themselves*. Ideally we should develop such an internal feeling of wellbeing that excess consumption either becomes uninteresting or is eschewed out of respect for environmental ethics. Improved parenting, personal development and also organizing business, education and government to support rather than control people can all contribute to developing genuine happiness and wellbeing.

From a systems point of view, other high leverage points of change for achieving ecological sustainability include improving industrial design and reducing the amount of advertising – perhaps by not regarding advertising as a tax deductible business expense.

To complete the Core Briefing we need to consider the core values that will lead to either a positive or a dismal future

The partnership-dominator contrast

In *The Chalice and the Blade* (1987) futurist Riane Eisler describes two quite different modes of relating to people. She calls them *partnership/respect* relating and *domination/control* relating. People who are using a partnership mode of relating take pains to understand other people, and they work collaboratively for the good of the whole. People who are using a dominator mode of relating work for their own aggrandizement (or that of their group), and characteristically use force to achieve their ends. They are quite willing to hurt other people, communities and the environment in the process. For example, think of the invasion of Iraq and of corporations who knowingly sell disease producing products and 'externalize' the medical and environmental costs,

Arguably it is people with dominator attitudes who shape the policies that are inducing increasing environmental and social deterioration. If so, then *the fate of the world depends upon partnership values coming to set the tone.*

The world has put an amazing amount of creative resources into building an astonishing industrial civilization. Now it is time to shift our emphasis to creating a world where *magnificent well-being* is our primary goal. And the good news is: we know a vast amount about how to actually do this. We have the technical design strategies that can vastly reduce adverse ecological impact (Lovins *et al.*, 1999). Proposals to change the economic system to better support community well-being have been put forward (Hawken, 1993). We know how to support parents in becoming more nurturing (McFarland, 2005). We have effective means of helping adults resolve their childhood hurt and become more loving (and inwardly contented) people (Feinstein, *et al.*, 2005). Experiential methods of personal growth including Aikido, the Feldenkrais method of body awareness and interactive creativity training can help people become more comfortable and skilled at partnership relating – a shift that makes us less hurtful to others and which can make businesses work better (Hull & Read, 2004). There are examples of running successful institutions on partnership values (Niel,1960; Semler, 1993). All in all these approaches can take us far towards an ideal of becoming so inwardly contented (and ecologically responsible) that excess consumption simply becomes uninteresting.

To sum up, we have two options. One is to continue on a path that is becoming increasingly dismal. The other is to get excited about building a world of magnificent wellbeing – and to invest in the intellectual, emotional and practical things necessary to actually build that world. The option of muddling along making small piecemeal environmental improvements is not an option that will lead to future well-being for young people alive today. The near future (within the lifetime of our children and grandchildren) will either be unspeakably horrible or, through a fundamental shift in social direction, surprisingly wonderful. It will either be utopia or oblivion, and creating utopia requires committing to a massive shift to partnership values and community well-being. So we complete our diagram by adding *Dominator* and *Partnership* as core values that drive the entire system in one direction or another.

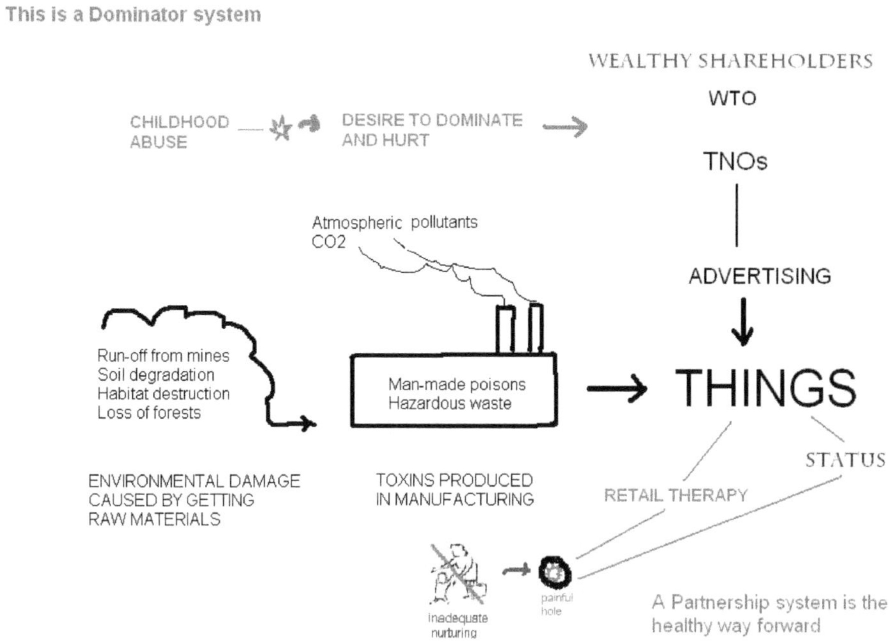

Figure 9

Our times call for an exceptional order of citizen leadership. Creating a healthy society is not just 'up to the government'; it is up to all of us. I believe that our entire culture must wake up and make massive goodwilled changes in every sphere from child-rearing to industrial design, economics, business and personal development. And 'everybody' must come to understand this.

To stimulate this understanding I have set up the Project to Make Well-being a National Priority through the Futures Foundation (www.futuresfoundation.org.au). We will start thousands of discussion groups around Australia enabling people to think through what is needed in more detail, thus developing the background understanding necessary for initiating and supporting the healthy changes we need. I would be pleased if you would join us. You can contact me at andrew.gaines@futuresfoundation.org.au.

Yours for a world that works.

Andrew Gaines
Katoomba, NSW

References

Eisler, Riane (1987), *The Chalice and the Blade* London: Harper Collins, ISBN 004440627.

Feinstein, D., Eden, D. and Craig, G. (2005) *The Healing Power of EFT and Energy Psychology*, London: Platkus ISBN 0749926791.

Grille, Robin (2005), *Parenting for a Peaceful World* Alexandria, NSW: Longueville Media, ISBN 1921004142.

Hawken, Paul (1993), *The Ecology Of Commerce,* New York: HarperCollins ISBN 0887307043.
Lovins, A., Lovins, L.H. and Hawkins (1999), *Natural Capitalism*, London: Earthscan ISBN 1853837636.
Hull, D., and Read, V. (2004), "*Simply the best: workplaces in Australia*", Working Paper, No. 88, ACIRRT, University of Sydney, Sydney.
McFarland, Robert B (2005), "Creating a Community Parenting Center" *Journal of Psychohistory* ISSN: 01453378, 32(4) (Spring).
Neill, A.S. (1960) *Summerhill*, Harmondsworth, Middlesex, Eng. : Penguin Books ISBN 0140209409.
Semler, Ricardo (1993), *Maverick: the success story behind the world's most unusual workplace*, New York, NY : Warner Books ISBN 0446516961.
Slater, Philip (1980), *Wealth Addiction*, New York: Dutton, ISBN: 0525230734.
Wolff, Edward N (2000), *"Recent Trends in Wealth Ownership, 1983-1998", Working Paper No. 300* Annandale-on-Hudson, NY: Jerome Levy Economics Institute.

Andrew Gaines (BA) is Coordinator of the Project to Make Wellbeing a National Priority. The goal of the Project is to contribute to Australia's cultural transformation to ecological sustainability and social health. The Project is sponsored by the Futures Foundation. Andrew maintains a private practice as a creativity trainer, Feldenkrais practitioner, psychotherapist and Emotional Freedom Technique practitioner in Katoomba, NSW. He is trained in the Natural Step and as a sustainability educator, and is a proponent of integrated design a la *Natural Capitalism.* He is the author of *Creativity Games* and *Evolving a World That Works,* and has published articles suggesting that a cultural shift to respectful 'partnership relating' is essential for achieving ecological sustainability in the *Journal of Corporate Citizenship* and the *CSIRO Sustainability Netletter.* His special contribution to sustainability discussions is powerful insights into how personal development disciplines can help us become the kinds of people who can make and enjoy a sustainable world.

Theory, Identity, Language and Practice: The Value of Participatory Practices to Sustainability

J. Fisher[1], S. Goff[1], J. Lavarack[2], L. Porter[3] & P. Whitecross[4]
[1]Centre for Cultural Research, University of Western Sydney, AUS
[2]Sustainability Strategy, University of Sydney, AUS
[3]Murray Mallee Community Health Service, AUS
[4]Northern Sydney and Central Coast Area Health Service, AUS

Introduction

How do we know sustainability, and what is our relationship to such knowledge?

Between 2004 - 2006 twelve participatory practitioners in public, academic and community sectors in three Australian states collaboratively explored the value of participatory practices to sustainability.

The research was a doctoral candidature supported by the Centre for Cultural Research at the University of Western Sydney. The task of this Inquiry was to explore the nature of "participatory practices" being developed by each person within their range of disciplinary and inter-disciplinary approaches to aspects of "sustainability". A central working assumption was that "participation" was more than "doing" in an observable and quantifiable sense; that "participation" and those practices that manifest it embrace being, knowing, doing and becoming.

Some of us work to a "triple bottom line" concept of sustainability addressing environmental and energy footprints; others work to social sustainability where the question of the environment remains relatively disassociated in public policy; some explore our experiences of sustainability and participation from personal, philosophical and cultural perspectives.

As a result of this still ongoing work, and through a powerful three-day interactive event in a cave in the Blue Mountains, we came to understand that when we approach the questions of sustainability without presuming an epistemological stance new ways of sensing, knowing and being in relationship to the sustainability quest arise, giving new forms to participation.

In this paper four of the practitioners reflect briefly on their various insights. Susan Goff describes how Torbert's (Torbert, 2001; Torbert & Reason, 2001) proposition of Action Inquiry, while not used as a research mechanism, was a frequently discerned pattern in our work. John Lavarack reflects on the splitting and joining of self-identity between work, community and virtual environments. Peter Whitecross explores the liberating and disturbing influence of myth in re-minding us to be open to the multiple forms of knowing that take shape outside conventional holds. Lesley Porter considers the power of poetics to shift ways of knowing within a community health service's team of therapeutic practitioners. Jane Fisher completes the text with critical reflection and editorial review.

We speak autonomously, partially and together, the hallmarks of interdependent freedoms that characterise the social-ecology of our collaboration.

Torbert's Action Inquiry in our hands: Ebbs, flows and flourishes

"We cannot know everything through rational intelligence and must accept incompleteness" (Marshall, 2004: 306).

Our Inquiry offers a small "step to the side" of conventional ideas about "change". This convention is governed by pre-determined "targets" undertaken through political and economic legitimacy and resourcing. It tends to use an instrumental form of knowing which is conducive to prediction and control, measure, and notions of "implementation".

Our concern with this convention of change is that it is adaptive to existing powers rather than transformative of and with them. With its origins in the Western enlightenment, luxuriating in colonial and class-based exploitation, our knowing, doing, being and becoming within the narrowing scope of this mono-epistemology can only prolong our un-sustainability possibly to the point of violent extinction:

"There can be no true solution until we turn away from the 'empty world' vision that underlies much of our political and economic thought, and commit ourselves to aspirations consistent with a world of finitude and limits... such a step is undeniably momentous..." (Bradford, 2000:9)

Our Inquiry considered how we could know differently as a matter of survival; how to experience the "undeniably momentous" steps of being, doing and becoming differently. We wanted to do this recognizing the utility of instrumental approaches to, say, reducing ecological footprints and also recognizing the dangers of instrumentalism as the only or "inappropriately appropriated" form of knowing. We wanted to emerge into and through an "epistemologically free zone"; to know whether such an experience informed our understandings of sustainability and participation; and whether "the Institution" could be persuaded.

While entertained by it, with the exception of my university, none of the participants' institutions could conceive of a formal association with the Inquiry any more than we were ready to conceive of one ourselves. Our Inquiry was about understanding this persistent disconnection, not from a strategic perspective, but from an epistemological one. We shared the problem with our institutions and took responsibility for it. We struggled with the same silences and inaction that our institutions have been struggling with - and we found new ground. The activism, community building, social learning and development of governing principles that we needed to do this work, with ourselves as much as any institution, located it in the field of "participatory practice".

Throughout this experience we kept seeing glimpses of Torbert's (and Reason's) Action Science as a pattern, rather than a methodology. His idea is that inquiry unfolds over three dimensions of personhood. His proposition is startling for its suggestion that we reverse the conventional sequence of "change" from one that is sourced in, and is often limited to a-critical, legitimate authority (law, regulation, policy and strategy for example), to one that is sourced in our selves and becomes a "mutually inter-penetrating first-, second- and third-person action inquiry in the present" (Ibid: 251). For Torbert such inquiry practice begins with instilling appreciation in our illegitimate (free), subjective self, each other and our experiences of the present.

Our Inquiry provides a *complementarity* to those who hold that formal shifts towards sustainability need to occur at any cost, including that of our personal and societal freedom. Our experience is that the two ideas of freedom and sustainability are coevolutionary of each other.

Through a glass brightly: Translating self between work and home

I have written emails from my workplace, an inner city university facility management office, to my home account as notes to self. These have served as the means of sending some piece of my work home: a problem to solve, a report to review, an opinion to consider. I have been struck, on reading them at home, by the difference between my work self and home self. In a strange moment of narcissism I treated the one at work as a friend generally addressing timidity and resolving issues from the peace of my home.

I am intrigued by this transition and have had conversations about it over the phone with participants in our research circle. The strangeness of the workplace is something that constantly confronts me as I live in the same neighbourhood and consider myself part of the communal reach of the University. On walking from home to work I come to the workplace door where I must swipe a card to get in. There is something that disturbs me in this, as much as I appreciate the security concerns that brought the door into being. The sense of transition in this manner is felt as unwarranted as an outsider and warranted as an insider.

I have worked in environments where strangers could, and sometimes did, come right up to my desk to pull up a chair and have a chat. One was a community education centre based in Sydney's Inner West, in which I participated as a researcher, learner and, sometimes, mentor. There was a strong spirit of participatory practice in this environment and many remarkable and sustainable things followed, not least of which has been its continued existence beyond its original grant-funded status.

While the security of employees is something that I entirely applaud, I also believe that to address the apprehension that comes with measures such as security cards, we must find ways to a greater degree of participation in political, socio-cultural and aesthetic realms. These are the very generation of social capital - shared identities arising with shared actions. The optimal experience of moving in and out of this kind of work space can be characterized as 'watery' in the sense that while a sense of boundaries (i.e., of the sensed difference between "being" and "not being") may be essential to identity, the experience of transition between identities, as of diving into water, should be a pleasurable experience of new possibilities and not an abandonment of them.

I believe that workplaces need to become kinder environments and I believe that there may be a connection between the stifling of creativity in a workplace and the presence of security controls. Inequity, lack of transparency and exclusion may become the hallmarks of such an organization. To make a general observation, I am conscious of destructive relationships that form in organizations in which people conflate risk to their person with that to the corporation. The technical problem is revealed as a political one that is revealed as the minutiae of daily existence – things perceived or apprehended that act like fields of knowing, ways of being, plays in three acts with characters caught up in cycles of meaning and resisting.

Instrumental ways of knowing need to be subject to discipline within organizations to ensure that ethical being is accountable in deliberations. Qualitative approaches to evaluation exist in many frames but what distinguishes participatory practice is that it is infused with knowing, gracing and being in as many strands as possible, as fits. This can amount to simple but powerful moments of reassuring colleagues of something unspoken. Tacit.

"We serve tea at our desks with a tray and pot to which more distant colleagues may come and pause briefly over a cuppa while we get their news. We kick our shoes off and prop ourselves on top of desks to sit cross legged for animated conversations. We create moments of freedom in which to consider action and sense the next event" (Lavarack, J. Inquiry Evaluation 3, 2005).

Mythology: A deeper code of discourse

What attracts me to mythology as a system is its power to dissolve and resist instrumental epistemological conventions and structures. Participation is by its nature emergent, full of surprises and unbidden memories. Systems that place a high value on certainty, predictability and security are inimical to the messiness of participation. Myth on the other hand can provide a narrative/quest framework for us to recognise and give voice and shape to an emergent pattern of being, knowing, doing and becoming. Myth can come in the form of a full blown classical narrative, e.g., Orpheus and Eurydice or a line from a pop song, e.g., 'stairway to heaven' or 'stand by me'. Insofar as it gives voice, shape and beauty to a relationship it helps propel that experience in a meaningful direction, helping us to remain engaged and committed.

In parallel to this Inquiry, I had the privilege of working with a group of older women who wanted to evaluate their 'program' of wellness activities in a participatory way; largely because the essence of their wellness activities is participation. While creating an innovative social care modality they were simultaneously recreating (and re- storying) the myth of activist crone. Pre-sensing this enabled us to move with the grain of the relationships rather than against them. We were enabled and empowered to take imaginative leaps that would not have been available to us if a more pedestrian and programmatic evaluative pathway had been taken. A consequence of these choices is that the report struggles to be 'useful' as a piece of advocacy writing while charming all who read it.

Metaphor, myth and ritual can infiltrate and illuminate discourses through time, virtual space and real life space. This happened both in the cave and with the older women. As a deep code of discourse, epistemology is a simultaneously limiting and enabling framework; framing possible questions/ solutions and creating tools; providing a bridge in the journey. Epistemology is a self-fulfilling prophecy - in both senses of the term; that is - a prophecy that fulfils itself (comes true) and a prophecy that fulfils (and becomes) your self. Recognizing this, I feel more at home with such a storied system of sense making than with a more abstract system. Myth is full of choices and bifurcations, and while always recognizable it is never predetermined; no two tellings being the same.

Nature has traditionally spoken to us in mythic forms – perhaps Aesop was the world's first biosemiotician? It is basic and obvious that the first step in this journey is the acknowledgement that such a conversation with Nature is possible, without condescension or arrogance. Humility, love and wonder seem to be keystones to the foundation of this more integrated way of being with Nature. For it's fortieth anniversary the magazine "Resurgence" asked many luminaries of the environment movement to reflect on the past forty years. Jonathon Porritt (2006) made the point that while the movement could hardly be faulted on its record of rational evidence based science and rhetoric, it had not engaged in any meaningful way with people's personal narratives and identities. Myth is one promising tool for reaching out to engage with the stories that are important to people everywhere.

Poetics and power: The lyrics of dissonance in transformative experiences

"Life is a creative, intimate and unpredictable conversation if it is nothing else, spoken or unspoken, and our life and our work are both the result of the particular way we hold that passionate conversation" (Whyte , 2001: 6).

Whyte suggests that life and work are more closely related than most of us realize (Whyte: 2001). He believes that work is a pilgrimage where people shape and develop identity, and life's purpose. With this assertion he states that there is a grander scheme in operation during everyone's working lives, and contends that this grander scheme requires closer attention and connection. Furthermore he states that work involves a 'conversation with grander, more eternal, more essential parts of ourselves' (*Ibid.*: 8). With the sensibility of a poet Whyte believes that working relationships are closely connected to the metaphysical and poetic realm of life. I am strongly drawn to these ideas, in my role as a manager of a rural primary health care team. In recent years I have experimented with opportunities that allow for the inclusion of the poetic realm in team work, while also daring to question some of the limitations of normative professional boundaries in our work with community.

In early 2003 my team and I became involved in managing and providing a community based therapeutic program for a group of asylum seekers from Afghanistan living in rural Murray Bridge, South Australia. The program focused on keeping the men well, alive and socially connected. These men had fled their country fearing for their lives following a long history and experience of persecution and marginalization in Afghanistan. All of the men had been detained for several months on arrival into Australia and they were all separated from their families and loved ones. In early 2003 a member of this community, a highly respected man Dr. Wahede took his own life. This tragic loss shocked the Murray Bridge community. We reached out to these men by inviting them to attend the community health service every Friday night to have a shared meal and swim in our hydrotherapy pool. On the first night over 50 men arrived accompanied by local community people who had supported them by offering English classes every week on a Saturday, in a voluntary capacity.

The men's stories were rich in pathos, poetry, sorrow and confusion. For several years we supported the men in this way, and we still hold our Friday night groups today. We also involved the men in art therapy workshops which culminated in the highly successful "Pictures in My Heart" exhibition which continues to tour regional South Australia with Country Arts SA. Nearly all of the men now have permanent protection visas, and many of them have managed to bring their families safely to Australia. I am telling this story here to say that this work was characterized by multiple difficulties and although it was based on participatory practice, we all needed to draw on that which was yet to be spoken, in terms of the stories that were hidden, the stories that were consumed by political agendas, the stories designed to reinvent and reshape realities and truths, and the stories of how we were all impacted by this experience in ways that can only be described as connected to the universal themes which web and weave poetry into our lives and our work.

As a therapist, a team manger, a researcher and participatory practitioner Whyte inspires me to consider the intersection between poetry's sacred and universal narratives and participatory practice. If we consider participatory practice in relationship to universal themes of: belonging, compassion, ethics in practice, overcoming isolation, engagement with despair and hope, negotiating loneliness and connectedness, responding to subjugation and injustice, a possible bridge forms metaphorically and literally to themes and possibilities for sustainability. From this position, standing on this bridge I am able to connect with the poetics and aesthetics' of the participatory endeavour. In this connection I am interested in poetics and aesthetics as a relational resource for sustaining and enriching my participatory practice.

Whyte makes reference to delving into the unknown 'to bring out words' that makes sense and gives meaning to day-to-day existence (Whyte, 2001). Bird considers 'talk that sings' in her linguistic and relational approach to making meaning within the therapeutic relationship (Bird, 2004). In poetry, therapy and participatory practice there is an intentional exploration with what is yet to be said and yet to be known. I believe that participatory practice and poetics attempt to connect people and communities, including practitioners with thoughts and feelings of hope, nurture and the possibility of a desired or longed for future, while also navigating the depths and experiences of despair, loss and disconnection. In my world participatory practice is interested in exposing the intersection between poetry and therapy, the poetry and song of the therapeutic relationship, and to make more visible that, which is considered sacred and poetic.

Conclusion

In writing this text we are encouraged by the other co-researchers to resist the felt convention to order our reflections with common themes, findings, principles, models or any homogenization that would make digestion familiar. And yet our disparate experiences hold within them some sense of a shared context even if perceived differently, and constructed of different elements through each person's form of participation.

Torbert's construction hovers behind our words, a gentle permeating architecture of being in solitude, together and of the world. It holds the gravity of the first person in reflexive appreciation that relational practices generate, allowing the personal, solitary and inward knowings to cautiously reach towards someone else's fragile and uncertain hold on "what is". As we found our "sites" for question, reflection and renewed intentions, we found new caves in the depths of our writings, the silences and appreciations of our conversations.

It is not that we changed any aspect of infrastructure as a recognizable third person inquiry would expect; but the infrastructures changed around us as we brought new and different qualities of valuing first and second person inquiries together and into them. We did not change our behaviors in a mechanical sense, but we did change the context of behaviors in a meaningful sense. We learned to let connections come to us, rise within us and not be stitched together; in so doing energy formed through us and was not taken from us.

We grew to appreciate, as this paper follows, ways of being with theory that were not about compliance or elitism, but about pattern and recognition. We learned about identity by ripening our sense of person and appreciating the depth of being in others. We let rise within each of us not only new words, but new forms of communication, an extra pause in a conversation, a question where previously we might close a thought down. We let myth and poetry stir, reframed emails as artefacts of cultural practices, and gathered and worked images of many kinds. We found within us unexplored sensitivities that opened whole worlds of freedom, hurts, confusions and leaps of understanding hitherto hidden from our life worlds.

We learned to walk into such waters and take their dampening disturbances into other locations, not as formal contracted works gearing shifts from one way of doing to another, but as shifts in qualities of our presence with others, as subtle lightening bolts of understanding:

"Before I sought to create space; now I am to draw attention to its emergence, reflect my being within it" (Fisher, 2006).

These were invisible journeys actually undertaken, travelling through the spaces in between (Weil, 1997: 380) theory, identity, language and practice, opening ourselves up to "what and how" we know, the "monstrous knowledges" (Foucault, 1976: 224 in Hodge, 1995: 36) of emergent knowings. They are those forms of knowing that as Hodge sees, must paradoxically recognise and reject convention if they are to spring into being. In these ontological spaces are homeless knowings and home making beings, built on the principle of:

"Normal and other ascribed by society. Many caves; much participatory work. The antithesis of control; people being what people are, not what I [we] want them to be. Not like us. The zenith of respect in humanity" (Fisher, 2006).

It is into these spaces that we invite you to participate in the quest that is sustainability.

"The shift from the idea to the being is not even
I am suddenly the shifting sand the undulating body of grace;
I see a perfect portion of the white soft cloud moving slowly across the clear blue;
The sky framed perfectly by the overhang of an enormous cave-like a mouth;
Shouting out the joy, the pain, the love and the blessing of the creative spirit.
I am the sand, I am the woman, I am the earth, and I am myself at grace"
(Iuliano, 2006)

Jane' commentary: Reflecting on finalizing this paper for printing

When this paper began I chose not to participate in its collaborative authorship. I watched its development and agreed with the authors that, following its peer review, we each use the paper as a reflective tool in co-writing in participatory practice. In time, I took my role in progressing the paper towards submission for printing. To do so is my practice. How then to be with the two issues that emerged for the authors from the reviews of our paper? How many times have I been here before? Experiencing the reality of collaborative action that emerges from rigorous and reflected-on participatory practice, and facing the dismissal of the emergent knowledge by those reading but not having experienced the action. Like Peter I have experienced that the...

"...consequence of these choices is that the report struggles to be "useful" as a piece of advocacy writing while charming all who read it" (Whitecross, P., this paper).

Again: language as a barrier, not as an enabler. Again: I see the fear of dismissal of our knowledge rising in the eyes of my fellow practitioners.

"Can 'the Institution' be persuaded?" (Goff, S., this paper).

I observe myself, and others, dismissing what we dislike, that which does not fit with the knowledge we have created in our Inquiry. I observe us reaching for the standard tools – translation, explanation. I find myself exploring different ways of laying out the paper, to accommodate the non-linear knowings into some acceptable form. If only a poster, a multi-dimensional e-tool. I get an email from a friend asking me how I work on such papers as these - am I 'very organized' and work from beginning to end, or ... And so I ask myself, where is my practice now? In facilitating the final draft of a co-written work about our experience of participatory practice, am I practising what I have learnt?

I cease reflecting alone. I email the authors:

"Hooray! The reviewer finds participatory practice and how it is spoken of difficult to come to terms with. Also the reviewer has noticed paradox. We have done well" (email communication: 3/10/06).

I re-read the paper and listen anew to my colleagues' words:

- *"... simple but powerful moments of reassuring colleagues of something unspoken"* (Lavarack, J., this paper);
- *"The point of resistance – our sharp edges and our porous skin"* (Goff, S., this paper);
- *"...a possible bridge forms metaphorically and literally to themes and possibilities for sustainability"* (Porter, L., this paper)."

A reviewer challenged the notion of an 'epistemologically free zone'. We pondered on this, collectively and individually, and accept that, rather than standing outside what we know, we observe that there are many ways in which people know, many of which are unnamed by science and philosophy. Peter has proposed that we sought to emerge through an 'epistemologically fresh zone'.

Another reviewer suggested that we shift the first paragraph of the conclusion into the introduction to help readers, who have not experienced what the authors have in the Inquiry, to engage with the paper. We decided to leave the paper to unfold to you. However, now you can choose to read it with the first paragraph of the conclusion as the opening paragraph of the paper, and to reflect on how you have come to know.

"Into these spaces we invite you ..." (Goff , S., this paper).

References

Bradford, W. (2000): "Global capitalism and sustainable development". Background paper for *International Conferencc on Global Ethos at United Nations University*, Tokyo Japan; October, 2000.

Bird, J. (2004): *Talk that sings*. Auckland, New Zealand; Edge Press.

Fisher, J. (2006): *Journey, 3 Sept, 2006,* contribution to Susan Goff's thesis, unpublished.

Foucault, M. (1976): *The archaeology of knowledge*. New York; Colophon Books.

Hodge, B. (1995): "Monstrous knowledge: doing PhDs in the new humanities". *Australian Universities' Review*; 38 (2), 35 – 39.

Iuliano, J. (2006): "My grace". *Writings from the cave, October* (Jane Fisher, Ed): Coda to

"The value of participatory practices to sustainability". Goff, S. thesis, in progress.
Marshall, J. (2004): "Living systemic thinking", *Action Research*, 2(3), 305-325.
Porritt, J. (2006): "Edging closer to meltdown". *Resurgence*, 238: 10-11. (http://www.resurgence.org/contents/238.htm. Retrieved, 8/10/06).
Reason, P. and Torbert, W. (2001): "The action turn: toward a transformational science: a further look at the scientific merits of action research". *Concepts and Transformations;* 6 (1): 1-37.
Senge, P., Scharmer, C.O., Jaworski, J., Flowers, B.S. (2005): *Presence – exploring profound change in people, organizations and society*. London; Nicholas Brearley Publishing.
Torbert, W.R. (2001): "The practice of action inquiry". Peter Reason and Hilary Bradbury, (Eds): *Handbook of action research*. London; Sage Publications.
Weil, S. (1997): "Social and organisational learning and unlearning in a different key: an introduction to the principles of critical learning theatre (CRT) and dialectical inquiry (di)". In F. S. Stowell; R. Ison and R. Armstrong et al, (Eds): *Systems for sustainability: people, organisations and environments*. New York; Plenium Press.
Whyte. D. (2001): *Crossing the unknown sea -work and the shaping of identity*. London; Penguin Books.

Proceedings of the 12th ANZSYS conference - Sustaining our social and natural capital
Katoomba, NSW Australia, 3rd-6th December, 2006

Learning to be a "Transdisciplinary" Sustainability Researcher: A Community of Practice Approach

Juliet Willetts & Cynthia Mitchell
Institute for Sustainable Futures, University of Technology, Sydney, AUS

This paper utilizes a 'community of practice' model to reflect on the post-graduate research program at the Institute for Sustainable Futures, UTS. Our work at the Institute involves resolution of complex problems in today's society, a task which requires insights generated through multiple disciplines. Over the last five years we have conducted an evolving program of activities for our post-graduate students to equip them with the necessary skills for this challenge. This program has been transformational for both individuals and the group, which now operates as a cohesive, mutually learning team. In this paper we look to the 'community of practice' model as a critical lens to examine our program and assist in identifying new opportunities to improve our approach to transdisciplinary research training.

Background and introduction

The purpose of this paper is to use the model of communities of practice as a lens for reflecting on our evolving transdisciplinary postgraduate program. Our program has developed organically, drawing on diverse inputs and built up through action research cycles that have responded to voiced and observed needs of the students. We believe the program to be comprehensive and working well for students and supervisors alike, and are looking for a theoretical frame to guide its further development and encourage translation by others. In this paper we utilise the 'communities of practice' model to provide such a frame.

The Institute for Sustainable Futures (ISF) at the University of Technology Sydney takes as its mission 'to create change towards sustainable futures'. Sustainability throws up the challenges of solving complex real-life problems that transcend traditional disciplinary boundaries and knowledge systems, making a PhD in the area of "sustainable futures" a distinctly challenging journey. Our postgraduate program is small (typically 8-10 students at once) and delightfully diverse, for example at the moment they range from a study of sustainable change in the Australian building industry, to the application of learning theories to international aid utilizing a case study of agricultural reform in Mozambique.

ISF's mission is mirrored in the ideology and action of our research students, and implies a strong change agenda in their research processes. The challenge in this passion is to find the balance between rigour and relevance.

As well as a commitment to create change, the research conducted by the Institute's students seeks to adhere to concepts of transdisciplinarity. Transdisciplinarity differs from multi-disciplinarity and inter-disciplinarity in that it actually calls into question disciplinary knowledge, and attempts to offer ways to articulate between different forms of knowledge (Thompson Klein, 2004). Transdisciplinarity is an emerging field with a slowly forming discourse and theoretical basis. As yet, literature is focused primarily on contending the definition of the term, in search of a coherent identity (Horlick & Jones, 2004; Balsiger, 2004; MaxNeef, 2005), with only a small number of papers reporting on the realities and challenges of conducting such research with integrity (Depres et al., 2004; Bruce et al., 2004), and even fewer attempts to propose evaluative structures (Wickson, *et al.*, 2005; Bergmann, *et al.*, 2005). As such, the guidance available to students venturing into such territory is limited.

A further challenge for our students is find the balance point in their research for their original discipline and/or experience base: it needs to inform and be informed, but neither dominate nor disappear. A dominant original discipline transgresses transdisciplinary intent because it fails to make space for other discourses and epistemological perspectives, whilst a disappearing original discipline probably fails to capture the contributions from the depths of experience, and may suggest a lack of reflective practice, once again transgressing transdisciplinary intent. The integration of knowledge and methodologies from different fields requires our students to question the values and underlying assumptions in their habitual thought processes and modes of perception. In other words they must learn about their own particular epistemological perspective before they can meaningfully engage with new disciplinary fields. As MaxNeef (2005) points out, in transdisciplinary research, 'an integrating synthesis is not achieved through the accumulation of different brains. It must occur inside each of the brains...'

Our post-graduate program has developed since 2001 in response to these challenges, and has transformed our postgraduates from a group of disparate disconnected researchers into a mutually learning team. We now seek to reflect upon this program in the light of 'communities of practice' as a meta-model to guide the program's on-going development. We start below with an introductory description of our activities followed by their analysis in this frame.

Activities of our post-graduate program

Our post-graduate community activities have evolved over time, through action research cycles of "learn, plan, do". They include annual retreats, small peer groups, special workshops and fortnightly roundtables, all coordinated through meetings to plan and reflect on such activities.

Our yearly residential retreats have become a focal point. We began this initiative five years ago with the intent of creating bonds between the disparate individual students. Nowadays, the retreats are coherently themed, and have become the time when we take an intellectual leap together.

The idea for small peer groups for accountability and support (GAS groups) came from the experiences of Kath Fisher and her colleagues (Fisher, *et al.*, 1998). The students have mostly decided the size and constitution of GAS groups, based on congruent research areas, personalities and stages in the research process. Each group then articulates the structure and focus of their activities. They also consider how to build reflection into their practice, how their meetings will be facilitated and when to stop and review their outcomes and processes.

The yearly retreat often results in identification of common questions, so we have held a variety of research training workshops, using internal and external experts, covering topics like epistemology and theoretical frameworks, validity frameworks in the social sciences, research writing and focusing on specific methodologies such as Actor Network Theory.

Our final area of endeavour is more typical of postgraduate research training programs generally, and consists of weekly or fortnightly roundtables in which the students present, seek feedback on, or debate issues raised in their research. These provide a forum for interaction with the rest of the Institute (who are involved in research consultancy in sustainability) and keep everyone up to date on developments in each other's thinking and topics.

Communities of practice as a critical lens for reflection

What is a community of practice and why use it as an analytical frame?

The discourse of communities of practice began with Lave and Wenger's work (1991) on the learning processes that occurred within apprenticeships and the way in which the community surrounding an apprentice acts as a living curriculum. They focused on "what kinds of social engagements provide the proper context for learning to take place" (Lave & Wenger, 1991: 14) and described learning as "an evolving, continuously renewed set of relations" (ibid.: 50). Wenger (1998) went on to define the unique characteristics of a community of practice as being "mutual engagement, joint enterprise and shared repertoire", where learning processes focus on the negotiation of meaning through two critical inter-relating processes: reification and participation.

In their work on postgraduate supervision and training, Pearson and Brew (2002) propose the use of communities of practice to conceptualise research communities. They suggest the model would provide insight to the complex social relationships in supervision and research practice and would offer "guidance for the creation of productive learning environments" (p. 142). In addition, the strongly "situated" nature of our post-graduate students' research and the reification inherent in disciplinary research makes this social learning theory an appropriate frame to analyze and reflect on our practice. Below we begin with Wenger's characteristics of mutual engagement, joint enterprise and shared repertoire, examining how they relate to and might inform our program. We then utilise Wenger's ideas on learning design to provide additional ideas to further our approach.

Mutual engagement

Mutual engagement is dependent on "being included in what matters" (Wenger, 1998: 74). In the context of our community of practice, our challenge was and is to provide multiple opportunities for interaction between the students; to create space for dialogue to occur and relationships to be built and to facilitate this process of mutual engagement. The full range of activities mentioned above has helped serve this purpose, through both occasional intense development spaces (retreats and workshops) and ongoing, 'maintenance' spaces (roundtables and GAS activities). Our annual retreats have been particularly important in achieving what Wenger describes as "a community of practice [that] can become a very tight node of interpersonal relationships". For example, one student's comment in reflecting upon the retreat was:

"At the broadest (should that be deepest?) level, it was the mental stimulation, journey sharing, reality checking and companionship with a very special bunch of people. To be frank and honest, I may well have decided to discontinue my PhD studies by now if it was not for that sharing. Whilst retreats are many things at many levels, it is this one that is most important to me."

An interesting point Wenger makes about the dynamics and relationships within a community of practice are that they are not necessarily harmonious. Indeed he contends that "disagreement, challenges and competition can all be forms of participation" (*ibid*: 77). On reflection, it is true that both co-operation and alignment of people's outlooks and approaches as well as dissonance and the voicing of alternative perspectives have shaped our practice and have been essential to our joint negotiation of meaning. The quote below demonstrates how debate triggered a personal realization of a need to change views from their dominant disciplinary approach (positivist, scientific background) to accommodate the methods they were utilizing (born from social sciences):

"What changed for me ...? - I now know how I am going to deal with sustainability in my thesis. This change came as a result of Simon's forthright stance and Paul playing back my shrill 'you must have a position on sustainability' decree ... I also realized that I will need to disclose my own values (and how they construct/ underpin my conception of sustainability)"

The opportunity for us is to focus more on bringing perspectives from beyond the boundary of our group, including those who specifically support views that differ from our own.

Joint enterprise

Our joint enterprise centres on the two aspects of research common to our community of practice: sustainability and transdisciplinarity. According to Wenger joint enterprise is "defined by the participants in the very process of pursuing it". Sustainability and transdisciplinarity are both large, contended realms and thus offer plenty of scope for our community to evolve our own meanings and definitions of these terms. We have used a mixture of participation and reification centred learning approaches, which according to Wenger, are both essential to progressing learning on an issue (Wenger, 1998: 55).

Our pursuit to explore sustainability has resulted in extensive dialogue within our community about its definition, worth and relationship to each student's research. In general, our activities focused on sustainability have focused on participation. Reified forms of this knowledge already exist in the academic literature and the students already spend much time alone in reading and reflecting on these. So what has been important for our community has been to test what this means in our practice, share our experiences and debate variations in our perspectives.

In facing the challenge of conducting transdisciplinary research with little prior exposure to, or experience in such a realm, the students have needed structured exposure to various forms of reified "knowledge' and "competence" in the required skills, as well as participatory activities to trial our use of such materials. During our retreats we have conducted a range of such learning activities. These activities have touched areas of peer review methods and frameworks, development of critical thinking and research writing skills, exploration of epistemological stances, theoretical frames and their influence on research methodologies, and sharing of personal accounts of managing a transdisciplinary PhD journey. The result is that many of the students are now at ease in working across a range of disciplines, demonstrated by reports of two of Chris Reidy's thesis examiners, a leader in futures thinking and a leading environmental academic:

"The candidate has obviously ventured beyond any disciplinary 'comfort zone' and stretched himself to cover the various 'bases' of the chosen methodology."

and,

"I can only say that I am in awe of Chris' abilities in researching and synthesizing over this broad field – a breadth which in my view is essential if we are to make progress in sustainable development – but one which few would be game to take on!"

Within the pursuit of a joint enterprise, Wenger refers to the formation of a "communal regime of mutual accountability" (Wenger, 1998: 81) which has a broad definition as follows:

"what matters and what does not, what is important and why it is important, what to do and not to do, what to pay attention to and what to ignore, what to talk about and what to leave unsaid, what to justify and what to take for-granted, and what to display and what to withhold, when actions and artefacts are good enough and when they need improvement or refinement"

Here we examine three types of accountability regimes and how they have played out over time. They are peer evaluation, accountability within GAS groups and our roles as designers of the learning processes as well as participants in the community.

We have initiated forms of peer evaluation at various points in time, and have provided frameworks to the students to inform such evaluation. While some students have taken up this practice, particularly in their GAS groups, for others it is not yet a habitual activity. This provides a challenge for us to develop new ways and means to embed peer evaluation in our practice.

The accountability regime established in the GAS groups has turned out to be broader than we had first anticipated. We had initially understood the idea of accountability in these groups to mean that the students would hold each other accountable for progress or completing tasks in their studies. In reality, this kind of accountability has not been a strong focus in the groups. Instead the review processes for these groups revealed mutual support and sharing of their ideas with one another as the established "accountability regime". For example the needs and benefits voiced by the students included emotional support, social contact, a reduced sense of isolation, development of trust between students, motivation to get things done, validation, and sharing of the journey. In addition, at an intellectual level, the students noted the value of peer validation and critique. Accountability for progress in their work instead features more predominantly in the supervisor-student relationships.

Our own role within the community of practice rests on a balance between leadership and participation. Wenger mentions the establishment of different roles within a joint enterprise, such that the responsibility of different parties becomes implicit and understood (Wenger, 1998: 81). This is something visible with regard to our role in guiding the community of practice, and initially we were much relied upon to lead the group. However, as Wenger states, "the defining of a joint enterprise is a process, not a static agreement" (ibid: 82). Indeed we now have moved to a different dynamic in which the students take greater responsibility within the community. For example at the last retreat, the structured sessions were both planned and led by the students.

Shared repertoire

A shared repertoire defines the "resources for negotiating meaning" (Wenger, 1998: 82). One of the outcomes of the series of retreats and our other activities has been the building of a shared language and understanding about our endeavours. Concurrently, we are indeed creating a growing set of resources generated to utilise useful tools invested by others, and to summarize and synthesise the thoughts which we have crystallized together. Some examples are a set of critical thinking resources, a framework for reviewing research writing (the meta, macro, micro framework), an evolving document covering different epistemologies and theoretical frameworks, documented reflections of our retreat and the GAS groups, and documented tips and tricks to doing a PhD ("The accumulated wisdom of ISF PG students"). The most recent addition is an emerging paper on what constitutes quality in transdisciplinary research. This paper demonstrates new intellectual advances in transdisciplinarity by engaging both with theory and with the constraints and practicalities of conducting such research. Wenger's focus on creating a shared repertoire reminds us that it will be important to keep persisting with the effort to document such new developments within the thinking of our community.

Informing our design of future learning activities

Wenger presents a framework for guiding the design of social learning processes, with the caution that "learning cannot be designed", that it "moves on its own terms. It slips through cracks; it creates its own cracks." (Wenger, 1998: 225). His conceptual framework suggests a focus on three elements: engagement, imagination and alignment (Wenger, 1998: 237).

In examining his framework, some opportunities for future directions in our community arise. Under engagement we are reminded of the importance of "peripherality" (*ibid*: 237) and the need for our community to look outwardly as well as inwardly in order to keep our practice "current". This area also directs us to consider ways to create "continuity" (*ibid*: 238) such as through ensuring encounters between different generations or cohorts of our students. Within the concept of imagination, we might consider activities that allow us to locate our own meanings within a larger context, and to push boundaries through visioning exercises. A focus on alignment and infrastructure to support alignment would likely help us converge again on our common purpose, and think consciously about how we coordinate and jurisdict our processes.

Conclusion

In conclusion, there are important lessons to be learnt through reflecting on our activities through the lens of communities of practice. These lessons stand us in good stead to continue to develop new ways to embed the skills for transdisciplinary research in our students. In addition, it becomes clear that the 'community' we have created has been an essential ingredient in broadening the perspective of our students, such that they are able to engage and achieve integration across different disciplines meaningfully.

Finally, the 'communities of practice' model has provided a means for us to describe elements of our program in a coherent way such that the ideas may be transferred to other settings. The community of practice model could likely offer other research training programs important additional dimensions. Indeed interpretivist researchers report on the need for "apprenticeship" to learn the required competencies in interpretivist research (Angen, 2000). The approach could also be useful to a wider set of activities that require people to cross disciplinary boundaries and backgrounds.

References

Balsiger P.W. (2004) Supradisciplinary research: history, objectives and rationale, Futures 36 (2004)

Bergmann, M., Brohmann, B., Hoffmann, H Loibl, M C., Rehaag, R and Schramm E. (2005) *Quality Criteria of Transdisciplinary Research: A Guide for the Formative Evaluation of Research Projects.* Institute for Social-Ecological Research GmbH; Frankfurt am Main ISSN 0947-6083

Bruce, A., Lyall, C., Tait, J., Williams, R., (2004) Interdisciplinary integration in Europe: the case of the Fifth Framework Programme, Futures 36 (2004)

Despres, C., Brais, n., Avellan, S., (2004) Collaborative planning for retrofitting suburbs: transdisciplinarity and intersubjectivity in action, Futures 36

Fisher, K., Bennett-Levy, J., and Irwin, R. (1998) *What a GAS! Action research as a peer*

support process for postgraduate students (unpublished)
T. Horlick-Jones,T., Sime, J., (2004) Living on the border: knowledge, risk and transdisciplinarity, Futures 36
Max-Neef, M. A. (2005) Foundations of Transdisciplinarity, *Ecological Economics* 53:5-16
Lave, J., & Wenger, E. (1991). *Situated learning: Legitimate peripheral participation*. New York: Cambridge University Press.
Pearson M and Brew A (2002) Studies in Higher Education 27(2) 'Research Training and Supervision Development'
Snyder, W and Wenger, E 2003, Our world as a learning system a community of practice approach, In Clawson, J. and Conner, M. (eds.) *Creating a Learning Culture: Strategy, Practice, and Technology*. New York: Cambridge University Press
Thompson Klein, J (2004) Prospects for transdisciplinarity, Futures 36: 515-526
Wenger, E. (1998). Communities of practice: Learning, meaning, and identity. New York: Cambridge University Press.
Wenger, E (2002) Communities of practice *Encyclopedia of the Social Sciences*. Volume 1.5, Article 5. Elsevier Science, Amsterdam.
Wickson, F., Carew, A Russell W (2006) Transdisciplinary Research: characteristics, quandaries, and quality, *Futures* (in press)

Acknowledgements

The postgraduate students, supervisors, and other staff at ISF, are thanked for their participation and contribution that make our community of practice such fertile ground for our mutual learning.

Juliet Willetts (BSc, BE, PhD) is a Research Principal at the Institute for Sustainable Futures, UTS. Juliet works towards sustainable change in practice and policy through learning. She has assisted with coordination of the Institute's postgraduate program since 2002, and has a keen interest in transdisciplinary research and what it takes to achieve integration across disciplines and knowledge systems. Her particular areas of focus and expertise are sustainable water management and international development, particularly monitoring and evaluation. Her previous publications span technical fields (for example wastewater management processes) and the social sciences (for example innovative qualitative monitoring systems for international aid).

Cynthia Mitchell (BE, PhD): As Research Director and Associate Professor at UTS' Institute for Sustainable Futures, Cynthia's goal is to create change towards sustainable futures. Her project brief includes strategic analysis, policy advice, and learning/change creation materials development in the water and building sectors. She also coordinates the ISF postgraduate program. Even though her practice-based work for change overlays the principles from her science and engineering training with futures thinking, systems thinking, diffusion of innovation, learning, and ecological economics, Cynthia is still wondering whether transdisciplinarity really exists ☺. She has published 4 monographs, about 25 international journal articles and 65 refereed conference papers, and has supervised 13 postgraduate students, including 7 in transdisciplinary sustainability studies.

Learning Together for Sustainability: The Value of Group Based Peer Learning

Jade Herriman, Juliet Willetts & Emma Partridge
Institute for Sustainable Futures, University of Technology Sydney, AUS

This paper reflects on a particular approach to community education for sustainability that has emerged in Australia in recent years, following successful examples elsewhere. Group based peer learning sees people coming together for extended periods to learn from each other, and in the process, inspiring one other, and acting as witnesses to commitments made to change. We examine a selection of examples of such learning programs both in Australia and overseas, and identify some of their predominant features. We consider the benefits of this learning approach, examine some of the challenges to using this technique in public programs, and consider how to overcome barriers and maximize benefit. The paper highlights the value of this approach as one that provides less formal, less 'expert'-based, less extensive compared with other more conventional learning approaches, but possibly more significant learning experiences.

Introduction

"Environmental Education must be holistic and about connections. In order to address environmental challenges, people need to think broadly and understand systems, connections, patterns and causes. The challenges themselves frequently have social, scientific, cultural, economic and ethical aspects, all of which must be considered for their effective management." (Environment Australia, July 2000)

Local government in Australia has had, over the last few decades, an emerging interest in supporting residents to 'live more sustainably'. Waste and environment focused programs[1] have primarily relied on incentives (such as discounted compost bins), compliance (enforcing legislation against illegal dumping or stormwater pollution) or structural change (implementing domestic recycling services, introducing greenwaste collection). Education or social marketing campaigns have tended to be primarily information based, mass communication campaigns (brochures, material on Council websites) or 'face to face' programs such as lectures or workshops delivered by experts on single issues or topics.

In contrast, much of the literature on effective processes for changing values, habits, and attitudes highlights the value of 'word of mouth' messages, and the norming of new behaviors by uptake and role modeling by early

1 A distinction between waste and environment programs is often made due to the structural arrangements of Councils in Australia, where environmental management and waste management are commonly different operational units, environment often located within planning and waste within in capital works

adopters (Rogers, 2003) Work on voluntary behavior change for sustainability stresses the value of people having 'trusted others' to discuss change. Similarly, health promotion work has shown that programs that use peers as teachers or facilitators are effective in influencing change (e.g., Baldo, 1998).

This paper is written from a practitioner perspective, reflecting on practice in environmental education (EE) and education for sustainability[2] (EfS) within local government. The objective of this paper is to investigate a range of group based peer learning programs that cover the spectrum of approaches available. To this end we utilize a set of case studies of group based peer learning programs to examine the ways in which these programs have brought people together to learn, reflect and discuss their lifestyles through the lens of sustainability. We discuss the typical characteristics of the case study programs and highlight the range of possible benefits they have delivered. In doing so, we also acknowledge some of the barriers faced, including those related to the culture within institutions which tend to pay for and deliver community education programs. We conclude with a discussion of a range of ways to maximize the benefit of group based peer learning programs as tools for change towards sustainability.

What is group based peer learning for sustainability?

"Specialist discipline-based knowledge, while contributing critically, is no longer adequate by itself - an holistic appreciation of the context of environmental problems is essential." (Environment Australia, July 2000)

"Australian society is diverse and apparently ever changing. As such a participatory process that is learning and responsive to community needs is a step in the right direction" (Voronoff, 2005: 21).

Group based peer learning for sustainability is a term used loosely in this paper to mean a style of learning in which small groups of people come together to learn with and from each other on some topic of mutual interest, with an underlying objective of change – either personal change (attitudinal, conceptual, behavioral), social change or both. Peer group education is a tool that has been used effectively in public health and health promotion, especially in sexual health (Baldo 1998), sexually transmitted disease (Norr, *et al.*, 2004), and women's health (Haider, *et al.*, 2000).

Public programs designed to achieve environmental education objectives have historically taken many forms – from mass communication campaigns to small skill-development workshops for householders (on composting or green cleaning for example), to public and community art projects (such as murals, sculptures, or theatre). In the last 10 years or so group based peer learning has also been used for environmental education, or more recently, education

2 It is recognized that Education for Sustainability (EfS) is a distinct approach to community education and possesses features that some traditional environmental education may not have – the terms are by no means automatically interchangeable. For a discussion of the differences in these approaches, see Tilbury and Cooke (2005) and ARIES 2006.

for sustainability objectives.

Clearly, people learn about sustainability in a range of formal and informal ways, both in groups and individually. Many situations exist where people learn from peers, including through informal interactions like significant conversations in social groups or homes. This paper does not refer to all of these situations, but specifically to those community education programs which have been designed to feature ongoing group engagement and that value the exchange that happens between participants.

For the purpose of discussion, it is assumed that a program designed to facilitate learning for sustainability may have change objectives along a spectrum from simple exposure to information, to measurable personal change in a desired direction (e.g., water conservation), to more systemic personal change with an empowerment focus and no targeted specific behavioral change. In this context, group based peer learning methodologies are considered as tools for achieving all of these objectives. They are considered from the perspective of case studies and also from theory about learning, group processes and reflection.

Case study examples

The examples of the peer based learning approach to sustainability education have been selected with a view to demonstrating the breadth of possible approaches. In addition, they include two programs (Living Waters Living Communities, Leichhardt Cool Communities) with which the authors have had direct experience. The case studies included in this paper are by no means an exhaustive list. Appendix 1 contains a description of these programs and others, providing relevant details and references. The case studies included in this paper fall into two main categories, the first focused on sustainable approaches to living, and the second are more narrowly focused on a particular environmental issue.

The 'sustainable living' case studies combine a broad range of thematic perspectives such as waste, energy, water, toxics in the home and household purchasing. Our case study examples include:

- International programs;
    * Global Action Plan (GAP) EcoTeam Program, and;
    * the New Zealand Sustainable Households Program;
- Australian programs,
    * Living Waters Living Communities program in NSW;
    * Living for Our Future in Victoria, and;
    * Sustainability Street in Victoria and NSW).

The case studies with an environmental theme or aspect of sustainable living are:

- Leichhardt Cool Communities program (energy and greenhouse)
- Watermark Australia, (water).

How do they work? Characteristics of the programs

Reflecting on the Leichhardt Cool Communities program, Tilbury and Ross (2006) give a sense of how group based peer learning programs operate:

"After the group decided on their joint answer, each group would tack their answers up for everyone to see, and the entire group would perhaps laugh and discuss the variety of answers. This type of process – placing questions, discussing, and sharing - is important for exchanging ideas and creating understanding of other people's thought processes and assumptions. This process also allows participants' ideas to evolve and be expanded on."

Some of the case studies identified above have a greater emphasis on action, others on exposure to information, others on creating a space for reflection and inspiration for personal change. There are also differences in the degree of inward or outward focus of the group – whether seeking to change some structural aspect of society or focused more on enriching the experience of the individuals present. However, there are a range of common characteristics to these programs. These commonalities and differences are described below based on our case study examples. Broadly, group based peer learning programs may:

Focus on a single theme or more broadly, on sustainability

Leichhardt Cool Communities program had a specific 'greenhouse' focus. Whilst this enabled topics such as sustainable transport, water efficient appliances, and composting to be covered, it tended to have a strong 'in the house' focus, and did not engage with broader questions of consumption, or complex issues such as embodied energy, focusing rather on direct, measurable greenhouse gas reductions. In comparison, Living Waters Living Communities covered a broad range of themes and enabled participants to spend time considering global as well as local issues of concern.

Meet regularly

Living Waters Living Communities participants met for eight sessions (including a wetland walk and celebration event). Participants in the Sustainable Households Program in NZ met for between six and ten 2-hour sessions. Watermark Australia suggests teams meet monthly.

Recruit and support groups over a short or longer period of time

A feature of the Watermark program is that it is national and groups can be set up anywhere and run by any interested community member. It is an ongoing program and so new groups can be established at any time to suit participant interest. The Leichhardt Cool Communities program on the other hand had a more tightly defined timeline, due to the grant funding it received. The evaluation noted that these types of programs require long timeframes for implementation (Hole 2004).

Be existing or purpose-built using existing facilitators or training community members to facilitate
In some programs community volunteers are recruited and trained to be team leaders/ facilitators in the program and recruit participants (e.g.. Living Waters Living Communities, GAP). Some programs are delivered by program leaders/ facilitators and advertised to the broader community (Sustainable Households, Sustainable Living at Home). These may result in groups of people who do not share a geographical region or an existing social connection. In some cases existing groups may be used – in the Leichhardt Cool Communities program existing community groups were offered the program as a 'guest speaker' arrangement.

Reflect on experiences, emotional responses and underlying values, not just knowledge
Peer based groups typically consider information, values, attitudes, behaviors and feelings. The format is open to members articulating 'big picture' issues, exploring connections or emotional responses. People can share their personal experiences of changing lifestyles – both successes and challenges. This results in strengthened relationships between participants and supports critical thinking, dialogue, self diagnosis and action.

Be supported with information, activities and recognition
Programs run with centralised support and varying degrees of devolved responsibility. Examples of the range of support offered by organisers or facilitators include expert speakers, handbooks, information materials, access to utility data, feedback on the quantifiable impacts of changes, templates for 'commitments' or eco-goals, ongoing email communication email between meetings, case studies, media articles, quizzes and surveys, comparison data and tools to benchmark consumption (e.g., the ecological footprint approach – http://www.footprintnetwork.org/). Field trips (for example wetland walks in Living Waters Living Communities, industry visits in Sustainable Living at Home) are often offered as a way to make abstract issues 'real' and emphasise the local context – revealing aspects of the local environment which may have been hidden to participants previously. Recognition may also offered, with profiles of participants, 'good news stories' in local media, prizes, certificates and presentation and celebration ceremonies being examples of tools used.

Provide opportunities for and encourage 'outreach'
Often participants identify opportunities for extending their learning, sharing their learning with others or creating structural changes in their local area (e.g.. Living Waters Living Communities participants organizing a public seminar on sustainable investment, or lobbying for an area of park to be converted into community garden).

What are the benefits of these programs for sustainability?

"The program gave our family the feeling that we really can help improve the quality of life in our family, our neighborhood and our city." (Columbus Ohio GAP Eco Team participant)[3]

"There is a continual expression of a desire for 'community'. This may or may not be a false hope or idea of the past –but it does point to something perceived as missing – perhaps: local connections." (Paine, 2004)

Group based peer learning programs have delivered a number of benefits in the communities in which they have been used. General social benefits such as 'involvement' and social cohesion, or 'knowing the neighbors' are often cited, and are perhaps signs that such programs fulfil the 'desire for community' and local connection that Paine notes is so common. Other benefits cited are behavioral change outcomes related to reduced resource consumption, and individual benefits such as development of group skills, critical thinking or facilitation.

Most programs are reported on favorably by their organizers, although it must be said that many have not been subject to independent evaluation. Further, the causes of long term social or behavioral change are notoriously difficult to measure and the complex relationships between cause and effect are often overlooked in the monitoring of project outcomes. However, longitudinal study that was carried out in the Netherlands measured the durability of the short term effects of the EcoTeam Program into the long term, with particular attention to improvements in environmentally relevant behavior, intensified environmental investments in the household, and quantitative savings on environmental resources. This analysis found that effects seen just after the EcoTeam Program was completed were maintained for up to two years after the intervention had ended.

The acceptance of these programs in communities speaks of the value of informal as well as more formal styles of education. Time for discussion, questioning, reflecting and thinking about material between sessions provides opportunities for shifts in thinking. They also demonstrate value in the 'non-expert' as well as the expert opinion - where community members are recognized to possess a broad range of insights that can be the basis of learning for others. As such, they value the importance of relationships between participants in facilitating changes in perspective.

Another benefit of these programs for sustainability is that they feature a high degree of integration. For example, through integrating learning across 'topics' (which may otherwise be tackled in separate programs and campaigns) participants are able to draw connections between different processes (social, economic, technical, environmental, political) that impact on environmental problems and influence their own behaviors. The programs also integrate learning with doing, providing opportunities for a range of learning styles and learn-

3 Empowerment Institute website: *http://www.empowermentinstitute.net/files/Studies.html*

ing experiences. Importantly they also integrate social outcomes with environmental outcomes – combining objectives for community development with objectives related to reducing ecological 'footprints'.

Discussion: How can the benefits of peer group based learning be maximized?

This paper has shown that there are a range of characteristics common to group based peer learning programs, and a range of possible benefits to this approach. However, if these processes are to be used to best advantage there are a number of barriers that may need to be overcome. Barriers and challenges to this approach include organizations perceptions and expectations of education programs based on their organizational culture, the achievement of appropriate timing, flexibility and resourcing, and the lack of an overarching sustainability initiative across Australian councils. Opportunities to maximise benefit include ways to overcome such barriers as well as methods such as collaborative approaches across traditional organizational 'silos', an ongoing timeframe to allow successive waves of community involvement, realistic staffing and integrating group processes from other fields such as community consultation and deliberation.

Barriers, sources of resistance and their resolution

One of the most significant barriers to using this approach to learning for sustainability or incorporating this type of program into formal environmental education programs may be the culture within institutions. The flexible, responsive nature of support needed for such a program does not always align with the work planning and outcomes focus of public institutions. It also challenges the single issue focus for many programs, where staffing and funding is often tied to a single issue (such as 'water conservation', 'social isolation for aged residents') so partnerships or innovative programs which cross boundaries require more coordination and commitment from individual staff to justify new approaches. Quantitative evaluation of programs, as is often required, may result in performance indicators such as the number of participants, the number of meetings, the number of direct, short-term measurable changes in consumption, the duration of the program. In such an evaluation, more subtle community development outcomes, or long term shifts in environmental behaviors are likely to be overlooked.

One way to address this may be to deliver such programs as a joint initiative between environmentally focused and community focused organizations and sections of Council. This may result in broadening the range of performance indicators for the project, and providing useful social science or qualitative research perspectives to be applied to what could otherwise be seen merely as a vehicle for technical, quantifiable change.

Timing could also pose a challenge, as change towards sustainability requires programs to be supported over a long-term. Ampt (2003) suggests that 6 months might be the minimum time frame needed to run an effective voluntary behavior change program on transport. The US GAP EcoTeam program runs for only 4 months, but the infrastructure for support is ongoing, with staffed offices to support the Eco Program. The challenge in organizations such as councils, is that there is often interest in running short-term programs with demonstrable immediate impacts. Maintaining education programs for a period of years is potentially a challenge in the face of pressure to deliver 'new' outputs and take part in the many newly devised programs and projects which Councils are exposed to.

One of the case study examples demonstrates a way to overcome this barrier regarding timing. A feature of the Watermark program is that it is designed to allow for successive waves of involvement rather than only recruiting for groups at the start of a program dictated by grant or program funding (a common feature of participatory education programs in the public sector). Programs that are designed to capitalise on the interest that participant involvement will generate in the community may maximise involvement and ease of recruitment compared to those that run intensely over a shorter period.

Flexibility of program content may be both a potential strength and challenge for programs like these. Education programs focused on voluntary behaviour change in transport reinforce the importance of an approach where 'people can identify their own problem or frustration and find a solution to it rather than trying to align their own value system with that of the client' (Ampt, 2003). Developing and sourcing content in response to the needs of participants requires ongoing support throughout the project, but means that the program can be most meaningful for the participants involved. In the Leichhardt Cool Communities program for example, this involved conducting further research into Green Power options for residents.

Resourcing may also be a barrier, with significant establishment and preparation time required to create materials, structure sessions, train community volunteer 'team leaders' or staff facilitators, provide incentives, and provide ongoing communications and support for teams. These outlays may be a large proportion of total project cost especially when projects are running for a short duration. This often seems to be the case with grant-funded programs, despite the best intentions of initiating staff that may see the initial phase as a pilot, to be built on and incorporated into the organization's core programs.

One of the key challenges facing practitioners in the local government or community sector wanting to implement group-based peer learning programs for sustainability, in NSW, may be the lack of an overarching program structure to 'tap into'. Whilst the NSW Government initiated EarthWorks program provides some structure and resources for those wishing to carry out waste focused community workshops, the broader mantle of sustainability or sustainable living is not addressed through this program. As a result, many individual sustainable living programs are being designed and implemented without coordination or effective sharing of lessons learnt. This may be resulting in lost opportunities for shared learning and common resources to draw on. The NZ experience of pooling financial contributions between Councils to fund the creation of a centrally designed and evaluated program, with flexibility for regional differences in implementation may be a useful model for consideration in Australia.

Conclusions

We have demonstrated the group-based peer learning approach to be a useful tool to add to the set of approaches that sustainability practitioners might use to engage with the community. The case study scenarios show the approach to be valuable in achieving significant changes in both how people interact with the natural environment and how they interact with each other in neighborhoods and communities. Based on the importance of this type of learning approach for sustainability, there is a need for further research into such interventions. Some suggestions would be to situate them on a spectrum depending on their structure, to consider factors such as the degree to which learner experience is leader-dependant, sources of resources and openness to changes in direction.

Conceptually we have shown that group-based peer learning is consistent with many existing action-based approaches to community development and social change. The learning is guided to a greater or lesser extent by participant-identified priorities, the strong role that social relationships play in influencing behavioral norms is recognized, and the connections and interplay between different themes of concern in relation to sustainability are recognized and explored rather than being considered in isolation through rigidly defined and separated topics. The way that this approach recognises community and individual expertise is aligned with a decentralised model of 'authority' and information transfer; where agencies facilitate and support citizen action rather than dictating and directing.

We conclude that there is scope for this approach to be used more widely in Australia, and that use of the approach will benefit from the reflections presented in how to overcoming likely barriers and to maximise effectiveness.

References

Ampt, E.S, 2003. Voluntary Household Travel Behaviour Change – Theory and Practice, paper presented at the 10th International Conference on Travel Behaviour Research, Lucerne, August 2003.

ARIES, 2006. What is the relationship between EfS and EE?' Australian Research Institute for Education for Sustainability EfS Portal website. http://www.aries.mq.edu.au/portal/about/relationship.htm [August 22 2006]

Baldo, M. 1998. Peer Education: a successful strategy with some constraints. Sexual Health Exchange 1998;(4):1-3.

Calvert, F. 2004. *The role of research in Cool Communities.* Proceedings from the Effective Sustainability Education Conference http://www.environment.nsw.gov.au/resources/calvert.pdf

City of Port Phillip. 2003. Sustainable Living at Home Workbook

Donlen, J., Condon, L. and A. Tourney. 2005. Measuring community engagement with sustainability: 'Living for our Future'. Paper presented to the Social Change in the 21st century Conference. Centre for Social change research. Queensland University of Technology, Carseldine. October 2005.

Environment Australia. 2000. Environmental Education for a Sustainable Future – National Action Plan. http://www.ea.gov.au/education

Haider, R, Ashworth, A., Kabir, I. and SRA Huttly. 2000. Effect of community-based peer counselors on exclusive breastfeeding practices in Dhaka, Bangladesh: a randomized

controlled trial. The Lancet, Vol 356, Nov 11, 2000.
Harland, P and Staats, H.J. 1997. Long term effects of the EcoTeam Program in the Netherlands. Study 4: The situation two years after participation. Leiden: Centre for Energy and Environmental Research, Leiden University. E&M/R-95-57.
Hole K., 2004. Community Education Programs for Reducing Greenhouse Gas Emissions, Discussion Paper, Prepared for Sydney Coastal Councils Group, Rockdale City Council, and the Institute of Environmental Studies, UNSW, Sydney.
Leichhardt City Council. 2004. *Cool Communities at Leichhardt Council: Project Report.* Unpublished.
McDonald, A. and Taylor, R (eds) 2004. Sustainable Households Pilot Program 2001-4: Final Evaluation Report. Published by Marlborough District Council (as project manager) on behalf of the 12 Partner Councils. New Zealand. www.sustainablehouseholds.org.nz/issuepdfs/Sustainable_Households_Pilot_Evaluation_2004.pdf
McKenzie-Mohr, D. and W. Smith, 1999. Fostering Sustainable Behaviour Change. New Society Publishers. ISBN: 0865714061
Meadows D., 2002, Dancing with Systems, The Systems Thinker Vol 13 No 2, Pegasus Communications
Marilyn Mehlmann, 2006. Global Action Plan (GAP) International& Fenixgruppen, Stockholm, Sweden; personal communication. 9 March 2006.
Norr, K.F., Tlou, S. and Moeti, M.R. 2004. Impact of Peer Group Education on HIV Prevention among women in Botswana. Women's Health XXX ?
Owen, H. 1997. Open Space Technology: A user's Guide. 2nd edition, Berrett-Koehler.
Owen, H. Website: http://www.openspaceworld.com/ Accessed 2006.
Paine G., 2004, Wholes, Patterns and Sustainable Development: Closing the Gap Between Concern and Action. Unpublished PhD thesis. University of Western Sydney.
Pickens, P.M. 2002. Community-based social marketing as a planning tool. Community and regional Planning masters project. University of Oregon Architecture and Allied Arts Program. http://www.empowermentinstitute.net/files/CBSM_MastersProject.pdf
Reinhold, J. 1993. Working with Rural Communities in Nepal: Some Principles of Non-Formal Education Intervention. UNESCO – YCF Action Research in Family and Early Childhood.
Rogers, EM, 2003. Diffusion of Innovations, 5th Ed, Free Press, New York., NY
Staats, H.J. and Harland, P. (1995). The EcoTeam Program in the Netherlands. Study 4: A longitudinal study on the effects of the EcoTeam Program on environmental behaviour and its psychological backgrounds. Summary report. Leiden: Centre for Energy and Environmental Research, Leiden University. E&M/R-95-57.
Taylor R., 2005, Promoting Behaviour Change Through The Sustainable Households Programme: (Work in progress), Paper presented at National Zero Waste Conference, Stream 2, Thursday 7 April 2005, Kaikoura, New Zealand.
Tilbury, D. and Ross. K. 2006, Living Change: Documenting good practice in Education for Sustainability in NSW. Framework: http://www.aries.mq.edu.au/pdf/frameworkJan06.pdf, Case study: Cool Communities http://www.aries.mq.edu.au/pdf/Cool_CommunitiesJan06.pdf
Tilbury, D and Cooke, K, 2005. ' A National Review of Environmental Education and its Contribution to Sustainability in Australia: Frameworks for Sustainability' Canberra: Australian Government Department of the Environment and Heritage and the Australian Research Institute in Education for Sustainability (ARIES) .
Voronoff D., 2005, Community Sustainability, A review of what works and how it is practiced in Victoria, prepared for Environment Victoria
Warringah Council & OzGREEN, 2001. 'Living Waters Living Communities. A Sustainable living program – the first of its kind in Sydney!'. An initiative of Oz GREEN

and Warringah Council assisted by the NSW Government through its Stormwater Trust. (Project booklet.) pp 32.

Acknowledgements

Thank you to Suzanne Cronan for research support and Ann Hobson for review and to two anonymous reviewers for their constructive comments.

Jade Herriman has 6 years experience in sustainability research and project management within Local Government. She has worked with leading NSW Councils to deliver innovative sustainability programs across many technical disciplines. She has worked on municipal greenhouse gas reduction and energy efficiency programs, urban biodiversity conservation, water management planning and community engagement projects. She has designed and delivered a variety of community education and engagement programs such as the Whites Creek Wetland community education program for Leichhardt Council and the Youth Environment Forum for Warringah Council. Her current research interests include decision-making frameworks for sustainability, participatory processes and social, institutional and economic drivers for consumption.

Juliet Willetts is a Research Principal at the Institute for Sustainable Futures, UTS. Juliet works towards sustainable change in practice and policy through learning. She has assisted with coordination of the Institute's postgraduate program since 2002, and has a keen interest in transdisciplinary research and what it takes to achieve integration across disciplines and knowledge systems. Her particular areas of focus and expertise are sustainable water management and international development, particularly monitoring and evaluation. Her previous publications span technical fields (for example wastewater management processes) and the social sciences (for example innovative qualitative monitoring systems for international aid).

Emma Partridge is a qualitative social researcher with extensive experience in policy, research and consulting contexts. Her work with a range of stakeholders focuses on the social and cultural dimensions of creating change towards sustainability. Emma has worked in various senior research and policy development roles in the state and commonwealth public sector prior to joining ISF. This work focused particularly on issues relating to women, children and young people, juvenile justice and community services. Emma's research interests focus on the social dimensions of sustainability, the interrelationship between environmental issues and social and cultural issues, the links between sustainability, health and social justice, and strategies for achieving sustainability in the urban environment.

Appendix 1 – Community Sustainability education programs which use or have used group based peer learning (Alphabetical by title)

Community Sustainability education programs which use or have used group based peer learning			
What	**How**	**Where, Who**	**More information**
Living Waters Living Communities	**Sustainable living focus** GAP Eco Team style program delivered by Warringah Council and OzGreen funded through NSW Government Stormwater Trust. Eco teams were lead by facilitators who had volunteered from the community, the group met on an ongoing basis over a period of months, and a full range of sustainability topics were covered in a group setting, and supported with materials such as information, surveys, guest speakers, group exercises and opportunity for reflection and discussion. Participants: EcoTeam Leaders are recruited and trained from the community. Teams included both existing groups (workplaces, churches, neighbourhoods) and custom made teams (in response to advertisements and information nights)	Australia: NSW, Warringah Council. Organisers: Warringah Council, OzGREEN, NSW Stormwater Trust	Warringah Council & OzGREEN (2001), Paine 2004
Living for Our Future	**Sustainable living focus** Vic sustainable living program run in the City of Boroondara using community based social marketing principles, as a pilot, with 120 residents in small groups over 6 months (Donlen, Condon and Tourney 2005).	Australia: Victoria, City of Booorondara	Donlen, Condon and Tourney 2005
Sustainability Street	Sustainable living focus The program is run is various places: Penrith, Wollongong, Willoughby, (NSW), Moonee Valley, City of Melbourne, (VIC)	Australia: Victoria and NSW	http://www.voxbandicoot.com.au/SustainabilitySt/the%20street%20report%20no2_scn.pdf

Community Sustainability education programs which use or have used group based peer learning			
What	**How**	**Where, Who**	**More information**
Sustainable Living at Home	Sustainable living focus A series of 5 workshops with guest speakers, surveys and the opportunity for participants to make short, medium and long term plans to carry out sustainability actions in the home. The project is staffed by one full-time environment projects officer supported by the Environmental Programs Team and other City of Port Phillip officers as required. Six monthly "SLAHminars" are held at council offices featuring a range of guest presenters and council staff who share their knowledge with participants about the topic of the month. Following the monthly SLAHminar, teams of up to 20 households meet to learn from each other and develop practical solutions to environmental problems from the SLAH workbook. This resource book provides easy to follow guidelines on environmental sustainability and the five program themes. A range of practical exercises in the workbook helps participants audit their waste and travel habits, detect water leaks, and change their energy consumption habits. The SLAH program has involved 364 City of Port Phillip households (representing 875 people) since December 2001.(source: ICLEI global – members in action http://www.iclei.org/index.php?id=1192)	Australia: Victoria, Port Phillip Project partners: Department of Sustainability and Environment, Australian greenhouse Office Cool Communities, Melbourne Water and South East Water.	http://www.portphillip.vic.gov.au/slah.html City of Port Phillip. 2003. Sustainable Living at Home Workbook ICLEI global – members in action http://www.iclei.org/index.php?id=1192
Sustainable Households Pilot Program	Sustainable living focus The Sustainable Households Program was developed as an action research project at multiple locations in NZ. The program consisted of a facilitator managing either a series of individual community study groups and tutored classes on different sustainability issues such as implementing insulation, improving green waste management and reading product labels. The programme design uses adult education and social marketing techniques to bring groups of interested adults together repeatedly for a timetabled study series (2hours, between 6 and 10 sessions). These are led by tutor-facilitators and supported by reference material, quizzes, activities for the group and home, measurement and questionnaires (Taylor 2005). The program was linked with existing media and environmental marketing campaigns in New Zealand. The three year pilot ended in 2004, and a transition to national availability was underway in 2005.	New Zealand Organiser: Marlborough District Council on behalf of 12 partner Councils Participants: interested individuals coming together in groups in a community college environment	MacDonald and Taylor 2004, Taylor 2005

Community Sustainability education programs which use or have used group based peer learning

What	How	Where, Who	More information
Global Action Plan	**Sustainable living focus** The Global Action Plan (GAP) EcoTeam Program is probably the most well-known example of group-based peer learning for sustainability. Founded in 1989 in the USA, Global Action Plan for the Earth (GAP) is a nonprofit organization . The EcoTeam Program's primary goal is to reduce resource consumption at the household level by promoting individual behaviour change. The program supports community members who are interested in bringing together a team of people from the same neighbourhood to work through a workbook and conduct surveys of resource use. The focus is on empowerment and features the use of feedback to stimulate and reinforce change. This program originated in the USA and spread to Europe, including the Netherlands and the UK. GAP International is a network of independent organizations and each chooses its own approach, with some commonality. All are for instance committed to supporting measurements and feedback. Where no statistics are readily available, from utilities or elsewhere, program participants are encouraged and helped to make their own measurements. GAP Belgium (EcoLife) has for example become a world leader in helping households to calculate their ecological footprint (Mehlmann 2006) GAP now operates in seventeen countries with a combined participation of 150,000 people to date (Pickens 2002). The focus is on empowerment and features the use of feedback to stimulate and reinforce change.	USA, UK, Netherlands	www.globalaction-plan.org, Harland, P and Staats, H.J. (1997), Staats, H.J. and Harland, P. (1995), Pickens 2002, Mehlmann 2006 pers. comm.
Reality Bites	**Sustainable food focus** Anti Apathy's sustainable food program is focused on food choices and the environmental and social impacts of these choices. The Food program involves a small handful of people committing to learn more about and reflect on their food habits, and doing so in the public domain by sharing their experiences online. Although each individual is physically separate from the others, it could be considered that they are coming together in a 'virtual group' both with the others in the group, and also with the much broader group of people who are engaging with their journey and interacting with them through the website.	UK: London Organisers: AntiApathy Participants:	http://www.anti-apathy.org/social-experiments/index.php?i=71

Community Sustainability education programs which use or have used group based peer learning			
What	**How**	**Where, Who**	**More information**
Watermark Australia	**Water focus** Watermark is an Australian water education and action project which uses centralized support – education materials, website – to support small groups of people who want to meet regularly for group based learning on water issues and specifically achieving reduced water footprint. The essence of the Watermark Australia process is that ordinary people bring together small groups of people, meeting monthly to discuss water, then sharing their thoughts, anecdotes and 'folk-wisdom' , bringing it all back to the Watermark Australia team. Funding: To date, a small number of private citizens have contributed their own money towards the project. As well, The Myer Foundation has made two grants at key stages of the project's development - supporting the early concept work, and helping to implement the project Participants: Over 200 group convenors spread throughout Australia have joined the project and are bringing other citizens into the process.	Australia Organisers: Watermark Australia operates out of the office of the Victorian Women's Trust in Melbourne, and is comprised of: the Watermark Australia team, the Victorian Women's Trust staff, a small group of volunteers, and a Reference Group. Partner organizations: http://www.watermarkaustralia.org.au/info.asp?pg=partners	Website: http://www.watermarkaustralia.org.au/home.asp Results: http://www.theage.com.au/news/national/cutting-consumption-street-by-street/2006/08/04/1154198331857.html Reports: http://www.watermarkaustralia.org.au/info.asp?pg=convenor_submissions
Leichhardt Cool Communities	**Greenhouse and energy focus** Wanting to engage a 'non-green' audience in Leichhardt the project focused on working with existing community groups to increase understanding of the connection between energy use and greenhouse gas emissions, and the options available to householders for change. The workshops were delivered to five pre-existing community groups, usually at their regular time and place of meeting Participating groups included a seniors group and Spanish women's group. A series of 3 workshops featured quizzes, games, discussions and prizes. The Cool Communities commitment sheet had boxes where residents committed to ways they personally would reduce greenhouse gas emissions by changing their actions involving hot water, energy use, car fuel use, and waste.	Australia, NSW, Leichhardt Council Organiser: Leichhardt Council, NSW Cool Communities facilitator, funded through the AGO's Cool Communities Program (round 1) Participants: existing community groups	Leichhardt Council 2004, Calvert 2004, Tilbury and Ross 2006. http://www.aries.mq.edu.au/pdf/Cool_Communities-Jan06.pdf

How the Common Language of 'Child Abuse' Makes Enemies of Parents

Suzanne Regan[1] & David Thorpe[2]
[1]Centre for Collaborative Intervention in the Public Sector, Lancaster University Management School, UK
[2]Sør-Trøndelag University College, NOR

In his book "The Social Construction of What? Hacking (1999) points to a growing list of things which are said to be socially constructed. Included among the list is "child abuse." Noting its 'discovery' as an object of medical science by authoritative child paediatricians in Denver, Colorado in the 1960's, Hacking speculates as to how and why "child abuse" has become a ubiquitous feature of social and personal realities even though the numbers of avoidable child deaths and seriously harmed children have remained stable across the developed world over many years. In this paper we are specifically concerned with the ways in which the language of "Child Abuse" has been unproblematically deployed to define the aims of public policy in the field of child welfare programmes and as well, to organize the activities of individuals engaged in transdisciplinary children's services activity. Drawing selectively on research data on the organization and outcomes of Child Protection programmes undertaken by these authors in eighteen local authorities in the UK and in West Australia, we first describe briefly how the discourse of Child Protection came into being. Secondly we consider the state of relationships between parents and social workers arguing that inherent in the dominant discourse of Child Protection are perceptions of parents as potentially dangerous 'others.' These oppositional positioning results in actions which simultaneously fail to recognize the complexity of situations in which intervention occurs and any possibility for improvement.

Introduction

The discourse of Child Protection has dominated the childcare social work field for well over twenty years. Its negative effect on professional and transdisciplinary practice and learning in the field of children's services has been considerable. Prior to the institution of Child Protection the activities of social workers in particular were based on the collectivist principles enshrined in the welfare state legislation of the post-war period. The role for social workers in assisting people who were deemed to be at a social disadvantage in a capitalist economy due to poverty, disability or some other form of ill-health or adversity was more straight forward during this period.

Programmes designed only and specifically to "protect" children from avoidable harm had their genesis in the 1970s. At this time a series of scandals involving the avoidable deaths of children who were known to local authorities prompted far-reaching enquiries and recommendations for changes in practice to avoid such events re-occurring (Thorpe, 1994). As a result of the first public enquiry into the death of a child called Maria Colwell in 1974, a direct connection was made for the first time between the failures of health and welfare staff and a fatal outcome. The conclusion was that "a better functioning system would have prevented Maria's death (Gibbons, 1997: 79). The imperative for government to take action to prevent such tragedies was further fuelled by subsequent enquiries into more child deaths (Beckford, 1985; Henry ,1987; Carlisle, 1987; Neve, 1996 and most recently Climbie, 2001). These deaths came to signify all that was thought to be wrong in the child welfare system.

Before the term "child abuse" was coined the Denver child paediatricians labelled the healing fractures they observed in X-rays of young children as illustrative of the "battered baby syndrome" but later this reference was rapidly extended and new connotations were acquired. Hacking notes that:

"The idea [of child abuse] became embedded in legislation, incorporated in practices and changed a wide range of professional activities involving social workers, police, school teachers, parents, busybodies. It acquired new moral weight: child abuse became the worst possible vice" (Hacking, 1999: 126).

The effect of the repeated media scandals in the UK also led to similar developments in state-funded services for children. Notwithstanding the extensive measures put in place to protect children, research into the outcomes of the Child Protection programme commissioned by the Department of Health in the UK (Department of Health, 1995) revealed that Child Protection procedures and protocols failed not only to prevent avoidable child deaths but also served to alienate families and others wrongly accused of 'child abuse.' Most worryingly, the research found that following an intrusive investigation, usually carried out in conjunction with police officers, many families needing help often refused assistance from Social Services departments. (Department of Health, 1995; Thorpe, 1994). The government report noted:

"Many investigations are undertaken, many families are visited and case conferences called but [...] in the end, little support is offered to the family. In such situations, it is unsurprising that participants become angry, alienated and bewildered. Furthermore, the children are not helped and a chunk of valuable child care resource has been consumed with little apparent benefit" (DoH, 1995: 54-55).

Despite an official catalogue of failure and repeated attempts to curb the negative effect of overzealous investigations, programmes of Child Protection persist and continue to cultivate powerful backing from various quarters (King, 1997). Children's rights groups, media, medical practitioners, government officials, academics and children's charities such as the NSPCC uncritically deploy the discourse of "child abuse" to explain various kinds of adversities confronting children and their generally impoverished carers.

In the following excerpt from an interview conducted by these authors as part of their research (Regan, 1997-2006; Thorpe, 1988-2006), social workers describe how their practices changed as a result of the introduction of child protection procedures:

"If it [a referral] was about a child you pinked it . . . the two teams [north and south] were pinking them [child protection referrals] differently. Everything was logged as a CP [Child Protection]. The scene was set for CP... One time a teacher reported a child with a red mark on it's face. The teachers would not ask any questions they were told not to . . . . We went out to find that it was raspberry jam... Social workers started to act like policemen. You couldn't ask the child any questions that the police should ask. We knew it was stupid but that's what we had to do... I remember in one joint investigation we had to put the child up to the window to see the bruise. . . I was horrified . . . it was going to court. . . When I first arrived [in this authority] it made me rethink my views on Child Protection. We'd always had a strong Family Support base... I started to think that I wasn't any good as a social worker because of this. . . You were just investigating and didn't have time to do social work ... I didn't want to practice this way ... working long hours and the chaos!"

The account paints a disturbing picture of an ideology which has more in common with the pre-enlightened, medieval world of witch-hunts and inquisitions than it does with post-war welfare reform. Listening to these and other stories told by social workers on post–qualifying training programmes one could be led to believe that childcare social work is engaged in a contemporary form of social engineering designed to enforce aspects of middle-class norms of child rearing. When encouraged to describe childcare social work in their organizations, conversations invariably focused on either the assessing or investigating of parents and accumulating developing evidence against them using court procedures or finding or fixing foster placements, wrangling with managers over funds to pay for an out- of-county placement or the long distances travelled to carry out mandatory placement reviews. Very few childcare social workers even those working in so-called 'family support' teams ever mentioned their role as 'resourceful friends' for families in need of assistance (Holman, 2000).

Somewhere along the line the idea of providing the sort of interventions which would enable more parents to better care for their own children no longer existed as a viable reality in the minds of many social workers nor indeed in the minds of those government policy makers responsible for raising the standards of practice. The steadily rising numbers of children being accommodated by the state, the setting of targets for adoption and polices for the 'twin-tracking' of children for adoption suggest that policies which privilege a conception of the 'child's world' (Howarth, 2001) as being the object of practice interventions, parents have become the enemy

The failure of childcare social work to genuinely believe in change as a possible outcome of intervention cannot be seen in isolation from the restructuring of organizational arrangements and changes to practice interventions (Regan, 2006). The fragmentation of childcare social work into short term/ initial assessment and long term family support/permanency work has engendered a moral vacuum where accountability for the consequences of decisions made about the future of a child and its family has resulted is strangely absent. Many social workers claim that the material resources and the requisite skills they need to assist families in situations where poverty, unemployment, poor housing, substance misuse and other forms of ill-health strongly feature are no longer available to them.

In addition to this the quasi market model of commissioning make resources more difficult for social workers and other professionals to access. In some local authorities contracts for service provision are established with voluntary or private sector agencies in situations where workers and volunteers often lack the necessary training to intervene in families facing multiple long-term problems. (In our research we find these families being re-referred to social services). More to the point the government's policy of 'fast-tracking' children for adoption pointedly enacts a negative view of parents who often love their children but who do not have the emotional or material resources to fight the courts for their care(1). At the same time they can be denied regular assistance to enable them to demonstrate their capacity to care. Often these financial and practical resources are subsequently offered to more 'worthy' foster and adoptive parents.

Removing children from their parents and local communities in the hope of producing a better outcome is an intervention fraught with risk. Forty-three percent of placements break-down in the first year and the chances of children who remain in care after six weeks of returning home are dramatically reduced by two-thirds (Thorpe, 1988). Many children lose contact with their natural families and siblings (Bilson, 1995). The statistics produced by the DfES on movements in the 'looked-after system,' the educational attainment of children in care and performance on targets for adoption fail to convey the human cost of interventions in terms of fractured lives, distorted realities, disrupted attachments and the loss of a child's capacity over time to constitute an identity which is connected in meaningful ways with their wider birth family and place of origin. Social workers discover that children even temporarily removed from their homes often return with a new set of problems caused by the anxiety of separation from their familiar surroundings and exposure to unforeseen events (such as the breakdown of the placement which was supposed to produce stability in the child's life). The outcomes for many children who are looked after are notoriously bleak, as noted recently in the Society pages of the UK Guardian newspaper (4/10/06):

"You only have to look at the shocking figures. Only 1% of care leavers go to University, while half of all prisoners under 25, and 80% of Big Issue sellers have been in care"

Of the eighteen local authorities where these authors have carried out research only two local authorities were found to support practices which result in social workers maintaining their involvement with families over a period of time. The first, Cheshire Social Services had also retained family support workers as integral members of their generic childcare social work teams, while the second and most notable exception, North Lincolnshire Children Services, supported a range of specialist resources which remained within the firm control of that local authorities child care managers (Regan, 2003). Of course there are circumstances where substitute care is the only alternative but nonetheless if some local authorities like Cheshire and North Lincolnshire can organize to keep their entries to care to a minimum then why are these outcomes considered as exceptions rather than the rule?

The answer to this puzzle seems to reside in the connecting together of a clearly stated management philosophy of enabling parents to provide for their children rather than a philosophy of child rescue, the translation of this philosophy into a practical reality in terms of flexible, accessible and locally available resources (both for childcare workers and those they assist) and the belief of workers that they can make a difference even in the most distressing of circumstances.

In our view the national programme of Child Protection and its language of "risk" and "abuse" has steadily eroded the moral necessity for social workers to see parents as 'others' worthy of professional engagement. In the discourse of Child Protection expressions such as 'perpetrator' 'abuse' 'risk' 'safety' 'protective factors' 'threshold' 'domestic violence' and 'danger' form a common language amongst childcare professionals which in a philosophical sense casts the 'others' of intervention as either perpetrators or potential victims of some form of 'abuse' or 'risk' (Regan & Thorpe, 2005). Rescuing child victims from parents who pose 'risks' of various kinds has conferred an unassailable moral legitimacy on social work action which can simultaneously fail to recognize the complexity of situations in which intervention occurs nor any possibility of improvement. As a result social policy more generally is bereft of ideas as the knowledge which informs practice derives from a world view which no longer values the advancement of humanity and the relief of poverty and adversity through programmes of social justice and welfare reform.

Note

[1] Social services departments in the UK run a "Twin Tracking" system when children are taken into care. This means that when a child gets taken into care, plans for adoption are run alongside (twin tracked) against any other plans, such as the child coming home or being placed with another family member. The apparent logic behind this is that if for any reason it is judged that the child cannot be returned to the birth family, then they have adoptive parents ready and waiting for them and do not have to "languish" in the care system. The government are also encouraging Local Authorities to drastically reduce the time it takes from a child coming into care, to being adopted. This is giving the parents hardly any chance to get a case together and fight for their child (see http://www.fassit.co.uk/fassit_uk_reforms)

References

Bilson, A. (1995) Parental Contact with Children Fostered and in Residential Care after the 1989 Children Act in *British Journal of Social Work Vol 25, No 36.* ISSN 0045-3102.

Conn, D.(2006) I Survived By Accident: interview with Maxine Wrigley *The Guardian*,4/10/2006 (accessible at http://society.guardian.co.uk/children/story/0,,1886537,00.html.

Department of Health (1995) *Messages from Research.* London: HMSO.

Gibbons, J. (1997)*Relating Outcomes to Objectives in Child Protection Policy* in N. Parton (ed) Child Protection and Family Support: Tensions, Contradictions and Possibilities, London: Routledge, ISBN 0515142245.

Hacking, I. (1999) *The Social Construction of What?* USA: Harvard College, ISBN 067481200X.

Holman, B. (2002) *Champions for Children,* Bristol: The Policy Press, ISBN 1861343426.

Howarth, J. (2001) *The Child's World Assessing Children in Need,* London: Jessica Kingsley Publishers, ISBN 1417553642.

King, M. (1997) *A Better World for Children,* London: Routledge, ISBN 041550175.

Regan, S. (1999) *Transformations in the Organization of Social Work in the Public Sector.* PhD thesis published by NTNU-Norway (2006), ISSN 0802-8974.

Regan, S. (2003) *Organizing Assistance to Children and their Families in North Lincolnshire1998-2003* (Available at http://www.lums.lancs.ac.uk/files/ccips/6387/download/).

Regan, S. and Thorpe, D. (2005) *Reflecting on Boundaries in Statutory Child Welfare Programmes: An Example of Critical Systems Thinking in Practice.* Proceedings of the 11th ANZSYS/Managing the Complex V conference Christchurch, New Zealand,5th-7th December,2005.

Regan, S. (1997-2006) *Video-recorded studies on the organization of referral-taking in social services,* Unpublished research reports to eighteen UK local authorities.

Thorpe, D. (1988) Career Patterns in Childcare-Implications for Service *British Journal of Social Work* (18-2), 137-53, ISSN 0045-3102.

Thorpe, D. (1994) *Evaluating Child Protection,* Buckingham: Open University Press ISBN0335157521.

Thorpe, D. (1988-2006) *Reports on the Outcomes of Child Protection Referrals.* Unpublished research reports to eighteen local authorities.

Suzanne Regan has a background in child welfare work and began her career as a district officer for the NSW Department of Youth and Community Services in the mid 1970s. After arriving in the UK from Australia in 1984 she studied for her MA and PhD in Organizational Behaviour and Analysis from Lancaster University. As an organizational research and development analyst specializing in health and welfare organization, Suzanne has researched practice in over twenty UK local authorities. Her research interests involve the development of new methods of researching practice which involve the use of edited video-recorded data to improve practices in local work settings.

After leaving Lancaster University in 2006, **David Thorpe** has taken up a post as a Professor of Applied Social Science at Trondheim University Norway. Much David's work also entailed the development of new methodological approaches to evaluating child welfare programmes which took him in the early 1980s to the United States (Harvard Law School and the Universities of Southern California and Ann Arbor, Michigan) and Australia where he has developed a continuing relationship with child welfare agencies and academic institutions. In 1987-90 he directed the Australian study into the outcomes of child protection services on secondment from Lancaster University and then in the late 1990s, was responsible for the UK component of a nine-nation European Union research programme on child protection under the BIOMED II programme which was based on the methodology originally developed for eight Australian states. His other international activities have include those of visiting Research Fellow at the University of Western Australia a continuing research collaboration programme with the Norwegian Child Research Centre at the University of Trondheim, L'Office Nationale de la Naissance et de L'Enfance in Brussels.

How to Teach Oral Ecology Using Complexity Approach?

Vladimir D. Dimitrov[1] & Harald M. Eriksen[2]
[1]University of Western Sydney, AUS
[2]University of Tromsø, NOR

Oral ecology is a recently emerged branch of inquiry that studies the oral world - the 'world of human mouth'. When applying the approach of complexity in teaching oral ecology, we seek to explore the 'oral world' in relationships with the world of the whole organism existing in unbreakable unity with the social and natural environment in which we live. This is our rational for suggesting an oral ecology based approach in dental education.

Complexity and ecology

The word "complexity" has roots in two Latin words: *complexus* which means "totality" and *completere* which means "embrace". The science of complexity seeks to embrace totality of the studied phenomena and processes. And as far as the totality of reality is sustained by dynamics – energies and forces which never cease to manifest, complexity focuses on them; it is their interactions and interrelationships, their emergent phenomena and self-organization, their evolution and transformation that complexity tries to understand, describe and explain (Bak, 1996; Dimitrov, 2003; Eriksen & Dimitrov, 2003a; Gleick, 1987; Mandelbrot, 1982; Solé & Goodwin, 2000; Waldrop, 1992)

Ecology, on the other hand, originates in the Greek word *oekos* meaning "house"; the house is a place where its inhabitants closely relate to one another and interact dynamically. When 'the house' of the humans expands so as to encompass the planet, the universe and the all-embracing totality of existence responsible for the emergence, unfoldment and transformation of human lives, then ecology and complexity converge into one indivisible field of inquiry; to which we refer as a study of self-organization (Dimitrov, 2003). The impetus towards self-organization is inherent in the dynamics of nature; every living form embodies this impetus. More than a century ago, the pioneer of social ecology, the French researcher Elisée Reclus defined the human beings as "nature achieving self-consciousness" (Reclus, 1905); it is the self-organizing impetus of nature that makes it seek to achieve self-consciousness through us.

Oral ecology

The mouth is 'the house' of the oral dynamics and the oral ecology focuses on these dynamics. Complexity theory explains that the oral dynamics cannot be understood only by keeping them locked in the mouth. Through the prism of complexity, dynamical processes in human mouth are seen as nested in dynamical processes of the individual's body, nested in social dynamics, which are nested in the dynamics of nature and the planetary and universal dynamics (Bak, 1996; Gleick, 1987). Therefore, the first thing that needs to be underlined when we teach oral ecology using the approach of complexity science is its inseparable connectedness with human ecology – a study of the ever-changing, evolving and transforming interrelationships and interactions operating at different level of human nature - physical, emotional, mental, spiritual, social, environmental, universal. Human ecology is organically imbedded in the ecology of society (social ecology) (Dimitrov, 2003) and the ecology of nature and planet (environmental ecology) (Sole & Goodwin, 2000; Bak, 1996). This implies that the understanding of oral ecology is impossible without developing understanding of the essentials of human, social and environmental ecologies. This is a direct consequence of applying the complexity approach. One can never grasp the nature of a complex dynamic entity when studying it in isolation from the entities which it relates to - depends on, exercise influence to, and coevolves with.

If we focus on the micro-world of the human mouth, we can observe more than 400 distinct species of micro-organisms, mainly bacteria. In addition, recent investigations have documented that a number of non-cultivable species might be added, interfering in the balance between oral health and disease (Aas, 2006). Billions of interacting bacteria live on the surface of each tooth, in the crevices of the tongue, on the inner sides of the cheeks, on the gums and the palate. In this humid environment, saliva participates both in positive and negative feedback loops in relation to the whirling bacterial dynamics. The proteins which the saliva contains have positive impact on the growth of the bacteria, as they provide them with nutrients, while the antibacterials agents in the saliva (lysozyme, immunoglobulins, antifungal and antiviral components) have negative impact impeding their growth. Saliva also contains acid-buffering substances. Bicarbonate ions in the saliva neutralize the tooth-decaying acids produced by a variety of acidogenic bacteria and the phosphatic and calcium ions in the saliva act both as buffering and re-mineralizing agents repairing microscopic initial de-mineralized area.

Inseparability of oral and human ecologies

According to the principles of complexity, changes in human ecology – the ecology of the human organism as a whole – bring forth changes in oral ecology. When we are under severe stress or intensive emotional, mental, or spiritual involvement, we lose control over the quality and quantity of the food and drinks which we consume and which directly may affect our oral health.

It is not only the substances, which we put in the mouth, that may affect its dynamics. Any experience classified by the human ecologists as negative life events (fear, anxiety, anger, jealousy, grief, animosity, envy, hatred, anguish, suicidal urges, etc.) have the potential to interfere with the balance of the oral dynamics (Breivik, *et al.*, 1996). Everyone is familiar with a sudden emergence of dryness or bitterness or an emission of bad breath, when some stressful personal or social conditions trigger a rupture of negative experience. The Hindu sages used to say that our words can pollute the mouth much more than the substances we put in it. Harsh, cruel and offensive words directed to another person 'pollute' the mouth of those who use it. They are heavily impregnated with negative thoughts and emotions – indirect triggers of serious disturbance to the dynamical processes in human mouth.

Complexity approach to oral dynamics

When teaching topics related to oral health and disease by applying the principles of complexity, the whole complex of factors behind oral ecology is emphasized instead of the search for a fragmented scientific description of the oral phenomena studied (a reductionist approach). The search for answers of this and similar kind of questions starts with recognition that the functioning of the mouth cannot be isolated from the functioning of the organism, and the organism cannot be isolated from its environment. In "The Tree of Knowledge", Maturana and Varela wrote: "When we speak of living beings, we presuppose something in common between them... Our proposition is that living beings are continually self-producing. We indicate this process when we call the organization that defines them an autopoietic system... The most striking feature of an autopoietic system is that it pulls itself up by its own bootstraps and becomes distinct from its environment through its own dynamics, in such a way that both things are inseparable" (Maturana & Varela, 1978). Although distinct from our environment, we are inseparable from it.

An important concept in the study of human complexity is the concept of multiple attractors: human dynamics are drawn towards many attractors. Often these attractors are unwholesome - habits like smoking, excessive consumption of alcohol or (mis)use of drugs. It is clear that the oral dynamic are severely disturbed by these kinds of habits not only directly, but also indirectly as a result of the destructive effects which the unhealthy habits have on the ability to control one's own behavior, to be responsible for one's own actions, to take care for one's own health. If individuals lack this ability, the ecology of their entire lives is destroyed: they may become helpless victims of the unwholesome habits.

Complexity approach to dental education

When applied to the teaching of oral ecology, complexity makes us clear that the oral health cannot be sustained by caring for the mouth only (Eriksen & Dimitrov, 2003a, 2003b). The whole complex of the three vital constituents of human nature – body, mind and soul – must be healthy. Wholesome life ecology that is based on complexity theory aims at exploring the secrets of the wholesome - healthy and fulfilling - human life (Dimitrov & Naess, 2005).

What connects and centers the body-mind-soul dynamics is the human spirit. When life is lived in a wholesome way, it seems that the human spirit takes care for its unfolding, and the power of spirit is irresistible; any attempt to suppress it undergoes fiasco. It does not matter that the science cannot understand the power of spirit. In analogy with the chaos theory, where the mathematical proof of a chaotic attractor is in the emergence of the effects which the attractor manifests in the phase space of the chaotic dynamics, the proof of the power of human spirit is in the emergence of the effects this power manifests in the experiential space of the human dynamics. To live wholesomely is *conditio sine qua non* for keeping the oral health at its highest possible level. And to study oral health ecology through the lens of complexity means to learn how to live in a wholesome way.

There is another key reason for applying complexity thinking in dental education. It is not only the organism and its environment that affect oral dynamics. Changes in oral dynamics may affect the whole organism and its environment in various ways (Bergdahl, 2002). For example, smoking pollutes not only the mouths of the smokers and thus affects directly their oral ecology, but has also serious negative effects on the functioning of their entire organisms. One whole issue of the journal *Oral Health and Preventive Dentistry* was devoted to that topic (Watt *et al.* 2006). Such holistic negative effects often strike back on the health of the mouths of the smokers and thus again affect (this time indirectly) their oral ecologies. Ancient thinkers considered the mouth as "a sacred gate to the temple of the body" – a gate that permanently needs attention, vigilance, and care. Any worsening of the oral ecology implies worsening of the functioning of the body, that negatively affects human health and the perception of quality of life (Inglehart & Bagramian, 2002). Health has strongly manifested social dimensions. The increase in the number of sick people in society has a serious impact on its social dynamics.

During the last years, efforts have been made towards a more comprehensive, "holistic" dental education (Shanley, 2004; Plasschaert, *et al.*, 2005) emphasizing an evidence-based approach (Haynes & Haines, 1998). So far this has mainly been a pragmatic approach. What has been missing in dental and oral education is an underlying conceptual theory supporting and framing this way of thinking. In previous publications, we have tried to emphasize this way of thinking (Eriksen & Dimitrov, 2003a; Dimitrov, 2003; Eriksen, *et al.*, 2006; Loesche 1997; Dimitrov, 2001). The present paper is an attempt to propose a "complexity" approach in dental curriculua and to promote this view from a "complicity" perspective.

References

Aaas J.A. 2006 Microbial Flora in Oral Health and Disease Studied by Molecular Genetics, *PhD Thesis*, University of Oslo, Norway.

Bak, P. 1996 *How Nature Works. The Science of Self-organized Criticality,* NY: Copernicus.

Bergdahl. M. *et al.* 2006 Natural teeth and cognitive function in humans. *Scandinavian Journal of Psychology*

Brevik T. *et al.* 1996 Emotional Stress Effects on Immunity, Gingivitis, and Periodontitis, *European Journal of Oral Sciences*, 104: 327-334.

Dimitrov, V. 2003 Complexity of Human Life, in *A New Kind of Social Science: Study of Self-organization of Human Dynamics*, Morrisville: Lulu.press, pp. 35-44.

Dimitrov, V. and T. Naess 2005 *Wholesome Life Ecology*, Morrisville: Lulu.press.

Dimitrov, V. 2001 Bridging Complexity and Ecology: An Outline of Health Ecology, *Complex Systems* , 13: 393-413.

Eriksen, H. and V. Dimitrov 2003a Ecology of Oral Health: A Complexity Perspective,

European Journal of Oral Sciences, 111: 285-290.
Eriksen, H. and V. Dimitrov 2003b The Human Mouth: Oral Functions in a Social Complexity Perspective, *Acta Odontologica Scandinavica*, 61: 172-177.
Eriksen, H. *et al.*, 2006 The Oral Ecosystem: Implication for Education, *Eur J Dent Res* (in press)
Gleick, J. 1987 *Chaos*, London: Abacus.
Haynes, B. and A. Heines 1998 Barriers and Bridges to Evidence Based Clinical Practice, *Br. Med. J.*, 317: 273-276.
Inglehart, M. and. R. Bagranian 2002 *Oral Health-related Quality of Life*, Chicago: Quintessence Books.
Loesche, W. 1997 Association of the Oral Flora with Important Medical Diseases, *Current Opinion Peridintol*, 4, pp. 21-28.
Mandelbrot, B. 1982 *The Fractal Geometry of Nature*, NY: Freeman
Maturana U. and F. Varela 1978 *The Three of Knowledge*, London: Shambala.
Plasschaert, A. *et al.*, 2005 Profile and Competences for the European Dentists, *Eur. J. Dent. Ed.*, 9, pp. 98-107
Reclus, E. 1905 *L'Homme et la Terre*, Paris: Librarie Universelle.
Shanley, D. 2004 Convergence Towards Higher Standards in International Dental Education, *NY State Dent J.*, 70:35-39.
Solé, R. and B. Goodwin 2000, *Signs of Life*, NY: Basic Books.
Waldrop M. 1992 *Complexity. The Emergent Science at the Edge of Order and Chaos*, NY: Simon&Schuster.
Watt R. *et al.* 2006 Public Health Aspects of Tobacco Control: Setting the Agenda for Action by Oral Health Professionals Across Europe. *Oral Health Prev Dent* 4:19-26.

Vladimir Dimitrov, PhD, DSc is with the School of Education, University of Western Sydney. He is author of many publications in the field of complexity and fuzzy logic, including the books: "A New Kind of Social Science" (2003) and "Introduction of Fuzziology: Study of Fuzziness of Human Knowing" (2005); he is co-author of "Wholesome Life Ecology"(2005) and "Social Fuzziology" (2002), and co-editor of "With Fuzzy Logic in the New Millennium" (2002) and "Fuzzy Systems Design" (1998).

Harald M. Eriksen is Professor and Head of Institute of Clinical Dentistry at the University in Tromsø (Norway). He has numerous publications in the field of dentistry and oral health.

THEME EIGHT

ENGINEERING SYSTEMS AND INFRASTRUCTURE DESIGN

Stephen C. Cook & Timothy L.J. Ferris,
University of South Australia

We are engineers who have approached systems thinking through involvement in the problem of designing suitable products to fit in their context. Early engineering work (up to the end of the 19th century) tended to demonstrate awareness of the practical need to take a context aware view of engineering design. In the mid 20th century (particularly in electronics) the need for a holistic view of the product had been lost (with unfortunate consequences). Systems engineering was developed in the telecommunications and defence industries to help address this problem. Systems engineering has gradually broadened its scope of concern and methodology to include a growing range of systems methodologies developed in the systems thinking community.

In our paper (Ferris and Cook) we evaluate the role of systems engineering (SE). This study was prompted by two opposing positions in the SE community: some regard SE very narrowly as concerning the methods to design technological systems; others claim that any process of investigating a problem and proposing a creative solution is an application of the SE methodology. We consider both these positions to be inappropriate, so we use the approach of Ulrich's twelve critical heuristics questions to explore the bounds of the appropriate scope of SE. The benefit of this work is to 'right size' the claims of coverage of SE.

In the paper by Asproth, the prevalence of IT systems has led to an increase in the amount of information, AND to an increase in the amount of information that seems to be worth archiving, BUT 10 years is a long time in the evolution of computing equipment and data formats, WHILST much data requires long-term preservation. This situation presents a major dilemma for IT systems administrators. Changing hardware standards and software can make data unreadable. The problems identified in the paper include the difficulty of maintaining: readability of data; and legal integrity of data records for use as legal evidence. The paper follows a general description of the problems confronted with a program of matters for research related to the broad problems of data storage for long-term purposes, such as government operations.

The paper by Andrews and Petkov returns us to one of the technical fields which prompted the development of systems engineering – telecommunications. However – unlike the focus of the original SE, the focus of this paper is not solely (or even primarily) on the technical system. The paper describes a framework for rural telecom planning using the Interactive Planning methodology (associated with SE methods to address a combination of hard and soft issues). The framework draws on Mingers' multimethodology to provide a means to mix methods from different paradigms in one intervention. The framework methodology used workshops involving the range of stakeholders to develop plans for action. This created a problem of the difficulty of having the community stakeholders present at every workshop meeting because of the need of these people to work. This contrasted with the availability of the professional solution providers because of workshops being part of their work. The solution sought relied on the formulation of a description of the 'mess' – the situation to be improved by the proposed intervention – and the community at large was part of this process. The formulation of the 'mess' was not solely the responsibility of the professional solution providers.

Remington and Pollack' paper considers the problem of major civil infrastructure projects, investigating the reasons for the limited success of many such projects. Infrastructure project are generally managed according to the Project Management paradigm – which in turn is based on the conceptualisation underlying control theory. Control Theory regards control as an action performed in a situation with a mathematical relation of inputs and outputs. The authors assert that the control paradigm is poor because infrastructure projects may have responses which are sensitive to events in the project context. Two kinds of complexity are described, following Williams (1999): Many independent elements – 'structural' complexity; Unclear goals and methods – 'uncertainty'. The Project Management role is not generic: some projects require good legal or schedule skills; others require good ambiguity handling skills. Any particular project manager is unlikely to be good in both sets of skills. Identification of the two kinds of complexity enables the characterisation of the project, and thus the necessary project manager skills. This conceptualization is consistent with seeing infrastructure projects as Complex Adaptive Systems.

In the final paper by Schauenberg, Sinha and Atyeo, the issue addressed in the work is the complex problem of fleet management of rotary wing aircraft in military fleets. The authors state that there is a need for an holistic analysis because of the disparate concerns of operations and logistics. The authors used an SE method to design a fleet management system; a bespoke management tool (software) is required because existing fleet management tools are designed to suit the much simpler concerns of a surface fleet situation. The fleet management elements include: scheduling and planning; maintenance; operations; platform; and personnel. The environmental issues include: regulatory bodies; weather; human resources; foreign affairs; information; economy.

The Bounds of Systems Engineering

Timothy L.J. Ferris & Stephen C. Cook
Systems Engineering and Evaluation Centre, University of South Australia, AUS

One of the themes in systems thinking is how to determine the boundary of the system under consideration. The possibility of achieving insightful investigation of the system is strongly affected by the placement of the boundary. Systems engineering is a branch of engineering concerned with the whole of life cycle and holistic development of technical and socio-technical systems. However, ideas of systems engineering within the professional community differ markedly with respect to the boundary. Within the systems engineering community there is an expressed need for a strongly formed theory of the nature of systems engineering. Any attempt to determine a philosophy of a field must appropriately demarcate the field from other fields so that neither too much nor too little is included in the field. The traditional, narrow, understanding of the role of systems engineering is that it concerns the development of the technical product system reflecting whole of lifecycle concerns, and a desire to achieve a balance across technical facets, cost and delivery schedule. However, technical systems are both developed and deployed in social environments resulting in socio-technical systems. The question arises as to the extent to which systems engineering should be concerned with the social side of the use and impact of the systems developed, and the extent to which the social system of deployment should be considered in the design process. The work of systems development is conducted in the human environment of the engineering company or consortium of companies constituting another system aspect of the task of systems engineering. Further, the engineering project is conducted within an organisation, which itself is part of a broader socio-political and cultural context. This paper explores the reasonable boundaries for systems engineering, including broad enough but not excessive scope, in order to support practice in the profession and to enable participants in other fields to know what is not claimed by systems engineers as part of their field.

Introduction

Systems engineering is a relatively recent field of engineering, having opened up in the early post World War II period to address certain problems that had arisen with the design and development of large, complex technical systems. These problems concerned the problems sometimes encountered when large complex technical systems projects failed to deliver a satisfactory system on schedule with the budgeted cost. The response to this problem was the development of a methodology for defining the intended deliverable system and for managing the wide range of project decisions required to complete the project and to ensure that the project was completed in a manner ensuring traceability of the decisions made to the most appropriate decisions for the technical success of the project.

Since the origin of systems engineering the methodologies have been extended and refined to address needs arising from several sources. These sources are the lessons learned through the many projects which have been addressed using systems engineering methodologies, using the performance of actual projects to identify weaknesses in prevailing practices. In addition, the projects planned have become more ambitious than the projects prompting the development of systems engineering, leading to further problems of complexity and scale challenging the ability of engineers to deal with the complexity and increasing the risk associated with the projects because of their scale. Changes in the social and political environment in which major projects, usually for government acquisitions, are performed have resulted in a need to improve the probity related aspects of systems engineering. These changes in systems engineering have led to modified practice.

The context of practice of systems engineering, particularly in USA, has resulted in the development of a significantly large professional community of practitioners. As always happens in large communities of practice, there has developed a diversity of views concerning the nature and scope of systems engineering. The divergence in the view of the scope of systems engineering ranges from a narrow position reflecting the concerns of systems engineers working in contracting organizations, who tend to seen systems engineering as concerning the processes for performing large complex projects, where the overall vision of the purpose of the project has been defined, but including the analysis required to make wise judgements concerning technical decisions within the project to this plus all of the processes and analysis required to determine what product should be developed. The latter position reflects a broad view that systems engineering concerns all the decisions required in order to make the product system delivered the most appropriate solution to the need presented as the stimulus for the project.

There are some practitioners who take this breadth a step further, and using a perspective that systems engineering is primarily a methodology for analyzing problems and transforming the analysis into a design for a solution to the problem are inclined to hear of a problem leading to a project to provide a solution in almost any sphere of human endeavour, and respond saying "that's just systems engineering". We contend that this attitude is too simplistic, and does not provide significant respect to the specifics of particular situations in the diverse areas in which people may set out to solve problems, and does not adequately respect the diverse fields of endeavour of others who do not claim to be systems engineers. In addition, we are concerned that anything which claims to be the solution to all problems is unlikely to provide deeply thought out and developed solutions to any problem, or at best to a sub-class of problems. If not addressed, the panacea claim of these systems engineers could lead to unnecessary personal conflict that is not helpful in achieving improvement in society.

This concern has led us to study the issue of the appropriate bounds of systems engineering in order to delineate those concerns and methods which are properly classed as within the range of concerns of systems engineering, whilst also delineating those that are outside the range of effective application of systems engineering methods. Our interest in bounding the field of systems engineering will use Ulrich's work providing twelve questions for bounding a system. In our case the system of interest is the field of systems engineering (Ulrich, 1987). The benefit of rationally determined boundaries to systems engineering includes that the communities of practice in other fields, and most particularly those in other systems methodologies, will have a claim to a space addressed appropriately by systems engineering, which in turn leaves a space which is not claimed as the preserve of systems engineering. Where there is overlap in the claims of different fields there will be a need to resolve the issue of overlapping claims, but this can most effectively be considered in an atmosphere of the participants making finite claims which only partially overlap.

Definition of systems engineering

Systems engineering is variously defined, which is not surprising for a living and dynamic field of practice, and especially one in which there a several sub-communities who see the activity from different perspectives. In the case of systems engineering the major different perspectives pertain to the supplier and acquisition communities, whose systems engineering responsibilities are significantly different. The major specialist professional association is International Council on Systems Engineering, INCOSE. In a recently released publication INCOSE define systems engineering:

"Systems engineering is a profession, a process, and a perspective as illustrated by these three representative definitions. "Systems engineering is a discipline that concentrates on the design and application of the whole (system) as distinct from the parts. It involves looking at a problem in its entirety, taking into account all the facets and all the variables and relating the social to the technical aspect." Ramo in (Federal Aviation Agency, 2004). "Systems engineering is an iterative process of top-down synthesis, development, and operation of a real-world system that satisfies, in a near optimal manner, the full range of requirements for the system." (Eisner, 2002) "Systems engineering is an interdisciplinary approach and means to enable the realization of successful systems." (INCOSE, 2004) Certain keywords emerge from this sampling – interdisciplinary, iterative, socio-technical, and wholeness." (International Council on Systems Engineering, 2006)

Whilst there is some diversity of definition, the common theme in the definitions is expressed in the list of keywords provided in the final sentence of the quotation. These keywords distinguish systems engineering from other branches of engineering, which are largely focused on the use of particular kinds of technology or the development of technical solutions to issues arising in a particular application context. These definitions do show systems engineering as belonging in the total family of engineering disciplines, because they indicate that the expected solution to the problem involves some significant technical component. However, unlike other engineering disciplines, in which the focus is usually on technology of a certain kind, systems engineering places the technological focus on the determination of the appropriate technical strategy, and so is open to the use of technology from, any family of technologies. Further, although it is normal for the problem solution to involve a significant technological component the systems engineer is open to obtaining a solution involving significant non-technical components, such as organizational restructuring or variation of operating procedures.

Project centric nature of systems engineering

At root systems engineering is a methodology of dealing with projects in a holistic manner. The idea of holism has been introduced above in relation to the span of interest of systems engineering. This concern indicates, particularly, that the manner of address of an issue performed in systems engineering is to consider the whole of the solution offered and its context so that the solution accepted as suitable is demonstrated to be the most appropriate according to a broad, multi-dimensional evaluation of a variety of performance and context interaction measures.

Systems engineering was developed to deal with projects. Therefore it is important to provide clarity concerning the idea of project. A project is an activity of work which addresses a particular identified need, and seeks to perform some intervention in the current state of affairs such that the project brings effect to some transformation of the state of affairs between the initial condition and a final condition, which is expected to become an established steady state which will last for a significant duration before any further intervention action is required. In turn, this concept indicates that a project is an activity expected to take a finite duration, and so projects are finite interventions in the current state performed in order to effect a transformation to some envisaged potential future state which is believed to be better than the current state.

Since systems engineering has developed as a professional practice to overcome recognized actual and potential difficulties pertaining to product system performance, development project schedule, development budget and probity it has been found helpful to codify practice through the development of standards to define the matters to address, the processes to be used and the process products to be developed as the means to both perform the work of the processes and to document the work done and the fact that the work was done. Through the history of systems engineering there have been several standards developed to define the processes and process products that should be developed through the performance of the project. The earlier standards emphasized the itemization of the processes and process products required. The ISO 15288 standard has taken a different approach and identified the matters about which it is necessary to have a process with which to deal (ISO/IEC, 2002).

ISO 15288 divides the matters to be addressed in the life-cycle processes into Technical Processes, Project Processes, Agreement Processes and Enterprise Processes. Technical Processes refer to those processes that are used to perform the technical work of the project, and so include such matters as the technical aspects of the design, construction and testing of the product system.

Project Processes refer to the processes of administering the project and managing activity. The combination of technical and project processes are the processes required in order to do the work of performing the project. Agreement Processes concern the processes required to ensure that agreements between the entity performing the project and all other stakeholder entities are managed appropriately. Stakeholder entities include, but are not limited to, suppliers, customers and regulators. Enterprise processes are the processes of the entity performing the project that facilitate and sustain its existence and capability to perform the present and future projects, and enable the decision to pursue particular projects.

In presenting the system life cycle processes of systems engineering as a set of matters about which effective processes, policies and procedures are required ISO 15288 has shifted thinking about systems engineering from process and process product centric views to the higher level of purpose and means to achieve desired purposes. All the systems engineering standards (Department of Defense, 1974; IEEE, 1998; Electronic Industries Alliance, 1999), including ISO 15288, have a strong element of interest in technical processes, either by stating that effective policies, procedures and processes are required to deal with a range of technical matters, or by actually specifying processes to deal with those matters, which indicates that the typical systems engineers perception of systems engineering involves the performance of projects delivering some significant piece of technology as at least a part of the total system produced as a result of the activity. This point is important in addressing the question of the bounds of systems engineering. Whilst superficially similar methods may be used to address projects which do not contain any technology, that is the project commences with an investigation of the perception of the problem held by each of the stakeholder groups, and an analysis of the proposed end state is made, with comparison to the current state, so that means to effect the transition from current to desired states are developed, where there is no newly made or acquired technology of any kind involved in the solution the project demands skills outside the normal range of engineers working as engineers, and the various analytical and modelling methods and tools of engineering will be found to be inapplicable, and so the solution cannot be designed and developed using methods that correlate well to traditional engineering methods, and therefore projects involving situations not requiring the acquisition or deployment of technology as a significant proportion of the project demand such different development methods that systems engineering methodology is unsuitable.

The paragraph above has considered the situation of projects in which there is very little or no technical component. Where there is a situation in which it appears likely that the solution will require significant technical work, either in the development of a technical entity or the selection, installation and commission of technical entities, as a part of the project the systems engineering methodology is appropriate. The systems engineering methodology provides means to identify the need in a manner such that the details of the technical specification of the technical aspect of the project can be defined. The systems engineering method recognizes that the appropriate technical solution must be cognisant of the human aspect of the situation affected by the project, both at the detailed, ergonomic level, and at the organizational or social level, to ensure that the technology proposed by the project is suitable for the situation into which it will be deployed, and will contribute to the establishment of a sound final product system. The judgement that the product system is effective can only be obtained where the design of the system incorporates a holistic view of the technology and the human use of, and interaction with the technology, and the effect of the technology on the modes of interaction of the people in the system. The design process used in systems engineering includes consideration of all these factors as relevant to the success of the proposed system. For example, if users require specific training to be able to use the proposed system, then the design of training programs would be considered as part of the project. If the proposed project would change the work procedures in the organization where the system is deployed, then consideration of the effect of those changes in work procedures is considered properly part of the systems engineering project for the development of the solution.

By way of example, suppose a large organization requires an information system to support its activity. The information system itself is a significant piece of technology, involving hardware and software. However, the work procedures required after implementation of the system are different than the procedures in place prior to system development. The change of work procedures will have significant effects on organizational relationship matters such as the organizational level at which decisions are made, work flow and the devolution of responsibility for action.

Systems engineering methodology

The authors have elsewhere argued that the systems engineering methodology is a methodology for harnessing other methodologies with the definite intent of achieving at least satisfactory solutions for the matters of concern addressed in the projects (Ferris & Cook, 2005). Thus, the systems engineering methodology does not, by itself, claim to be a unified methodology for performing entire engineering projects, but rather it is fundamentally a meta-methodology which enables the appropriate selection of particular methodologies to support particular aspects of its cases, the projects. However, within this construct there are some general things which can be asserted about systems engineering and are true of the meta-methodology as a whole.

The first fundamental issue is that systems engineering is a particular branch of engineering, and so it shares the underlying ethos of engineering as a goal directed activity seeking to deliver technical products that satisfy particular needs arising in the community. The driving force in engineering project is awareness of the objective, and other things are set aside in order to bring effect to that objective. The nature of engineering is to be goal directed, and focused on the technological aspect of the project that seeks to attain a solution to the identified need that prompted the initiation of the project. In turn, because of the nature of engineering work, and the organizational environments in which the work is done engineers tend to develop work practice traits that are goal focused. This is a contributing factor in the popular social impression of engineers as people suffering some measure of challenge in relation to sensitivity to social issues which are commonly regarded as important.

Systems engineering has the goal directed focus, as described above, even though it seeks to take an holistic perspective on the systems which it seeks to establish. The holistic perspective is taken in systems engineering out of recognition that it is fundamentally important to achieving a genuinely successful overall product system to respect the full range of concerns of all the stakeholders and to achieve a balanced technical solution. The concern for holism, which manifests itself in systems engineering, as a distinguishing mark from other engineering disciplines, in the goal of achieving a well rounded overall solution to the need presented rather than an optimized technical solution focused particularly on one of the technology focused branches of engineering. The latter perspective is common among projects developed primarily from the interests of the discipline providing the technology in a project driven by a technology rather than drawn by an identified need. In the systems engineering concern for an holistically desirable product system the dimensions of interest include the dimensions of technical performance measures of the technology components of the final product system, but also concern measures of effectiveness of that system as a system for delivery effects which are the satisfaction of the identified need. The measures of effectiveness are a formal test and evaluation means to determine whether the product system as developed is able to perform in a manner that is suitable for addressing the originally identified need in the context in which the product system placed. This evaluation focus on effectiveness, rather than on technical performance measures *per se* is an important expression of the holistic focus of systems engineering, and in turn demands that a thorough investigation of the situation or situations of in which the product system is expected to be effective, that is work in a manner delivering the capabilities required in order to effectively address the identified need.

As an engineering discipline systems engineering is fundamentally concerned with making things that achieve goals, but the things themselves are or include artefacts which provide certain functionality under constraints. As such, the means of achievement of the concerns of systems engineering are bound by the laws of nature, and by the basic constraint that anything which is one thing cannot be something else. Therefore, whilst systems engineering is concerned with developing product systems which respect the constraints imposed by the laws of nature and therefore must implement a mechanistic view of the world, it is also pluralist, in that it is concerned with respecting the positions of diverse stakeholders and seeking a solution which is reasonable given a rational analysis of the values, interests and concerns represented by the various stakeholders. This statement is made with respect to the matrix of problem classifications and applicable systems methodology presented by Flood and Jackson (1991), except that, for the reasons discussed here we hold that the placement of systems engineering by Flood and Jackson is inappropriately in the 'simple' and 'mechanistic' cell, and should rather be in the 'pluralist' and 'mechanistic' cell. Our contention concerning the appropriate placement of systems engineering reflects a combination of changes in the field of systems engineering, and our interpretation of what is meant by the ideas represented by the columns of Flood and Jackson's matrix, that is 'simple', 'pluralist' and 'coercive'. The change in the positioning of systems engineering in the Flood and Jackson Total Systems Intervention matrix is justified by the broad perspectives on the role of systems engineering presented by two recent documents published by INCOSE, the major professional body of systems engineers, the *Technical Vision* (INCOSE, 2005) and *Systems Engineering Handbook*, version 3 (International Council on Systems Engineering, 2006). However, although much of the early practice of systems engineering supported the Flood and Jackson position, for example (Gifford, 1954; Schultz, 1956; McMullin ,1958), there were also early works in the field supporting the present contention that systems engineering is a means to deal with complex pluralist situations (Collins, 1957).

These two positions, of systems engineering representing a mechanistic and pluralist perspective, present a tension. Since the product system must be something in particular and the stakeholders may hold divergent and even contradictory ideas of what is 'good' in the case of the proposed product system and the situation into which it will be deployed the product system must represent a balance, or possibly a compromise, between the positions of the stakeholders. It is most likely that not all stakeholders can be completely satisfied by any decision related to any particular aspect of the product system. In addition, there is finite time within which to make decisions about the product system and the effect that it will have.

Therefore the systems engineering processes of trade-off analysis, in which various possible approaches to the design of the product system are compared with a view to enabling a choice to adopt the approach with the best overall benefit, determine the most appropriate product system design through weighing the various stakeholder interests with respect to what is achievable, given the constraints, in the product system. This process is one which implements a consequentialist view of ethics. That is, the possible alternative designs are compared using a filter which weights particular performance quantities in each of a number of dimensions of interest in order to produce a 'figure of merit' for each of the candidate design strategies, and then leads to the selection of the design strategy with the highest 'figure of merit'. In systems engineering, this approach to trade-off analysis is taken because systems engineering is used as a means to obtain the best possible solution given that a decision to proceed with a project to develop a solution to a certain need situation has already been taken.

Thus, the systems engineering method has a means to determine what is 'good' in the context of a project, and once this 'good' has been transformed through the trade-off analysis process to a 'most desirable' design strategy the design is proceeded with, through to the development and delivery of the solution. Once the trade-off analysis has been done it may appear that the systems engineering method rides roughshod over the concerns of stakeholders other than the developers and the 'customer'. However, this appearance only arises because once a satisfactory design strategy has been found through the trade-off process there is no further reason to further delay the project through continuing discussion with sundry stakeholders. However, systems engineering is not as imperialist as this description appears. Firstly, the trade-off analysis process itself includes extensive efforts to ensure that the concerns of all stakeholders are incorporated in the determination of the weights that enter into the trade-off analysis process. Therefore the driving through of the decisions made is done only after the concerns of the stakeholders have been included in the process of deciding the design strategy that should be taken. Secondly, the systems engineering methodology itself permits changes in objectives and design to be decided, when necessary, through the design of the product system. Systems engineering processes include particular processes for suggesting and evaluating the appropriateness of possible changes. Therefore, should it become evident that the concerns of a particular stakeholder are very inadequately represented in the arrangements leading to the current version of the design a change can be effected. However it is well known that the cost of changes in both budgetary and schedule dimensions is considerable, and so the systems engineering manager will require strong justification for any proposed change, resulting in changes to design being managed to minimise their number and magnitude. Most definitely, changes arising from conflicting stakeholder positions are likely to be treated as "chasing after the wind" and therefore judged as not adequately justified.

The systems engineering methodology is focused on delivering a system which addresses the concerns articulated in the statement of need as expanded and elaborated in the needs and requirements analysis phases at the commencement of the project. The exploration of the needs and requirements involves detailed evaluation of the concerns of the various stakeholders and the *sitz im leben* of the intended product system so that that product system, when delivered will represent a reasonable balance of the concerns of the plurality of stakeholders and a means to support the intended purpose as an entity capable of the dynamism genuinely to provide the ready-to-hand (Heidegger, 1973) capability sought in the development of the entity.

The perspective taken in the systems engineering task is to translate the difficult to describe qualities of the intended product system related to the intended *sitz im leben* and capability into a statement of requirements of the product system. This process of translation transforms the understanding of the intended entity from a concept of an entity situated within life and performing a function that supports a desired lifestyle to a description of the entity in terms of the material of which it is to be comprised, which is a translation to the present-at-hand properties of the entity.

The engineering task design task of engineering concerns the arrangement of material, logical and human elements, all viewed as technical elements, thus emphasizing only that of which they are comprised, but de-emphasizing the dynamism in life of those elements as they gain through their actual entering into relations with other elements of the various kinds comprising the whole system. Thus the purpose of the engineering task is to take the elements denuded of their life setting and assemble them in combinations such that the resultant combination obtains properties which transform the elements from being just elements to being an entity which belongs in and supports its intended *sitz im leben*.

Whilst regarding the human *sitz im leben* aspect of the intended product system as of crucial importance, and therefore taking this part of the human interaction with the proposed product system as of crucial importance there is another aspect of the human side of systems development which the systems engineering methodology gives a low priority. The interaction with people in the methodology concerns determination of the properties of the system which must interface effectively to the people, but the methodology itself does not ask questions about relationships between people such as whether those relationships are hierarchical, exploitative or supportive, and there is no explicit place in the systems engineering methodology for addressing such 'imbalances', or otherwise deliberately modifying existing relationships except in so far as the parties commissioning the project have determined that such a change is their objective in their decision to commission a project to achieve a certain end. To illustrate, a government may determine that it has a problem with housing for the socio-economically disadvantaged, and in response establish a project to address this political concern. At the political level the government makes the decisions about the transformation of relationships between people that the government wishes to effect, for some primarily political purpose of its own. This political objective leads to a determination of the goals in the political arena and the life setting arena for the project. The systems engineering role commences with these objectives, and then uses whatever combination of technical and social science skills are necessary to determine a technical and social solution to the need identified by the political group commissioning the project. Once the project arrives in the systems engineering sphere the decisions about the objectives have already been made and the remaining work is to determine means to satisfy the objectives but not to address questions of whether the objectives set are the right ones, because that role is one for the political masters of the project to determine.

To put the above discussion more succinctly, it can be said that the systems engineering methodology may regard some issues related to human interactions as impediments to be overcome rather than as the matter of particular focus. This could be illustrated from a situation arising in a fairly typical systems engineering application. In the design of a warship the systems engineering methodology would be applied to the design of an operations room. The focus of attention is the processes to be performed by operators at each of the processing stations and communication required between participants. The focus is on the efficient and accurate performance of the tasks required because in the extreme situation of the ship being under attack it is likely that the participants may have at most a minute from first detection of an actual threat to the last possible time to deploy a response, during which considerable analysis of the nature of the threat must be performed so that the most appropriate response is made. The systems engineering is concern with the effective performance of the task required, but is not interested in any of the social subtlety of the relationship between the participants because such a concern is not helpful in obtaining a good response in the threat situation.

Ulrich's 12 boundary questions

Ulrich (1987) considered the issue of a general social system for which some kind of systemic intervention is required. The systems envisaged by Ulrich are primarily social systems, such as organizations in which some intervention is required in order to achieve some outcome considered desirable. These systems contrast with those for which systems engineering is designed, which are those with a significant technical component expected as part of the solution. However, Ulrich's questions are addressed here because many of the question provide some insight into the bounding of systems engineering relevant issues, and so assist us in providing bounds to the space in which systems engineering is the appropriate methodology.

1. Who ought to be the client (beneficiary) of the system S to be designed or improved?

In situations where systems engineering is suitable there is a defined client of the project. This is either a contractually identified client, who has commissioned the work, or an organizational manager who, for reasons of pursuing organizational strategy has commissioned a particular project. These parties have reasons why they seek the development of the product system, but the project must address their interests in commissioning the project in order to be identified as a success. However, this question is not as simple as this. Within systems engineering it is recognized that identification of the customer, user and owner of a product system is important because it is likely that these three will be different. In particular the user refers to particular people who directly interact with the system in order to perform some activity, whereas the owner may be some organization whose obvious measures of effectiveness may well be different that the concerns of operators. In defence acquisition in particular, in many countries one organization is established to administer the project to acquire the system whilst another becomes the owner and deployer of the system. A similar tension in the answer to this question may arise in many other situations, where for example an organization acquires the product system in order to provide a service or effect some change in a social context. The systems engineering task includes exploring this issue so that the right questions can be asked in the needs and requirements phases, in order to determine the correct objectives for the technical project. If this is not identified the product system is likely to be severely impaired in its ability to deliver the service intended.

2. What ought to be the purpose of S, i.e., what goal states ought S be able to achieve so as to serve the client?

This question is crucial to the systems engineering endeavour because the development of a product system that addresses the concern expressed in the original articulation of the need will only be regarded as successful by any of the stakeholders concerned with the system in so far as the product system satisfies concerns of that stakeholder. However, since there are usually multiple stakeholders concerned with any product system the concerns of the range of stakeholders must be understood in order to determine the appropriate goal state to be sought through the detailed design phase. The systems engineering methodology determines the balancing of the concerns of the various stakeholders through a trade-off process in which for each dimension of interest a weighed contribution from the achievable performance is added to provide an overall measure of the merit of a particular proposed system architecture.

3. What ought to be S's measure of success (or improvement)?
This issue is addressed in systems engineering through the development of measures of effectiveness which will be used to validate the product system developed as a means to satisfy the need as articulate in the original proposal that something be done, and the expansion of that original expression of desire performed in the needs analysis phase of the project.

4. Who ought to be the decision taker, that is, have the power to change S's measure of improvement?
In the systems engineering methodology the chief systems engineering is responsible for decisions made concerning the project. Because of the magnitude and diversity of the analysis required in order to provide foundations for the decisions the systems engineering is dependant on the detailed analysis of other project participants with relevant expertise to analyze problems of certain classes. The systems engineer must receive the reports of the work of the other staff providing analysis and design work and make decisions about the response which should be made, to accept or modify design work, and for the project directions that should be taken in response to the findings of investigations. Thus the systems engineer is responsible for the conduct of the project, but has the opportunity and responsibility to seek for and to use advice obtained by experts in other fields.

5. What components (resources and constraints) of S ought to be controlled by the decision taker?
The systems engineering methodology regards the systems engineer as the party responsible for the use of resources in the conduct of the project and also for the determination of the final form of the product system. Thus the resources are all ultimately controlled by the systems engineer, with the usual kinds of delegated authorities which are present in most large organizations to facilitate efficient action within the constraints of the framework decided by the systems engineer. In addition, the resources, and use of resources is similarly governed by the systems engineer.

6. What resources and conditions ought to be part of S's environment, i.e., should not be controlled by S's decision taker?
The preferred position in systems engineering is that the maximum possible range of resources and other factors associated with the project should be under

the control of the systems engineer. This view is held because it provided the systems engineering with the maximum flexibility in dealing with the project, and making decisions concerning what should be made and ensuring the most effective use of resources, and the best possible opportunity to make most effective product system. The systems engineering methodology recognizes that each project will have a particular range of constraints imposed by the laws of nature, the resources that can be made available for the product systems and the systems engineering team, either at development and production time, or during deployment, and that there may be various constraints imposed for reasons related to the context of the project for reasons other than the inherent nature of the product system and its context of deployment. These constraints limit the freedom of the systems engineer to make decisions relevant to the design of the product system, and so to provide the best opportunity to product the most effective product system possible it is generally believed that the systems engineer should have control to make decisions about all remaining resources and constraints, rather than to have further impediments to the free ranging responsibility.

7. Who ought to be involved as the designer of S?

The issue is important to systems engineering. Systems engineering adopts the position that the product systems design should follow an architecture developed because it is the most appropriate architecture to address the objectives of the project. Therefore, at the initial architectural stages of design the systems engineering methodology is agnostic concerning the most appropriate technology, or other factors to deploy in the design, with the trade-off process used to assist these decisions. Once the appropriate architecture has been identified people with the appropriate expertise to perform the detailed design of those portions of the product system are engaged and applied to those parts of the design. Thus, it is unclear who should design the system until the initial system architectural work has been done and it emerges what technical or systems strategies are to be applied to the problem, and then appropriate people are engaged to perform the design work.

8. What kind of expertise ought to flow into the design of S, i.e., who ought to be considered an expert and what should be his role?

The systems engineering methodology asserts that each part of the work should be performed by a specialist or expert in that facet of the work. Thus each technical aspect of the design should be performed by a suitable expert. This call for expertise requires that there be a teaming of design staff with expertise in facets of the task, such as design in the technical specialty, manufacturing and maintenance, amongst other fields of possible expertise.

9. Who ought to be the guarantor of S, i.e., where ought the designer seek the guarantee that his design will be implemented and will prove successful, judged by S's measure of success or improvement?
The systems engineer is the authority who can give assurance of the final design to be implemented, subject to their negotiation with other stakeholders, most particularly the buyer of the proposed system. However, at no time during design is there the possibility of guaranteeing that there will be no future change to any part of the design, let alone the whole. This is because there is no guarantee that the external circumstances will remain constant, resulting in continued need for the proposed system in exactly the form that has so far been negotiated. Indeed, the proposed system may no longer be required in the event of some potential changes in the environment. Further, when the proposed system is placed under evaluation defects may be identified. Because in a system there is interaction of all the components to provide the range of whole of system effects any defect in the whole could result in a need to modify any, some, many or all of the components in the product system. Consequently no guarantee can be given at any time that any part of the design has become 'frozen' without risk of considerable detriment to the whole. The nearest there is to a 'guarantee' of frozen design is that the customer acquiring the product system sign-off for acceptance of the product system in its current form. However, even with this sign-off there is the possibility of later need for modification.

In systems engineering, once the major design decisions have been made, through the architectural trade-off decision process it is normal that the design changes which are likely to happen to the product system will be changes within a general theme of elements. Thus, it is unusual for the product system architecture to be changed, with results such as a change of kind of technology to be used. The design changes in systems engineering tend to be refinements within the general range of solutions permitted or expected by the earlier, whole of system design decisions. This contrasts with design in social systems, where a problem is likely to be addressed by a significant change of design away from that proposed in the original design to one implementing quite a different analysis of the situation addressed.

10. Who ought to belong to the witnesses representing the concerns of the citizens that will or might be affected by the design of S? That is to say, who among the affected ought to get involved?
In the systems engineering methodology there is considerable attention paid to the interests and concerns of diverse stakeholders. Where possible the actual people who are stakeholders, or a representative sample of the stakeholders is used to obtain input. This is possible where the stakeholders can be clearly identified, and where communication with them about the intended product system can be done in a way that does not lose competitive advantage or provide some other significant disadvantage to the principal in the project. Where the sensitivities of the project are such that it is possible to use real or representative stakeholders this is done. Where the sensitivities preclude the use of real or representative stakeholders the alternative position is to model the concerns of

the stakeholder class and to deal with the matter indirectly. When stakeholder concerns are identified indirectly, this is a practical manifestation of the systems engineering concern with the goal of providing a product system which provides satisfactory consideration of all stakeholders concerns, but also that it is more important to have a reasonable compromise solution product system than to try so hard to please all the stakeholders that nothing tangible happens.

11. To what degree and in what way ought the affected be given a chance of emancipation from the premises and promises of the involved?

Within the systems engineering methodology there is opportunity for either direct or indirect involvement of the affected, at the discretion of the systems engineer, to contribute to the trade-off analysis. Often the principal in the project will give consideration to the position of the affected as well as the involved, because the principal's public image and reputation will be affected by the effect that they have on the affected through their actions. Thus, although there is no formal place to incorporate the views of the affected in the systems engineering process, only the unwise would totally ignore the views of the affected. In addition, where the affected will be involved in interactions with the system, the final judgement of the success of the system will include the effect that it has on the affected, and any potential action that they may take as a result of this effect. For example, if a large organization introduces a new information system, where the involved are the employees of the company, the customers of the company, who may never directly interface with the information system are affected. The customers may be benefited if the staff of the company are enabled to serve them in ways that are more convenient, but may be disadvantaged if the information system makes it more difficult for company staff to access or interact with customer information. If the affected become sufficiently frustrated with the change in quality of service they may shift their custom to another provider of similar services, to the loss of the company introducing the information system.

12. Upon what world-views of either the involved or the affected ought S's design be based?

The systems engineering concern with the worldviews of the involved and the affected is with understanding the *sitz im leben* of the product system, with the consequent concern that the product system design will be suitable for the purpose for which it is intended, and that the product system will provide a desirable enhancement of the life of the involved and affected.

Our discussion of Ulrich's questions concerning the boundary of a systems issue has presented systems engineering as a functionalist methodology concerned with the development of tangible solutions to the problems addressed, and as such the consideration given in response to each question has emphasized the engineers concern with developing workable solutions to the problems confronted. The engineers concern has resulted in the approach of consultation and participation with stakeholders where possible or practical, but otherwise presenting a coercive view of systems engineering as concerned more with the delivery of solutions which have been determined through the methods of systems engineering and then imposed on the surrounding human and physical environment.

What is not systems engineering

The discussion above has presented the position that systems engineering is concerned with the definition and design of systems involving at least the expectation of some significant technological component. A rather vague expression like this is required because at the beginning of a project to address a need it is not necessary clear whether the solution will be obtained through a design, development or selection of technological components, in association with the existing and potentially creatable organizational, political and social environment of use of the proposed system. Thus most systems will involve some balance of social and technical aspects. The systems engineering methodology has been developed to address the processes of development of large complex technical systems, which have been found, in the natural course of the systems engineering work, to involve people. The first view of people in systems engineering work was at the mechanistic level with the development of 'human factors' a field largely concerned, at least initially, with the mechanics of the interaction of people and equipment, addressing such matters as size, activation forces, and other quantifiable factors in the interaction of people and equipment, driven by an interest in safety and effectiveness of the interaction of the people and the technical aspect of the product system.

Much more recently, and particularly noticeably since 2002/2003[1], the systems engineering professional community has become significantly interested in the human aspect of the social community of system use, with a view to placing emphasis on the need for the system to be designed in order to either enhance or influence the manner in which the users interact with each other in order to better bring effect to their intentions in particular missions of the use of the product system. This is seen to be a desirable field of consideration in order to improve the usefulness of the systems developed.

Even in the emphasis placed on the human social influence of the product system the emphasis in systems engineering the interest is primarily a goal directed focus of making product systems that more effectively support the intended functionality and capability of the entity that owns and deploys the product system as means to bring effect to some objective that it has.

It is useful to contrast the positive statements of what systems engineering properly concerns, expressed above, with some examples of things which systems engineering does not claim to address, and for which the systems methodology is poorly suited.

---

1 This remark is based on the authors' personal observations of the growing emphasis of this made evident in the annual symposia of the International Council on Systems Engineering. The year when the most obvious change from the previous year was evident was 2004. The trend is observed to be continuing, as at the International Symposium, of July 2006.

1. Systems engineering does not claim to address fundamentally social and political concerns. Systems engineering is used in the development of product systems that are sponsored by someone, such as a government, to deal with social and political concerns, but the contribution of the systems engineering methodology is in the exploration of the need, as identified by the principal, so that the principal's objectives can be achieved, and the design and production of a product system that enables the principal's social and political objectives to be achieved. The principal, in this case, is the activist entity that decides what must be done, and the systems engineer is a servant bringing effect to those objectives.
2. Systems engineering is not designed to be a means to provide on-going management of operations of an entity. On-going operations should be managed using establishment management methods applicable to the kinds of issues presented by the situation. Systems engineering is a methodology developed to deal with the issues of projects to define, design, develop and deliver product systems which themselves will become structures requiring on-going management. The systems engineering concern with the operations phase is to design the product system so that it requires on-going operation interventions of a kind, magnitude and intensity which have been agreed by the principal to be reasonably supportable given the anticipated circumstances of the principal, or the circumstances that it is anticipated that owners and users will face during deployment.
3. The systems engineering methodology is not designed to deal with the organizational aspect of the continual renewal process in business operations because many of these matters are purely social, or concern the design of work, but do not concern the development or acquisition and deployment of technology into the organization. Such activities, even if organized as project with defined commencement and object states address matters which are outside the range of matters for which the systems engineering methodology was developed, and so it is more appropriate to investigate and address these projects using other systems methodologies identified in the literature. In these cases the people leading the projects should be expert in those other fields.

Conclusions

This paper has discussed the nature of systems engineering, including the purpose for which it was developed, and issues that it has come to address more recently as a result of confronting those issues in the course of its normal activity. The broadening of the application of systems engineering which has developed has led some systems engineers to view the world of intervention in situations to establish desirable new states as being the sole preserve of systems engineering.

We have asserted, and argued a defence for our view, that systems engineering is suitable for use in a broad range of cases where there is a need which is to be addressed by means of a project, that is an activity with a defined beginning and end to deliver a new, product, system which will address the need identified. We have narrowed the set of projects to include those for which, at the time of project commencement, it is reasonable to expect that the product system which provides the solution is likely to include a significantly large technological component. The solution may be solely technical, or may be a sociotechnical system. However, this claim to the applicability of systems engineering also declaims applicability to situations in which it is expected or planned that the solution will not involve any of the design, development, selection, installation or deployment of significant technology to contribute towards the intended product system. The scope of this claim, being limited to situations in which it is expected that the reasonable intervention action is to develop or deploy some technology in order to address the core issue is important in constraining the claim of some systems engineering that the systems engineering methodology applies to everything. The constraint of claim presented in this paper is important negatively in rejecting the excessively broad claims of a small number of systems engineers who regard any situation requiring an investigation of the problem, proposal of alternative possible courses of action, investigation of the alternatives leading to selection of one, and implementation of the solution according to that method as an implementation of systems engineering, because such a broad claim is unhelpful in providing sufficient insight into classes of situation which could be presented to facilitate improved solution development. The constraint of claim is helpful positively in disclaiming the space which is the primary concern of many other systems intervention and investigation methodologies which have been developed to obtain insight and effective interventions in situations, normally not expected to involve the development or deployment of a significant item of technology.

We have also argued that the system engineering methodology is suited to projects, activities with a defined start and objective. This claim, correspondingly declaims the space of on-going management of operations as a concern of systems engineering.

Our claim to the space delineated above indicates that we view systems engineering as a discipline with a broad-ranging area of concern and application, but that that area is characterized by some clearly identifiable characteristics which distinguish the area of application of systems engineering from other areas of interest which we state are best addressed using other systems methodologies.

Even within the area of application that we have claimed above, we recognise that within the systems engineering led projects to address identified needs there will often be a place for application of other systems methodologies as the issues associated with the social aspect of the system solution are investigated to ensure that the solution provides the most satisfactory outcome possible. However, the systems methodologies may be constrain in their application by the over-arching concern of the lead discipline of systems engineering to deliver a real product system that effectively addresses the identified need fro a perspective balanced across all identified stakeholders.

References

Collins, W. C. (1957). "Electronics application in the County of Los Angeles." IRE transactions on vehicular communications **8**(1): 45-46.

Department of Defense (1974). MIL-STD-499A Engineering management. Washington D.C., Department of Defense.

Eisner, H. (2002). Essentials of project and systems engineering management. New York, N.Y., Wiley.

Electronic Industries Alliance. (1999). Processes for engineering a system. Arlington, VA, Electronic Industries Alliance.

Federal Aviation Agency (2004). Systems engineering manual.

Ferris, T. L. J. and S. C. Cook (2005). Away from. International engineering management conference, St Johns, Newfoundland.

Flood, R. L. and M. C. Jackson (1991). "Total systems intervention: a practical face to critical systems thinking." Systems practice **4**(3): 197-213.

Gifford, R. P. (1954). "The knee of the nose." Transactions of the IRE professional group on vehicular communications(4): 40-51.

Heidegger, M. (1973). Being and time. Oxford, Basil Blackwell.

IEEE (1998). IEEE standard for application and management of the systems engineering process. New York, Institute of Electrical and Electronics Engineers.

INCOSE (2004). Systems engineering handbook, version 2.0a, International Council on Systems Engineering.

INCOSE (2005). INCOSE Technical Vision, Version 1.1, May.

International Council on Systems Engineering (2006). Systems engineering handbook a guide for system life cycle processes and activities. Seattle, Washington, International Council on Systems Engineering.

ISO/IEC (2002). ISO 15288 systems engineering - system life cycle processes, ISO/IEC.

McMullin, T. J. (1958). "The end of the line." IRE transactions on vehicular communications **11**(1): 20-32.

Schultz, C. (1956). "Spectrum compression and its problems." IRE transactions on vehicular communications **6**(1): 76-82.

Ulrich, W. (1987). "Critical heuristics of social systems design." European journal of operational research **31**: 276-283.

Long Term Preservation of Electronic Information

Viveca Asproth
Mid Sweden University, SWE

Information technology has lead to that much of the information that earlier was produced on paper is now being produced in electronic form. The e-government services, where the whole idea is to exchange information electronically, will radically redirect paperbound information towards electronically generated and managed information and documents. The challenges associated with digital preservation are strategic, organizational and structural as well as technological. In this paper some research directions connected to digital preservation are presented.

Introduction

Information technology in combination with changes in organizational structures and forms for work has led to an increasing amount of information. In addition totally new forms of information and records are created. Much of the information that earlier was produced on paper is now being produced in electronic form, for example as e-mail and in databases. The Dutch archive theoretician Eric Ketelaar (1999) has invented the word "archivalisation" to describe the phenomenon that more and more seems to be worth to document and archive. This means that a considerably larger amount of information needs to be handled. Electronically generated and stored information is also exposed to change and manipulation in quite another way than paper-bound information.

A 10-year period is a long time when it comes to preserving electronic records. This is of course due to the rapid technological development, for example the problems with obsolete software and file formats. Goldstein (2004) stresses the need for a quick solution to the problem of archiving electronic information for the future. Further, he says that the major threat to archived material today is not the fire hazard but the rapid development of different file formats for documents, sound and images. "The last decade has generated more recorded information than any previous decade of human activity. The fact that the majority of these data is less accessible than ever before is one of the ironies of the modern information age" (Duranti, 2000). This implies a risk for a situation where the societal memory no longer can be guaranteed. In data mining and data warehousing the issue of archiving data is addressed, but the time horizon for this is limited and the use of archived data is for a specific purpose.

Vitale functions in society such as the social insurance and national registration have to be accessible for long periods of time. Without authentic, accessible, searchable, and reliable records these functions in society are in fact impossible to maintain. The technical development and transition to electronic media has led to that traditional methods have become obsolete or at least insufficient.

Problem domains

There still remain a lot of problems to solve before secure and efficient long-term preservation will be brought about. It seems that although much of the challenges associated with digital preservation is strategic, organizational, and structural and not only technical, the research concentrates at solving the technical issues. Digital preservation requires, in addition to the technological development, elaborated strategies, new workflows and organizational structures, standards and common metadata, new specific competences, and close cooperation across different professions from traditional preservation management to computing science. (Asproth, 2005)

Technical problems

Issues at stake in this domain cover hardware, software and formats. The large amount of electronic records to store, for active use, demands new mass storage media. The fragility of the media leads to low durability and there is also often a lack of knowledge about the durability of different media for storage. The software problem concerns the rapid development of new versions of the software product and occasionally changeovers to new software products. Connected to the problem with rapidly changing software, there is the problem with changing formats, which in the long run can lead to the consequence that archival information become unreadable (Gibbs & Heazlewood, 1999).

Legal issues

One important requirement on records is evidence. The problem of how to ensure authenticity and how to preserve evidence is emphasized by, among others, Bearman (1994), Duranti (2001) Rothenberg (2000), and Wilson (2000). The rapidity and the ease with which changes can be made mean that there are challenges associated with ensuring the continued integrity, authenticity, and provenance of digital materials (Beagrie, 2002).

Organizational problems

Most of the challenges associated with digital preservation are organizational – not technical. The first line defense against loss of valuable digital information rests with the creators, providers, and owners of digital information. (Marcum, 1997). There is also a matter of inter-organizational problems to cope with. For example, governmental organizations want to offer electronic services to the citizens that requires collaboration as the services assume information from more than one organization (Asproth, 2006a).

Context and metadata

How should it be possible for someone in the future to understand the context of the information and the processes that created it? This is a question of metadata, a question that is considered as more important in the digital age. How can we determine what amount of evidential historicity is required and how to capture and retain it (Bearman, 1994)?

Long-term preservation of information is also uncertain as we can not fully foresee who the users are and for what purpose they want to use the information.

Current state in Swedish organizations

Here some studies carried through by the research group "Archives of the Future" at Mid Sweden University are presented.

The first study aimed to investigate how different governmental departments managed electronic information both in a short and long run. The ambition of many governmental departments is to become 24/7 agencies. Of the studied organization, four have e-services taking care of at least some part of their activities. Two others plan to start e-services. The county council will not, for obvious reasons, supply any e-service for case records. All the studied organizations handle electronic records in the studied cases. They saved the electronic records in XML or other formats that is said to be appropriate for long-term preservation. Most of them also scanned paper documents and used them as electronic information. Much information is also stored as database records (Asproth, 2006b).

The active time for the electronic information vary quite a lot, but mostly the information is active for decades, in some cases for several decades. In some of the organizations not all information is preserved during that time, some sorting out is performed. Information like notes, e-mails, etc. is sorted out. As the amount of information is huge, just a selection is obliged to be preserved. However several of the organizations chose to save more than they were obliged to. Many of the organizations chose to save all the database records and did not sort out any of them. In some cases scanned documents were chosen to be the original and thus preserved. Other chose to print out the information and preserve paper documents. It deserves to be mentioned that electronic records has not existed for so long time and a feeling of uncertainty about how to do was expressed (Asproth, 2006b).

The second study concerns how police officers use electronic records in their operational work, reacting to calls from the dispatch centre. Information technology use aims to increase the efficiency of police work. The Swedish police have several computer-based information systems, and several of those information systems have become a necessary tool for police work. Some information systems manage records and their contents should be preserved long-term. There is a widespread use of information retrieved from electronic records. This has changed the possibilities for police officers to get reliable and authentic information for both tactical and legal decisions, which increases the officers' ability to make correct decisions within operational work. Information technology and information systems now handle many administrative tasks, and allow access to and searching of electronic information independently of the physical location of the officer. This opens up a possibility for mobile access to trustworthy information that supports police work. (Borglund, 2005)

In the third study the characteristics of records was empirically identified and compared to archival theory. Four organizations and their records were analyzed. The study shows that there are discrepancies between practical use and archival theory. There are also differences between the organizations. There are examples of information systems that include records that do not meet record-keeping requirements as well as manual systems with low maturity. The conclusion is that some of the organizations have a lot to do before reaching good record-keeping (Borglund & Öberg, 2006).

Ongoing research

There is a lot of ongoing research aiming to solve the technical problems associated with long-term preservation of digital information, although the projects very seldom are directed especially towards this area. Examples of research areas are mass storage media, format conversation, migration of technical environment (hardware and software), emulation of technical environment, digital signatures and other security issues. There is also research going on to develop models for long term preservation. As an example the Record Continuum Model (Upward, 2005) can be mentioned, which the Australian School on record-keeping is based upon.

The project "Archives of the Future" at Mid Sweden University has concentrated on the organizational and metadata issues in long-term preservation of digital information.

A problem with many organizations and IT architectures is that they are built up in a stove-piped rather than in an integrated way. In Sundberg and Wallin (2005a, 2005b) a model for integration of IT strategy, information resource strategy and costumer/e-services strategy is presented.

Governments, both local and national, want to offer electronic services to the citizens. As an example can be mentioned the application of building permits. In Sweden an application can include information and partial permits from several departments within the local government but also from other governmental organizations. To facilitate for the citizen the e-service will include the whole chain as one single e-service disregarded the organizational boundaries. Approaches for handling inter-organizational problems are described in Asproth (2006a)

By linking documents to the processes where it has been created and changed, an organization will have the ability to follow the information flows afterwards. Capturing of connections between information and the processes will lead to trust worthier information, because of the fact that it serves as evidence of an organization's actions. Studies and suggestions of how to capture the components that makes it possible to follow the processes are described by Öberg (2005, 2006).

Trust worthy information is a key criteria for information that have to be preserved for long time. A predictive model for attaining quality in record-keeping is suggested by Borglund (2006).

Concluding remarks

To be able to preserve information for a long time and to guarantee access to it the technical issues have to be solved or managed. Furthermore the information must be kept understandable over time, which requires a lot effort on context and meta-data. Supposedly new context information must be added over time. Methods for analysis of needs for new context information must be elaborated. The information must be kept independent, not according to to-days organization as the organizations changes or even disappear over time. A new model for organization independency must be developed. There is also a question about what to preserve as it is not possible to preserve everything. What is a representative selection? And finally, the information must be able to access now and in the future. How can the needs be foreseen or how can a completely independent access be accomplished.

References

Asproth. V. (2005) Information Technology Challenges for the Long-term Preservation of Electronic Information. *International Journal for Public Information Systems*, no 1, pp 27-37.

Asproth, V. (2006a) Inter-organizational management and decision-making. Proceedings of the UKSS 10th International Conference, 11th - 13th September, 2006, Oxford, UK.

Asproth, V. (2006b) Long-term preservation of electronic information – a study of seven Swedish governmental organizations, Proceedings of the 15th International Conference on Information Systems Development, 31 August - 2 September 2006, Budapest, Hungary

Beagrie, N. (2002) A Continuing Access and Digital Preservation Strategy for the Joint Information Systems Committee (JISC) 2002-2005. [On-line] Retrieved 2005-11-22 from: http://www.jisc.ac.uk/index.cfm?name=strategy_preservation

Bearman, D. (1994) Electronic evidence: strategies for managing records in contemporary organizations. Pittsburg: Archives and Museum Informatics

Borglund, E. (2005) Operational use of electronic records in police work. *Information Research*, 10(4), paper 236 [Available at http://InformationR.net/ir/10-4/paper236.html]

Borglund, E. (2006). *A Predictive Model for Attaining Quality in Recordkeeping.* Mid Sweden University, Sundsvall.

Borglund, E., Öberg, L-M. (2006) What are the Characteristics of Records? *International Journal for Public Information Systems*, no 1, pp 55-76.

Duranti, L: (2000) The InterPARES Project. The complete proceedings from the February 2000 InterPARES Symposium. [On-line] Accessed 2006-06-19 from http://www.interpares.org/display_file.cfm?doc=ip1_symposium_2000.pdf

Duranti, L. (2001) The impact of Technological Change on Archival Theory. *Archival Science, no 1, pp* 39-55.

Gibbs, R., Heazlewood, J. (1999) Electronic Records: Problem solved? A report on the Public office Victoria's Electronic Records Strategy. *The Journal of the Australian Society of Archivists, Vol. 27 , No 1,* pp 1-13

Goldstein, H. (2004) The Infinite Archive, *IEEE Spectrum*, Vol. 41, p 74.

Ketelaar, E. (1999) Archivalisation and archiving, in: *Archives and manuscripts* 27, pp 54-61.

Marcum, D. B. (1997) A Moral and Legal Obligation: Preservation in the Digital Age. *International Information and Librarian Review* 29, pp 357-365

Öberg, L-M. (2005). To Capture and Preserve Documents in Processes. In K. S. Soliman (Ed.), *Information management in modern enterprise: Issues & solutions. Proceedings of the 2005 international business information management conference July 5 - 7, 2005 Lisbon, Portugal.*

Öberg, L-M. (2006). *Needed components to reconstruct processes.* Paper presented at the Archiving 2006, May 23-26, Ottawa, Canada.

Rothenberg, J. (2000) Preserving Authentic Digital Information. *Authenticity in a Digital Environment,* pp 51-68.

Sundberg, H., Wallin, P. (2005a) Co-ordination of Business and IT Development Processes - Managing Stovepiped Organisations. *International Journal of Public Information Systems, 2005*(1), 53-69.

Sundberg H., Wallin, P (2005b). Icebergs drifting apart? –Customer strategies affecting systems development and information management. Proceedings of e-Society Conference, 27-29 July, Valetta, Malta

Upward, F. (2005) The records continuum. In S. McKemmish, M. Piggot, R. Barbara & F. Upward (Eds) *Archives: Recordkeeping in society* (pp 197-222). Wagga Wagga: Charles Stuart University, Centre for Information Studies

Wilson, I. (2000) The test of time. *Authentic records in the electronic age,* pp 24-34

Viveca Asproth received the doctor (PhD) and the reader (Docent) degrees from the University of Stockholm. Currently she is an Associate professor in Computer and Systems Science at Mid Sweden University in Östersund, Sweden. She has published papers on visualization, spatial systems, decision support, anticipation and fuzzy systems. She is member of the board of Graduate Education and deputy head of the department of Information Technology and Media at Mid Sweden University. In her current research she is focusing on Spatial Decision Support Systems: modelling, simulation and visualization but also in long-term preservation of digital information.

Implementation of a Multimethodology Rural Telecommunications Planning Framework: Reflections on the Case Study and Some Lessons Learned

Theo Andrew & Don Petkov
Faculty of Engineering and the Built Environment, University of Johannesburg, ZAF
Dept of Business Administration, Eastern Connecticut State University, USA

A systemic framework for the planning of rural telecommunications infrastructure in developing countries was developed by the authors in 2002. The framework as a whole was founded on the theory of critical systems thinking, and consists of a mix of soft systems approaches using Interactive Planning as the dominant methodology. A case study based on a rural area in South Africa was used for the practical validation of the framework. The paper discusses the practical implementation of our framework and some lessons in the light of Midgley's critical reflection model.

Introduction

A systemic framework for the planning of rural telecommunications infrastructure (RTI) in developing countries was developed by the authors in 2002. The framework as a whole was founded on the theory of critical systems thinking, and consists of soft systems approaches using Interactive Planning as the dominant methodology. At various stages of the intervention Interactive Planning was enhanced with rich pictures, stakeholder analysis, Interpretive Structural Modelling and Critical Systems Heuristics. Techniques from Systems Engineering practice were also included. To the best knowledge of the authors this is the first approach for planning of rural telecommunications incorporating a suitable combination of soft and hard systems engineering techniques, taking into account the interests of all stakeholders. The full theoretical and philosophical justification of the framework and its components, and the entire development and implementation of the framework cannot be discussed in this paper due to limited space. Further details on the techniques involved can be found in Ackoff (1981, 1993); Ulrich (1996); Warfield (1994); Mingers (1997); and Checkland & Scholes (1990), while the justification of our framework is presented in Andrew & Petkov (2002; 2003).

The planning framework is essentially a soft systems framework and learning is fundamental if the framework is to be regarded as useful. In dealing with methodological pluralism in a systemic intervention Midgley(2000) has produced a model for critical reflection when learning about methodology and methods. The model pays attention to three challenges to methodological pluralism: paradigm incommensurability, the cultural problem, and the psychological problem. It consists of several layers, one of which is "reflections on practice" that emphasizes the importance of feedback from practice to methodology. The contribution of this paper is in the discussion of the practical implementation of our framework in the light of Midgley's critical reflection model and some of the lessons learnt.

The paper continues with an overview of the planning framework, highlighting certain of its features. An overview of the implementation and validation of the framework follows, setting out the various interventions in the form of workshops with stakeholders, during which the various methods and techniques were applied. The paper continues with reflections on these interventions. Due to the limitations of space this paper provides a discussion on the overall learning that was achieved during the implementation of the framework, and related issues.

The salient features of the rural telecommunications planning framework

Andrew & Petkov (2002, 2003, 2005) have shown that planning for rural telecommunications is really planning for a complex dynamic socio-technical system. This system, termed the rural telecommunications system (RTS), is not just a technological system. It is a complex system of people and technology interdependent on other systems/subsystems such as the economic, political and regulatory, physical environment, and socio-cultural.

The analysis of the RTI and its interaction with the environment led to the need to apply a systems thinking approach to the planning of RTI. Figure 1 presents the process of the proposed planning framework. The framework was developed along the lines of critical systems thinking (see Flood and Jackson, 1991; Jackson, 2000; Midgley, 2000). A recent strand of it, Multimethodology, Mingers (1997), was chosen as a foundation for the framework because of the flexibility that it allows in mixing methods from different paradigms in one intervention. Multimethodology draws partially its philosophical justification on Critical Social Theory (see Habermas, 1984) and hence corresponds to the idea that RTI should promote development and empowerment of the rural communities.

Interactive Planning (IP) (see Ackoff, 1981; 1993) is used as the dominating systems approach guiding the framework. However, IP on its own is insufficient to address the complexity of the problem situation but it allows for the incorporation of other relevant methods within its process. Therefore IP was enhanced at its various stages with other appropriate methods and techniques as shown in Figure 1. The techniques included elements from Soft Systems Methodology (see Checkland & Scholes, 1990), Interactive Management (see Warfield, 1994) and from other methodologies.

An overview on the implementation of the framework

The preparation for the implementation of the framework took more than 18 months of continuous interaction with the relevant stakeholders. A number of meetings with individuals took place as well as several more formal events happened. Two preliminary seminars were given by the authors to Telkom SA (the sole fixed-line network operator in South Africa) planners on

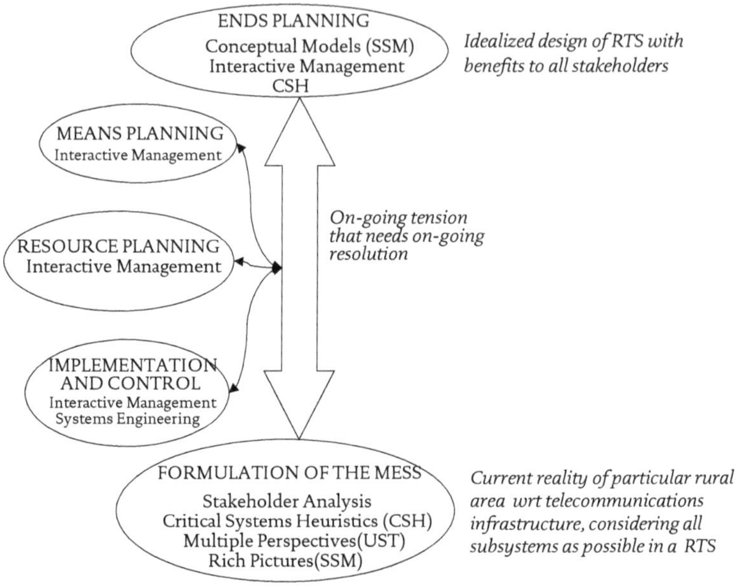

Figure 1 *A systemic planning framework for rural telecommunications infrastructure showing the mix of various methods/techniques for the different phases of the process (Andrew & Petkov, 2005)*

various aspects of the research as it unfolded. It was decided to have four workshops for the implementation of the framework, with the understanding that more would be scheduled if it were necessary for the validation of the framework. The implementation was done in Mapumulo, a typical rural area in KwaZulu Natal (KZN), South Africa. It involved staff from the planning sections at Telkom SA and community representatives from the Mapumulo rural area concerned. The Telkom participants came from different regions within KZN and represented different sections within the planning department. The first author facilitated all the workshops.

The initial workshop focussed on the preparation for the implementation of the formulation of the mess phase. Apart from the primary reason of implementing the framework, the intention was also to use this workshop to gauge the commitment of the participants to non-traditional processes such as those involved in stakeholder analysis and rich pictures, so that one could address any shortcomings before the next workshop. The second workshop was dedicated to the ends planning. In terms of the requirements of the framework the community based stakeholders should have been present at this workshop. This did not materialise due to Telkom's reservation at this stage to interact directly with other stakeholders before assessing the framework.

Keeping in line with the principles of systemic intervention, a third workshop was subsequently arranged with stakeholders from the Mapumulo area and a few representatives from Telkom. The purpose of this workshop was to compensate for the lack of stakeholder representivity in the second workshop. It focussed on issues covered in the latter and produced the stakeholders' response to the inputs made by Telkom.

The fourth workshop dealt with the leverage phase of the framework and this was carried out with the Telkom planners only as this is the phase that focuses on action in the real world with respect to the final plan for design and deployment of the infrastructure. The workshops provided a lot of material for reflections which is discussed below.

Reflections and some lessons learnt

The output itself of this research could be regarded as a systemic intervention into the problem situation of providing telecommunications infrastructure to rural communities. There are two fundamental but related aspects to this intervention: the nature of the infrastructure planners and the planning situation, and the development of the rural community itself as a spin off from the RTI. Attention will be focussed on these two aspects.

Midgley (2000) points out that learning is a process over time and increases with practice. His reflection model consists of several layers, one of which is "reflections on practice" that emphasizes the importance of feedback from practice to methodology. The authors shall now attempt to reflect on the way the various components (methods and techniques) of the framework worked together (or did not) in the case study. This will serve two related purposes: to provide a basis for validation of the proposed approach for planning of RTI as a soft learning framework and to outline possibilities for the further refinement of the framework in future implementations.

On paradigm incommensurability and the mixing of methods in our framework

A solution to the first issue, that of *paradigm incommensurability*, is discussed in depth by us in Petkov, *et al* (2006) and will not be treated here in detail. In summary, we propose an integrated treatment of the paradigm incommensurability problem. It combines the views of Deetz (1996), Midgley (1997), Jackson (2000) and Hassard (1990) in their separate arguments for certain forms of paradigm mediation. Our complementary explanation of the issue of paradigm incommensurability resolves what is defined as the main problem in a multimethodology intervention: the legitimacy of transferring a technique developed within one paradigm to another.

The complexity of rural telecommunications infrastructure planning demonstrated the need to apply a mix of methods following critical systems thinking principles and along the lines of its recently proposed strand of multimethodology (see Mingers and Gill, 1997). The implementation justified the need to provide a planning framework promoting rural development based on critical systems thinking because of its emancipatory nature. The mixing of methods was based on another principle of critical systems thinking, related to acknowledging awareness of methodological pluralism. The methods chosen here were each contributing according to their strengths to the framework for planning of RTI suggested here. The facilitator felt that the participants in the workshops considered as natural the links between the various stages of the planning framework no matter that they employed techniques from different paradigms.

Addressing the psychological problem in working across paradigms

In attempting to address the *psychological problem*, Midgley (2000) emphasizes the importance of learning over time, starting from whatever knowledge the intervener/s have at the start of the intervention. The intervention demonstrated how the softer issues raised during the formulation of the mess stage were interpreted and allowed to inform the harder issues in planning, during the leverage phase.

The chosen dominant systems methodology of IP proved to be an appropriate systemic guide to the planning process. The methodology of IP kept the processes intact and connected. It paved the way for moving from an initial situation of uncertainty, characterized by a system of problems to a stage where the complexity was managed to the extent of facilitating appropriate action for the purposes of appropriate RTI deployment. One of the practical advantages was that the workshops could be arranged around the different phases of IP. This is an advantage because the formulation of the mess workshop for example may be regarded as an intervention in itself, and the learning that took place among the planners during this stage influenced their activities in the rest of the planning phases.

The stakeholder analysis further enabled the planners to delve deeper into the RTS. Although the original intention of the stakeholder analysis was to identify who should be involved in the rest of the planning, it served as a means of boundary critique in terms of the extent of the systems of concern and the complexity of the problem. One must assess this in the light of the fact that most of the planners came from a technology planning background.

The authors wonder whether the categorisation of stakeholders as "standard, fiduciary and future" served the same purposes as the boundary questions. This needs to be investigated further as the framework is refined in practice. If this is the case then the inclination would be to retain the boundary questions rather, due to their emancipatory connotations. The rich pictures generated did include many perspectives on the problem situation. This could also be attributed to the comprehensive stakeholder analysis. The general thrust of the next phase of the workshop, ends planning, followed the idealized design as recommended by Ackoff (1981, 1993). It included the constrained and unconstrained idealized design.

ISM/IM enhanced the methodology of IP with respect to the leverage phase. It facilitated the transition from the complex messy issues in the previous phases to intervention in the real world with respect to the provision of RTI. Whereas in the previous stages the systemicity was at a conceptual level (refer also to SSM) ISM/IM helped to transfer this systemicity into the real world i.e. the Mapumulo area. An interesting observation during this phase of planning was that the ends planning and means planning were inextricably linked in certain cases. For example, the desire to address the health issues immediately led to the notion of a telemedicine system. This suggests that the boundary between the ends planning and the means planning is actually arbitrary.

On the cultural difficulties when applying multimethodology: Some lessons from this project

Little is said in the same source about the cultural problem but an important point raised is that cultural acceptance is related to whether or not methodological pluralism is perceived as adding value to the intervention. With respect to the psychological (and perhaps cultural) barriers to multi-paradigm multimethodology the consultations and seminars held with the RTI planners over a period of 18 months served as part of the education/learning process that contributed to their acceptance of methodological pluralism. In addition, the responses to evaluation questionnaires demonstrate further the tendency of the RTI planners towards methodological pluralism.

The authors suggest that one of the enabling factors that led to results at the initial phase of planning was the defocus on the technology issues. In particular, the concept of the RTS as a complex sociotechnical system which was presented to the planners for consideration, helped to focus their minds on the other subsystems of the RTS.

The outputs of the ISM/IM session paved the way for the design and implementation of the telecommunications technology using traditional methods such as Systems Engineering and in-house processes. In addition, one was now able to assess the appropriateness of these in-house processes in the light of the structures generated by ISM/IM. More specifically, the particular structure that was generated (problematique) informs the activities and processes that should follow for the design and implementation of the various components of the network.

One of the reasons for choosing a critical systems thinking approach to the framework was the emancipatory aim (which is linked to development of the community) of the planning exercise and the RTI. How did the fact that not all stakeholders were present in all the workshops impact on the learning and the envisaged aims of the planning framework? This was a practical constraint imposed by the realities of the previously existing planning practices at Telkom and by the impossibility to involve physically representatives of the local community in every workshop due to their availability. It can be stated that this aim was met through the consultation with the community in the third workshop and the report back to the Telkom planners.

It is difficult to envisage what would have happened, with respect to learning, if the community stakeholders participated with the planners in all the workshops. One can only speculate. However, significant learning did take place throughout the workshops and the authors argue that the conducted workshops did have some emancipatory benefits both for the planners and the community. The results of the formulation of the mess, and the results of the ISM/IM session indicate the recognition by the planners of the need to plan RTI for the development of a socio-technical system (and the implications of such a system).

The community was given an opportunity to contribute to the formulation of the mess and the ends planning and, more importantly, these were taken into consideration by the Telkom planners (in the fourth workshop). This knowledge, together with the satisfaction of being part of the planning process, could also lead to more ownership of the infrastructure, and therefore resolve some of the security issues with respect to telecommunications equipment.

Conclusion

The paper provided a reflection on the implementation of a systemic framework for planning of rural telecommunications infrastructure in light of Midgley's (2000) model of critical reflection. Paradigm incommensurability was addressed theoretically by adopting a complementary approach that integrates the existing previous treatments of that problem (see Petkov, *et al.*, 2006). The practical difficulties of working with methods across different paradigms were addressed through careful preparation of the facilitator and education of the stakeholders participating in the intervention on an ongoing basis and also the attention to detail in the transition from one method to the next in the framework shown in Figure 1. The actual experience of the authors with the intervention further justified their initial choice of the mix of different methods in this complex engineering and societal problem.

Close contact and interaction of the facilitator with the stakeholders were crucial for overcoming the cultural and psychological problems associated with pluralist systemic interventions, mentioned by Midgley (2000). The implementation of the framework was a learning process both for the infrastructure planners and also for the representatives of the local community and other government departments. Our work showed that it could be possible to simplify a little the planning framework at some stages by avoiding the use of methods that may lead to similar insights.

Rural telecommunications infrastructure planning is a large, complex societal problem and hence we cannot claim that our framework is the ultimate answer to all issues related to. We hope that it provides a viable path for further work towards its improvement. Our overall conclusion from the reflection over the implementation of the systemic planning framework is that it served not only the narrow technical goal of planning rural telecommunications infrastructure but went beyond and facilitated the development of a collaborative understanding between the telecommunications planners and the local community, while creating a common ground for better interpretation of the role of telecommunications in rural development.

References

Ackoff, R. L. (1981). Creating the Corporate Future, J. Wiley, New York-Chichester, ISBN 0471090093

Ackoff R. L (1993). "Idealized Design: Creative Corporate Visioning", OMEGA, ISSN 0305-0483, (21)4: 401-10.

Andrew, T. and Petkov, D. (2002). Critical Systems Thinking Applied To Rural Telecommunications Infrastructure Planning, *ISSS Conference Proceedings*, China, ISBN 0966418387.

Andrew, T. and Petkov, D (2003). "The Need for a Systems Thinking Approach to Planning of Rural Telecommunications Infrastructure", Telecommunications Policy, ISSN 0308-5961, 27(1-2): 75 - 93.

Andrew, T. N. and Petkov, D. (2005). "A case study on the initial enquiry stage in a framework for improved planning of rural telecommunications infrastructure in developing countries, Int. J. Technology Management, ISSN: 0267-5730, 31(1/2):64-77.

Banville, C. Landry, J-M Martel and C. Boulaire, (1998). "A Stakeholder Approach to MCDA", Systems Research and Behavioral Science, ISSN 1092-7026, 15(1): 15-32.

Checkland, P. B. (1993). Systems Thinking, Systems Practice, J. Wiley, Chichester, ISBN 0471279110.

Checkland, P. & Scholes, J. (1990). Soft Systems Methodology in Action, J. Wiley, Chichester, ISBN 0471927686.

Deetz, S. (1996). Describing Differences in Approaches to Organizational Science: Rethinking Burrell and Morgan and their Legacy, *Organization Science*, ISSN 10477039, **7**(2), 190-207.

Flood, R. L. and Jackson, M. C. (1991). Creative Problem Solving, J. Wiley, Chichester, ISBN 0471-93052-0.

Habermas, J. (1984). The Theory of Communicative Action Vol.1: Reason and Rationalization of Society, Heinemann, London, ISBN 0807015075.

Hassard, J. (1990), An Alternative to Paradigm Incommensurability in Organization Theory, in Hassard, J. and Pym, D. (Eds). (1990), *The Theory and Philosophy of Organizations: Critical Issues and New Perspectives*, Routledge, London, ISBN 0415063132.

Jackson, M. C. (2000). Systems Approaches to Management, Kluwer Academic / Plenum, New York, ISBN 0-306-46500-0.

Midgley, G. (1997). Mixing Methods: Developing Systemic Intervention, in Mingers J. and A. Gill (Eds) (1997), *Multimethodology: the Theory and Practice of Integrating OR and Systems Methodologies*, Wiley, Chichester, 249-290, ISBN 0471974900.

Midgley, G. (2000). Systemic Intervention: Philosophy, Methodology, and Practice, Kluwer Academic / Plenum Publishers, New York, ISBN 0-306-46488-8.

Mingers, J. (1997). Multi-paradigm Multimethodology, in Mingers J. and A. Gill Eds) (1997), *Multimethodology: the Theory and Practice of Integrating OR and Systems Methodologies*, Wiley, Chichester, 1-22, ISBN 0471974900.

Mingers, J. and Brocklesby, J. (1996). Multimethodology: Towards a Framework for Critical Pluralism, Systemist, **18**(3), 101-131, ISSN 0961-8309.

Mingers, J. and Gill, A. (1997). (Eds.), Multimethodology: the Theory and Practice of Integrating OR and Systems Methodologies, Wiley, Chichester, ISBN 0471974900.

Petkov, D., Petkova, O., Nepal, T. and Andrew, T. (2006). "Paradigm Issues in Critical Systems Thinking and Their Interpretation in Three Developmental Systemic Interventions", in proceedings of 50th annual conference of ISSS, Complexity, Democracy and Sustainability, J. Wilby, J. K. Allen and C. Lourriro-Koechlin (eds.), ISBN 0-9740735-7-1.

Ulrich, W. (1996) *A Primer to Critical Systems Heuristics for Action Researchers*, Centre for Systems Studies, University of Hull, South Humberside, United Kingdom.

Warfield, J. N. (1994). A Science of Generic Design: Managing Complexity through Systems Design. Iowa State University Press, Iowa, ISBN 0-8138-2247-5.

Theo Andrew is the Executive Dean of the Faculty of Engineering and the Built Environment at the University of Johannesburg. Theo is a Fellow of the South African Institute of Electrical Engineers (SAIEE). He is also a member of the

International Society for the Systems Sciences. Professor Andrew holds a PhD in the area of rural telecommunications planning from the University of Natal. He serves as a reviewer for the National Research Foundation (NRF). He was the leader of the NRF funded Research Unit in *ICT and Development*, and of a *Centre of Excellence in Rural Telecommunications*. His research interests include rural telecommunications, systems thinking, developmental engineering, and engineering education.

Don Petkov is an associate professor and Coordinator for BIS at ECSU, USA. Previously he was a professor at the University of Natal. He is a member of IEEE CS, AIS, Intl Society on MCDM and other organizations. He serves on the editorial board of Systems Research and Behavioural Science and also of Scientific Inquiry and is the co-editor of IJCSS. He is the chair of the systems track for the 2007 ISOneWorld conference. His publications and research are in IS management, software engineering and systems thinking applied to complex technical and societal problems.

Complex Infrastructure Projects: A Systemic Model for Management

Kaye Remington & Julien Pollack
University of Technology Sydney, AUS

Large infrastructure projects exhibit characteristics that might be able to be explained in terms of complex adaptive systems. Size, technical challenges, unclear or unshared goals and changes to context over time are the major sources of complexity. These sources of project complexity have important implications for management of large infrastructure projects. Drawing upon findings from the research literature in combination with the authors' personal industry experience and evidence from other practitioners, this paper uses ideas from complex systems thinking to inform the applied discipline of project management in order to explore the challenges offered by such projects. The paper advances the idea that such projects are more effectively managed systemically, as interrelated groups of projects or programs, with discreet management strategies chosen to address the specific sources of complexity. Synthesized in the form of a model, the aim is to stimulate discussion about strategies for managing complex infrastructure projects.

Introduction

The complexity inherent in large-scale infrastructure projects means that conventional project management approaches are often ineffective. The paper begins by introducing the idea that complex projects exhibit characteristics that can be readily understood through the terminology of complex adaptive systems. A brief discussion follows outlining how understanding the nature of project complexity can positively influence decisions about how the projects can be managed. Finally we present a model for thinking systemically about complex infrastructure projects.

Too many large infrastructure projects are failing (Williams, 1999, 2002; Flyvberg, 2003). Evidence is mounting that complex infrastructure projects exhibit qualitatively different characteristics to projects from traditional project-based disciplines, such as construction and engineering projects, on which much project management thinking has been formulated (Koskela & Howell, 2002; Costello, *et al.*, 2002). Processes for managing projects have been based on control theories (Yeo, 1993) and 'hard' systems thinking (Winter & Checkland, 2003). These methodologies assume that outcomes can be clearly defined, agreed and adhered to, that technological issues can be solved early in the life cycle of the project and that the project can be managed within a relatively stable context (Hobbs & Miller, 2002; Remington & Crawford, 2004). With large infrastructure projects conditions such as these are often the exceptions (Williams, 2002; Flyvberg, 2003).

Complex projects and complex adaptive systems

The literature on complex adaptive systems and systems thinking can help describe many of the characteristics of complex infrastructure projects which make their management problematic. Complex projects often exhibit a rapid change from one state to another. They can continue apparently according to plan for a very long period and then suddenly reach a point where change seems to be occurring in all parts of the project at once. This can be thought of as a phase change. Like complex adaptive systems complex projects exhibit different degrees of non-linearity (Beeson & Davis, 2000). This can be caused by positive feedback loops occurring when the risks are triggered that mutually reinforce other risk outcomes. In complex projects we can also observe the propensity for self-organization or adaptiveness that may be out of the direct control of those managing the project. During such projects, new and unexpected pairings and groupings of people might occur, unanticipated cultural territories and political coalitions can become apparent presenting challenges for management that are extremely difficult to predict. Sensitive dependence on initial conditions is also a feature shared by complex adaptive systems and complex projects. A small change in initial conditions can have a profound effect on system outputs (Mumby, 2005). Traditional risk management approaches, based on linear thinking, are not very effective as predictors of emergent characteristics, particularly as complexity increases during implementation.

We argue that people responsible for planning and implementing complex infrastructure projects should be able select from a multitude of methodologies or paradigms in order to manage the projects successfully. This can be referred to as 'systemic pluralism' – *systemic* in that we need to recognise that these projects have many characteristics in common with complex adaptive systems – *pluralism* in that we cannot restrict ourselves to one paradigm when thinking about appropriate management strategies (Munro & Mingers, 2002; Mingers & Brocklesby, 1997; Midgley, 1997). 'Hard' systems thinking, 'soft' systems thinking and thinking in terms of 'complex adaptive systems' are all appropriate depending upon the aspect of the project under consideration (Jackson & Keys, 1984; Flood & Romm, 1997; Flood & Jackson, 1991). For example, traditional project management methodologies provide very efficient ways of managing those parts of a complex project that can be clearly defined and for which firm contracts can be negotiated. However when characteristics of complexity are present the manager needs to be able to move freely between paradigms and build methodologies based on his/her interpretation of the situation (Midgley, 2000; Gregory, 1996).

Managing project complexity through a program

Project complexity is not a single, unvarying attribute. The nature of complexity can vary in different parts within the project. It is helpful therefore to think of any large project as a group of related smaller projects, commonly referred to as a program (Thiry, 2006). Turner and Cochrane (1993) classified projects into four types based on their degree of clarity or certainty about goals and methods. Following this, one of the first to explicitly recognise that projects exhibited different kinds of complexity was Williams (2002) who identified two kinds of complexity found in projects. He described projects with many interdependent elements as having 'structural' complexity and projects having unclear goals and methods as being characterized by 'uncertainty'. Remington, *et al.* (2006) have further developed Williams's categories, identifying four distinct categories of complexity encountered in projects. Any project might exhibit any one or all of these aspects of complexity. The larger the project the more likely it is to exhibit many aspects of complexity. We argue that it is helpful to be able to recognise the nature of the complexity being encountered when designing the management structure for these projects.

Infrastructure developments are typically too large to be managed as a single project or initiative. In order to manage systemically, which implies taking both an overview and detailed views of the individual initiatives, large projects are best conceptualized as integrated group of projects, or programs (Thiry, 2004). This enables the program to be designed so that each project within the program is small enough to be the responsibility of a single manager or management team. Within a program, the projects are typically interconnected. Their impacts on each other, in terms of risks and resource use, can be tracked at the program level and appropriate action can be taken to manage risks and distribute resources effectively. Most importantly, changes in the environment can be tracked, the impact on any project assessed and action taken to continue, change or eliminate that particular project from the program. This model for management requires a manager or management team responsible for managing the program. It also requires that the program management team be in direct contact with those responsible for organizational strategy. The model is summarized in Figure 1.

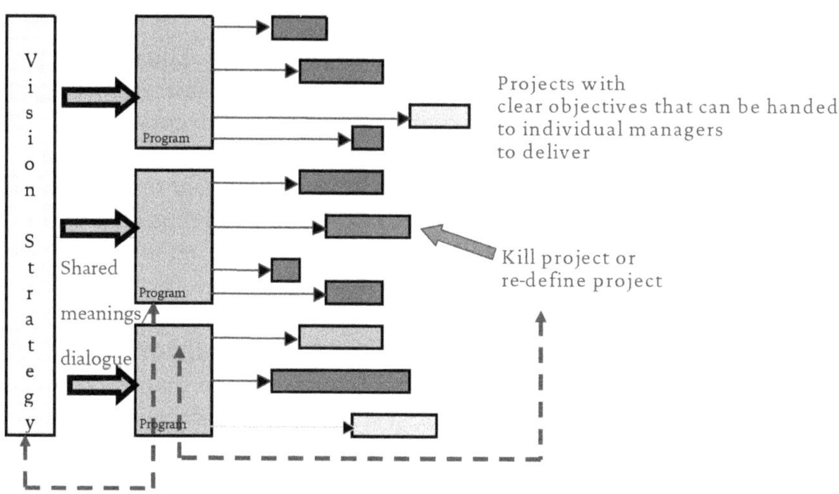

Figure 1 *Translation of the strategic vision through programs of projects that can be monitored and redefined as necessary in response to changes in strategic direction.*

Decisions about how best to implement the projects within the program relate specifically to the nature of project complexity expected. For instance there might be some projects within the program that exhibit very little complexity. On a spectrum from control to chaos they exist at the control end. They have relatively few parts, objectives can be clearly defined and agreed, there are few technological surprises and they will be unlikely to be affected by political or environmental change during implementation. These projects are most efficiently managed using traditional project management methods based on 'hard systems' thinking. Other projects within the same program may exhibit varying aspects of complexity. Understanding the nature of the complexity that characterizes each project has important implications for a number of key decisions to be made about the projects, such as the project governance structure, role capabilities and the selection of key personnel, approaches to risk management, finance and procurement.

Governing complex programs and projects may offer challenges that are quite different from those found in functional organizations. The organizational structure for the program and each project within defines how information is exchanged formally and authority is allocated. Recent research (Helm & Remington, 2005) strongly indicates that the capabilities needed for role of the executive sponsor in complex projects may not be sufficiently understood. Evidence is also mounting about the importance of the role of the executive sponsor and the governing board which are now considered to be fundamental to project success in any large complex project (Crawford, *et al.*, 2006).

Appropriate selection of key personnel is critical for complex projects. Lack of appropriate resources and divided loyalties of project personnel are well recognized sources of project failure (Dinsmore, 1993). Project teams are often expected to work much longer hours than required by functional roles. Many personnel suffer from stress-related symptoms as a result (Birch and Paul, 2003). Finding appropriate personnel who have the ability to work with complex issues while under pressure is not easy. We often assume that a project manager's role is generic. However a person who is able to cope with political uncertainty and ambiguity, may not also be able to be sufficiently hard-nosed to manage multiple contracts and ensure strict time deadlines, or motivate designers and technical experts to deliver within a timeframe. Being able to identify the source and nature of the complexity may assist the program director to fit the nature of the complexity with people's capabilities.

A model for thinking about complex infrastructure projects

In the model below (Figure 2) we offer an approach to managing complex infrastructure projects that demands a systemic perspective and accommodates a pluralistic approach to methodologies selected for implementation. It must be stressed that although certain initial decisions, made when developing the first iteration of the program structure, may look like a top-down approach to planning, the model will not work unless it is iterative (Thiry, 2006). In response to information coming from the project teams, regular re-evaluation and revision during the entire life cycle of the initiative should cause the program to expand and contract organically during the life cycle of the initiative.

The model suggests that the senior management team first identifies the nature of project complexity expected and devise a program of separate projects that reflects the various aspects of that complexity. This will have implications for the structure and capabilities of the governing board and for selection of key people to direct the program and manage each project within the program. Once the projects within the program have been defined, approaches to managing risk, finance and procurement can be determined for each project. Key to decision-making also is knowledge of the cultures and constraints inherent in the organizations involved together with a detailed understanding of the roles and capabilities of the key people associated with the project.

Conclusion

Many projects exhibit characteristics that can be explained in terms of complex adaptive systems. This is a result of their size, technical challenges, unclear or unshared goals and unpredictable changes to the political and social context over time. These aspects of project complexity have implications for management. The model presented in this paper is intended to stimulate thinking about management strategies for complex infrastructure projects. We argue for the development of management strategies that specifically target the nature of the complexity, which may vary substantially,

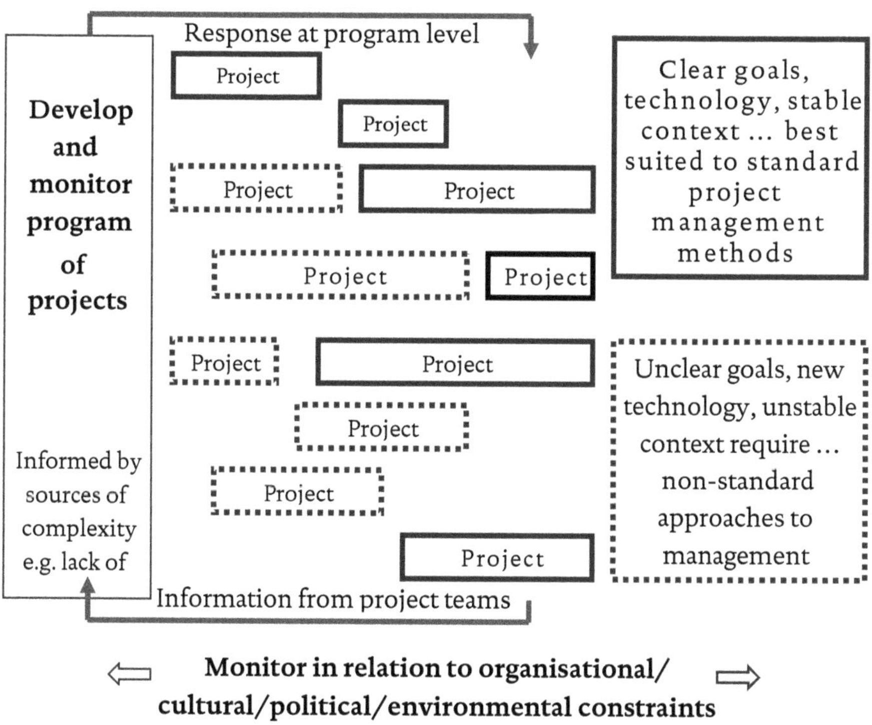

Figure 2 *A systemic model for thinking about complex infrastructure projects*

in different parts of a large infrastructure project. We therefore argue that any large infrastructure project be organized within an integrated framework of a program to allow for management approaches informed systems thinking. By adopting a pluralistic perspective management strategies may targetted specifically for each project within the program permitting more effective monitoring and strategic responses to contextual changes.

References

Beeson, I. and Davis, C. (2000) "Emergence and accomplishment in organizational change," *Journal of Organizational Change Management.* **13**(2), pp. 178-190, ISSN 0953-4814.

Birch, C. and Paul, D. (2003). *Life and work: Challenging economic man.* Sydney, Australia: University of New South Wales Press, ISBN 0-89840-670-8.

Crawford, L., Cooke-Davies, T., Labuschagne, L., Hobbs, B. and Remington, K. (2006) "Exploring the Role of the Project Sponsor," *PMI Research Conference, 18 July, Montreal,* ISSN 0961-3218.

Costello, K., Crawford, L., Bentley, L., Pollack, J. (2002) "Connecting Soft Systems Thinking with Project Management Practice: An Organizational Change Case Study," in Ragsdell et al. (Eds.) *Systems Theory and Practice in the Knowledge Age.* New York: Kluwer Academic/Plunum Publishers, ISBN 0-306-4724-3.

Dinsmore, P. C., (1993) *Human Factors in Project Management.* New York, Amacon, ISBN 0-309-04848-6.

Flood R., Jackson M. (1991) *Creative Problem Solving: total systems intervention.* New York: John Wiley and Sons, ISBN 0-471-93052-0.

Flood, R. and Romm, N. (1997) "From MetaTheory to 'Multimethodology'," in Mingers J, Gill, A. (Eds.) *Multimethodology: The Theory and Practice of Combining Management Science Methodologies.* Chichester: John Wiley & Sons, pp 291-322, ISBN 0-471-9760-0.

Gregory, W. (1996) "Discordant Pluralism: A New Strategy for Critical Systems Thinking," *Systems Practice,* **9**(6), pp 605-625, ISSN 0894-9859.

Hall, M, Holt, R, Purchase D. (2003) "Project sponsors under New Public Management: lessons from the frontline," *International Journal of Project Management,* **21** pp495-502, ISSN 0263-7863.

Helm, J. and Remington, K. (2005) "Effective Sponsorship, Project managers' perceptions of the role of the project sponsor," *Project Management Journal,* **36**(3), pp 51-62, ISSN 8756-078X.

Hobbs, B. and Miller, R. (2002) "The Strategic Front End of Large Infrastructure Projects: A Process of Nesting Governance," *Proceedings of the PMI Research Conference,* Seattle, USA, pp 14-17, ISSN 0961-3218.

Jackson, M. and Keys, P. (1984) "Towards a System of Systems Methodologies," *Journal of the Operational Research Society,* **35** pp 473-486, ISSN 0160-5682.

Koskela, L. and Howell. G., (2002) "The Underlying Theory of Project Management is Obsolete," *Proceedings of PMI Research Conference,* pp293-301, ISSN 0961-3218.

Midgley G. (1997) "Mixing Methods: Developing Systemic Intervention," in Mingers, J., Gill, A, (Eds.) *Multimethodology: The Theory and Practice of Combining Management Science Methodologies.* Chichester: John Wiley and Sons, pp 249-290, ISBN 0-471-9740-0.

Midgley G. (2000) *Systemic Intervention: Philosophy, Methodology, and Practice.* New York: Plenum Publishers, ISBN 0-306-46488-8.

Mingers, J. and Brocklesby, J., (1997) "Multimethodology: Towards a Framework for

Mixing Methodologies," *Omega, International Journal of Management Science,* **25**, pp 489-509, ISSN 0894-069X.
Mumby, D. (2005) "Theorizing resistance in organization studies: a dialectical approach," *Management Communication Quarterly.* **19**(1), pp 19-44, ISSN 0893-3189.
Munro, I. and Mingers, J. (2002) "The use of multimethodology in practice - results of a survey of practitioners," *Journal of the Operational Research Society,* **54**, pp 369-378, ISSN 0160-5682.
Remington K. and Crawford L., (2004) "Illusions of Control: Philosophical foundations for Project Management," *Proceedings of IRNOP VI Conference,* 27 August, Turku, Finland: Abo Akadami University Press, ISBN 952-12-1389-2.
Remington, K., Paul, D. and Pollack, J. (2006) "A systemic approach to managing complex change projects in the public sector, "*Proceeeings of the ANZSOG Conference,* February, 2006, Canberra, Australia: ANZSOG, ISSN 1442-8458.
Thiry, M. (2006) "Recent Developments in Project-Based Organisations," *Proceedings of IRNOP VII Project Research Conference,* October, 2006, Xi'an, China: IRNOP, pp 2-15, ISBN 7-121-03252-X.
Turner J.R. and Cochrane, R.A. (1993) "Goals-and-methods matrix: coping with projects with ill defined goals and/or methods of achieving them." *International Journal of Project Management,* **11**, p. 93, ISSN 0263-7863.
Williams T., (1999) "The need for new paradigms for complex projects." *International Journal of Project Management,* **17**, pp 269-273, ISSN 0263-7863.
Williams, T. (2002) *Modelling Complex Projects,* Sussex, UK: John Wiley & Sons, ISBN 0-471-89945-3.
Winter M. and Checkland P. (2003) "Soft Systems: a fresh perspective for project management." *Civil Engineering – Proceedings of ICE,* **156**, pp187-192, ISSN 1751-7672.
Yeo K. T., (1993) "Systems thinking and project management - time to reunite." *International Journal of Project Management,* **11**, pp111-117, ISSN 0263-7863.

Until very recently **Kaye Remington** was Course Director of the post-graduate Project Management program at the University of Technology Sydney. She teaches regularly overseas with recent appointments as a visiting academic at universities in France, Sweden and China. Prior to joining UTS in 1996 Kaye's 25-year career in project, program and senior management began in the UK where she worked on large-scale engineering and construction projects. Her career took her to Europe, the Middle East and back to Australia, where she founded her own consultancy practice. During that period she extended her expertise to projects in fields such as education, IT, HR and organizational change. Kaye consults regularly to a wide range of industries, including engineering, rail, housing, public works, water supply, power, banking, telecommunications and product manufacture. Recently she has also completed development and delivery of a large infrastructure program of 54 cross-organizational change projects encompassing specialist disciplines such as IT, HR, finance and facilities management, deemed a success by all key stakeholders. Together with Julien Pollack she is coauthoring a book, 'Tools for Complex Projects', due to be published by Gower Publishing in 2007, ISBN 0 566 08741 3. Kaye now balances a consultant practice with research and writing.

Julien Pollack has pursued a career that balances practice, research and teaching. He has taught project management at both post-graduate and under-graduate levels and has been an active member of the Project Management Research team at the University of Technology Sydney since 2000. In 2003 Julien won the Australian Institute of Project Management student medals for research into project management at both State and National levels and in 2005 he won the International PM Tage student research award, Vienna, Austria. From 2002-2005, while completing his PhD, he was Administrative Manager for the Health Professionals Registration Boards (HPRB) of NSW with responsibility for a large and complex infrastructure project involving formulation and implementation of a five-year plan for the strategic development of HPRB in relation to changing Government initiatives, management of information systems and information management projects and staff development and training in project management and information systems use. He has authored and coauthored several refereed journal papers combining the fields of project management and systems thinking. Together with Kaye Remington he is coauthoring a book, 'Tools for Complex Projects', due to be published by Gower Publishing in 2007, ISBN 056608741 3. Julien is currently working in Japan.

Fleet Management System for an Advanced Helicopter Platform: A Research Program

Axel Schauenburg, Arvind Kumar Sinha & Simon Atyeo
The Sir Lawrence Wackett Centre for Aerospace Design Technology, RMIT University, AUS

The defence helicopter fleet will be undergoing a major acquisition process including upgrade under a 20+ year strategic plan. This program develops aims to rationalise the number of helicopter types operated, simplify operational requirements and reduce through life support costs. The helicopter fleet presently comprises of nine helicopter types, to be grouped to three to four fundamental types. A fleet management methodology needs to be developed and modelled for each fundamental type of helicopter. The research will assess current practices in aerospace technology management of military aircraft fleets and establish requirements for both, civil and military rotary-wing platforms and design a specific methodology for the fundamental Multi-Role Helicopter (MRH) of the defence forces. The outcome will be a technology management methodology, demonstrated through simulation for application on the MRH.

Introduction

Rotary wing technology provides major benefits to the community and the industry in the civil and military sectors. It has proven its capabilities in service for emergency and military support over the years. Rotary wing platforms cover a variety of missions, conducting passenger and emergency transport, search and rescue tasks, heavy lift duties and military roles. The defence forces operate and maintain a large fleet of rotary wing platforms that require advanced fleet management practices.

Presently fleet management is mainly based on commercially available tools to manage the maintenance and movements of vehicles (Vijayakumar, 2006). Companies specializing in the development of softwares, customise the tools based on fleet types, resulting in several commercially available tools. These tools assist the operators to increase the availability and enhance service life of the vehicle. The tools are specialised on engineering maintenance support management including operations to cover the allocation of resources (Vijayakumar, 2006).

The providers of fleet management tools and their services are for application on surface vehicle fleets and limited for aviation applications. The complexity of fixed- and rotary-wing systems and their associated support structure makes it challenging in the application for fleet management. The service life of surface vehicles and its type of associated support differ from aircraft on several issues; to illustrate - military aircraft undergo life-extension through mid-life upgrades to enhance mission capability. This requires an in-built flexibility in the technology insertion management system. None of the tools support management of life-extension.

Studies have been initiated in fleet management to investigate issues related to increase in average age of military aircraft fleets (Crowley & Mutzman, 2003), and its associated costs on operators. The research aims to extend the service life (Tiedeman, *et al.*, 2002) of the airframes to enhance operational effectiveness (Brewer, 2002) and predict fleet reliability (Meyer, *et al.*, 2001). To address issues of aircraft scheduling, performance and cost management; probabilistic approaches (Kaczor, 1999 and Oore, 2005) and statistical models (Rumph, 2006) are being considered; including single or multiple type fleets to optimize maintenance and training (Scott & Gaerke, 2002).

A fleet management methodology needs to be developed and modelled specific to aerial platforms. In this paper the MRH helicopter platform is considered to demonstrate the methodology. The research will assess current practices in aerospace technology management of military aircraft fleets. It will investigate the requirements for both, civil and military rotary-wing platforms and design a specific methodology for fleet management of the MRH for the defence forces.

Systems approach

The operation, maintenance and support of a rotary-wing fleet are complex. The investigations on fleet management systems need a holistic analysis to capture the disparate requirements of operations and logistics support. The objective of a fleet management is to ensure safety in operation and enhance the availability of vehicles to downsize the fleet. An efficient fleet management system provides a cost efficient mode of operation and support. To adopt a systems approach for the development of a methodology to establish a fleet management system, the various practices of fleet management are studied to formulate the system elements, attributes and their relationships which are defined as follows (Blanchard & Fabrycky, 2006; Cavaleri & Krzysztof, 1993):

- **Components (Elements):** The operating parts of a system consisting of input, process and output;
- **Attributes:** The properties of the system's components; and
- **Relationships:** The links between components and attributes.

The system elements, attributes and their relationships will provide avenue to further identify the system structure of fleet management. Each element within the overall 'fleet management' layout acts as a parent for the underlying subsystems. The components are further classified as systems where each section is part of an overall larger system, forming a logical hierarchy (Blanchard, & Fabrycky, 2006 and Cavaleri & Krzysztof, 1993). The research established the following core elements in fleet management:

1. Scheduling & Planning;
2. Operation;

3. Platform;
4. Maintenance; and
5. Personnel

The purpose of the fleet management system as a whole is achieved through the functional relationships linking components and attributes (Blanchard & Fabrycky, 2006). A diagrammatic representation of the system concept is presented in Figure 1. Fleet Management is organized through the following components and their relationships are represented in a defined environment:

The relationship between the elements indicates information flow direction where the dependency of the elements in relation to each other is clearly formulated.

1. **Scheduling and Planning**: The central element of fleet management governs the utilisation of vehicles, designation of tasks amongst the crew and the involvement of personnel for the stipulated mission. Maintenance intervals based on feedback gained through manufacturers and regulatory bodies is at the core of the methodology and is designed to address the issue of fleet availability.
2. **Operation**: The working environment is taken into consideration such that the mission requirement can be stipulated accordingly based on specific time and location. As a result, mission scheduling and planning together with the allocation of vehicle and crew can be undertaken, to obtain total system effectiveness based on the previously identified environmental conditions.

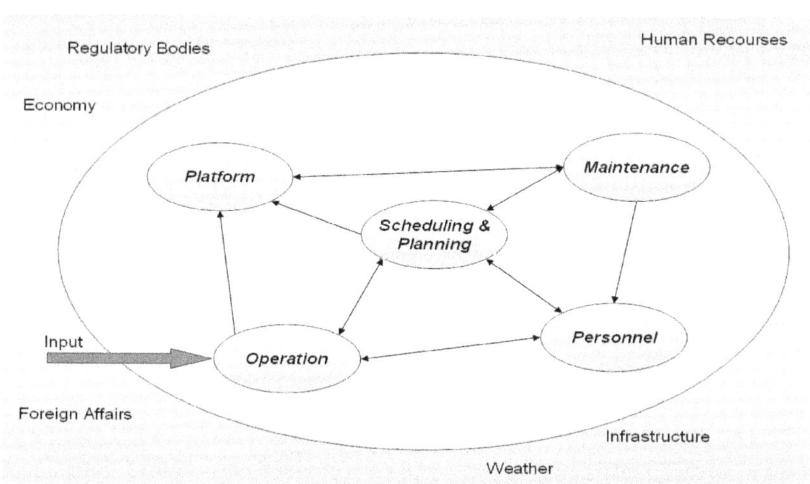

Figure 1 *The general concept of a fleet management system*

3. **Platform**: Provides a detail fleet operational perspective within which vehicles under operational and maintenance status are identified. The movement of helicopters can be identified accordingly thus, creating a bi-directional dependency. The requirement of conducting a specific time dependent mission leads to the assignment of a number of available airframes through the element of 'Scheduling and Planning'.
4. **Maintenance**: Scheduled repair work is up to 80% of all work performed (Dossier Fleet Management Software, 2006) and is a key element within the overall functionality of rotary-wing aircrafts. Maintenance comprises of classification of subsystems through the documentation of repair schedules, providing logistic support and guidance together with the identification and tracking of parts.
5. **Personnel**: All human recourses in the fleet management system are grouped in this element. It therefore interacts with all elements, including assigning vehicles to appropriate crew members through scheduling and planning such that a feasible design solution is obtained.

The fleet management system is classified as an 'open' system, based on the interaction between the system's elements and its environment. Open systems interact with their environment, allowing information, energy and matter to cross its boundaries. The system 'fleet management' significantly interacts with its environment, where the fleet operates in various environments including weather and terrain, thus exposing it to corrosion and fatigue. Economical issues are further taken into consideration, where changes in fuel price and stock fluctuations with regards to the implemented inventory is further included in the analysis. Input from the environment into the fleet management system is achieved by establishing the requirement to conduct a specific mission in a specific location and time. The direct output of the system is the successful or unsuccessful accomplishment of the mission requirement. However the main objective of the system fleet management is to obtain a cost effective, safe and reliable solution that is conducive to rotary-wing operations, through the interaction of the identified system components.

Fleet management concept

The development of a fleet management methodology involves systems engineering principles to ensure an interdisciplinary, comprehensive and structured approach. Systems engineering is a life-cycle approach concerned with the overall process of defining, developing, operating, maintaining, and ultimately replacing quality systems such as the investigated MRH helicopter. The customer needs and requirements are documented early in the development cycle, where the proceeding design synthesis and system validation process required to complete the problem is formulated.

The Systems Engineering Process (SEP) is iterative and applied continuously in the fleet management application (INCOSE, 2002; Blanchard & Fabrycky, 1990). The research establishes a set of requirements in the rotary wing application leading to rotary wing operations and support requirements for the identified operators. Similarly civil and military rotary wing technology management requirements are developed through a literature review. Based on these requirements, a functional analysis is generated to apportion the appropriate system-level requirements down to the subsystem and lower levels of the system hierarchy. System analyses are accomplished to evaluate the various alternatives that are considered feasible in meeting the identified needs (Blanchard & Fabrycky, 1990). The research will accomplish the following SEP tasks during the first two phases (INCOSE, 2002):

1. Define the fleet management objectives (needs of the rotary-wing operators);
2. Establish functionality of the objectives (functional analysis);
3. Establish performance requirements (requirements analysis);
4. Evolve design and operations concepts (architecture synthesis);
5. Select a baseline (through cost/benefit trades); and
6. Verify that the baseline meets the stipulated requirements (the rotary-wing operators' needs).

To establish the needs and requirements of the operator, the development of a fleet management methodology is investigated through a 'Life-Cycle Analysis' (Blanchard & Fabrycky, 2006 and Chapman, *et al.*, 1992) perspective. The adoption of the above mentioned scheme within the design and development process is undertaken to ensure design compatibility with the related physical and functional requirements. The system life cycle can be classified into the following seven phases (Chapman, *et al.*, 1992):

1. Requirements Development;
2. Concept Development;
3. Full Scale Engineering Development;
4. System Development;
5. Test and Integration;
6. Operations, Support, and Modification; and
7. Retirement and Replacement Phase

The fleet management system design process is initiated through investigations on the helicopter operator's needs and requirements. The client's needs form a set of slated requirements while the design and development process evolves through a) conceptual design (the establishment of performance parameters, operational requirements and support policies), b) preliminary systems design, and c) detail design.

All elements of the system 'fleet management' are considered on an integrated basis from the start. Emphasis is not only directed to elements that relate directly to the mission performance (including crews, operating facilities, prime equipment and operational data and information), but also to the system maintenance and support elements, involving large financial recourses (Blanchard & Fabrycky, 2006). The prime system elements of the helicopter fleet are designed to be effectively and efficiently supported during the entire system life cycle. The maintenance and support infrastructure concept is designed to be responsive to this requirement and evolves from the system operational requirements. The fleet management system framework therefore follows the SEP and is structured in four phases covering a period of three years (*Figure 2*):

The operations and support requirements are developed from the fleet needs and requirements of the defence forces that operate the platform. The attributes of the system 'MRH' (including but not limited to technical, financial or operative specifications) are integrated into the research, forming the framework for the initial phase of the research program. Phase two will analyze the helicopter platform systems technology design and support management. The requirements of current aerospace practices in technology management of civil and military rotary-wing platforms will be investigated and established in this part of the research. Phase three will present a novel methodology to support management needs and cover MRH platform technology needs. The rationalization program's specific fleet management needs are further addressed within the context of this phase. The results of phase one and two will support phase three investigations. Following, the last phase will simulate the design technology based on the outcomes of phase three. Furthermore the application to helicopter platform technology, design, operation and management is investigated. A best practice model to include the MRH platform technology and a simulation demonstration model will be established in this last part of the research.

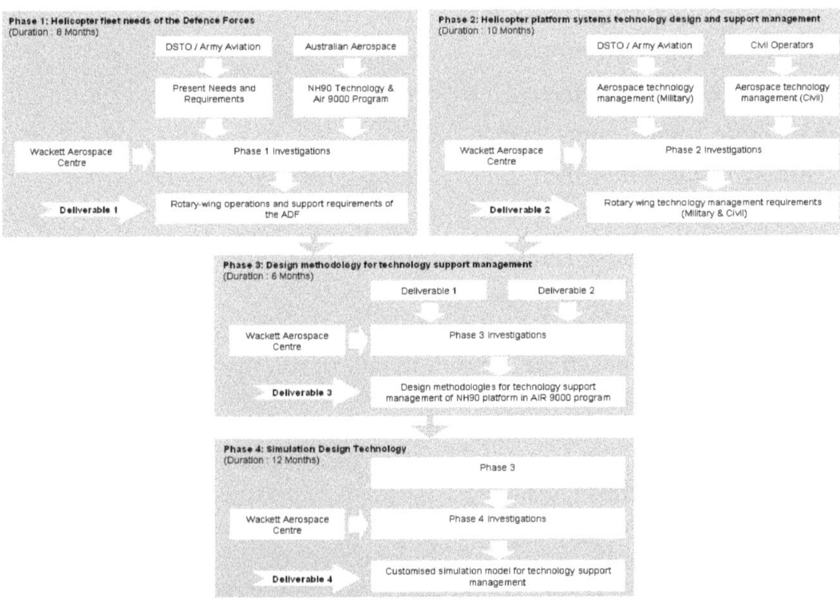

Figure 2 *Research Structure – Tasks (Phases 1 to 4)*

The investigations on fleet management tools will form the base for the development of a customized software simulation model, specifically designed for application on MRH technology.

Technology management tools

Fleet management is based on commercially available tools to monitor and track movements, expenses and maintenance of vehicle fleets. The tools are tailored to the application on surface vehicles, with no flexibility to be applied on rotary- or fixed wing fleets (Vijayakumar, 2006). The described high level of complexity is partly caused by the maintenance system involved due to the high number of parts (Ogden, 2003) as well as by the regulatory framework stipulating the certification procedures.

Presently, the research investigates technology management tools and its degree of applicability within the rotary-wing platform. The fleet management software market is served by a number of specialised companies, focussing in the development and implementation of fleet management tools (Vijayakumar, 2006). The investigated tools are designed for application on surface vehicle fleets, representing the bulk of the market. Surface vehicle fleet management is of lower complexity compared to airborne vehicles, allowing a wide application on surface vehicle types.

A low degree of applicability on rotary-wing technology was found in all of the ten investigated tools (Vijayakumar, 2006). The specific design of the tools for application on surface vehicles can be related to the following design features:

- **Performance Parameters**: Mission performance is governed by several parameters. Parameters of interest in surface vehicle applications include but not limited to fuel consumption, mission duration, speed, distance and delivery of cargo. Although selected performance parameters match with the parameters of interest in rotary-wing aviation, they can not be addressed due to the disparate units of the metric and imperial system.
- **Missions**: The rotary-wing mission spectrum is versatile. Surface vehicle fleets perform passenger and cargo transport, patrol and emergency missions. Further civil tasks include agricultural duties and special tasks. None of the investigated tools support military missions nor aerial civil tasks.
- **Maintenance**: Surface vehicle maintenance differs from the applied aviation principles of fail-safe or safe-life. The required maintenance efforts in aviation exceed the capabilities of the tools, with platforms requiring over 30h maintenance per flight hour (Wikipedia, 2006). The aviation certification procedures are either not addressable or different in surface vehicle fleet applications.
- **Involved Personnel**: A significant difference between aerial and surface vehicle platforms is in the amount of personnel involved and their required training. The expenses in crew training are higher due to the complexity in rotary-wing operation and maintenance.

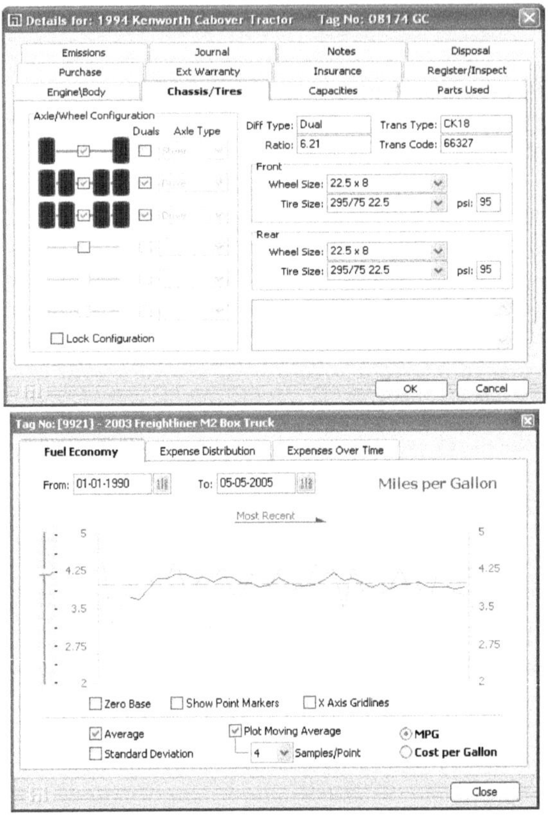

Figure 3 *Fleet management tool 'Fleetmate', vehicle details (left) and fuel economy (right)*

- **Environment**: Differences in environment between surface and aerial vehicles include weather dependency and terrain, which is in most applications limited to roads and urban environments for surface vehicles. Airborne operations require a home base which is essential to support equipment, provide maintenance including crew and refuelling facilities. Both, surface vehicles and aircraft are dependant on economical aspects influencing the fuel price.
- **System Life-Cycle**: Civil and military aircraft exceed the life expectance of surface vehicles due to extensive maintenance requirements and mid-life upgrades. Because of the comparable low unit cost of surface vehicles, replacement becomes more economic than upgrades, resulting in a comparable low average age.
- **Regulations**: Civil and military safety regulations are complex due to the involved risk of airborne operations. Regulatory bodies require full documentation of all maintenance and operational tasks performed. In contrast, surface vehicles are undergoing technical roadworthiness inspections in intervals of two to three years with lower maintenance requirements.

Figure 3 illustrates the properties to be entered by the user in a tool designed for application on surface vehicle fleet management (Fleetmate®, 2006). The properties of interest include insurance, engine/body, chassis/tires, warranty, purchase and parts used. The tool returns the fuel economy graph, displayed in miles per gallon, leaving no flexibility for applications on rotary-wing technology management.

Conclusions

The fleet management methodology under development will provide an avenue to address the operational and logistical challenges of rotary-wing technology management. The systems engineering design process applied in the development of the fleet management methodology is well established and demonstrated in various applications. The fleet management concept will support cost effective and safe operations of the MRH platform within the defence forces, to be demonstrated through a simulation model for application on MRH technology.

Glossary

ARH	Armed Reconnaissance Helicopter
ASW	Anti Submarine Warfare
MRH	Multi-Role Helicopter
SEP	Systems Engineering Process

References

Blanchard, B.S. and Fabrycky W.J., (2006), '*Systems Engineering and Analysis*', 4th ed. Prentice Hall International Series in Industrial and Systems Engineering, ed. W.J. Fabrycky and J.H. Mize, Upper Saddle River: Prentice Hall

Blanchard, B.S. and Fabrycky W.J., (1990), '*Systems Engineering and Analysis*' 2nd ed., New Jersy: Prentice Hall

Brewer, T., (2002), '*Organic Depot Maintenance Re-capitalization*', 6th Aging Aircraft Conference, San Francisco, CA, USA, September 2002

Cavaleri, S. and Krzysztof O., (1993), '*Management Systems: A Global Perspective*', Belmont, CA: Wadsworth Publishing Company

Chapman, W.L., Bahill A.T., and Wymore A.W., (1992), '*Engineering modelling and design*', Boca Raton, Florida: CRC Press, Inc.

Crowley F., Mutzman R., (2003), '*Air Force Fleet Viability Board (AFFVB)*', 7th Joint Council on Aging Aircraft, New Orleans, LA, USA

Dossier Fleet Management Software, (2006), Arsenault Associates, http://www.truckfleet.com/ (Accessed 28.08.2006)

Fleetmate®, Fleet Maintenance Management System, (2006) <www.fleetmate.com> (Accessed 27.04.2006)

INCOSE, International Council on Systems Engineering: the Systems Engineering (SE) Handbook Working Group, (2002), '*Systems engineering handbook: a "how to" guide for all engineers*', Version 2.0, ed. J. Whalen, R. Wray, and D. McKinney. 2002, Seattle: International Council on Systems Engineering

Kaczor S., (1999), '*CC-130 Optimum Maintenance Scheduling using Probabilistic Damage Tolerance Methods*', Bombardier Aerospace Defence Services, AA Conference

Meyer E.S., Fields S.S., Reid P.A., (2001), '*Projecting Aircraft Fleet Reliability*', The Boeing

Company, St. Louis Missouri, USA
Ogden, M., (2003), Aviation Today Newsstand, '*PacRim*', Issue: 1, 1 June 2003
Oore M., (2005), '*Aircraft Life Cycle Cost Management Using a Probabilistic Approach*', IMP Aerospace Group, Joint Council on Aging Aircraft, USA
Rumph F., (2006), '*A Statistical Model for Predicting Probable Future Performance*', US Army Aviation Engineering, 9th Joint FAA/DoD/NASA Conference on Aging Aircraft, Atlanta, GA, USA
Scott, J.P. and Gaerke J., (2002), '*CC-130 Hercules Individual Aircraft Management*', Aging Aircraft Conference
Tiedeman T., DeSalle R., Gaskin A., Hamlin D., (2002), '*Aging Fleet Integrity & Reliability Management (AFIRM)*', Lockheed Martin Aeronautics Company and WR-ALC
Vijayakumar S., (2006), *Fleet Management Systems – A Market Study*, Technical Report, The Sir Lawrence Wackett Centre for Aerospace Design Technology, RMIT University, Melbourne, VIC
Wikipedia, (2006), <http://en.wikipedia.org/wiki/SH-2> (Accessed 25.08.2006)

Axel Schauenburg is a researcher and doctoral candidate at The Sir Lawrence Wackett Centre for Aerospace Design Technology, RMIT University in Melbourne. He holds a Bachelor and Masters Degree in Aerospace Engineering from the Technical University of Munich, Germany and has experience on several industrial projects, including work for Boeing Commercial Airplanes USA, Boeing Australia and EADS.

Arvind K. Sinha has a service record of 29 years, which includes defence forces, industry and academic institutions. He has several qualifications, scholarships, awards, industrial research projects, research papers and public presentations to credit. He is presently the Director of The Sir Lawrence Wackett Centre for Aerospace Design Technology, Royal Melbourne Institute of Technology, Melbourne, Australia.

Simon Atyeo is presently a doctoral research candidate at The Sir Lawrence Wackett Centre for Aerospace Design Technology, RMIT University in Melbourne. He graduated in Aerospace Engineering from RMIT University with honours and is currently pursuing his doctoral thesis on an intelligent system for helicopter emergency medical service operations. During his time at the centre he has also been involved in several key industry research projects.

APPENDIX ONE
CONVERSATION MAPPING WORKSHOP

Based upon the panel discussions, theme chairs were asked to identify a 'trigger issue' – some aspect for which there would be a range of views and useful as a beginning point for a broader discussion.

These statements were used as the basis of a conversation mapping workshop facilitated by Bruce McKenzie. Below are brief summaries of the conversations mapped during this exercise. One of these conversation maps has been reproduced in full along with background comments to this particular issue.

Trigger statements and summary statements drawn from the exercise

The place we are forced to take on sustainability is individual responsibility.

- Interconnection between individual and community. Institutional – it's about the structure;
- It is necessary but not sufficient. We live in a technologically and institutionally mediated world which often constrains responsibility;
- Institutional / structural conditions need to be challenged and need to change;
- Sustainability is about dynamics and relations beyond the individual;
- Target marketing, e.g., not just home owners' different demographics – Reuters can't buy water tanks, install solar panels, change fixtures and fittings;
- Forced? Is this because the political systems and institutions are not taking up the challenge, therefore it is up to the individual;
- The individual acting on their own could be the critical mass, or '100th monkey' to tip the scales of sustainability;
- Sustainability is a collective activity that individuals contribute to, initiate...

Considerations in the use of cultural knowledge in sustaining our social and natural capital (special considerations for indigenous cultural knowledge).

- Isn't all knowledge cultural?
- So are we just saying that all knowledges deserve special consideration and privileging;
- Why do we want to understand cultural knowledge?

- I value this knowledge and know it is worth knowing;
- Does this map onto the current society;
- Not to understand but to respect cultural knowledge – indigenous people (many not all) know how to care for country;
- We are all indigenous to this planet and it will take all of our collective knowledge, cultural wisdom to create a sustainable planet;
- It is about heritage 'honouring the past' and facing up to what we have inherited – including untruths. It is about our present and how we hold such knowings within our modern constructions of community. It is about our future – the courage and wisdom we make available to all regardless of race and any other constructed separation (fragmentation);
- Opportunity for mutual learning;
- Culture is what I seek to sustain the ability to think and to range over our thought, to be sane in a sane world, to love and to put my feet up and to look at useless and beautiful things;
- It may be that in a truly egalitarian society intellectual property would be less of an issue;
- There are class, race, age gender, ability/disability, sexuality, cultural considerations and they all require our attention;
- Intellectual property rights are a western capitalist construct and it's as absurd as the idea that you can patent 'genes';
- Value of cultural knowledge, and arguments over conserving it;
- At this time and space there is an issue in (trust, respect) in sharing cultural knowledge;
- (Perhaps we should avoid conserving heritage from the past. It suited the past not present nor perhaps the future;
- Existing cultural knowledge is insufficient;
- Cultural knowledge and protection of cultural diversity is as critical as the protection of biodiversity;
- Yes, but given more attention and resources already (if not universally, i.e., generally Eurocentric), i.e., Mona Lisa is valued and protected, tigers and orang-utans are on their way out and no-one seems to care;
- Fire burning regimes are important ways of protecting biodiversity but indigenous and non-indigenous disagree often;
- We need to consider intellectual property (traditional knowledge);
- Whose knowledge? In NZ Maori assert a right to decision-making over the use of their traditional knowledge. This is a power issue;
- Cultural knowledge is something from the past and not sufficient to answer the problems of the present;
- Moreover, it is false to privilege cultural knowledge over protection of biodiversity;

- Indigenous cultural knowledge is important, but indigenous peoples assert their rights over this and therefore non-indigenous peoples may not have access to it.)

Conflicting paradigms are a recipe for organizational chaos or creativity.

- Not all paradigms conflict. Both the synergies and antagonisms happen – both are useful;
- It is a source of strength and learning: out of conflict resolution or finding a balance comes new understanding;
- Are creativity and chaos necessarily independent? Either/or outcomes or they can be simultaneous, i.e., feed each other;
- There is a dynamic continuum between order and chaos that is important for creativity;
- Conflicting paradigms and conversations between persons holding them will force new ways of looking at issues and creativity;
- Creative tension – a great motivator;
- Conflict may be essential to move forward;
- It is where paradigms conflict and interact that some of the most interesting developments can occur;
- Do you necessarily need to know where you have conflicting paradigms? In practice, conflicting paradigms are often not relevant;
- On a personal learning journey we can enable a creative tension between coherence and openness to new ideas.

Systems studies should be recognized as a separate discipline in University environment.

- Institutional forms:
    - Perhaps a new kind of entity - whose focus is on the much vaunted concept of transdisciplinarity.
    - So, then, what if in Humanities where arts, history, story and other creative skills reside?
    - If it were separate you would still need to position it within a faculty.
    - Not so, there are models (e.g., SOLAR at UWE) of 'independent' or multiplayer positioned research centres placed within universities and on their edges.
    - I have a feeling that universities are precisely where this will not happen, in their current form, conception and operation.
    - That's why universities have to change too.
    - The thing that stops this from happening in universities may not be

that different from what stops them anywhere else – inquire into our immediate spheres of action to generate systemic transformation – before we expect others to do it in theirs.
    - Universities are hierarchical, status and silo …
    - Why not high school?
    - Can't there be multiple forms? Separate and integrated elsewhere, as and when needed and feasible.
    - Perhaps it belongs outside the universities as a special kind of institute?
    - Within mathematics.
    - The search for viable institutional form(s) for systemic studies is not yet over, but multiple forms are possible, and we should not always expect institutions to persist indefinitely.
- Scope of systems thinking / the nature of the beast.
    - Systems studies need to exhibit scholarship. Otherwise it will not remain part of a university.
    - What is scholarship in a world on the edge of extinction?
    - There needs to be a systemic discussion on what constitutes systems studies.
    - When I told a friend I was coming to a 'systems' conference she replied "…you mean it's a discipline now, not just a way of thinking?"
    - The knowledge foundations must be integrated into curricula no matter what the course.
    - Is big picture a part of the mechanistic mindset – we account for the relevant pushes and pulls on the issue.
    - Oh no! There are non-mechanistic big pictures too?!
    - No because to be a discipline we need a recognised framework of ideas, methodology, area of concern; area of concern covers much of human endeavour. Rather teach systems in every degree.
    - Systems is more than a discipline – it is a way of knowing differently in all disciplines.
    - Systems studies do not display the rigor of scholarship or framework to stand alone as an area of study in university.

Systemic practitioners are experts for change.

- Who determines that the intervention is systemic?
- If we think we are experts for change then we are probably not.
- Agents for change does not necessarily equate to systemic practitioners.
- I have a problem with a notion of expert.

- Are they sufficiently trained for the task?
- Are systemic academics also agents of change and is this relevant for them?
- Experts for change is an oxymoron.
- Practitioners should be open about their agendas.
- Everyone is an expert and everyone is a lay person. It is OK to bring expertise as long as it is open and transparent.
- Experts / agents have a role in challenging linear, positivist approaches.

Nurturing systems thinking within ourselves and our constituencies requires new innovations.

- 'New' innovation is one possibility of systems thinking; 'affirming' is another.
- Ways of nurturing ourselves include mentoring, providing proof for sceptics, engaging others, going to places where leaders meet, and recognizing the relevance of what we know already.
- Why is systems thinking the minority paradigm? Because it challenges the current disciplinary world views, methodologies of command and control and economic rationalism. Because other systems have a history of at least 500 years and is embedded throughout our other systems. Uncomfortable but life changing when exposed to it. Challenges notions of power and control
- Systems thinking is the minority paradigm because it challenges the current disciplinary world views, methodologies and power and control structures.
- We need to be more active and courageous in encouraging the systemic.
- Requires both individual and shared learning journeys.
- Let us be strong about our position and point of view and the tools we can bring.
- We need a set of champions.
- We need to be more active and courageous in encouraging the systemic view.

Our social and natural capital is sustained by learning and articulating transdisciplinarity.

- Transdisciplinarity is understanding the concept necessary for progress to be made.
    - Is there an opportunity for de-jargonising?
    - The quality and validity of the articulation is a determinant
    - Accessibility is another determinant to those who don't know about 'transdisciplinarity'.

- There is power in naming what is happening (and not naming).
- Can we say it's really happening? How do we know?
- Efficacy of language and engagement at this level for different contexts / places / communities (comes from a very theoretical position).
- This is about language and engagement.

- What is a metaphor for?
  - Metaphors like this can be useful for helping people start to engage with the issue.
  - Learning and meaning making are relational.
  - Role of the heart.
  - Learning and meaning making – in community: connected to a specific place; 'a-disciplinary' – the meaning-making / knowledge of indigenous people.
- Meaning making
  - Is an innate human desire or feeling.
  - Is an emotional process.
  - Does communication precede meaning making or meaning-making precede communication?

Synthesizing 'hard' and 'soft' approaches we can engineer positive changes in socio-technical systems.

- Why do we need the duality of hard / soft and who decides?
- My point is that the hard/soft duality distinction reifies the dualism. Rather than the duality, it is also phalocentric.
- Synthesizing may be the wrong term as each stands on different assumptions.
- It implies that people can synthesise. It is a capacity that needs development before we can address the hard / soft duality.
- It may be possible (and I think it has been done) to develop new assumptions that allow us to transcend hard / soft labels / distinctions.
- In practical application the distinction is false, but may be a useful abstraction for analysis.
- Integration
- 'Hard' and 'soft' are different conceptualizations of the same phenomenon.
- Omitting either the hard or soft builds in the seeds of failure.
- You need knowledge in both fields but many persons have just knowledge about either of them.

- Necessity for addressing Australia's salinity issues with water! Many issues are like that.
- Perhaps it is really a practice/theory continuum in multi-directions.
- Insight: Need a systemic practice of creative design of methods that can incorporate the range of hard and soft systems approaches.
- Issue: Need knowledge of a range of approaches.

Conversation map background statement

Bruce McKenzie
Systemic Development Institute
Vincentia

One issue that has engaged members of ANZSYS in vigorous debate over the last twelve years has been the status of systems studies within the academy.

At the 2006 ANZSYS Conference delegates participated in building a 'conversation map' triggered by the statement "Systems Studies should be recognized as a separate discipline in the University environment". The conversation map is formed by participants recording on the map material (four attached sheets of newsprint), their responses to the trigger statement and to the responses of others. Through continuous interaction with comments of fellow participants the map quickly emerges as a 'rich picture' of this complex issue. Comments from any and every perspective are incorporated with no comment being inappropriate. Participants in this process can all speak at once and it is very difficult for anyone to direct the conversation. Further every comment is of equal significance.

When additional contributions have ceased participants explore the map to identify emergent patterns of themes and/or issues. These emergent patterns invariably confront the participants with new ideas of how to approach the 'trigger' issue or ways to take a new pathway to address associated problematics. Conversation mapping is widely used for dealing with complex management issues in conjunction with SSM.

Three themes were identified in the time available at the conference from the 'systems studies as a discipline' conversation map:

- The search for viable institutional form(s) for systems studies is not yet over, multiple forms are possible, and we should expect institutions to persist indefinitely;
- Systemic approaches are more than a discipline they are ways of knowing differently within all disciplines;
- Systems Studies does not exhibit the rigor of scholarship or an adequate coherent internal framework to stand alone as an area of study in a University.

These themes can be used as the basis for 'Transformation statements' within an SSM root definition to leverage further understanding of the topic.

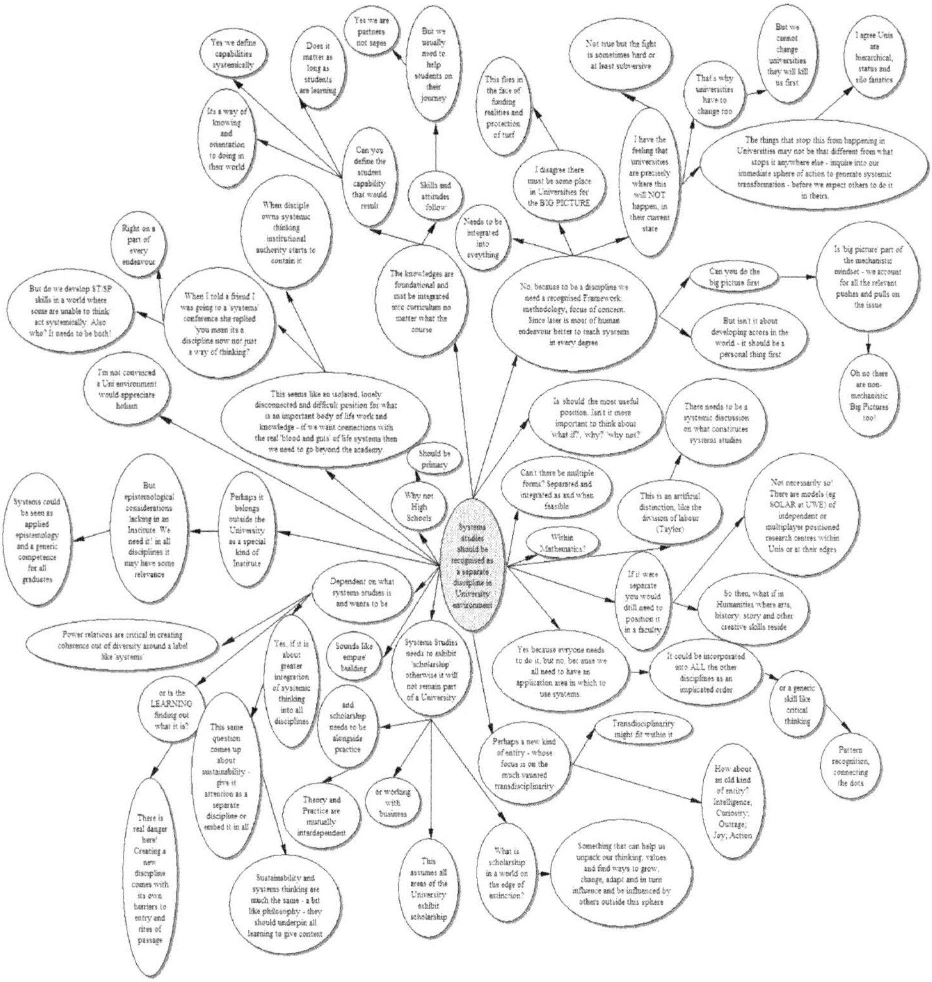
Systems studies should be recognised as a separate discipline in University environment
The knowledges are foundational and mst be integrated into curriculum no matter what the course
Can you define the student capability that would result
Yes we define capabilities systemically
Its a way of knowing and orientation to doing in their world
Does it matter as long as students are learning
Skills and attitudes follow
But we usually need to help students on their journey
Yes we are partners not sages
No, because to be a discipline we need a recognised Framework, methodology, focus of concern. Since later is most of human endeavour better to teach systems in every degree
Needs to be integrated into everything
I disagree there must be some place in Universities for the BIG PICTURE
This flies in the face of funding realities and protection of turf
I have the feeling that universities are precisely where this will NOT happen, in their current state
Not true but the fight is sometimes hard or at least subversive
That's why universities have to change too
But we cannot change universities they will kill us first
The things that stop this from happening in Universities may not be that different from what stops it anywhere else - inquire into our immediate sphere of action to generate systemic transformation - before we expect others to do it in theirs.
I agree Unis are hierarchical, status and silo fanatics
Can you do the big picture first
Is 'big picture' part of the mechanistic mindset - we account for all the relevant pushes and pulls on the issue
Oh no there are non-mechanistic 'Big Pictures' too!
But isn't it about developing actors in the world - it should be a personal thing first
Is 'should' the most useful position. Isn't it more important to think about 'what if?', why? 'why not?
Can't there be multiple forms? Separated and integrated as and when feasible
Within Mathematics?
This is an artificial distinction, like the division of labour (Taylor)
There needs to be a systemic discussion on what constitutes systems studies
If it were separate you would dtill need to position it in a faculty
Not necessarily so! There are models (eg SOLAR at UWE) of independent or multiplayer positioned research centres within Unis or at their edges
So then, what if in Humanities where arts, history, story and other creative skills reside
Yes because evryone needs to do it, but no, bec ause we all need to have an application area in which to use systems.
It could be incorporated into ALL the other disciplines as an implicated order
or a generic skill like critical thinking
Pattern recognition, connecting the dots
Perhaps a new kind of entiry - whose focus is on the much vaunted transdisciplinarity
Transdisciplinarity might fit within it
How about an old kind of entity? Intelligence; Curiosity; Outrage; Joy; Action
What is scholarship in a world on the edge of extinction?
Something that can help us unpack our thinking, values and find ways to grow, change, adapt and in turn influence and be influenced by others outside this sphere
Systems Studies needs to exhibit 'scholarship' otherwise it will not remain part of a University
This assumes all areas of the University exhibit scholarship
or working with business
and scholarship needs to be alongside practice
Theory and Practice are mutually interdependent
Sounds like 'empire' building
Dependent on what systems studies is and wants to be
Yes, if it is about greater integration of systemic thinking into all disciplines
This same question comes up about sustainability - give it attention as a separate discipline or embed it in all
Sustainability and systems thinking are much the same - a bit like philosophy - they should underpin all learning to give context
or is the LEARNING finding out what it is?
There is real danger here! Creating a new discipline comes with its own barriers to entry and rites of passage
Power relations are critical in creating coherence out of diversity around a label like 'systems'
Why not High Schools
Should be primary
Perhaps it belongs outside the University as a special kind of Institute
But epistemological considerations lacking in an Institute. We need it! in all disciplines it may have some relevance
Systems could be seen as applied epistemology and a generic competence for all graduates
I'm not convinced a Uni environment would appreciate holism
This seems like an isolated, lonely disconnected and difficult position for what is an important body of life work and knowledge - if we want connections with the real 'blood and guts' of life systems then we need to go beyond the academy.
When disciple owns systemic thinking institutional authority starts to contain it
When I told a friend I was going to a 'systems' conference she replied 'you mean its a discipline now not just a way of thinking?'
Right on a part of every endeavour
But do we develop ST/SP skills in a world where some are unable to think/ act systemically. Also who? It needs to be both!

APPENDIX TWO
Workshop Summaries

Applying the Cognitive Edge models to facilitation and problem solving
Bob Dick

In the whole group we discussed reasons for the growth in models for addressing complex problems. We examined the slow collapse of bureaucracy after many thousands of years of success. We addressed the increasing realization that rational problem solving models often aren't effective with very complex problems.

In small groups we then tried out one of the Cognitive Edge models for addressing complex problems, the ABIDE model. ABIDE is an acronym for attractors, boundaries, identities (that is, roles), diversity, and environment – (see Snowden, 2006).

Some of us then explored the notion of 'high reliability organizations' (Weick and Sutcliffe, 2001). HROs are organizations and groups (like bush fire fighters, for example) which experience low error rates despite the ever-present potential for disaster. In small groups we considered if the characteristics of HROs were also useful for other organizations.

Snowden, D. (2006). Stories from the frontier. Emergence: Complexity and Organization, 8(1), 85-88.
Weick, K., and Sutcliffe, K. (2001). Managing the unexpected: assuring high performance in an age of complexity. San Francisco: Jossey Bass.

Creating innovative conversations
Andrew Gaines

Innovative conversations are those wonderful conversations where people spark off each other and great new ideas arise. Can we get into creative space together intentionally? Yes!

Here's how. We play exciting improvisational games that teach thinking skills used by great innovators. They provide a way of tapping into our rarely used stores of knowledge experience and imagination – spontaneous lateral thinking! They also exemplify the fundamental values shift essential for a sustainable society – from domination/control relating to partnership/respect relating. Heavy stuff, eh? Actually it is all a lot of fun. You will be surprised by your friends' brilliance.

Theory, identity, language and practice: The value of participatory practices to sustainability

Susan Goff and colleagues

How do we know sustainability, and what is our relationship to such knowledge?

Between 2004 - 2006 twelve participatory practitioners in public, academic and community sectors in three Australian states collaboratively explored the value of participatory practices to sustainability.

The research was a doctoral candidature supported by the Centre for Cultural Research at the University of Western Sydney. The task of this Inquiry was to explore the nature of "participatory practices" being developed by each person within their range of disciplinary and inter-disciplinary approaches to aspects of "sustainability". A central working assumption was that "participation" was more than "doing" in an observable and quantifiable sense; that "participation" and those practices that manifest it embrace being, knowing, doing and becoming.

This workshop built upon the concepts developed by these authors in their paper.

Building workforce capacity to address complex health, housing and social inclusion issues (such as homelessness, drug misuse, domestic violence, unemployment) through improved collaboration and knowledge management

Janet McIntyre

"Shadow lines", writes Kinnane, are "wide lines of negotiation that we all use to make sense of our difference and interconnections "(379) ...Cuts leave scars. Scars leave tracks. Tracks can be followed." (Kinnane in Schaffer & Smith 2004: 118)

The project is called: "Building Workforce Capacity to Address Complex Health, Housing and Social Inclusion Issues through Critical Systems Thinking and Practice". This is a collaboration involving SA Health, Flinders University, University of South Australia and Neporendi Forum Inc, an Aboriginal NGO and in partnership with Anglicare. The multi-method research ensures that the service users, together with the providers design what works, why and how on the basis of their perceptions of successful outcomes that have achieved wellbeing.

The notion of experts making decisions on behalf of others undermines well being, removes a sense of control and fosters welfare dependency. In addition the project examines the effectiveness of established service delivery options. This relates to the policy and research areas that address Indigenous wellbeing, health, housing needs, aspirations and access to services. The research establishes the linkages across home tenure, engagement, capacity, health, education, employment and wealth accumulation. These are all issues that impact on wellbeing and it is accepted that all these issues have a role to play in ending the cycle of disadvantage experienced by Aboriginal Australians.

The issue is how do you do sustainable governance that is accountable to diverse people and to complex environments that moves from expert compartmentalized knowledge to systemic governance? Can the governance example funded through the ARC Linkage grant be applied more broadly to wider governance and risk management concerns? The ARC has funded a project to develop better health and wellbeing outcomes between the SA Dept of Health and an NGO. The idea is to find out how well the design process works and then to apply it more widely at the state level and to different domains of interest. In other words to any form of governance in the public, private and non government spheres.

This is an attempt to make so-called joined up governance work effectively and to apply the rule of subsidiarity as a way to address cross cultural and intercultural philosophies for governance.

Our approach makes it possible to avoid 'master narratives' (Hampden Turner, 1981) and to work across a number of knowledge domains so as to achieve participatory governance in a way that avoids the problems raised by Martin (2005).

Multiple perspectives in knowledge management: a sensemaking approach.

Jim Sheffield and Ricky Guo

Research in the critical pluralist tradition involves multiple perspectives, each of which provides a competing (but potentially complementary) epistemological framework. For example, knowledge management in the UK tradition recognizes hard (positivist) and soft (interpretive) perspectives. Conflict between these two frameworks is to be resolved using a critical perspective, frequently via the application of Habermasian theory. This workshop briefly describes principles for a critical pluralist approach to knowledge management interventions based on Jackson (2000). Four critical pluralist frameworks are presented. These identify critical heuristics based on Habermasian theory (and Churchman, Ulrich and Midgley). Finally, the critical pluralist approach to KM problems is demonstrated using selected teaching examples. The examples are characterized by three perspectives – objective fact, inter-personal dialogue, personal intuition - each of which provides a competing (but potentially complementary) view. The demonstration suggests that a critical pluralist approach to KM does illuminate complex issues in knowledge dynamics.

Notes

Notes

Notes

Notes

Notes

Notes

Notes

Notes

Notes

Notes

Notes

Notes

Notes

Notes

Notes

Notes

Notes

Notes

Notes

Notes

Notes

www.ingramcontent.com/pod-product-compliance
Ingram Content Group UK Ltd.
Pitfield, Milton Keynes, MK11 3LW, UK
UKHW020225290926
15470UKWH00001B/376

9 780979 168888